Zero Carbon Vehicles and Power Generation

Zero Carbon Vehicles and Power Generation

Editors

Wenbin Yu
Guang Zeng

MDPI

Basel • Beijing • Wuhan • Barcelona • Belgrade • Novi Sad • Cluj • Manchester

Editors

Wenbin Yu
Shandong University
Jinan
China

Guang Zeng
SPIC Northeast Energy Technology Co., Ltd.
Shenyang
China

Editorial Office
MDPI AG
Grosspeteranlage 5
4052 Basel, Switzerland

This is a reprint of articles from the Topic published online in the open access journals *Sustainability* (ISSN 2071-1050), *Energies* (ISSN 1996-1073), *World Electric Vehicle Journal* (ISSN 2032-6653), and *Designs* (ISSN 2411-9660) (available at: https://www.mdpi.com/topics/zero_carbon).

For citation purposes, cite each article independently as indicated on the article page online and as indicated below:

Lastname, A.A.; Lastname, B.B. Article Title. *Journal Name* **Year**, *Volume Number*, Page Range.

ISBN 978-3-7258-1943-0 (Hbk)
ISBN 978-3-7258-1944-7 (PDF)
doi.org/10.3390/books978-3-7258-1944-7

Contents

About the Editors . . . vii

Wenbin Yu and Guang Zeng
Zero-Carbon Vehicles and Power Generation
Reprinted from: *Sustainability* **2024**, *16*, 6447, doi:10.3390/su16156447 . . . 1

Haibo Wu, Xingwang Tang, Sichuan Xu and Jiangbin Zhou
Research on Energy Saving of PHEV Air Conditioning System Based on Reducing Air Backflow in Underhood
Reprinted from: *Energies* **2022**, *15*, 3183, doi:10.3390/en15093183 . . . 6

Min Zhang, Huiqiang Zhi, Shifeng Zhang, Rui Fan, Ran Li and Jinhao Wang
Modeling of Non-Characteristic Third Harmonics Produced by Voltage Source Converter under Unbalanced Condition
Reprinted from: *Sustainability* **2022**, *14*, 6449, doi:10.3390/su14116449 . . . 21

Andrew Yang Wu, Yui-Yip Lau and Ju-Ai Wu
Integration of Electric Vehicles into Microgrids: Policy Implication for the Industrial Application of Carbon Neutralisation in China
Reprinted from: *World Electr. Veh. J.* **2022**, *13*, 96, doi:10.3390/wevj13060096 . . . 36

Jiawei Guo, Chao He, Jiaqiang Li and Heng Wei
Slope Estimation Method of Electric Vehicles Based on Improved Sage–Husa Adaptive Kalman Filter
Reprinted from: *Energies* **2022**, *15*, 4126, doi:10.3390/en15114126 . . . 46

Junik Jo and Chul-Ho Kim
Numerical Study on Aerodynamic Characteristics of Heavy-Duty Vehicles Platooning for Energy Savings and CO_2 Reduction
Reprinted from: *Energies* **2022**, *15*, 4390, doi:10.3390/en15124390 . . . 63

Shaobo Ji, Ke Zhang, Guohong Tian, Zeting Yu, Xin Lan, Shibin Su and Yong Cheng
Evaluation Method of Naturalistic Driving Behaviour for Shared-Electrical Car
Reprinted from: *Energies* **2022**, *15*, 4625, doi:10.3390/en15134625 . . . 74

Ran Tao, Jingpeng Yue, Zhenlin Huang, Ranran An, Zou Li and Junfeng Liu
A High-Gain DC Side Converter with a Ripple-Free Input Current for Offshore Wind Energy Systems
Reprinted from: *Sustainability* **2022**, *14*, 11574, doi:10.3390/su141811574 . . . 97

Andri Ottesen, Sumayya Banna and Basil Alzougool
How to Cross the Chasm for the Electric Vehicle World's Laggards—A Case Study in Kuwait
Reprinted from: *World Electr. Veh. J.* **2023**, *14*, 45, doi:10.3390/wevj14020045 . . . 113

Mario Mišić, Marinko Stojkov, Rudolf Tomić and Mario Lovrić
Optimal Hybridization with Minimum Fuel Consumption of the Hybrid Fuel Cell Train
Reprinted from: *Designs* **2023**, *7*, 45, doi:10.3390/designs7020045 . . . 135

Marwan Al-Shami, Omar Mohamed and Wejdan Abu Elhaija
Energy-Efficient Control of a Gas Turbine Power Generation System
Reprinted from: *Designs* **2023**, *7*, 85, doi:10.3390/designs7040085 . . . 151

Nestor Asiamah, Kofi Awuviry-Newton, Whitney Nesser and Evelyn N. Alvarez
Carbon Footprints of Active and Non-Active Transport Modes: Hierarchy and Intergenerational Narrative Analyses
Reprinted from: *Sustainability* **2023**, *15*, 12795, doi:10.3390/su151712795 . . . . **170**

Fei Ma, Lingyan Guo, Zhijie Li, Xiaoxiao Zeng, Zhencao Zheng, Wei Li, et al.
A Review of Current Advances in Ammonia Combustion from the Fundamentals to Applications in Internal Combustion Engines
Reprinted from: *Energies* **2023**, *16*, 6304, doi:10.3390/en16176304 . . . . **186**

Guangze Li, Boxuan Cui, Chenglin Zhang, Liuyong Chang and Longfei Chen
Formulation of a Jet Fuel Surrogate and Its Kinetic Chemical Mechanism by Emulating Physical and Chemical Properties of Real Jet Fuel
Reprinted from: *Sustainability* **2023**, *15*, 13792, doi:10.3390/su151813792 . . . . **206**

Sumayya Banna, Andri Ottesen and Basil Alzougool
Reasons Why Only Kuwaiti Citizens Drive Electric Vehicles despite Being Only a Quarter of the Population
Reprinted from: *World Electr. Veh. J.* **2023**, *14*, 287, doi:10.3390/wevj14100287 . . . . **232**

Arief Nurdini, Rahmat Nurcahyo and Anton Satria Prabuwono
Waste from Electric Vehicle: A Bibliometric Analysis from 1995 to 2023
Reprinted from: *World Electr. Veh. J.* **2023**, *14*, 300, doi:10.3390/wevj14110300 . . . . **250**

Philipp Hoth, Ludger Heide, Alexander Grahle and Dietmar Göhlich
Vehicle-Integrated Photovoltaics—A Case Study for Berlin
Reprinted from: *World Electr. Veh. J.* **2024**, *13*, 113, doi:10.3390/wevj15030113 . . . . **268**

Jianan Chen, Hao Yu, Haocheng Xu, Qiang Lv, Zongqiang Zhu, Hao Chen, et al.
Investigation on Traffic Carbon Emission Factor Based on Sensitivity and Uncertainty Analysis
Reprinted from: *Energies* **2024**, *17*, 1774, doi:10.3390/en17071774 . . . . **288**

Junhong Zhang, Ning Wang, Jian Wang, Hui Wang, Xueling Zhang, Huwei Dai and Jiewei Lin
Low-Friction and -Knocking Diesel Engine Cylindrical-Tapered Bore Profile Design
Reprinted from: *Energies* **2024**, *17*, 2042, doi:10.3390/en17092042 . . . . **302**

Daniel Silva Cardoso, Paulo Oliveira Fael, Pedro Dinis Gaspar and António Espírito-Santo
An Innovative Mechanical Approach to Mitigating Torque Fluctuations in IC Engines during Idle Operation
Reprinted from: *Designs* **2024**, *8*, 47, doi:10.3390/designs8030047 . . . . **320**

Diyun Huang, Jiayong Wang, Minshuo Shi, Puze Yang and Binyang Wu
Combustion Mechanism of Gasoline Detonation Tube and Coupling of Engine Turbocharging Cycle
Reprinted from: *Energies* **2024**, *17*, 2466, doi:10.3390/en17112466 . . . . **340**

Yize Liu and Wanhua Su
The Influence of Injector Nozzle Diameter on High-Density and Lean Mixture Combustion in Heavy-Duty Diesel Engines
Reprinted from: *Energies* **2024**, *17*, 2549, doi:10.3390/en17112549 . . . . **356**

About the Editors

Wenbin Yu

Prof. Dr. Wenbin Yu is a full professor at the School of Energy and Power Engineering, Shandong University, China. He received his Ph.D degree from the State Key Laboratory of Engines, Tianjin University. From 2011 to 2013, he joined Great Wall Motor as their Chief Technical Expert. From 2013 to 2020, he joined the National University of Singapore as a Senior Research Fellow. In 2020, he joined the School of Energy and Power Engineering at Shandong University as a Professor.

Guang Zeng

Prof. Dr. Zeng has been a full professor at the SPIC Northeast Energy Technology Co., Ltd., since Jan. 2023, which is the support unit of the National Power Plant Combustion Engineering Technology Research Center. He received his Ph.D. from the Harbin Institute of Technology. From August 2008 to May 2016, he worked at the Northeast Electric Power Research Institute State Grid, China, as an Engineering Technologist. From June 2016 to March 2018, he joined the DSSP Power at Sinar Mas Group, Indonesia, as a Power Plant Technical Advisor. From April 2018 to January 2021, he joined the National University of Singapore as a Research Fellow. From January 2021 to January 2023, he joined the Institute of Engineering Thermophysics, Chinese Academy of Sciences, as a Research Professor. Currently, Prof. Dr. Zeng is a guest professor at the Shenyang Institute of Engineering. Furthermore, he is a postgraduate supervisor at the Harbin Institute of Technology, North China Electric Power University, and Shenyang Aerospace University.

sustainability

Editorial

Zero-Carbon Vehicles and Power Generation

Wenbin Yu [1,*] and Guang Zeng [2]

1 School of Energy and Power Engineering, Shandong University, Jinan 250061, China
2 SPIC Northeast Energy Technology Co., Ltd., Shenyang 110179, China; zengguang@spic.com.cn
* Correspondence: wbyu@sdu.edu.cn

In recent decades, traditional fossil fuels such as coal, oil, and natural gas have made the greatest contributions to the economic development of the industrial sector. However, the negative impact of fossil fuel-based energy has also resulted in the pollution of the natural environment, which presents a major threat to long-term growth and prosperity. In response to energy and environmental crises, zero-carbon vehicles and power generation technology are becoming hot topics, in both the industrial and academic communities. More experts and scholars have demonstrated significant and broad influence in their field, and the number of related technical papers is also rapidly increasing. Therefore, we are committed to providing a platform for high-quality papers on research topics including but not limited to the following: renewable energy vehicles, ammonia and hydrogen technologies in power systems, virtual vehicles, thermal power plant peak regulation technologies, etc. The Topic focuses on fundamental and applied research examining the specific impacts of automation on mobility, energy demand, and greenhouse gas emissions. Many up to date concepts in technologies designed to accelerate the race to carbon neutrality and sustainable development in vehicles and power systems are explored, discussed, and published as original research papers in this Topic, "Zero Carbon Vehicles and Power Generation".

Wu et al. [1] proposed a novel method characterizing the air backflow of the underhood, in order to improve the thermal efficiency of the air conditioning system (ACS) and reduce the energy consumption of Plug-in Hybrid Electric Vehicle (PHEV). Additionally, a 1D model for analyzing air backflow in the underhood was established and a Computational Fluid Dynamics (CFD) method for calculating air backflow rate and distribution was proposed. It was found that the decrease in the air backflow rate of the underhood helped to improve the refrigeration capacity of the ACS, and, when the backflow ratio cannot be reduced below 10%, the air backflow should be distributed as evenly as possible at the front end of the condenser.

Zhang et al. [2] introduced an analytical model of the voltage source converter (VSC) under the unbalanced condition through mathematical derivations, and the final model was a coupled Thevenin circuit. The proposed model allowed for the direct computation of non-characteristic third harmonics using harmonic power flow studies. The results showed that the VSC, under unbalanced conditions, emitted both positive-sequence and negative-sequence third harmonics, and that the positive-sequence third harmonic was much larger than the negative-sequence third harmonic. It also showed that the unbalanced level and the size of the dc-link capacitor were critical to the level of non-characteristic third harmonics.

Wu et al. [3] proposed and analyzed a multi-source microgrid economic dispatch model to reduce carbon emissions. The implemented carbon trading mechanism contributed to achieving carbon emissions control. Companies with surplus carbon emission quotas can increase their additional income by selling such quotas to reduce the total generation and operation cost of the power system. The low-carbon dispatch model enabled wind power, solar energy, electric vehicles, and other distributed power generation units

Citation: Yu, W.; Zeng, G. Zero-Carbon Vehicles and Power Generation. *Sustainability* **2024**, *16*, 6447. https://doi.org/10.3390/su16156447

Received: 11 June 2024
Revised: 18 July 2024
Accepted: 22 July 2024
Published: 28 July 2024

to benefit from carbon trading and to reduce their operation costs while expanding the microgrid system, which ultimately reduced the total cost of the power system.

Guo et al. [4] established a system state equation based on the longitudinal dynamics equation of a vehicle, to deal with the factors influencing electric vehicles when driving under complex conditions. Combined with the improved Sage–Husa adaptive Kalman filter algorithm, a road slope estimation model was established. After the driving speed and rough slope observation were input into the slope estimation model, the accurate road slope estimation at the current time could be obtained. The road slope estimation method was compared with the original Sage–Husa adaptive Kalman filter road slope estimation method through three groups of road tests with different slope ranges, and the accuracy and stability advantages of the proposed algorithm in road conditions with large slopes were verified.

Jo and Kim [5] analyzed the aerodynamic interactions between platooning moving vehicles under different platooning conditions on a freeway. It was found that the effect of the vortex generated by the forward vehicle reduced the value of the stagnation pressure generated at the front of the rear vehicle, which effectively reduced drag on the driving vehicle.

Ji et al. [6] proposed a quantitative evaluation method of driving behavior, based on Naturalistic Driving Studies (NDS) data collected from shared electrical cars and online car-hiring services. In this study, data acquisition, the treatment method, and the data volume verification were analyzed to ensure the effectiveness of the dataset. The distribution characteristics of the main driving behavior parameters were studied. On this basis, the evaluation method was proposed and verified.

Tao et al. [7] presented a novel single-switch high-gain dc/dc converter with a ripple-free input current. The structure consisted of two cells: a coupled inductor cell and a switched capacitor cell. The coupled inductor cell in the proposed converter provided a ripple-free input current. The switched capacitor cell provided a high voltage gain. The converter had a simple control strategy due to the use of a single switch. Moreover, the output capacitor was charged and discharged continuously by a 180° phase shift to eliminate the output voltage ripple.

Ottesen et al. [8] explored consumer preferences for Electric Vehicles (EVs) in Kuwait, in terms of which factors are influential for the 'early majority' (i.e., a part of the general market vs. a niche one) that could influence their purchasing behavior in favor of EVs. A comprehensive and up-to-date picture of the preferences regarding this market was provided, while a variety of valuable promotional tactics were discussed, which may be implemented in conjunction with public incentives and policy changes in the State of Kuwait.

Mišić et al. [9] proposed special energy management for a mountain railway with optimal power distribution and minimum hydrogen consumption. A simulation model was created in a Matlab/Simulink environment for the optimization of hybridized power systems on trains, and it can be easily modified for the hybridization of any type of train. Optimization was performed using sequential quadratic programming (SQP). The results showed that this hybrid train topology had the ability to recover battery and supercapacitor state of charge (SOC) while meeting vehicle speed and propulsion power requirements.

Al-Shami et al. [10] described a novel approach for modeling, identifying, and controlling a running gas turbine power plant. A simplified nonlinear model structure composed of s-domain transfer functions and nonlinear blocks represented by rate limiters, saturations, and look-up tables was proposed. The model was used to design a multiple PI/PD control to regulate the gas turbine via the inlet guide vane and fuel vales, so as to raise and stabilize the compressor's differential pressure or pressure ratio, as well as the raise the set-point of the temperature exhausted from the combustion turbine.

Asiamah et al. [11] developed a hierarchy for understanding the impacts of active and non-active transport modes on the environment and analyzed the adoption of active transportation between older and younger people. The review suggested that the only active transport modes with no or negligible carbon footprint are walking, running, and swimming, as they do not result in a product that adds to atmospheric greenhouse gasses.

Ma et al. [12] analyzed previous studies and current research on the current technical advances emerging for use in the assisted combustion of ammonia. It was highlighted that plasma-assisted combustion (PAC) was able to change classical ignition and extinction S-curves to monotonic stretching, which makes low-temperature ignition possible while resulting in moderate NOx emissions.

Li et al. [13] developed a new four-component jet fuel surrogate which could satisfactorily emulate the chemical and physical properties of real jet fuel, including the cetane number (CN), threshold sooting index (TSI), molecular weight (MW), a lower heating value (LHV), the ratio of hydrogen and carbon (H/C), and the liquid density, viscosity, and surface tension. Furthermore, a reduced and robust kinetic chemical mechanism (containing 124 species and 590 reactions) that could be directly employed in practical engine combustion simulations was also developed.

Banna et al. [14] used a quantitative descriptive method (with close-ended questions) to collect data from a sample of 227 Kuwaiti nationals who were representative of the owners of half a million internal combustion engine (ICE) cars, categorized as early majority consumers. The findings indicated that over 50 percent of the respondents would prefer to buy an EV in the subsequent three years, when certain criteria were satisfied, including government-controlled pricing policies and recharging point availability, high-speed roads, and free EV-dedicated parking spaces. Furthermore, over 40 percent of respondents stated that they would contemplate purchasing an electric vehicle if the price of gasoline or diesel increased by 19 to 50 percent. The findings also indicated that more than 40 percent of respondents believed that EVs are fire- and crash-safe, and roughly 50 percent of the respondents would be willing to pay between 6 and 20% more for an EV because they believed that EVs are eco-friendlier vehicles and are significantly faster than conventional petrol vehicles.

Nurdini et al. [15] used the open-source R language and the bibliometrix package to carry out a bibliometric analysis of data from 593 scientific publications, taken from Scopus, to investigate the research landscape of electric vehicle waste management. It was revealed that the research area of recycling electronic waste from electric vehicles is still experiencing annual growth that is accelerating rapidly. The findings also indicated that China stands out as the leading contributor to publications, with Tsinghua University being a prominent research institution in this field. In 2023, the most frequently occurring topic was "closed loop", while "recycling" was the dominant keyword.

Hoth et al. [16] evaluated the solar energy potential of parking spaces in Berlin, considering challenges like building and tree shading using digital surface models and weather data for solar simulations. Utilizing open datasets and software, the analysis covered 48,827 parking spaces, revealing that vehicle-integrated photovoltaics (VIPV) could extend vehicle range from 7 to 14 km per day, equating to a median annual increase of 2527 km. The findings suggested possible median yearly cost savings of 164 euros from reduced grid charging, and this study introduced a method to pinpoint parking spaces that are most suitable for solar charging.

Chen et al. [17] comprehensively studied dynamic vehicle data contributing to traffic carbon emissions in terms of data sensitivity and uncertainty. The active subspace method could identify which input parameters were the most important through magnitudes of the input parameter weights, while exploring how the combination of inputs was related to the output of interest, without the expense of multiple simulations. It was concluded that the CO_2 emission factor was most sensitive to the vehicle specific power (VSP). The method has great potential to readily derive the relationship between the combination of inputs and outputs in a complex domain without the expense of multiple simulations. Also, the relationship between the input parameters (i.e., the active variables) and the CO_2 emission factor could be formulated using a quadratic function.

Zhang et al. [18] investigated the effects of bore taper, the starting height of the tapered profile, and ellipticity on the friction power and knocking energy of a piston ring cylinder bore (PRCB) system based on the full-scale test method, and the optimization of the design of the bore profile was carried out with the objectives of minimizing the system's friction power and the peak knocking kinetic energy.

Cardoso et al. [19] proposed an active, purely mechanical solution to the problem of irregular torque production in an alternative internal combustion engine. This solution used an actuator built on a camshaft and a spring, which stored and returned energy during the engine operating cycle, allowing torque production to be normalized, avoiding heavy flywheels.

Huang et al. [20] investigated the synergistic coupling process between the detonation and diesel cycles using gasoline as fuel. A numerical simulation model was constructed to analyze the detonation characteristics of a pulse detonation combustor (PDC), followed by experimental verification, which showed that the generation of detonation waves was influenced by flame and compression wave interactions. Increasing the airflow did not shorten the deflagration-to-detonation transition (DDT) time, whereas increasing the blockage ratio (BR) caused the DDT time to decrease and then increase. Large BRs affected the initiation speed of detonation in the tube, while small BRs impacted the DDT distance and peak pressure.

Liu et al. [21] combined the approaches of bench experiments and numerical simulations to investigate the influence of the injector nozzle diameter on the in-cylinder fuel–air mixture and combustion process, based on a combustion strategy characterized by a high-density and lean mixture. The research provided guidance and suggestions for the selection of an injector nozzle diameter in the development of advanced engine combustion systems.

The papers in this Topic reveal an exciting field, namely "zero carbon vehicles and power generation", which is critical to energy utilization and sustainable development. Pursuing work in this field requires professional knowledge in fields such as energy, fuel, automotives, and power generation. We are delighted to be invited to be the Editors for this "Topic". We have received strong support from professional researchers from many renowned universities and research institutions. We will continue to grow our strengths in order to make a greater contribution to carbon neutrality goals. We firmly believe that, through continuous collaboration among all researchers, these steps will help resolve global challenges and address energy sustainability at the source. We hope that this "Topic" will help bring the research community closer together. Finally, we would like to thank all authors, reviewers, and editors who contributed to this publication. We believe that the readers of all the journals involved in this Topic will find these scientific manuscripts interesting and useful for their future research efforts.

Author Contributions: Investigation, W.Y. and G.Z.; Writing—review and editing, W.Y. and G.Z. All authors have read and agreed to the published version of the manuscript.

Funding: This research received no external funding.

Conflicts of Interest: Author Guang Zeng was employed by the company SPIC Northeast Energy Technology Co., Ltd. The remaining authors declare that the research was conducted in the absence of any commercial or financial relationships that could be construed as a potential conflict of interest.

References

1. Wu, H.; Tang, X.; Xu, S.; Zhou, J. Research on Energy Saving of PHEV Air Conditioning System Based on Reducing Air Backflow in Underhood. *Energies* **2022**, *15*, 3183. [CrossRef]
2. Zhang, M.; Zhi, H.; Zhang, S.; Fan, R.; Li, R.; Wang, J. Modeling of Non-Characteristic Third Harmonics Produced by Voltage Source Converter under Unbalanced Condition. *Sustainability* **2022**, *14*, 6449. [CrossRef]
3. Wu, Y.; Lau, Y.-Y.; Wu, J.-A. Integration of Electric Vehicles into Microgrids: Policy Implication for the Industrial Application of Carbon Neutralisation in China. *World Electr. Veh. J.* **2022**, *13*, 96. [CrossRef]

4. Guo, J.; He, C.; Li, J.; Wei, H. Slope Estimation Method of Electric Vehicles Based on Improved Sage–Husa Adaptive Kalman Filter. *Energies* **2022**, *15*, 4126. [CrossRef]
5. Jo, J.; Kim, C.-H. Numerical Study on Aerodynamic Characteristics of Heavy-Duty Vehicles Platooning for Energy Savings and CO_2 Reduction. *Energies* **2022**, *15*, 4390. [CrossRef]
6. Ji, S.; Zhang, K.; Tian, G.; Yu, Z.; Lan, X.; Su, S.; Cheng, Y. Evaluation Method of Naturalistic Driving Behaviour for Shared-Electrical Car. *Energies* **2022**, *15*, 4625. [CrossRef]
7. Tao, R.; Yue, J.; Huang, Z.; An, R.; Li, Z.; Liu, J. A High-Gain DC Side Converter with a Ripple-Free Input Current for Offshore Wind Energy Systems. *Sustainability* **2022**, *14*, 11574. [CrossRef]
8. Ottesen, A.; Banna, S.; Alzougool, B. How to Cross the Chasm for the Electric Vehicle World's Laggards—A Case Study in Kuwait. *World Electr. Veh. J.* **2023**, *14*, 45. [CrossRef]
9. Mišić, M.; Stojkov, M.; Tomić, R.; Lovrić, M. Optimal Hybridization with Minimum Fuel Consumption of the Hybrid Fuel Cell Train. *Designs* **2023**, *7*, 45. [CrossRef]
10. Al-Shami, M.; Mohamed, O.; Abu Elhaija, W. Energy-Efficient Control of a Gas Turbine Power Generation System. *Designs* **2023**, *7*, 85. [CrossRef]
11. Asiamah, N.; Awuviry-Newton, K.; Nesser, W.; Alvarez, E.N. Carbon Footprints of Active and Non-Active Transport Modes: Hierarchy and Intergenerational Narrative Analyses. *Sustainability* **2023**, *15*, 12795. [CrossRef]
12. Ma, F.; Guo, L.; Li, Z.; Zeng, X.; Zheng, Z.; Li, W.; Zhao, F.; Yu, W. A Review of Current Advances in Ammonia Combustion from the Fundamentals to Applications in Internal Combustion Engines. *Energies* **2023**, *16*, 6304. [CrossRef]
13. Li, G.; Cui, B.; Zhang, C.; Chang, L.; Chen, L. Formulation of a Jet Fuel Surrogate and Its Kinetic Chemical Mechanism by Emulating Physical and Chemical Properties of Real Jet Fuel. *Sustainability* **2023**, *15*, 13792. [CrossRef]
14. Banna, S.; Ottesen, A.; Alzougool, B. Reasons Why Only Kuwaiti Citizens Drive Electric Vehicles despite Being Only a Quarter of the Population. *World Electr. Veh. J.* **2023**, *14*, 287. [CrossRef]
15. Nurdini, A.; Nurcahyo, R.; Prabuwono, A.S. Waste from Electric Vehicle: A Bibliometric Analysis from 1995 to 2023. *World Electr. Veh. J.* **2023**, *14*, 300. [CrossRef]
16. Hoth, P.; Heide, L.; Grahle, A.; Göhlich, D. Vehicle-Integrated Photovoltaics—A Case Study for Berlin. *World Electr. Veh. J.* **2024**, *15*, 113. [CrossRef]
17. Chen, J.; Yu, H.; Xu, H.; Lv, Q.; Zhu, Z.; Chen, H.; Zhao, F.; Yu, W. Investigation on Traffic Carbon Emission Factor Based on Sensitivity and Uncertainty Analysis. *Energies* **2024**, *17*, 1774. [CrossRef]
18. Zhang, J.; Wang, N.; Wang, J.; Wang, H.; Zhang, X.; Dai, H.; Lin, J. Low-Friction and -Knocking Diesel Engine Cylindrical-Tapered Bore Profile Design. *Energies* **2024**, *17*, 2042. [CrossRef]
19. Cardoso, D.S.; Fael, P.O.; Gaspar, P.D.; Espírito-Santo, A. An Innovative Mechanical Approach to Mitigating Torque Fluctuations in IC Engines during Idle Operation. *Designs* **2024**, *8*, 47. [CrossRef]
20. Huang, D.; Wang, J.; Shi, M.; Yang, P.; Wu, B. Combustion Mechanism of Gasoline Detonation Tube and Coupling of Engine Turbocharging Cycle. *Energies* **2024**, *17*, 2466. [CrossRef]
21. Liu, Y.; Su, W. The Influence of Injector Nozzle Diameter on High-Density and Lean Mixture Combustion in Heavy-Duty Diesel Engines. *Energies* **2024**, *17*, 2549. [CrossRef]

energies

MDPI

Article

Research on Energy Saving of PHEV Air Conditioning System Based on Reducing Air Backflow in Underhood

Haibo Wu [1,2], Xingwang Tang [1], Sichuan Xu [1,*] and Jiangbin Zhou [2]

1 School of Automotive Studies, Tongji University, Shanghai 200070, China; wuhaibo@csvw.com (H.W.); tangxingwang@tongji.edu.cn (X.T.)
2 SAIC Volkswagen Automotive Co., Ltd., Shanghai 201800, China; zhoujiangbin@csvw.com
* Correspondence: scxu_tongji@163.com

Abstract: A novel method characterizing the air backflow of the underhood in order to improve the thermal efficiency of the air conditioning system (ACS) and reduce the energy consumption of PHEV is proposed in this paper. In addition, a 1D model for analyzing air backflow occurring in the underhood is established and a CFD method for calculating air backflow rate and distribution is proposed. It is found that the decrease in the air backflow rate of the underhood helps to improve the refrigeration capacity of the ACS, and when the backflow ratio cannot be reduced below 10%, the air backflow should be distributed as evenly as possible at the front end of the condenser. Moreover, in order to eliminate the impact of backflow on the underhood of PHEV, the gap between the radiator and the bracket is sealed and the gap around the air guide is reduced. Compared with the original structure, the backflow rate of the optimized structure is reduced from 32.7% to 9.3% and the cabin temperature can be reduced by 3–5 °C.

Keywords: PHEV; air conditioning system; air backflow; thermal management

Citation: Wu, H.; Tang, X.; Xu, S.; Zhou, J. Research on Energy Saving of PHEV Air Conditioning System Based on Reducing Air Backflow in Underhood. *Energies* **2022**, *15*, 3183. https://doi.org/10.3390/en15093183

Academic Editors: Wenbin Yu and Guang Zeng

Received: 30 March 2022
Accepted: 24 April 2022
Published: 27 April 2022

1. Introduction

With the proposal of "carbon peaking" and "carbon neutrality", a higher requirement for vehicle exhaust emissions is required. The shortage of non-renewable resources and the increasingly stringent pollutant emission regulations are driving vehicle manufacturers to lower fuel consumption and lower exhaust emissions [1–3]. Compared with traditional fuel vehicles, plug-in hybrid electric vehicles (PHEV) clearly reduce fuel consumption and CO_2 emissions [4,5]. The PHEV power sources rely on electric power generated by electric machines and mechanical power generated by internal combustion engines (ICEs) [6]. Due to the high efficiency of electric machines and the quick development of battery technology, the PHEV is a rational choice for a long period in terms of environmental issues and energy saving [7].

In the past decade, a lot of scholars have mainly focused on the energy management strategies in the research of energy saving for PHEV [8–11]. The author of [12] proposed a predictive energy management strategy (EMS) based on reinforcement learning for PHEV. The EMS combined the velocity prediction with the optimal power distribution between engine and electrical motor, which greatly improved the vehicle's fuel economy. The author of [13] obtained the optimized operating point of the engine based on the dynamic programming algorithm and proposed an improved rule-based energy management method. The results show that the energy management strategy can effectively reduce the fuel consumption per 100 km of the vehicle equipped with a diesel engine. 22.80 L/100 km. A PHEV-integrated optimization simulation platform based on Isight and MATLAB/Simulink software for bus is established in [14]. They proposed a multi-objective optimization and matching method and proved its effectiveness and superiority in fuel reduction and better performance. The optimization of the series fuel cell vehicle components by a single and a multi-objective genetic algorithm to improve the vehicle performance and the operational costs is also performed in [15]. A method for PHEV online energy management utilizing an evolutionary algorithm was developed in [16]. The results show that the proposed self-adaptive control strategy outperforms the conventional binary control strategy with an average of 10.7% fuel savings without considering charging opportunity and 31.5% fuel savings when considering charging opportunity. The author of [17] provided a comprehensive study regarding the PHEV's optimum powertrain design based on an interactive adaptive-weight genetic algorithm approach, which aims to simultaneously minimize the PHEV's fuel consumption, battery state of health, charging time, and costs.

Nowadays, in order to reduce the energy consumption of the PHEV, most research is limited to the energy management strategy. However, there is little research that pays attention to the energy-saving potential of the air conditioning system. As the core component of the PHEV, the air conditioning system needs to keep the cabin at a proper temperature. However, because it is also one of the most important energy-consuming components of PHEV, especially the compressor of the air conditioning system [18–20], a low-efficiency air conditioning system will bring great challenges to the endurance of the vehicle. The cooling load of an air conditioning system can cause an 18–37% reduction in the Urban Dynamometer Driving Schedule (UDDS) range depending on ambient conditions [21]. Therefore, in order to reduce the energy consumption of PHEV, it is vital to pay attention to the air conditioning system.

In this paper, in order to reduce energy consumption of PHEV, the efficiency of the air conditioning system is addressed. A detailed and quantitative analysis of the air backflow of the PHEV's underhood is carried out to improve the efficiency of the air conditioning system, so that the system energy consumption can be reduced. In fact, the air backflow phenomenon in the underhood of PHEV refers to when hot air from the underhood returns to the front-end cooling module, which increases the temperature of the windward air passing through the condenser, intercooler, low-temperature radiator, and electrical radiator, resulting in a decrease in the heat dissipation performance of the cooling module. Therefore, the air backflow in the underhood can cause the performance of the condenser to decrease, which in turn will affect the performance of the air conditioning system, resulting in increased energy consumption.

Based on the reviewed literature, optimizing the air backflow of the underhood is an effective way to improve the thermal efficiency of the air conditioning system and reduce the energy consumption of PHEV. Therefore, the contributions of this paper are threefold:

- A heat flux marking method is proposed to characterize the mechanism of air backflow and its distribution in the condenser, thereby quantifying the air backflow phenomenon in the underhood of PHEV.
- The performances of the air conditioning system, including the evaporator outlet temperature, cooling capacity, exhaust pressure of the compressor, and COP, are investigated under different ambient temperatures, air backflow ratios, and air backflow distribution.
- An optimization model of the underhood is proposed to eliminate the impact of air backflow on the air conditioning system of PHEV.

The rest of the paper is organized as follows. The research method, including the 1D model characterizing the effect of air backflow on the air conditioning system performances, the 3D model characterizing and quantifying the backflow phenomenon in the underhood of PHEV, and experimental test system verifying the air conditioning system performance, is presented in Section 2. The evaluation results, including the impact of the air backflow effect in the underhood of PHEV on the air conditioning system, mechanism analysis of air backflow phenomenon, and performance optimization, are reported in Section 3. Section 4 concludes the paper.

2. Research Method

2.1. Research Framework of the Air Backflow Effect in the Underhood

To quantify the backflow phenomenon in the underhood of PHEV and explore the impact of the air backflow in the underhood of PHEV on the air conditioning system, the research framework is constructed in this paper, as shown in Figure 1. The quantification and analysis of the air backflow phenomenon of the underhood of PHEV are realized based on the heat flux marking method demonstrated in this paper. The performances, including the evaporator outlet temperature, cooling capacity, exhaust pressure of compressor, and COP of ACS are investigated under different ambient temperatures, air backflow ratio, and backflow distribution based on one-dimensional simulation. Furthermore, the computational fluid dynamics (CFD) simulation of the underhood can characterize and reflect the air backflow flow field distribution and provide a basis for the structural parameter optimization of the underhood. Moreover, a real vehicle experiment is carried out in order to verify the performance of the air conditioning system after the underhood of PHEV is optimized.

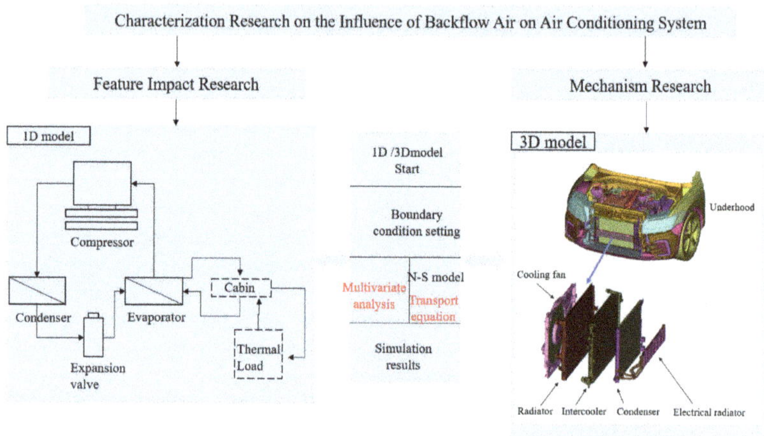

Figure 1. Research framework of the air backflow effect in the underhood of PHEV.

2.2. 1D Model

2.2.1. Compressor

A compressor is the core component of an air conditioning system;, so, the accuracy of its simulation model needs to be much higher than other components. Generally, a 1% error in the compressor simulation model will cause a 0.7% error in the air conditioning system approximately. In this paper, the compressor efficiency model is adopted and the physical model applied to the compressor efficiency model is shown in Equations (1)–(3) [22]:

$$m_{com} = \frac{\eta_v V_{th} r_{com}}{v_{suc}} \tag{1}$$

$$P_{comp} = m_{com} \left(h_{dis} - h_{suc}\right) / f_Q \tag{2}$$

$$h_{dis} = h_{suc} + \frac{h_{dis|s} - h_{suc}}{\eta_s} \tag{3}$$

where V_{th} is the displacement of the compressor, r_{com} is the rotary speed of the compressor, m_{com} is the mass flow of the compressor, η_v is the volumetric efficiency of the compressor, P_{comp} is the energy consumption of the compressor, h_{dis} is the exhaust enthalpy of the compressor, h_{suc} is the suction enthalpy of the compressor, η_s is the isentropic efficiency, $h_{dis|_s}$ is the exhaust enthalpy under the condition of isentropic compression, and f_Q is the heat loss of the compressor.

2.2.2. Condenser and Evaporator Model

In this paper, both the condenser and the evaporator of the ACS can be regarded as a model of the microchannel tube–fin heat exchanger; the working principle diagram of the heat exchanger is shown in Figure 2 [23]. The whole heat exchange process includes three parts: external convective heat exchange air–wall, heat conduction wall + fins, and internal convective heat exchange refrigerant–wall. In this model, the heat transfer correlation of the air side of the heat exchange can be expressed as follows [24]:

$$Nu = C_0 \times Re^N \times {Pr_t}^M \tag{4}$$

where Nu is the Nusselt number; Re is the Reynolds number; Pr_t is the Prandtl number; and M, N, and C_0 are the parameters that can be used to adjust the heat rejection performance of the heat exchanger.

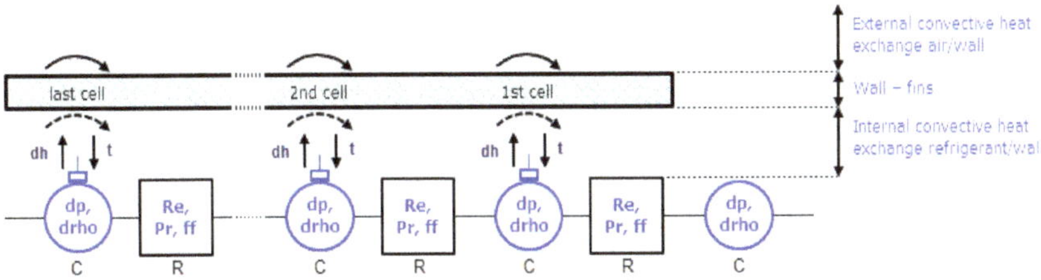

Figure 2. Diagram of discrete cell of the condenser component.

2.2.3. Expansion Valve

The expansion valve model is used in this paper to achieve different throttling effects by adjusting the opening degree of the expansion valve. The heat loss of the expansion valve is ignored in the model; so, the flow process of the refrigerant in the expansion valve is regarded as a constant enthalpy process. The mass flow rate is calculated as follows:

$$m_{exv} = \rho \times C_q \times A \times \sqrt{\frac{2 \times \Delta P}{\rho}} \tag{5}$$

$$\Delta P = P_i - P_O \tag{6}$$

where m_{exv} is the mass flow rate, kg/s; ρ is the density, kg/m^3; P_i is the inlet pressure, barA; P_O is the outlet pressure, barA; and C_q is the flow coefficient, which can be calculated by Formula (7):

$$C_q = C_{qmax} \times tanh\left(\frac{2 \times \lambda}{\lambda c}\right) \tag{7}$$

where C_{qmax} is the maximum flow coefficient; λ is the flow number; and λc is the critical flow number.

2.2.4. Fan and Radiator Model

The cooling fan is located at the front or rear of the radiator to increase the air flow through the radiator and improve the heat dissipation performance of the radiator. In this paper, the performance characteristic parameters of the fan are obtained based on the law of similarity, which mainly include the volumetric flow coefficient and pressure coefficient, and the calculation method is as follows [25]:

$$\phi = \frac{\dot{Q}}{A_{fan} \cdot V_{tip}} \tag{8}$$

$$\psi = \frac{\Delta P}{\frac{1}{2}\rho V_{tip}^2} \tag{9}$$

where ϕ is the volumetric flow coefficient; ψ is the pressure coefficient; $\dot{Q}$ is the volume flow of air; A_{fan} is the fan area; and V_{tip} is the linear velocity of the fan blade tip.

As for the radiator, the heat transfer and flow resistance of the cold side or hot side of the radiator can be calculated by the following formulas:

$$\Delta_p = \frac{1}{2} K \cdot \rho \cdot V_{\text{cool}}^2 \tag{10}$$

$$Q_{\text{rad}} = A_{\text{exch}} \cdot U \cdot (T_{\text{in}} - T_{\text{out}}) \tag{11}$$

where A_{exch} is the heat exchange area of heat exchanger; ρ is the air density; and K is the pressure loss coefficient.

2.3. 3D Models

2.3.1. Heat Exchanger

To characterize the backflow phenomenon in the underhood of PHEV, it is necessary to perform computational fluid analysis on the cooling module of the underhood. In the cooling module of the underhood, the condenser, intercooler and, radiator all belong to the heat exchanger. In this paper, the heat transfer model of the heat exchanger is established by NTU method, as follows [25]:

$$NTU = \frac{-\ln(1 - \varepsilon - \varepsilon C_r)}{(C_r + 1)} \tag{12}$$

where C_r is the heat capacity ratio and ε is the effectiveness.

The porous media model is adopted to characterize the pressure loss inside the heat exchanger; it also adopts the correction method of the standard momentum equation, which adds the momentum source term composed of viscous loss term and inertial to the momentum equations. The mathematical description is as follows [22]:

$$S_{\text{i}} = -\left(\sum_{\text{j}=1}^{3} D_{\text{ij}} \mu u_{\text{j}} + \sum_{\text{j}=1}^{3} C_{\text{ij}} \frac{1}{2} \rho |u| u_{\text{j}}\right) \tag{13}$$

where S_{i} is the source term in the momentum equation, $|u|$ is the velocity scalar, u is the velocity vector, μ is the aerodynamic viscosity, and D and C is the given matrix.

D and C are defined as diagonal matrices, and the formula can be expressed as follows:

$$S_{\text{i}} = -\left(\frac{\mu}{\beta} u_{\text{i}} + \eta \frac{1}{2} \rho |u| u_{\text{i}}\right) \tag{14}$$

where $\frac{1}{\beta}$ is the viscous resistance coefficient, and η is the inertial resistance coefficient.

The simplified momentum equation of porous media can be expressed in the form of pressure loss and source term as follows:

$$\Delta P = -\left(\frac{\mu \Delta n}{D} u_{\text{i}} + C_1 \frac{1}{2} \rho |u| \Delta n u_{\text{i}}\right) \tag{15}$$

The characteristic parameters of the heat exchangers are demonstrated in Table 1:

Table 1. The viscous resistance coefficient and inertial resistance coefficient of heat exchangers.

Heat Exchanger	Viscous Resistance Coefficient (m^{-2})	Inertial Resistance Coefficient (m^{-1})
Radiator	1192.5	160.1
Intercooler	813.6	74.5
Condenser	851.8	126.9
Electrical radiator	808.2	114.3

2.3.2. Meshing of the PHEV Underhood

The 3D solid model of the PHEV underhood, including the condenser, radiator, and other components, is shown in Figure 3a. In order to avoid the interference of the airflow disturbance in the limited space on the model accuracy, a rectangular air domain (10 L × 10 W × 6 H) is added, as shown in Figure 3b.

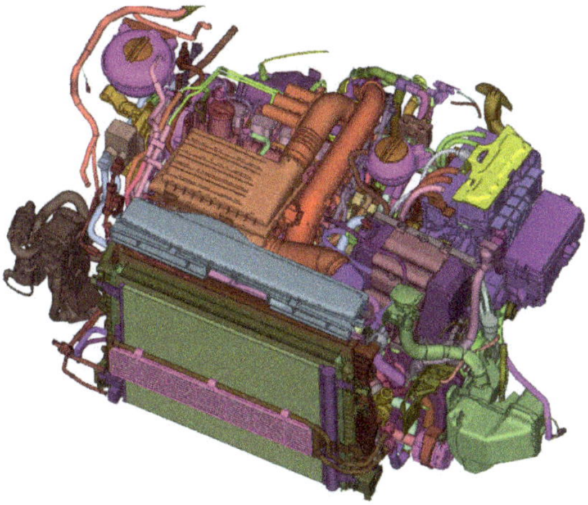

(a)

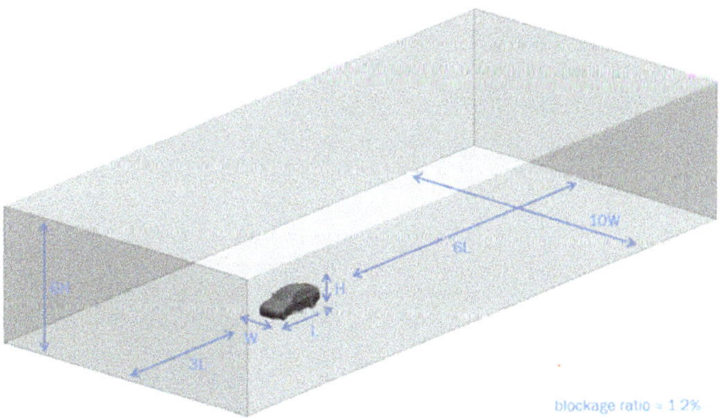

(b)

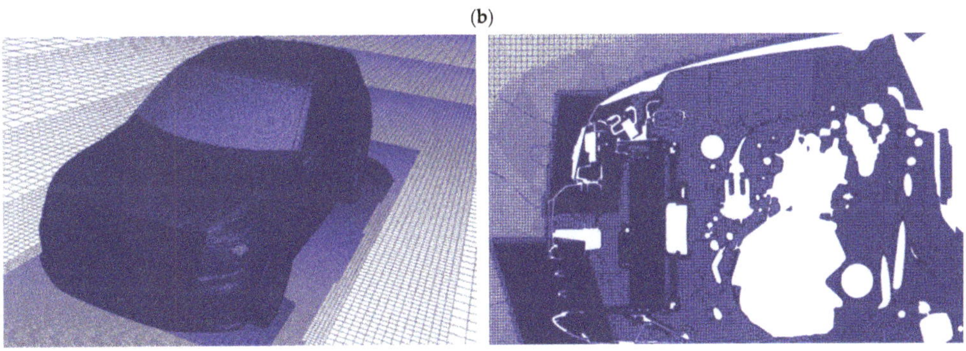

(c)

Figure 3. The 3D model of the underhood for PHEV. (**a**) A 3D solid model of the underhood; (**b**) the CFD calculation domain of the underhood for PHEV; (**c**) the mesh of the underhood for PHEV.

In view of the fact that the air backflow of the underhood is determined in this paper, the front half of the car body surface is also meshed to ensure that the flow details are captured. The mesh model with 80 million mesh is shown in Figure 3c.

2.3.3. Boundary Condition

The air backflow flow field distribution is investigated based on the CFD method. Some boundary conditions should be determined:

(1) RANS steady-state model is used for steady flow in this model;
(2) Pressure outlet and velocity inlet are selected for outlet and inlet boundary conditions, respectively;
(3) The condenser, intercooler, radiator, and electrical radiator are calculated based on the porous media model;
(4) The cooling fan is simulated based on the multiple reference frame (MRF) method.

2.4. Experimental Test System

In order to verify the effect of air backflow of the underhood on the cooling capacity of the air conditioning system for PHEV, a system bench experiment is established. The measurement points of the temperature sensor in the cabin are shown in Figure 4, and the cabin temperature value is obtained by calculating the average temperature of multiple measurement points.

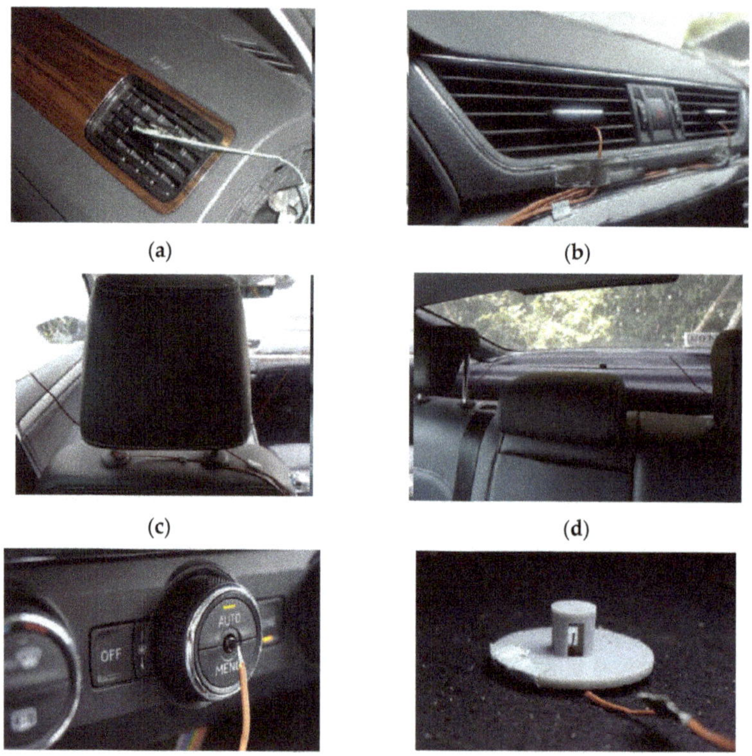

(a) (b) (c) (d) (e) (f)

Figure 4. The measurement points of temperature sensor in the cabin. (**a**) Side air outlet; (**b**) Middle air outlet; (**c**) Front seat passenger; (**d**) Rear seat passenger; (**e**) Dashboard area; (**f**) Cabin bottom.

In order to ensure more accurate simulation results in the 1D/3D calculation, the relevant key parameters of the condenser, evaporator, and other radiators are obtained through the enthalpy difference experiment. The laboratory is shown in Figure 5.

Figure 5. The heat exchanger performance test platform.

2.5. Heat Flux Marking Method

In order to solve the problem that the air backflow phenomenon of the underhood cannot be judged based on a streamline method, a heat flux marking method is proposed based on the mechanism and distribution of air backflow in the heat exchanger. The air flow passing the heat exchanger will be marked. In order to monitor the backflow air, a monitoring surface will be set at the front end of the heat exchanger. The schematic diagram is shown in Figure 6.

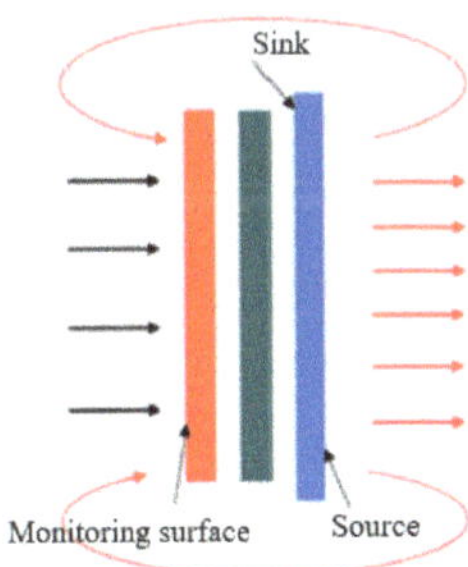

Figure 6. Schematic diagram of air backflow calculation method.

In this paper, in order to quantify the distribution of heated air and calculate the backflow rate, the scalar γ is used to express the proportion of the volume of hot air, referring to the momentum transport equation. The calculation method is as follows:

$$\frac{\partial \gamma}{\partial t} + u_j \frac{\partial \gamma}{\partial x_i} = \nu \frac{\partial^2 \gamma}{\partial x_j \partial x_j} - \nu_t \frac{\partial \gamma}{\partial x_j} \tag{16}$$

When the air backflow phenomenon is characterized based on the CFD method, the N-S equation for the entire fluid domain needs to be solved first. Then, the transport equation should be solved after the flow field is in a stable state. Finally, the distribution of hot air in the entire fluid domain can be obtained after the calculation converges.

3. Results and Discussion

In order to systematically analyze the air backflow phenomenon of the underhood of PHEV in this paper, the influence of the air backflow rate on the air conditioning system is explored under different ambient temperatures and backflow distribution. Furthermore, to eliminate the impact of air backflow of the underhood and improve the performance of the thermal management system, the underhood of PHEV is optimized in this paper.

3.1. Effect of Backflow Rate under Different Ambient Temperatures

In this section, the influence of air backflow rate on the air conditioning system is explored under different ambient temperatures. Figure 7 shows that under ambient temperatures of 30 °C, 35 °C, and 40 °C, as the backflow rate increases from 0% to 30%, the air conditioning system performance parameters, including evaporator outlet temperature, evaporator heat exchange, compressor discharge pressure, and COP, are changed accordingly.

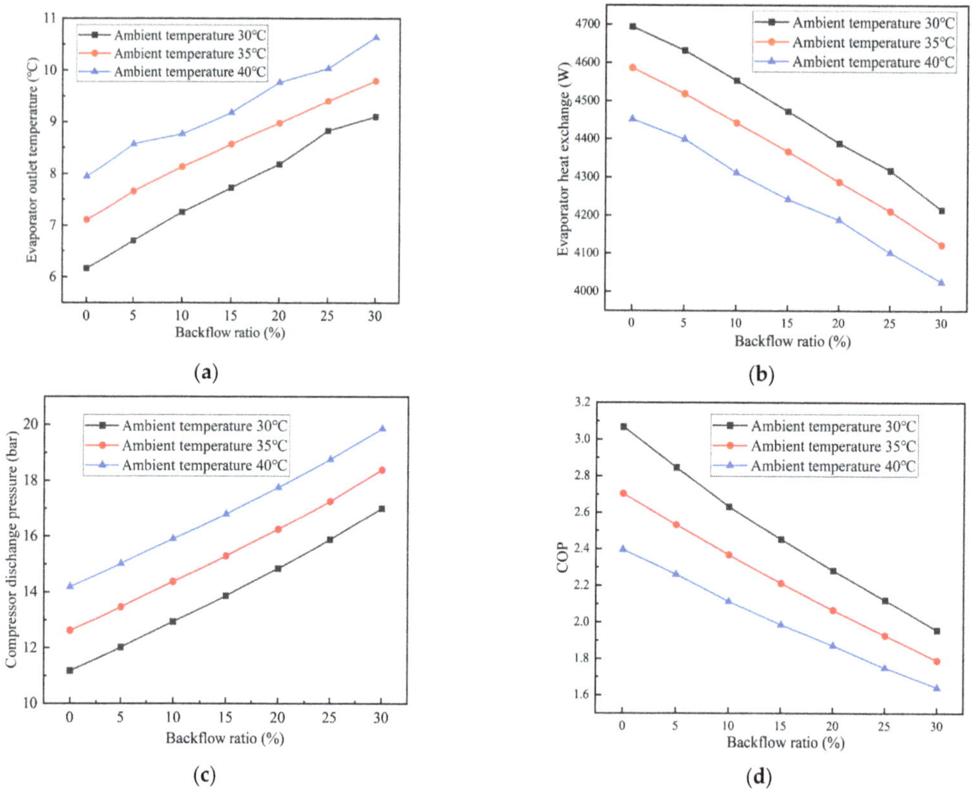

Figure 7. Influences of backflow rate on air conditioning performances under different ambient temperatures. (**a**) Evaporator outlet temperature; (**b**) evaporator heat rejection; (**c**) compressor discharge pressure; and (**d**) COP.

As shown in Figure 7, as the air backflow rate of the underhood is increased from 0% to 30%, the heat exchange performance and energy efficiency of the system is reduced. Specifically, when ambient temperature is 30 °C, the heat exchange of evaporator is reduced from 4693 to 4214 W and the reduction rate is 10.2%, at the same time, while the system COP is reduced from 3.07 to 1.96 and the reduction rate is 36.16%. Thus, more than a 35% improvement of COP could be achieved solely by eliminating air backflow of the underhood of PHEV, which indicates that the decline of the air backflow rate of the underhood enhances the cooling capacity in the air conditioning system. In fact, when the air backflow rate is enhanced, the oncoming air temperature of the condenser will be increased and the heat exchange capacity of the condenser will be decreased, enhancing the condensing pressure and the condensing temperature, thus increasing the discharge pressure of the compressor, as shown in Figure 2(C). Furthermore, when the compressor speed remains unchanged, the suction pressure of the compressor will increase, and the evaporation pressure and evaporation temperature will increase accordingly; thus, the cooling capacity will increase. Moreover, the increase in the suction pressure will cause the compressor power consumption to increase, and the system COP will decrease.

3.2. *Effect of Air Backflow Rate under the Different Air Backflow Distribution*

In fact, the air backflow phenomenon of the underhood occurs in different local areas of the condenser; thus, the influence of different air backflow distribution on the air conditioning system is analyzed in this section. In view of the fact that the distribution ratio of the cooling tube of the condenser analyzed in this paper is 27:13:7:5, the air backflow distribution form is set according to 27:13:7:5 (uniform distribution), 1:1:1:1, and 1:1:2:2. Figure 8 shows that under different backflow distribution, the air backflow rate has an effect on the performance of the air conditioning system.

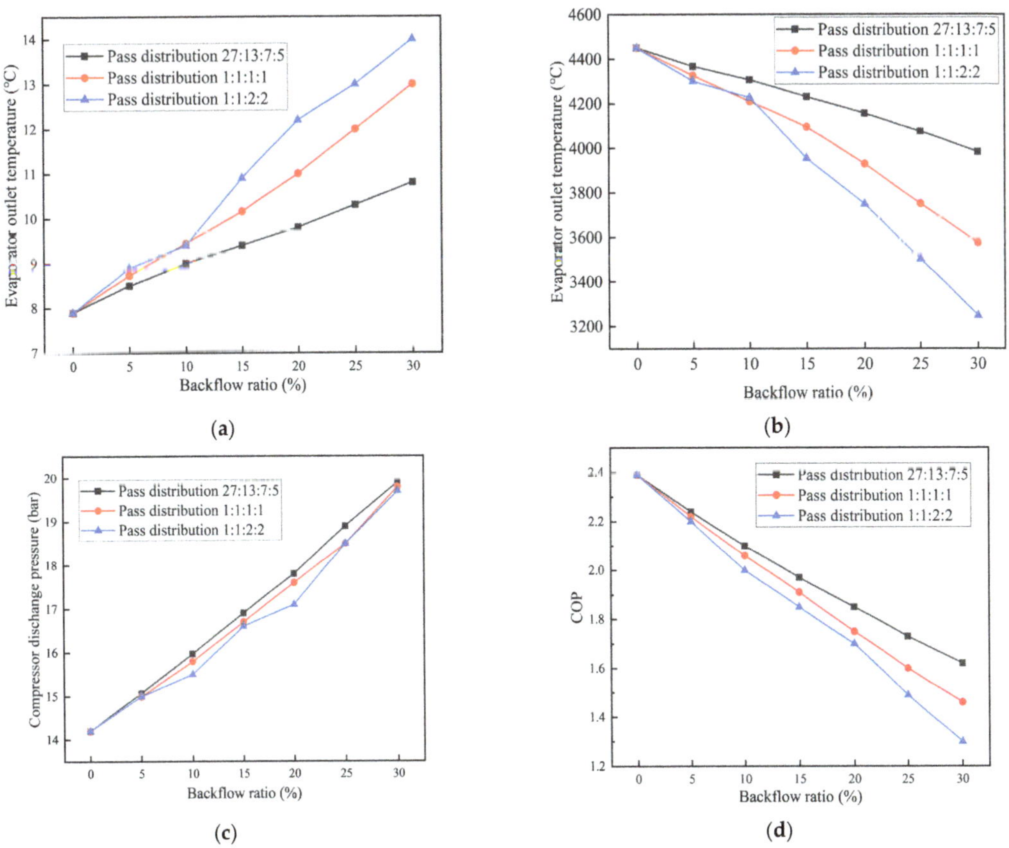

Figure 8. Influences of the backflow rate on air conditioning performances under different ambient temperatures. (**a**) Evaporator exit temperature; (**b**) evaporator heat rejection; (**c**) discharge pressure; and (**d**) COP.

As shown in Figure 8, when the air backflow rate is lower than 10%, the air backflow distribution has almost no effect on the performance of air conditioning system. However, when the air backflow rate is higher than 10%, compared to the air backflow concentrated under the condenser (Pass distribution 27:13:7:5) when air backflow is evenly distributed on the surface of the condenser, the cooling capacity of the air conditioning system will be higher and the system efficiency will improve. Specifically, compared to the air backflow concentrated under the condenser, when the air backflow rate is 30% and air backflow is evenly distributed on the surface of the condenser, the heat exchange performance increases from 3248 to 3983 W and the growth rate is 22.6%; at the same time, the system COP increases from 1.3 to 1.62 and the reduction rate is 24.6%, which indicates that the air backflow should be distributed as evenly as possible at the front end of the condenser.

3.3. Mechanism Analysis of Backflow Phenomenon

The analysis in the above section exhibits that the air backflow rate and air backflow distribution of the underhood have a great impact on the performance of the air conditioning system. In order to avoid the impact of air backflow on the air conditioning system of PHEV, the mechanism that causes the backflow phenomenon is analyzed in this section; thus, the performance of the air conditioning system can be optimized in the subsequent section.

The air backflow phenomenon of the underhood is analyzed and the distribution of air backflow on the surface of the condenser is shown in Figure 9.

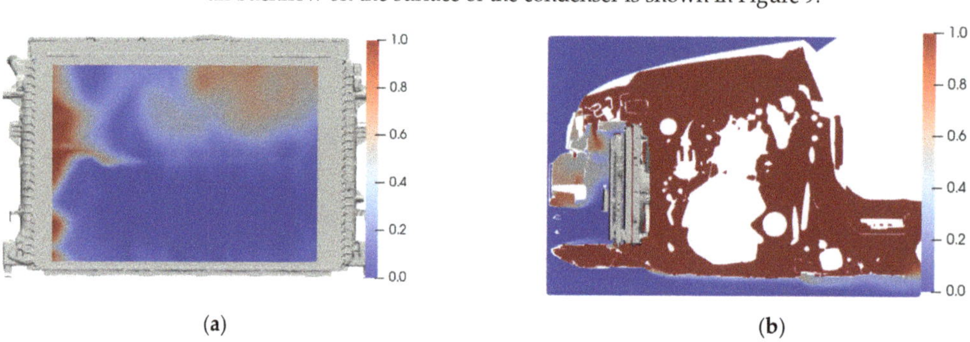

Figure 9. Backflow air distribution of original structure. (**a**) Backflow air distribution front view; (**b**) backflow air distribution side view.

Figure 9 shows that the distribution of the backflow hot air on the surface of the condenser is mainly concentrated on the left and upper right corners of the condenser. In order to trace the source of backflow hot air, as shown in Figure 9b, most of the backflow hot air is generated by the gap between the radiator and the bracket and the gap between the air guide parts.

3.4. Mechanism Analysis of Air Backflow Phenomenon

In order to eliminate the impact of air backflow on the air conditioning system, a cooling module of the underhood for PHEV is optimized in this section. The optimized structure is shown in the Figure 10. As shown in red in Figure 10, the gap between the radiator and the bracket is sealed and the gap around the air guide is reduced.

Figure 11 shows the optimized results of the PHEV underhood, and that the air backflow phenomenon in the left area is significantly improved and that there only is some backflow hot air around some pores that cannot be sealed.

In order to quantitatively analyze the backflow phenomenon of the optimized structure, the air backflow rate and the mass flow of backflow air are calculated based on the heat flux regression method, and the results are shown in Table 2.

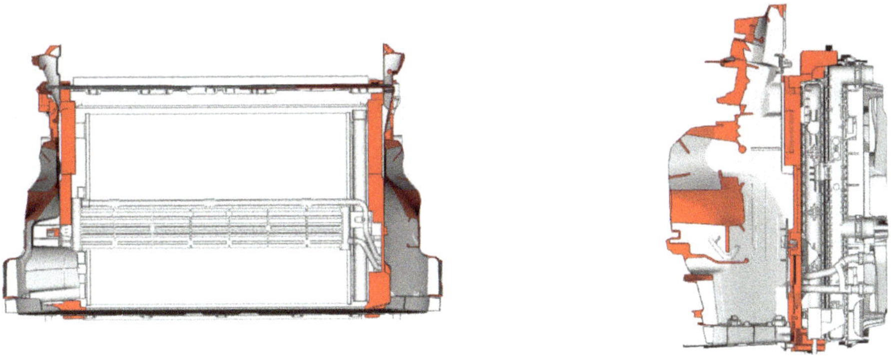

(a) (b)

Figure 10. Optimized cooling module of PHEV. (**a**) Optimized structure front view; (**b**) optimized structure side view.

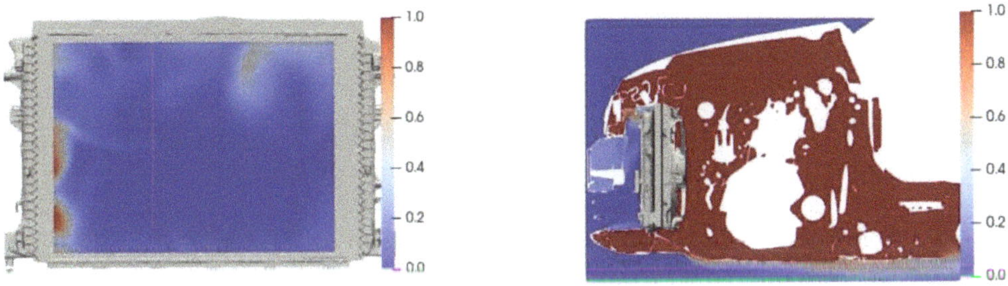

(a) (b)

Figure 11. Backflow air distribution of optimized structure. (**a**) Optimized structure front view; (**b**) optimized structure side view.

Table 2. Comparison of the backflow rate of the optimized cooling module and the original structure.

	Air Mass Flow (kg/s)	Backflow Ratio (%)	Backflow Air Mass Flow (kg/s)
Original Structure	0.518	32.7	0.169
Optimized structure	0.518	9.3	0.048

As shown in Table 2, compared with the original structure, the backflow rate of the optimized structure is reduced from 32.7% to 9.3%, the backflow air in the left area is significantly reduced, and the backflow air mass flow is reduced from 0.169 kg/s to 0.048 kg/s. The reduction rate is 36.16% and the backflow situation of underhood cooling module is significantly improved.

In order to verify results of the optimization, the performance of the air conditioning system installed with the original cooling module and the optimized cooling structure is tested in a real vehicle under driving conditions of 32 km/h. The performance results of the air conditioning system are shown in Figure 12.

As shown in Figure 12, when the air backflow phenomenon is weakened, the cooling capacity of the air conditioning system is significantly better. Compared with the original underhood cooling structure, the cabin temperature can be reduced by 3–5 °C.

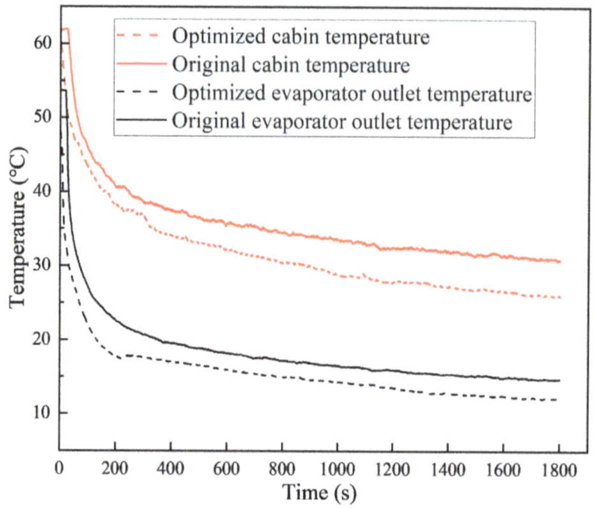

Figure 12. HVAC performances after reflux optimization.

4. Conclusions

The air backflow phenomenon of the PHEV underhood is quantitatively investigated in this paper and the relevant conclusions are as follows:

(1) The heat flux marking method proposed in this paper can accurately quantify the backflow phenomenon in the underhood of PHEV based on the momentum transport equation.
(2) The decrease in the air backflow rate of the underhood helps to improve the refrigeration capacity of the air conditioning system, thereby increasing the COP of the system
(3) When the air backflow ratio cannot be reduced below 10%, the air backflow should be distributed as evenly as possible at the front end of the condenser.
(4) In order to eliminate the impact of air backflow on the PHEV underhood, the gap between the radiator and the bracket is sealed and the gap around the air guide is reduced. Compared with the original structure, the air backflow rate of the optimized structure is reduced from 32.7% to 9.3% and the cabin temperature can be reduced by 3–5 °C.

In order to further investigate the effect of backflow phenomenon of the underhood on air conditioning system, digital twin technology should be considered in the future so that backflow phenomenon can be investigated in an all-round way without a large number of time-consuming and high-cost experiments and simulations.

Author Contributions: Funding acquisition, S.X.; investigation, H.W. and X.T.; methodology, H.W.; project administration, S.X.; visualization, J.Z.; writing—original draft, H.W.; writing—review and editing, X.T. All authors have read and agreed to the published version of the manuscript.

Funding: The authors wish to acknowledge for the financial support from the National Natural Science Foundation of China Grant No. 21776221, for the work reported in this paper.

Institutional Review Board Statement: Not applicable.

Informed Consent Statement: Not applicable.

Data Availability Statement: Not applicable.

Conflicts of Interest: The authors declare that they have no known competing financial interests or personal relationships that could have appeared to influence the work reported in this paper.

Nomenclature

PHEV	plug-in hybrid electric vehicles
ACS	air conditioning system
ICE	internal combustion engine
UDDS	urban dynamometer driving schedule
CFD	computational fluid dynamics
EMS	energy management strategy
COP	coefficient of performance
BTMS	battery thermal management system
MTMS	motor thermal management system
Acronyms	
V_{th}	displacement of the compressor
m_{com}	mass flow of the compressor
P_{comp}	energy consumption of the compressor
h_{suc}	suction enthalpy of the compressor
$h_{dis}\|_s$	exhaust enthalpy under the condition of isentropic compression
ρ	density
η	inertial resistance coefficient
γ	volume proportion of the hot air
ϕ	volumetric flow coefficient
Nu	Nusselt number
Pr_t	Prandtl number
V_{tip}	linear velocity of the fan blade tip
r_{com}	rotary speed of the compressor
η_v	volumetric efficiency of the compressor
h_{dis}	exhaust enthalpy of the compressor
η_s	isentropic efficiency
f_Q	heat loss of compressor
u	velocity vector
$\frac{1}{\beta}$	viscous resistance coefficient
C_r	heat capacity ratio
Re	Reynold number
$\dot{Q}$	volume flow of air
A_{fan}	Fan Area

References

1. Solmaz, H.; Ardebili, S.; Calam, A.; Yılmaz, E.; İpci, D. Prediction of Performance and Exhaust Emissions of a CI Engine Fueled with Multi-Wall Carbon Nanotube Doped Biodiesel-Diesel Blends Using Response Surface Method. *Energy* **2021**, *227*, 120518. [CrossRef]
2. Geng, L.; Xiao, Y.; Li, S.; Chen, H.; Chen, X. Effects of injection timing and rail pressure on particulate size-number distribution of a common rail DI engine fueled with fischer-tropsch diesel synthesized from coal. *J. Energy Inst.* **2021**, *95*, 219–230. [CrossRef]
3. Liu, J.; Wu, P.; Sun, P.; Ji, Q.; Zhang, Q.; Wang, P. Effects of iron-based fuel borne catalyst addition on combustion, in-cylinder soot distribution and exhaust emission characteristics in a common-rail diesel engine. *Fuel* **2021**, *290*, 120096. [CrossRef]
4. Zhang, Y.; Yuan, X.; Duan, L.; Xu, Y.; Lan, F. Environmental temperature effects on the energy flow of plug-in hybrid electric vehicles. *J. Power Sources* **2021**, *506*, 230231. [CrossRef]
5. Mahmoodi-k, M.; Montazeri, M.; Madanipour, V. Simultaneous multi-objective optimization of a PHEV power management system and component sizing in real world traffic condition. *Energy* **2021**, 233, 121111. [CrossRef]
6. Aljehane, N.; Mansour, R. Optimal allocation of renewable energy source and charging station for PHEVs. *Sustain. Energy Technol. Assess.* **2021**, *49*, 101669. [CrossRef]
7. Böhme, T.; Frank, B. *Hybrid Systems, Optimal Control and Hybrid Vehicles*; Springer: Berlin/Heidelberg, Germany, 2017.
8. Hemmati, S.; Doshi, N.; Hanover, D.; Morgan, C.; Shahbakhti, M. Integrated cabin heating and powertrain thermal energy management for a connected hybrid electric vehicle. *Appl. Energy* **2021**, *283*, 116353. [CrossRef]
9. Liu, J.; Chen, Y.; Li, W.; Shang, F.; Zhan, J. Hybrid-Trip-Model-Based Energy Management of a PHEV with Computation-Optimized Dynamic Programming. *IEEE Trans. Veh. Technol.* **2018**, *67*, 338–353. [CrossRef]
10. Fan Li Zhang, Y.; Dou, H.; Zou, R. Design of an integrated energy management strategy for a plug-in hybrid electric bus. *J. Power Sources* **2019**, *448*, 227391. [CrossRef]

11. Chen, Z.; Xiong, R.; Cao, J. Particle swarm optimization-based optimal power management of plug-in hybrid electric vehicles considering uncertain driving conditions. *Energy* **2016**, *96*, 197–208. [CrossRef]
12. Liu, T.; Hu, X.; Li, S.; Cao, D. Reinforcement Learning Optimized Look-Ahead Energy Management of a Parallel Hybrid Electric Vehicle. *IEEE Trans. Mechatron.* **2017**, *22*, 1497–1507. [CrossRef]
13. Peng, J.; He, H.; Xiong, R. Rule based energy management strategy for a series-parallel plug-in hybrid electric bus optimized by dynamic programming. *Appl. Energy* **2017**, *185*, 1633–1643. [CrossRef]
14. Song, P.; Lei, Y.; Fu, Y. Multi-objective optimization and matching of power source for PHEV based on genetic algorithm. *Energies* **2020**, *13*, 1127. [CrossRef]
15. Ribau, J.; Sousa, J.; Silva, C. *Multi-Objective Optimization of Fuel Cell Hybrid Vehicle Powertrain Design—Cost and Energy*; SAE Technical Paper; SAE: Detroit, MI, USA, 2013. [CrossRef]
16. Qi, X.; Wu, G.; Boriboonsomsin, K.; Barth, M.J. Development and evaluation of an evolutionary algorithm-based online energy management system for plug-in hybrid electric vehicles. *IEEE Trans. Intell. Transp. Syst.* **2017**, *18*, 2181–2191. [CrossRef]
17. Silva, S.; Eckert, J.; Silva, F.; Silva, L.; Dedini, F. Multi-objective optimization design and control of plug-in hybrid electric vehicle powertrain for minimization of energy consumption, exhaust emissions and battery degradation. *Energy Convers. Manag.* **2021**, *234*, 113909. [CrossRef]
18. Pan, L.; Liu, C.; Zhang, Z.; Wang, T.; Shi, J.; Chen, J. Energy-saving effect of utilizing recirculated air in electric vehicle air conditioning system. *Int. J. Refrig.* **2019**, *12*, 122–129. [CrossRef]
19. Zhang, Z.; Liu, C.; Chen, X.; Zhang, C.; Chen, J. Annual energy consumption of electric vehicle air conditioning in China. *Appl. Therm. Eng.* **2017**, *125*, 567–574. [CrossRef]
20. Oh, M.; Ahn, J.; Kim, D.; Jang, D.; Kim, Y. Thermal comfort and energy saving in a vehicle compartment using a localized air-conditioning system. *Appl. Energy* **2014**, *133*, 14–21. [CrossRef]
21. Jeffers, M.; Chaney, L.; Rugh, J. *Climate Control Load Reduction Strategies for Electric Drive Vehicles in Warm Weather*; SAE Technical Paper; SAE: Detroit, MI, USA, 2015. [CrossRef]
22. Liu, Y.; Gao, Q.; Zhang, T.; Cui, C.; Jin, S. Exploration of interactive thermal influence characteristics of power and air conditioning system based on 1D/3D coupling calculation in electric vehicle underhood. *Appl. Therm. Eng.* **2020**, *167*, 114717. [CrossRef]
23. Hofstadtera, R.; Jorge, A.; Kozek, M. Energy optimal control of thermal comfort in trams. *Appl. Therm. Eng.* **2018**, *143*, 812–821. [CrossRef]
24. Collier, J.; Thome, J. *Convective Boiling and Condensation*, 3rd ed.; Oxford University Press: Oxford, UK.
25. Shen, M.; Gao, Q. System simulation on refrigerant-based battery thermal management technology for electric vehicles. *Energy Convers. Manag.* **2020**, *203*, 112176. [CrossRef]

sustainability

MDPI

Article

Modeling of Non-Characteristic Third Harmonics Produced by Voltage Source Converter under Unbalanced Condition

Min Zhang *, Huiqiang Zhi, Shifeng Zhang, Rui Fan, Ran Li and Jinhao Wang

State Grid Shanxi Electric Power Research Institute, Taiyuan 030024, China; ypub75@163.com (H.Z.); sgwd53@163.com (S.Z.); ctwf99@163.com (R.F.); lydq47@163.com (R.L.); hkzd90@163.com (J.W.)
* Correspondence: ffwy26@163.com

Abstract: A three-phase three-wire voltage source converter (VSC) can produce third harmonics when it is operated under an unbalanced condition. It is essential to understand the mechanism of the production of this third harmonic and to assess its impact on power systems. Therefore, this paper presents an analytical model of the VSC under the unbalanced condition through mathematical derivations, and the final model is a coupled Thevenin circuit. The proposed model allows for direct computation of the non-characteristic third harmonics through harmonic power flow studies. The results show that VSC under unbalanced conditions emits both positive-sequence and negative-sequence third harmonics, and that the positive-sequence third harmonic is much larger than the negative-sequence third harmonic. It also shows that the unbalanced level and the size of the dc-link capacitor are critical to the level of non-characteristic third harmonics. The correctness of the proposed model and its application on noncharacteristic third harmonic evaluations have been verified using EMT simulations.

Keywords: voltage source converter; unbalanced condition; non-characteristic third harmonic; harmonic modeling

Citation: Zhang, M.; Zhi, H.; Zhang, S.; Fan, R.; Li, R.; Wang, J. Modeling of Non-Characteristic Third Harmonics Produced by Voltage Source Converter under Unbalanced Condition. *Sustainability* **2022**, *14*, 6449. https://doi.org/10.3390/su14116449

Academic Editor: Guang Zeng

Received: 25 April 2022
Accepted: 17 May 2022
Published: 25 May 2022

1. Introduction

In this context, this paper presents an analytical model of the VSC under unbalanced conditions for the direct computation of the non-characteristic third harmonics through rigorous mathematical derivations. The new model reveals the mechanism of the generation of third harmonics and the characteristics of third harmonics. The new model shows that a Thevenin circuit can represent the VSC for both the positive-sequence and negative-sequence third harmonics. In addition, there are coupling effects between the positive-sequence and negative-sequence third harmonics. The model also shows that the positive-sequence third harmonic is much larger than the negative-sequence third harmonic. The computation of the source and the impedance of the developed Thevenin model requires positive-sequence and negative-sequence fundamental frequency voltage and current on the ac side of the VSC, which can be obtained through load flow studies.

Power electronic devices have experienced a dramatic increase in modern power systems due to the initiative of renewable energies such as wind power generation and photovoltaic (PV) [1], and other applications such as flexible AC transmission system (FACTS) and high voltage DC (HVDC) transmission [2,3]. In recent years, the voltage source converter (VSC) has become the most popular basis of power electronic devices due to its advantages of bidirectional power flow and independent control of active and reactive power [4]. With the massive integration of these VSC devices, there is a concern about their harmonic impact on power systems [5–7]. Under the normal condition, VSC devices only produce very-high-frequency harmonics, which is produced by the PWM switching, but such harmonics can generally be impeded by a passive filter in front of the VSC [8,9]. However, VSC devices can generate low-order non-characteristic third harmonics under the unbalanced condition, which is a power quality concern of power systems. Such a phenomenon occurs because the unbalanced fundamental frequency voltage and current on the ac side of the VSC can induce second harmonic ripples in the dc-link voltage of the VSC through the switching process [9]. These second harmonic ripples in the dc-link voltage, in turn, will lead to a third harmonic voltage on the ac side of the VSC, again through the switching process. As a result, under the unbalanced condition, the VSC becomes a voltage source at the third harmonic. Therefore, it is essential to understand the mechanism of the production of the third harmonic voltage, especially in the distribution system, which can be unbalanced since single-phase loads are dominant [10]. A computation method of the third harmonics under the unbalanced conditions is also needed to assess their impacts.

To date, there are numerous papers presenting the studies of the VSC under unbalanced conditions, but most of these works focus on the control strategies that can suppress the unbalanced condition-induced third harmonics [11–14]. These papers have shown that the unbalanced fundamental frequency voltage and current will lead to double-frequency power oscillation on the ac side of the VSC. As a result, the dc side experiences second harmonic voltage, which can lead to third harmonics on the ac side of the VSC. Based on this finding, the control schemes that can be used to suppress the second harmonic on the dc-link are designed in [12,13]. However, these papers do not provide an analytical algorithm to compute the second harmonic on the dc side or the third harmonic on the ac side. A few other works are proposed specifically for assessing VSC's third harmonic emission under unbalanced conditions [15–17]. In [15], the switching function of the VSC and the harmonic interaction between the ac side and dc side of the VSC are analyzed to obtain the model that can be used to compute the third harmonic distortion. Nevertheless, the impact of the third harmonic on the switching function is not included in this method. Thus, the result could be inaccurate. In [16], the oscillation between the dc side and ac side of the VSC is analyzed based on the ac/dc power interaction, which reveals the mechanism of the production of the third harmonic, but the final analytical method of the uncharacteristic third harmonics is not presented. The reference [17] further provides a method for assessing the third harmonic distortions using the power interaction between the ac side and dc side of the VSC. However, the proposed model omits the effect of the control loop so that the model is not accurate. Moreover, this method is not able to fully show the differences between the positive-sequence and negative-sequence third harmonics and the coupling effect between these two kinds of third harmonics. References [18,19] propose a method that can be used to reveal the harmonic characteristics of the VSC by analyzing the control and AC/DC interaction, but the non-characteristic third harmonics are not analyzed. In summary, a full model that reveals the characteristics of third harmonics produced by VSC is still desired.

This paper presents an analytical model of the VSC under unbalanced conditions for the direct computation of the non-characteristic third harmonics through rigorous mathematical derivations. The development of the new model reveals the mechanism of the generation of third harmonics and their characteristics. The new model shows that a Thevenin circuit can represent the VSC at both the positive-sequence and negative-sequence third harmonics. In addition, there are coupling effects between the positive-sequence and negative-sequence third harmonics. The model also shows that the positive-sequence third harmonic is much larger than the negative-sequence third harmonic. The computation of the source and the impedance of the developed Thevenin model requires a positive-sequence and negative-sequence fundamental frequency voltage and current on the ac side of the VSC, which can be obtained through load flow studies.

The rest of this paper is organized as follows. Section 2 explains the mechanism of generating the non-characteristic third harmonics by the VSC under unbalanced conditions. Section 3 presents the development of the model of VSC under unbalanced conditions for a non-characteristic third harmonic. The proposed model is validated via time-domain simulations in Section 4. Section 5 shows some discussions on the proposed model.

2. Mechanism of the Production of Noncharacteristic Third Harmonics

A typical VSC system is illustrated in Figure 1a, and the control strategy of this VSC is presented in Figure 1b. As one can see, a two-loop control scheme is used, and the outer control loop is used to generate the reference current for the inner current control. The current is controlled using a PI regulator. A capacitor contains the dc-link voltage of the VSC. An inductor L_f is connected at the front end of the VSC to filter out the switching harmonics.

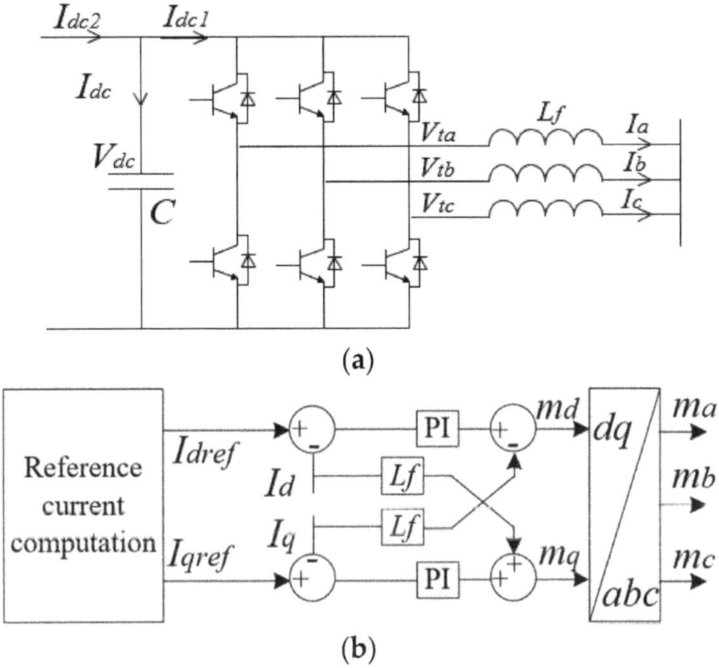

Figure 1. The VSC for unbalanced condition illustration. (**a**) VSC. (**b**) Control of the VSC.

Under the unbalanced operational condition, there are positive-sequence and negative-sequence fundamental frequency voltage and current on the AC side of the VSC, and the three-phase current can be expressed as

$$\begin{aligned} I_a(t) &= I_{1+}\cos(\omega_1 t + \delta_{1+}) + I_{1-}\cos(\omega_1 t + \delta_{1-}) \\ I_b(t) &= I_{1+}\cos(\omega_1 t + \delta_{1+} - 120^\circ) + I_{1-}\cos(\omega_1 t + \delta_{1-} + 120^\circ) \\ I_c(t) &= I_{1+}\cos(\omega_1 t + \delta_{1+} + 120^\circ) + I_{1-}\cos(\omega_1 t + \delta_{1-} - 120^\circ) \end{aligned} \tag{1}$$

It is noted that the modulation signal is a result of the inner current control, so the modulation signal contains the same types of components as those in the current. Thus, the three-phase modulation signals can be expressed as

$$\begin{aligned} m_a(t) &= m_{1+}\cos(\omega_1 t + \theta_{1+}) + m_{1-}\cos(\omega_1 t + \theta_{1-}) \\ m_b(t) &= m_{1+}\cos(\omega_1 t + \theta_{1+} - 120^\circ) + m_{1-}\cos(\omega_1 t + \theta_{1-} + 120^\circ) \\ m_c(t) &= m_{1+}\cos(\omega_1 t + \theta_{1+} + 120^\circ) + m_{1-}\cos(\omega_1 t + \theta_{1-} - 120^\circ) \end{aligned} \tag{2}$$

Since we focus on the study of the third harmonic, the high-frequency harmonics in the switching function can be omitted. Consequently, the modulation signal can be used to represent the switching function of the VSC. As a result, based on (1) and (2), the current on the DC side can be computed as

$$\begin{aligned} I_{dc}(t) &= m_a(t)I_a(t) + m_b(t)I_b(t) + m_c(t)I_c(t) \\ &= \tfrac{3}{2}m_{1+}I_{1+}\cos(\theta_{1+} - \delta_{1+}) + \tfrac{3}{2}m_{1-}I_{1-}\cos(\theta_{1-} - \delta_{1-}) \\ &+ \tfrac{3}{2}m_{1+}I_{1-}\cos(2\omega_1 t + \theta_{1+} + \delta_{1-}) + \tfrac{3}{2}m_{1-}I_{1+}\cos(2\omega_1 t + \theta_{1-} + \delta_{1+}) \end{aligned} \tag{3}$$

According to the above DC side current, the voltage on the DC-link capacitor can be written as

$$\begin{aligned} V_{dc}(t) &= \tfrac{1}{C}\int_0^t i_{dc}(t)dt + V_{dc}(0) \\ &= V_{dc0} + \tfrac{3}{4C\omega_1}m_{1+}I_{1-}\cos(2\omega_1 t + \theta_{1+} + \delta_{1-} - \tfrac{\pi}{2}) \\ &+ \tfrac{3}{4C\omega_1}m_{1-}I_{1+}\cos(2\omega_1 t + \theta_{1-} + \delta_{1+} - \tfrac{\pi}{2}) \end{aligned} \tag{4}$$

where V_{dc0} is the steady-state dc-link voltage. It can be seen that, due to the unbalanced condition, there are second harmonic ripples in the dc-link voltage. It is noteworthy that the expression in (4) is obtained on the basis that all dc-side harmonic currents pass through the dc-link capacitor. In practice, there will be a load (or source) on the other side of the dc-link that has an equivalent impedance Z_{load}, and the harmonic components in I_{dc} will be distributed between the load and dc-link capacitor (I_{dc1} and I_{dc2} in Figure 1a), that is, all of the harmonic components in (4) will be multiplied by $Z_{load}/(Z_{load} + Z_C)$. For simplification but not losing generality, Z_{load} is assumed to be very large so it can be ignored. The DC-link voltage in (4) is converted to AC voltage through the switching process, obtaining the expressions of the three-phase voltages, which are expressed in (5)~(7).

As one can see from (5)~(7), there are third harmonic components in the three-phase voltage of the VSC. In addition, it is found that the third harmonic voltages contain both positive-sequence or negative-sequence components, which are different from the conventional zero-sequence third harmonic. It can also be noted that the third harmonic is in the form of a voltage that is only affected by the positive-sequence and negative-sequence fundamental frequency components.

$$\begin{aligned} V_{ta}(t) &= m_a(t)V_{dc}(t) \\ &= \{m_{1+}\cos(\omega_1 t + \theta_{1+}) + m_{1-}\cos(\omega_1 t + \theta_{1-})\}V_{dc0} + \tfrac{3}{8\omega_1 C}m_{1+}{}^2 I_{1-}\cos(\omega_1 t + \delta_{1-} - \tfrac{\pi}{2}) + \tfrac{3}{8\omega_1 C}m_{1+}m_{1-}I_{1-}\cos(\omega_1 t + \theta_{1+} - \theta_{1-} + \delta_{1-} - \tfrac{\pi}{2}) \\ &+ \tfrac{3}{8\omega_1 C}m_{1-}m_{1+}I_{1+}\cos(\omega_1 t + \theta_{1-} - \theta_{1+} + \delta_{1+} - \tfrac{\pi}{2}) + \tfrac{3}{8\omega_1 C}m_{1-}{}^2 I_{1+}\cos(\omega_1 t + \delta_{1+} - \tfrac{\pi}{2}) + \tfrac{3}{8\omega_1 C}m_{1+}{}^2 I_{1-}\cos(3\omega_1 t + 2\theta_{1+} + \delta_{1-} - \tfrac{\pi}{2}) \\ &+ \tfrac{3}{8\omega_1 C}m_{1+}m_{1-}I_{1-}\cos(3\omega_1 t + \theta_{1+} + \theta_{1-} + \delta_{1-} - \tfrac{\pi}{2}) + \tfrac{3}{8\omega_1 C}m_{1-}m_{1+}I_{1+}\cos(3\omega_1 t + \theta_{1-} + \theta_{1+} + \delta_{1+} - \tfrac{\pi}{2}) + \tfrac{3}{8\omega_1 C}m_{1-}{}^2 I_{1+}\cos(3\omega_1 t + 2\theta_{1-} + \delta_{1+} - \tfrac{\pi}{2}) \end{aligned} \tag{5}$$

$$\begin{aligned} V_{tb}(t) &= m_b(t)V_{dc}(t) \\ &= \{m_{1+}\cos(\omega_1 t + \theta_{1+} - 120^\circ) + m_{1-}\cos(\omega_1 t + \theta_{1-} + 120^\circ)\}V_{dc0} + \tfrac{3}{8\omega_1 C}m_{1+}{}^2 I_{1-}\cos(3\omega_1 t + 2\theta_{1+} + \delta_{1-} - \tfrac{\pi}{2} - 120^\circ) \\ &+ \tfrac{3}{8\omega_1 C}m_{1+}{}^2 I_{1-}\cos(\omega_1 t + \delta_{1-} - \tfrac{\pi}{2} + 120^\circ) + \tfrac{3}{8\omega_1 C}m_{1+}m_{1-}I_{1-}\cos(3\omega_1 t + \theta_{1+} + \theta_{1-} + \delta_{1-} - \tfrac{\pi}{2} + 120^\circ) \\ &+ \tfrac{3m}{8\omega_1 C}m_{1+}m_{1-}I_{1-}\cos(\omega_1 t + \theta_{1+} - \theta_{1-} + \delta_{1-} - \tfrac{\pi}{2} - 120^\circ) + \tfrac{3}{8\omega_1 C}m_{1-}m_{1+}I_{1+}\cos(3\omega_1 t + \theta_{1-} + \theta_{1+} + \delta_{1+} - \tfrac{\pi}{2} - 120^\circ) \\ &+ \tfrac{3}{8\omega_1 C}m_{1-}m_{1+}I_{1+}\cos(\omega_1 t + \theta_{1-} - \theta_{1+} + \delta_{1+} - \tfrac{\pi}{2} + 120^\circ) + \tfrac{3}{8\omega_1 C}m_{1-}{}^2 I_{1+}\cos(3\omega_1 t + 2\theta_{1-} + \delta_{1+} - \tfrac{\pi}{2} + 120^\circ) + \tfrac{3}{8\omega_1 C}m_{1-}{}^2 I_{1+}\cos(\omega_1 t + \delta_{1+} - \tfrac{\pi}{2} - 120^\circ) \end{aligned} \tag{6}$$

$$\begin{aligned} V_{tc}(t) &= m_c(t)V_{dc}(t) \\ &= \{m_{1+}\cos(\omega_1 t + \theta_{1+} - 120^\circ) + m_{1-}\cos(\omega_1 t + \theta_{1-} + 120^\circ)\}V_{dc0} + \tfrac{3}{8\omega_1 C}m_{1+}{}^2 I_{1-}\cos(3\omega_1 t + 2\theta_{1+} + \delta_{1-} - \tfrac{\pi}{2} + 120^\circ) \\ &+ \tfrac{3}{8\omega_1 C}m_{1+}{}^2 I_{1-}\cos(\omega_1 t + \delta_{1-} - \tfrac{\pi}{2} - 120^\circ) + \tfrac{3}{8\omega_1 C}m_{1+}m_{1-}I_{1-}\cos(3\omega_1 t + \theta_{1+} + \theta_{1-} + \delta_{1-} - \tfrac{\pi}{2} - 120^\circ) \\ &+ \tfrac{3}{8\omega_1 C}m_{1+}m_{1-}I_{1-}\cos(\omega_1 t + \theta_{1+} - \theta_{1-} + \delta_{1-} - \tfrac{\pi}{2} + 120^\circ) + \tfrac{3}{8\omega_1 C}m_{1-}m_{1+}I_{1+}\cos(3\omega_1 t + \theta_{1-} + \theta_{1+} + \delta_{1+} - \tfrac{\pi}{2} + 120^\circ) \\ &+ \tfrac{3}{8\omega_1 C}m_{1-}m_{1+}I_{1+}\cos(\omega_1 t + \theta_{1-} - \theta_{1+} + \delta_{1+} - \tfrac{\pi}{2} - 120^\circ) + \tfrac{3}{8\omega_1 C}m_{1-}{}^2 I_{1+}\cos(3\omega_1 t + 2\theta_{1-} + \delta_{1+} - \tfrac{\pi}{2} - 120^\circ) + \tfrac{3}{8\omega_1 C}m_{1-}{}^2 I_{1+}\cos(\omega_1 t + \delta_{1+} - \tfrac{\pi}{2} + 120^\circ) \end{aligned} \tag{7}$$

3. Modeling VSC under Unbalanced Condition

The above analysis has shown that an unbalanced condition can introduce a third harmonic voltage for the VSC. Such a third harmonic voltage will further introduce a third harmonic current in the interconnected grid. As a result of the current feedback control strategy, the third harmonic current would lead to an additional third harmonic voltage on the ac side of the VSC via the control section. As a result, the third harmonic voltage of the VSC shall be computed as shown in Figure 2. As can be seen, there are two parts to be included. The first is the third harmonic voltage caused by the fundamental frequency voltage and current, and the second is the distribution of this third harmonic voltage. To compute the final voltage, it is necessary to include the third harmonic on the ac side of the

VSC, and, as a result, the steady-state current and modulation signal can be expressed as (8) and (9).

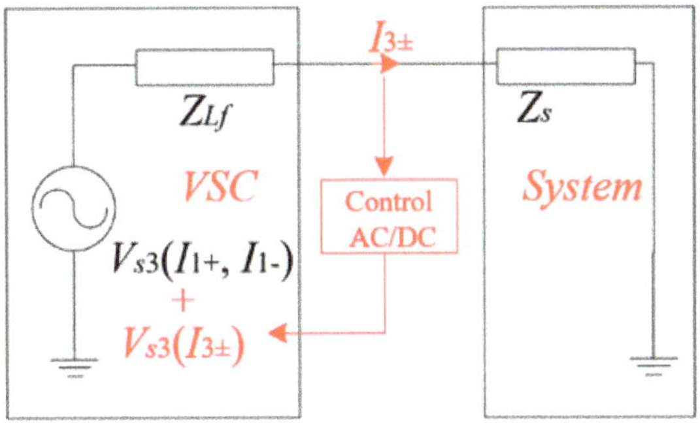

Figure 2. Illustration of modeling of the third harmonic.

$$\begin{array}{l} I_a(t) = I_{1+}\cos(\omega_1 t + \delta_{1+}) + I_{1-}\cos(\omega_1 t + \delta_{1-}) \\ +I_{3+}\cos(3\omega_1 t + \delta_{3+}) + I_{3-}\cos(3\omega_1 t + \delta_{3-}) \\ I_b(t) = I_{1+}\cos(\omega_1 t + \delta_{1+} - 120^\circ) + I_{1-}\cos(\omega_1 t + \delta_{1-} + 120^\circ) \\ +I_{3+}\cos(3\omega_1 t + \delta_{3+} - 120^\circ) + I_{3-}\cos(3\omega_1 t + \delta_{3-} + 120^\circ) \\ I_c(t) = I_{1+}\cos(\omega_1 t + \delta_{1+} + 120^\circ) + I_{1-}\cos(\omega_1 t + \delta_{1-} - 120^\circ) \\ +I_{3+}\cos(3\omega_1 t + \delta_{3+} + 120^\circ) + I_{3-}\cos(3\omega_1 t + \delta_{3-} - 120^\circ) \end{array} \quad (8)$$

$$\begin{array}{l} m_a = m_{1+}\cos(\omega_1 t + \theta_{1+}) + m_{1-}\cos(\omega_1 t + \theta_{1-}) \\ +m_{3+}\cos(3\omega_1 t + \theta_{3+}) + m_{3-}\cos(3\omega_1 t + \theta_{3-}) \\ m_b = m_{1+}\cos(\omega_1 t + \theta_{1+} - 120^\circ) + m_{1-}\cos(\omega_1 t + \theta_{1-} + 120^\circ) \\ +m_{3+}\cos(3\omega_1 t + \theta_{3+} - 120^\circ) + m_{3-}\cos(3\omega_1 t + \theta_{3-} + 120^\circ) \\ m_c = m_1\cos(\omega_1 t + \theta_{1+} + 120^\circ) + m_{1-}\cos(\omega_1 t + \theta_{1-} - 120^\circ) \\ +m_{3+}\cos(3\omega_1 t + \theta_{3+} + 120^\circ) + m_{3-}\cos(3\omega_1 t + \theta_{3-} - 120^\circ) \end{array} \quad (9)$$

The equations in (8) and (9) can be used to obtain the current on the DC side, which is given as (10):

$$\begin{array}{l} I_{dc}(t) = m_a(t)I_a(t) + m_b(t)I_b(t) + m_c(t)I_c(t) \\ = \frac{3}{2}\left\{ \begin{array}{l} m_{1+}I_{1+}\cos(\theta_{1+} - \delta_{1+}) + m_{1+}I_{1-}\cos(2\omega_1 t + \theta_{1+} + \delta_{1-}) \\ +m_{1+}I_{3+}\cos(2\omega_1 t + \delta_{3+} - \theta_{1+}) + m_{1+}I_{3-}\cos(4\omega_1 t + \delta_{3-} + \theta_{1+}) \\ +m_{1-}I_{1+}\cos(2\omega_1 t + \theta_{1-} + \delta_{1+}) + m_{1-}I_{1-}\cos(\theta_{1-} - \delta_{1-}) \\ +m_{1-}I_{3+}\cos(4\omega_1 t + \delta_{3+} + \theta_{1-}) + m_{1-}I_{3-}\cos(2\omega_1 t + \delta_{3-} - \theta_{1-}) \\ +m_{3+}I_{1+}\cos(2\omega_1 t + \theta_{3+} - \delta_{1+}) + m_{3+}I_{1-}\cos(4\omega_1 t + \theta_{3+} + \delta_{1-}) \\ +m_{3+}I_{3+}\cos(\theta_{3+} - \delta_{3+}) + m_{3+}I_{3-}\cos(6\omega_1 t + \theta_{3+} + \delta_{3-}) \\ +m_{3-}I_{1+}\cos(4\omega_1 t + \theta_{3-} + \delta_{1+}) + m_{3-}I_{1-}\cos(2\omega_1 t + \theta_{3-} - \delta_{1-}) \\ +m_{3-}I_{3+}\cos(6\omega_1 t + \delta_{3+} + \theta_{3-}) + m_{3-}I_{3-}{}^2\cos(\theta_{3-} - \delta_{3-}) \end{array} \right\} \end{array} \quad (10)$$

Accordingly, the DC-link voltage can be computed as

$$\begin{array}{l} V_{dc}(t) = \frac{1}{C}\int_0^t i_{dc}dt + V_{dc0} = V_{dc0} + \\ \left\{ \begin{array}{l} \frac{3}{4C\omega_1}m_{1+}I_{1-}\cos(2\omega_1 t + \theta_{1+} + \delta_{1-} - \frac{\pi}{2}) + \frac{3}{4C\omega_1}m_{1+}I_{3+}\cos(2\omega_1 t + \delta_{3+} - \theta_{1+} - \frac{\pi}{2}) \\ + \frac{3}{8C\omega_1}m_{1+}I_{3-}\cos(4\omega_1 t + \delta_{3-} + \theta_{1+} - \frac{\pi}{2}) + \frac{3}{4C\omega_1}m_{1-}I_{1+}\cos(2\omega_1 t + \theta_{1-} + \delta_{1+} - \frac{\pi}{2}) \\ + \frac{3}{8C\omega_1}m_{1-}I_{3+}\cos(4\omega_1 t + \delta_{3+} + \theta_{1-} - \frac{\pi}{2}) + \frac{3}{4C\omega_1}m_{1-}I_{3-}\cos(2\omega_1 t + \delta_{3-} - \theta_{1-} - \frac{\pi}{2}) \\ + \frac{3}{4C\omega_1}m_{3+}I_{1+}\cos(2\omega_1 t + \theta_{3+} - \delta_{1+} - \frac{\pi}{2}) + \frac{3}{8C\omega_1}m_{3+}I_{1-}\cos(4\omega_1 t + \theta_{3+} + \delta_{1-}) \\ + \frac{3}{12C\omega_1}m_{3+}I_{3-}\cos(6\omega_1 t + \theta_{3+} + \delta_{3-}) + \frac{3}{8C\omega_1}m_{3-}I_{1+}\cos(4\omega_1 t + \theta_{3-} + \delta_{1+}) \\ + \frac{3}{4C\omega_1}m_{3-}I_{1-}\cos(2\omega_1 t + \theta_{3-} - \delta_{1-}) + \frac{3}{12C\omega_1}m_{3-}I_{3+}\cos(6\omega_1 t + \delta_{3+} + \theta_{3-}) \end{array} \right\} \end{array} \quad (11)$$

The above dc-link voltage is converted to the ac-side voltage by the switching function, and the phase-A voltage is given as

$$
\begin{aligned}
&V_a(t) = m_a(t)\cdot V_{dc}(t) \\
&= \tfrac{1}{2}\left\{ \begin{array}{l} (m_{1+}\cos(\omega_1 t+\theta_{1+}) + m_{1-}\cos(\omega_1 t+\theta_{1-}) \\ +H_1(j2\omega_1)I_{3+}\cos(3\omega_1 t+\delta_{3+}) + H_1(j4\omega_1)I_{3-}\cos(3\omega_1 t+\delta_{3-})) \end{array} \right\} \\
&\cdot\left\{ \begin{array}{l} V_{dc0} + \frac{3m_{1+}I_{1-}}{4C\omega_1}\cos(2\omega_1 t+\theta_{1+}+\delta_{1-}-\frac{\pi}{2}) + \frac{3m_{1+}I_{3+}}{4C\omega_1}\cos(2\omega_1 t+\delta_{3+}-\theta_{1+}-\frac{\pi}{2}) \\
+\frac{3}{8C\omega_1}m_{1+}I_{3-}\cos(4\omega_1 t+\delta_{3-}+\theta_{1+}-\frac{\pi}{2}) + \frac{3}{4C\omega_1}m_{1-}I_{1+}\cos(2\omega_1 t+\theta_{1-}+\delta_{1+}-\frac{\pi}{2}) \\
+\frac{3}{8C\omega_1}m_{1-}I_{3+}\cos(4\omega_1 t+\delta_{3+}+\theta_{1-}-\frac{\pi}{2}) + \frac{3}{4C\omega_1}m_{1-}I_{3-}\cos(2\omega_1 t+\delta_{3-}-\theta_{1-}-\frac{\pi}{2}) \\
+\frac{3}{4C\omega_1}m_{3+}I_{1+}\cos(2\omega_1 t+\theta_{3+}-\delta_{1+}-\frac{\pi}{2}) + \frac{3}{8C\omega_1}m_{3+}I_{1-}\cos(4\omega_1 t+\theta_{3+}+\delta_{1-}) \\
+\frac{3}{12C\omega_1}m_{3+}I_{3-}\cos(6\omega_1 t+\theta_{3+}+\delta_{3-}) + \frac{3}{8C\omega_1}m_{3-}I_{1+}\cos(4\omega_1 t+\theta_{3-}+\delta_{1+}) \\
+\frac{3}{4C\omega_1}m_{3-}I_{1-}\cos(2\omega_1 t+\theta_{3-}-\delta_{1-}) + \frac{3}{12C\omega_1}m_{3-}I_{3+}\cos(6\omega_1 t+\delta_{3+}+\theta_{3-}) \end{array} \right\}
\end{aligned} \tag{12}
$$

and the phase-B and phase-C voltages can be computed similarly. As one can see, there will be numerous types of components in the ac voltage. Since we are only interested in the third harmonic components, the third harmonic components in (12) are extracted as in (13).

$$
\begin{aligned}
&V_{ta3}(t)=\tfrac{1}{2}\{m_{3+}\cos(3\omega_1 t+\theta_{3+})+m_{3-}\cos(3\omega_1 t+\theta_{3-})\}V_{dc0}+\frac{3m_{1+}^2 I_{1-}}{16C\omega_1}\cos(3\omega_1 t+2\theta_{1+}+\delta_{1-}-\tfrac{\pi}{2})+\frac{3m_{1+}I_{1-}}{16C\omega_1}\cos(3\omega_1 t+\theta_{1+}+\theta_{1-}+\delta_{1-}-\tfrac{\pi}{2}) \\
&+\frac{3m_{1+}^2 I_{3+}}{16C\omega_1}\cos(3\omega_1 t+\delta_{3+}-\tfrac{\pi}{2})+\frac{3m_{1+}m_{1-}I_{3+}}{16C\omega_1}\cos(3\omega_1 t+\delta_{3+}-\theta_{1+}+\theta_{1-}-\tfrac{\pi}{2})+\frac{3m_{1+}^2 I_{3-}}{32C\omega_1}\cos(3\omega_1 t+\delta_{3-}-\tfrac{\pi}{2})+\frac{3m_{1-}m_{1+}I_{3-}}{32C\omega_1}\cos(3\omega_1 t+\delta_{3-}+\theta_{1+}-\theta_{1-}-\tfrac{\pi}{2}) \\
&\frac{3m_{1+}m_{1-}I_{1+}}{16C\omega_1}\cos(3\omega_1 t+\theta_{1-}+\theta_{1+}+\delta_{1+}-\tfrac{\pi}{2})+\frac{3m_{1-}^2 I_{1+}}{16C\omega_1}\cos(3\omega_1 t+2\theta_{1-}+\delta_{1+}-\tfrac{\pi}{2})+\frac{3m_{1+}m_{1-}I_{3+}}{32C\omega_1}\cos(3\omega_1 t+\delta_{3+}+\theta_{1-}-\theta_{1+}-\tfrac{\pi}{2})+\frac{3m_{1-}^2 I_{3+}}{32C\omega_1}\cos(3\omega_1 t+\delta_{3+}-\tfrac{\pi}{2}) \\
&+\frac{3m_{1+}m_{1-}I_{3-}}{16C\omega_1}\cos(3\omega_1 t+\delta_{3-}-\theta_{1-}+\theta_{1+}-\tfrac{\pi}{2})+\frac{3m_{1-}^2 I_{3-}}{16C\omega_1}\cos(3\omega_1 t+\delta_{3-}-\tfrac{\pi}{2})+\frac{3m_{1+}m_{3+}I_{1+}}{16C\omega_1}\cos(3\omega_1 t+\theta_{3+}+\theta_{1+}-\delta_{1+}-\tfrac{\pi}{2})+\frac{3m_{1-}m_{3+}I_{1+}}{16C\omega_1}\cos(3\omega_1 t+\theta_{3+}+\theta_{1-}-\delta_{1+}-\tfrac{\pi}{2}) \\
&+\frac{3m_{1+}m_{3+}I_{1-}}{32C\omega_1}\cos(3\omega_1 t+\theta_{3+}+\delta_{1-}-\theta_{1+})+\frac{3m_{1-}m_{3+}I_{1-}}{32C\omega_1}\cos(3\omega_1 t+\theta_{3+}+\delta_{1-}-\theta_{1-})+\frac{3m_{1+}m_{3-}I_{1+}}{32C\omega_1}\cos(3\omega_1 t+\theta_{3-}+\delta_{1+}-\theta_{1+})+\frac{3m_{1-}m_{3-}I_{1+}}{32C\omega_1}\cos(3\omega_1 t+\theta_{3-}+\delta_{1+}-\theta_{1-}) \\
&+\frac{3m_{1+}m_{3-}I_{1-}}{16C\omega_1}\cos(3\omega_1 t+\theta_{3-}-\delta_{1-}+\theta_{1+})+\frac{3m_{1-}m_{3-}I_{1-}}{16C\omega_1}\cos(3\omega_1 t+\theta_{3-}-\delta_{1-}+\theta_{1-})
\end{aligned} \tag{13}
$$

The expression in (13) shows that the AC-side voltage of the converter contains third harmonic components, which means that the converter acts as a source at the third harmonic. Specifically, we can see that the terms of the third harmonic voltage in (13) can be categorized into two types. The first type is the terms that are determined by fundamental frequency components ($m_{\pm 1}$ and $I_{\pm 1}$). The second type is the terms that are affected by external third harmonic components ($m_{\pm 3}$ and $I_{\pm 3}$). The first type of third harmonic voltage is an independent voltage source and the second type of third harmonic voltage exhibits an impedance nature. Therefore, (13) indicates that the VSC can be represented as a Thevenin circuit at the third harmonic. In the following, the detailed Thevenin circuit will be presented.

In (13), it is noted that the third harmonic components in the modulation signal can be related to the third harmonic current on the ac side of the VSC by (14):

$$
m_{3\pm} = H(s \mp j\omega_1)(I_{ref3} - I_{3\pm}) \tag{14}
$$

It is noted that the outer control loop of the VSC generally has a very limited bandwidth (e.g., 100 Hz), so the reference current that is generated by the outer control loop can be considered as harmonic-free (i.e., $I_{ref} = 0$). Substituting (14) into (13) leads to an equation that contains the third harmonic voltage and current, which gives the following model

$$
\begin{bmatrix} V_{t3+} \\ V_{t3-} \end{bmatrix} = \begin{bmatrix} V_{st3+} \\ V_{st3-} \end{bmatrix} + \begin{bmatrix} Z_{3+} & Z_{3+,3-} \\ Z_{3-,3+} & Z_{3-} \end{bmatrix} \begin{bmatrix} I_{3+} \\ I_{3-} \end{bmatrix} \tag{15}
$$

where

$$
\begin{aligned}
V_{st3+}(t) &= \tfrac{3}{16C\omega_1}k_2 m_{1+}^2 I_{1-}\cos(3\omega_1 t + 2\theta_{1+} + \delta_{1-} - \tfrac{\pi}{2}) \\
&+ \tfrac{3}{16C\omega_1}k_2 m_{1+} m_{1-} I_{1+}\cos(3\omega_1 t + \theta_{1-} + \theta_{1+} + \delta_{1+} - \tfrac{\pi}{2})
\end{aligned} \tag{16}
$$

$$
\begin{aligned}
V_{st3-}(t) &= \tfrac{3}{16C\omega_1}k_2 m_{1-} m_{1+} I_{1-}\cos(3\omega_1 t + \theta_{1+} + \theta_{1-} + \delta_{1-} - \tfrac{\pi}{2}) \\
&+ \tfrac{3}{16C\omega_1}k_2 m_{1-}^2 I_{1+}\cos(3\omega_1 t + \theta_{1-} + \theta_{1-} + \delta_{1+} - \tfrac{\pi}{2})
\end{aligned} \tag{17}
$$

$$
\begin{aligned}
Z_{3+} &= \tfrac{1}{2}H_1(j2\omega_1)V_{dc0} + \tfrac{-j3}{16C\omega_1}k_2 {m_{1+}}^2 + \tfrac{-j3}{32C\omega_1}k_4 {m_{1-}}^2 \\
&+ \tfrac{-j3}{16C\omega_1}k_2 H(j2\omega_1) I_{1+} m_{1+} e^{j(\theta_{1+}-\delta_{1+})} \\
&+ \tfrac{-j3}{32C\omega_1}k_4 H(j2\omega_1) m_{1-} I_{1-} e^{j(\delta_{1-}-\theta_{1-})}
\end{aligned} \tag{18}
$$

$$\begin{aligned} Z_{3-} = &\tfrac{1}{2}H_1(j4\omega_1)V_{dc0} + \tfrac{-j3}{32C\omega_1}k_4 {m_{1+}}^2 + \tfrac{-j3}{16C\omega_1}k_2 {m_{1-}}^2 \\ &+ \tfrac{-j3}{32C\omega_1}k_4 H(j4\omega_1) I_{1+} m_{1+} e^{j(-\theta_{1+}+\delta_{1+})} \\ &+ \tfrac{-j3}{16C\omega_1}k_2 H(j4\omega_1) m_{1-} I_{1-} e^{j(-\delta_{1-}+\theta_{1-})} \end{aligned} \tag{19}$$

$$\begin{aligned} Z_{3-,3+} = &\tfrac{-j3}{16C\omega_1}k_2 m_{1+} m_{1-} I_{3+} e^{j(\theta_{1-}-\theta_{1+})} + \tfrac{-j3}{32C\omega_1}k_4 m_{1+} m_{1-} e^{j(\theta_{1-}-\theta_{1+})} \\ &+ \tfrac{-j3}{16C\omega_1}k_2 H(j2\omega_1) I_{1+} m_{1-} e^{j(\theta_{1-}-\delta_{1+})} \\ &+ \tfrac{-j3}{32C\omega_1}k_4 H(j2\omega_1) m_{1+} I_{1-} e^{j(\delta_{1-}-\theta_{1+})} \end{aligned} \tag{20}$$

$$\begin{aligned} Z_{3+,3-} = &\tfrac{-j3}{32C\omega_1}k_4 m_{1+} m_{1-} e^{j(-\theta_{1-}+\theta_{1+})} + \tfrac{-j3}{16C\omega_1}k_2 m_{1-} m_{1+} e^{j(-\theta_{1-}+\theta_{1+})} \\ &+ \tfrac{3}{32C\omega_1}k_4 H(j4\omega_1) m_{1-} I_{1+} e^{j(\delta_{1+}-\theta_{1+})} \\ &+ \tfrac{-j3}{16C\omega_1}k_2 H(j4\omega_1) m_{1+} I_{1-} e^{j(-\delta_{1-}+\theta_{1+})} \end{aligned} \tag{21}$$

The above final model gives the following key findings:

- A Thevenin circuit can represent the VSC at both the positive-sequence and negative-sequence third harmonics, and the analytical expression of the Thevenin circuit can be established.
- There is a coupling effect between the positive-sequence third harmonic and negative-sequence third harmonic. The coupling effect mainly originates from the interaction between positive-sequence fundamental frequency components and negative-sequence third harmonic components and vice versa.
- The source of the Thevenin circuit is determined only by the positive-sequence and negative-sequence fundamental frequency components, whereas the impedances are determined by both the inner PI regulator and the fundamental frequency voltage and current.

According to the above results, the fundamental frequency components are needed to complete the developed model. For this purpose, the load flow methods regarding the unbalanced power systems with VSC devices are needed. At present, numerous works have been conducted on this aspect [20–23]. These methods can be used to compute the positive-sequence and negative-sequence fundamental frequency voltage and current. The procedure to obtain the final model of the VSC can be summarized as the flow chart in Figure 3.

Compared with other existing models, the proposed model for third harmonic computation has the following advantages:

- The mechanism of the production of the non-characteristic third harmonics are fully revealed so that the contributions of the different components to the non-characteristic third harmonics emission can be easily analyzed. Accordingly, the control scheme that can suppress the third harmonic emission can be designed.
- The computation of the proposed model is very easy as the analytical expression of the proposed model is given. Compared with other methods, it is more straightforward for users.
- The proposed model can be integrated into the existing harmonic power flow tools. As a result, the non-characteristic third harmonic can be computed along with other harmonic components, which can be easily adopted by the users.

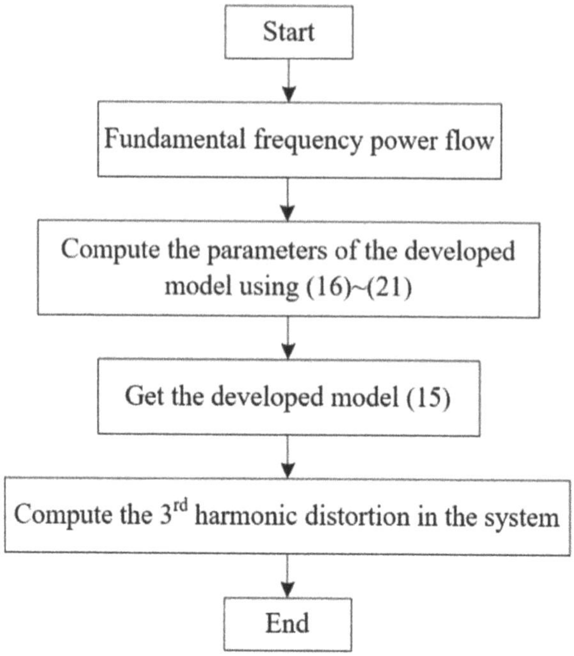

Figure 3. Procedure to compute the noncharacteristic third harmonics under the unbalanced condition.

4. Verification of the Proposed Model by Time-Domain Simulations

In this section, the developed model of VSC under unbalanced conditions is verified via time-domain simulations in the VSC system of Figure 4 in MATLAB/Simulink. The VSC has the same control strategy as that in Figure 1. The system parameters are listed in Table 1 [24,25]. The unbalanced condition is created by adding a negative-sequence fundamental frequency voltage in the AC source (V_s in Figure 4). Therefore, the ac voltage source can be expressed as

$$V_{sa}(t) = V_{s+}\cos(\omega_1 t + \phi_{s+}) + V_{s-}\cos(\omega_1 t + \phi_{s-}) \tag{22}$$

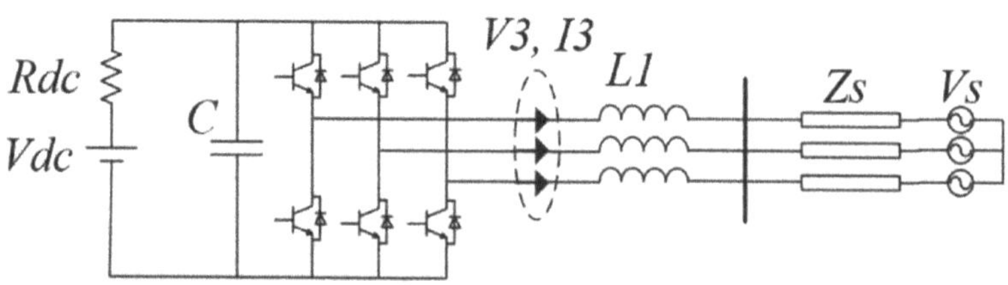

Figure 4. The VSC for unbalanced condition illustration.

Table 1. System parameters.

Parameter	Value	Parameter	Value
f_{switch}	3 kHz	Vs	200 V
L_1	5 mL	Ls	10 mL
Kp	0.023	Ki	43.15
C	50 µF	Vdc	600 V
R_{dc}	60 Ω	Nominal P	1 kW

For illustration, the unbalanced level of the system is defined as

$$l = \frac{V_{s-}}{V_{s+}} \times 100\% \tag{23}$$

The basic strategy to verify the proposed model is summarized as follows:

- Implement the simulation system in MATLAB/Simulink, then set different unbalanced levels to run the simulations. The simulation results give the third harmonic distortion levels.
- Based on the parameters of the simulation system and the unbalanced levels, the Thevenin circuit can be computed.
- Once the Thevenin circuit is obtained, the third harmonic distortion levels in the system can be computed for different unbalanced levels.
- The calculated third harmonic distortion levels are compared with the results obtained from the time-domain simulations in step (1). If these two types of results match well, then it verifies the correctness of the proposed model at third harmonics.

The developed third harmonic model is tested under an unbalanced level of 5%, 15%, 25%, and 35%. The simulated waveforms of the AC-side voltage and current of the VSC are presented in Figure 5. The waveforms of the VSC's voltage and current can be used to compute the third harmonic voltage and current using FFT, and the IHDs of the third harmonics for different unbalanced levels are shown in Figure 6. It can be seen that a larger unbalanced level leads to larger third harmonics. When the unbalanced level is only 5%, the IHD of the positive-sequence third harmonic current is around 2%, while it increases to 11% when the unbalanced level rises to 35%. Additionally, one can also see that the positive-sequence third harmonic is much higher than the negative-sequence third harmonic. The reason is that, according to Equations (16)~(21), the positive-sequence third harmonic is more relevant with the positive-sequence fundamental frequency components, whereas the negative-sequence third harmonic is more relevant with the negative-sequence fundamental frequency components. As a result of the large positive-sequence third harmonic current, the coupling between the negative-sequence third harmonic and positive-sequence third harmonic current cannot be ignored.

It has been mentioned previously that, to obtain the developed model, the positive-sequence and negative-sequence fundamental frequency voltage and current are needed. In the following, the calculated positive-sequence and negative-sequence voltage and current are compared with the simulation results first, and then these values are used to compute the developed model.

4.1. Computation of the Fundamental Frequency Components

The positive-sequence and negative-sequence fundamental frequency voltage and current can be obtained through load flow studies. However, it is noted from the derived model that both the positive-sequence and negative-sequence fundamental frequency components in the modulation signal are needed to compute the developed model. Due to the control and switching process features, the positive-sequence and negative-sequence fundamental-frequency components in the modulation signal can be computed using Equations (24) and (25), respectively.

$$m_{1+} = V_{1+}/V_{dc0} \tag{24}$$

$$m_{1-} = H(2j\omega_1)I_{1-} \tag{25}$$

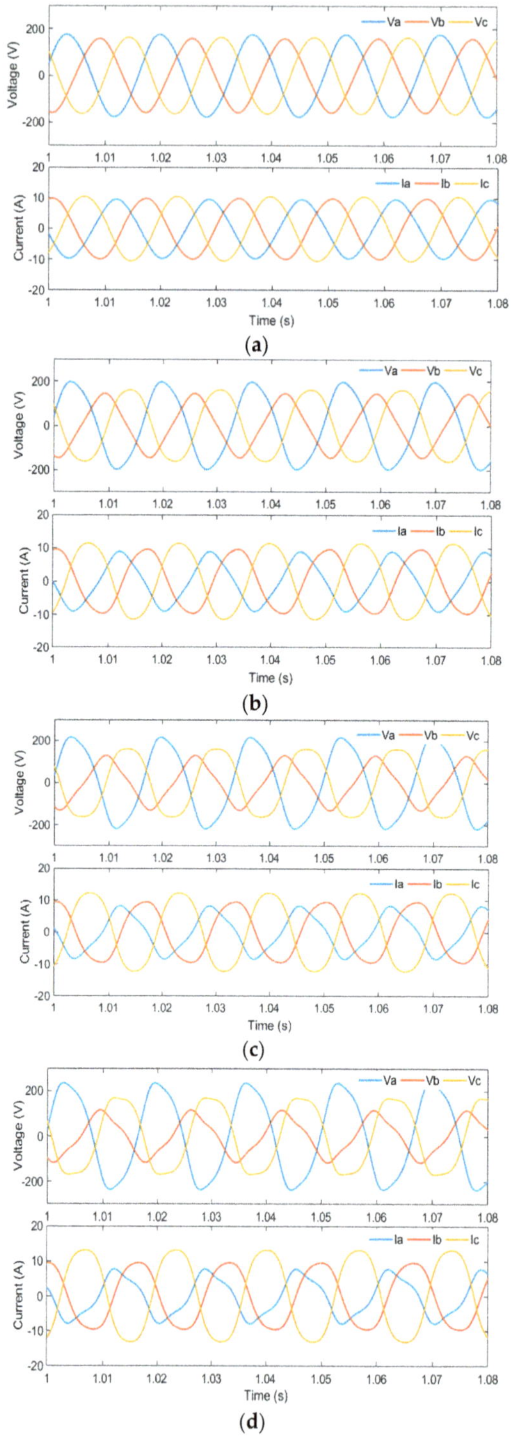

Figure 5. Voltage and current waveforms on the ac side of the VSC under different unbalanced levels. (**a**) l = 5%. (**b**) l = 15%. (**c**) l = 25%. (**d**) l = 5%.

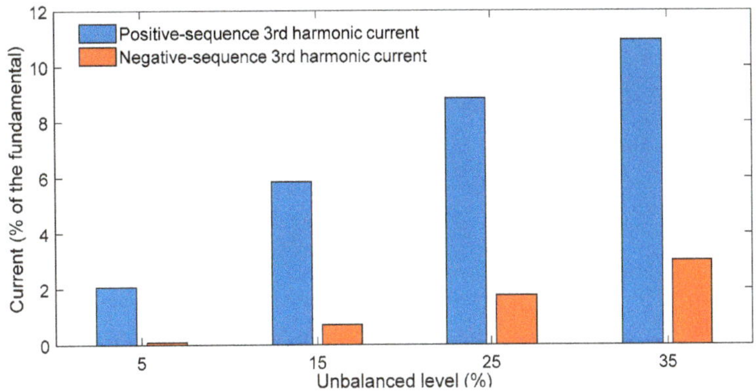

Figure 6. IHDs of the third harmonic on the ac side of the VSC.

For the studied four cases, the calculated ac-side positive-sequence fundamental frequency voltage and negative-sequence fundamental frequency current and modulation signals regarding various unbalanced levels are compared in Tables 2 and 3. It can be seen that the calculated fundamental frequency components under various unbalanced conditions match well with the simulated results. Based on the obtained positive-sequence voltage and the negative-sequence current, the corresponding modulation signal can be computed using (24) and (25).

Table 2. Results of the positive-sequence fundamental frequency voltage.

Unbalanced Level (l)	Simulated V_{t1+}	Calculated V_{t1+}
5%	$0.9815\angle{-70.9336^\circ}$	$0.982\angle{-70.9159^\circ}$
15%	$0.982\angle{-70.8655^\circ}$	$0.982\angle{-70.9159^\circ}$
25%	$0.9829\angle{-70.7298^\circ}$	$0.982\angle{-70.9159^\circ}$
35%	$0.9842\angle{-70.5274^\circ}$	$0.982\angle{-70.9159^\circ}$

Table 3. Results of the positive-sequence fundamental frequency current.

Unbalanced Level (l)	Simulated I_{1-}	Calculated I_{1-}
5%	−0.0137 − 0.0419i	−0.0137 − 0.0420i
15%	−0.0387 − 0.1233i	−0.0410 − 0.1259i
25%	−0.0586 − 0.1989i	−0.0540 − 0.1998i
35%	−0.0739 − 0.2677i	−0.0756 − 0.2797i

4.2. Verification of the Proposed Model

According to the obtained fundamental frequency components above, the parameters of the developed model in (15) can be computed, and the parameters of this model for different unbalanced levels are listed in Table 4. It can be seen that, since the positive-sequence third harmonic current is much larger than the negative-sequence third harmonic current, the coupling between the negative-sequence third harmonic voltage and the positive-sequence third harmonic current is more important. According to the calculated parameters, the third harmonic current can be computed using the following equation:

$$\begin{bmatrix} V_{s3+} \\ V_{s3-} \end{bmatrix} = \begin{bmatrix} Z_{3+} & Z_{3+,3-} \\ Z_{3-,3+} & Z_{3-} \end{bmatrix} \begin{bmatrix} I_{3+} \\ I_{3-} \end{bmatrix} + \begin{bmatrix} Z_{L1}+Z_{Ls} & 0 \\ 0 & Z_{L1}+Z_{Ls} \end{bmatrix} \begin{bmatrix} I_{3+} \\ I_{3-} \end{bmatrix} \quad (26)$$

and the third harmonic voltage on the ac side of the VSC can be computed as

$$\begin{bmatrix} V_{t3+} \\ V_{t3-} \end{bmatrix} = \begin{bmatrix} V_{st3+} \\ V_{st3-} \end{bmatrix} + \begin{bmatrix} Z_{3+} & Z_{3+,3-} \\ Z_{3-,3+} & Z_{3-} \end{bmatrix} \begin{bmatrix} I_{3+} \\ I_{3-} \end{bmatrix} \quad (27)$$

Table 4. Computed impedance and source.

l	Z_{3+}	Z_{3-}	$Z_{3+,3-}$	$Z_{3-,3+}$	V_{3t+}	V_{3t-}
5%	7.73 − 14.86i	6.44 − 7.87i	0.13 − 0.12i	0.03 − 0.22i	−1.58 + 0.50i	−0.06 + 0.09i
15%	7.84 − 15.26i	6.56 − 8.1i	0.37 − 0.35i	0.11 − 0.64i	−4.48 + 1.3i	−0.57 + 0.75i
25%	8.13 − 16.1i	6.84 − 8.55i	0.55 − 0.55i	0.21 − 0.98i	−6.82 + 1.76i	−1.53 + 1.79i
35%	8.07 − 16.3i	6.78 − 8.62i	0.69 − 0.71i	0.33 − 1.25i	−8.6 + 1.93i	−2.86 + 2.97i

The comparison results between the calculated harmonic voltage and current on the ac side of the VSC for the four studied cases are listed in Tables 5 and 6. One can see that both the calculated positive-sequence and negative-sequence third harmonic voltages match well with the simulated results, which verifies the correctness of the proposed model. In the calculations, the coupling between the positive-sequence and negative sequence third harmonics is included. This is very important for the computation for the negative-sequence third harmonics due to the large positive-sequence third harmonic current.

Table 5. Test results of the developed model for voltage.

Unbalanced Level (l)	Positive-Sequence Third Harmonic Voltage Using Time-Domain Simulations	Positive-Sequence Third Harmonic Voltage Using Proposed Model
5%	0.1742 − 0.113i	0.1745 − 0.1106i
15%	0.5131 − 0.284i	0.5141 − 0.2761i
25%	0.8169 − 0.3377i	0.8187 − 0.3189i
35%	1.058 − 0.2822i	1.0573 − 0.2508i
Unbalanced Level (l)	**Negative-Sequence Third Harmonic Voltage Using Time-Domain Simulations**	**Negative-Sequence Third Harmonic Voltage Using Proposed Model**
5%	0.0008 − 0.0086i	0.0008 − 0.0086i
15%	0.0088 − 0.071i	0.0088 − 0.071i
25%	0.0278 − 0.1733i	0.0286 − 0.1736i
35%	0.0574 − 0.2959i	0.0589 − 0.2956i

Table 6. Test results of the developed model for current.

Unbalanced Level (l)	Positive-Sequence Third Harmonic Voltage Using Time-Domain Simulations	Positive-Sequence Third Harmonic Voltage Using Proposed Model
5%	−1.9172 − 2.9548i	−1.9124 − 2.9669i
15%	−4.8182 − 8.7052i	−4.7920 − 8.7518i
25%	−5.7287 − 13.8587i	−5.6458 − 14.1981i
35%	−4.7867 − 17.9513i	−4.6459 − 17.5674i
Unbalanced Level (l)	**Negative-Sequence Third Harmonic Voltage Using Time-Domain Simulations**	**Negative-Sequence Third Harmonic Voltage Using Proposed Model**
5%	−0.1456 − 0.0139i	−0.1466 − 0.0136i
15%	−1.2053 − 0.1488i	−1.2154 − 0.1488i
25%	−2.9408 − 0.4724i	−2.9901 − 0.5058i
35%	−5.0207 − 0.9740i	−5.0250 − 0.9508i

4.3. Impact of the Dc-Link Capacitor

According to the derivations in Sections 2 and 3, the value of the DC-side capacitor plays an important role in the developed model. For the same unbalanced level, the larger the DC-side capacitor, the smaller the third harmonic. This can be verified by the simulation results in Figure 7, which are based on an unbalanced level of 25%. As one can see, when the DC-side capacitor is increased to 100 μF, the third harmonic voltage and current are significantly decreased. The results can also be supported by the harmonic voltage of the dc side, which is shown in Figure 8. It is found that the harmonic ripples (mainly second harmonics) in the dc-link voltage for a dc-link capacitor of 100 µF are much smaller than those for a dc-link capacitor of 50 µF. In summary, increasing the DC-link capacitor's value can resist the generation of the uncharacteristic third harmonic.

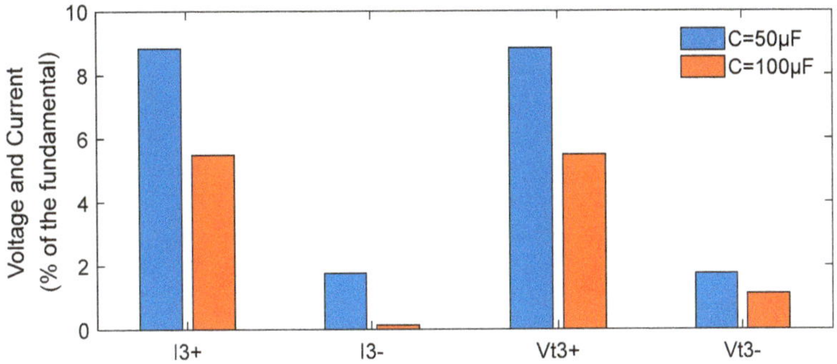

Figure 7. Impact of the size of the dc-link capacitor.

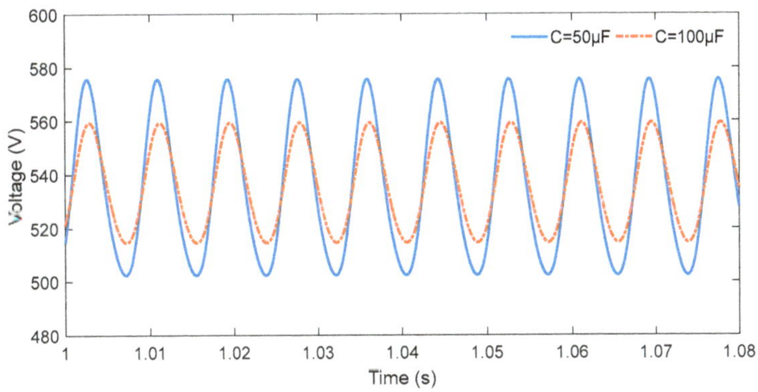

Figure 8. Waveforms of the dc-link voltage for different dc-link capacitors.

5. Discussion

The above results have shown the correctness of the developed model, but it has to be pointed out that the new model can only be developed with the accurate information of the positive-sequence and negative-sequence fundamental frequency voltage and current. If these parameters cannot be provided, then the model cannot be accurately developed. It is noteworthy that the positive-sequence third harmonic distortion is much larger than the negative-sequence third harmonic. Therefore, for the evaluation of the harmonic impact, the level of the positive-sequence third harmonic is sufficient. Moreover, since the negative-sequence harmonic is quite small, its coupling effect to the positive-sequence voltage can be ignored as well. As a result, for the evaluation of the non-characteristic third harmonics, the model can be practically simplified as a Thevenin circuit, and the computation can be simplified accordingly.

The proposed model is able to show the effect of different components on the Thevenin circuit, and, thus, an optimal parameter design can be achieved to reduce the third harmonic distortion. Additionally, the proposed model can help to implement the control schemes for suppressing the third harmonic distortion caused by the VSC under unbalanced conditions.

6. Conclusions

An analytical model of VSC for calculating the non-characteristic third harmonics generated by the VSC operated under unbalanced conditions has been developed. The model is a coupled Thevenin circuit, which can be used to compute the noncharacteristic third harmonics. The correctness and effectiveness of the proposed model for computing noncharacteristic third harmonics have been verified by EMT simulations. Based on the proposed model, the findings regarding the noncharacteristic third harmonic caused by the VSC can be summarized as follows:

- The VSC generates both positive-sequence and negative-sequence third harmonics under unbalanced conditions. The VSC can be modeled as a coupled Thevenin circuit at third harmonics. The source of the Thevenin circuit is determined by the fundamental frequency voltage and current. The impedances are determined by the control parameters and the fundamental frequency components.
- The positive-sequence third harmonic distortion is much larger than the negative-sequence third distortion. As a result, only computing the positive-sequence third harmonic is sufficient to evaluate the impact of the VSC's third harmonic emission under unbalanced condition.
- The larger unbalanced levels and smaller size of the dc-link capacitor introduce a larger third harmonic current.

Author Contributions: Methodology, M.Z. and H.Z.; validation, M.Z. and H.Z.; writing, M.Z. and H.Z.; funding acquisition, H.Z. and R.F.; resources, M.Z. and S.Z.; data curation, M.Z., R.L. and J.W. All authors have read and agreed to the published version of the manuscript.

Funding: This work is supported by State Grid Shanxi Electric Power Company Science and Technology Project Research: Research and application of fusion perception and data mining technology of distribution network power quality based on smart IOT (52053020000X).

Institutional Review Board Statement: Not applicable.

Informed Consent Statement: Not applicable.

Data Availability Statement: The data are available upon request from the corresponding author.

Conflicts of Interest: The authors declare no conflict of interest.

Nomenclature

$V_{(.)}$	Voltage
$I_{(.)}$	Current
$m_{(.)}$	Modulation signal
ω	Angular frequency
$(.)_{a}$	Quantities for phase A
$(.)_{b}$	Quantities for phase B
$(.)_{c}$	Quantities for phase C
$(.)_{1+}$	Quantities at positive-sequence fundamental frequency
$(.)_{1-}$	Quantities at negative-sequence fundamental frequency
$(.)_{3+}$	Quantities at positive-sequence third harmonic
$(.)_{3-}$	Quantities at negative-sequence fundamental frequency
δ	Angle of current
θ	Angle of modulation signal
$(.)_{dc}$	Quantities on the dc side
C	DC-link capacitor
L_f	Front-end passive filter
$H_1(s)$	Transfer function of VSC's inner control loop
i	The imaginary part of the plural

References

1. Carrasco, J.M.; Franquelo, L.G.; Bialasiewicz, J.T.; Galvan, E.; Portillo, R.; Prats, M.M.; Leon, J.I.; Moreno-Alfonso, N. Power-electronic systems for the grid integration of renewable energy sources: A survey. *IEEE Trans. Ind. Electron.* **2006**, *53*, 1002–1016. [CrossRef]
2. Segundo-Ramirez, J.; Medina, A. Modeling of FACTS Devices Based on SPWM VSCs. *IEEE Trans. Power Del.* **2009**, *24*, 1815–1823. [CrossRef]
3. Pan, E.; Yue, B.; Li, X.; Zhao, Z.; Zhu, Q. Integration technology and practice for long-distance offshore wind power in China. *Energy Convers. Econ.* **2020**, *1*, 4–19. [CrossRef]
4. Arazi, M.; Payman, A.; Camara, M.B.; Dakyo, B. Bidirectional Interface Resonant Converter for Wide Voltage Range Storage Applications. *Sustainability* **2022**, *14*, 377. [CrossRef]
5. Sirjani, R. Optimal Capacitor Placement in Wind Farms by Considering Harmonics Using Discrete Lightning Search Algorithm. *Sustainability* **2017**, *9*, 1669. [CrossRef]
6. Rao, S.N.V.B.; Kumar, Y.V.P.; Pradeep, D.J.; Reddy, C.P.; Flah, A.; Kraiem, H.; Al-Asad, J.F. Power Quality Improvement in Renewable-Energy-Based Microgrid Clusters Using Fuzzy Space Vector PWM Controlled Inverter. *Sustainability* **2022**, *14*, 4663. [CrossRef]
7. Iqbal, M.N.; Kütt, L.; Daniel, K.; Asad, B.; Ghahfarokhi, P.S. Estimation of Harmonic Emission of Electric Vehicles and Their Impact on Low Voltage Residential Network. *Sustainability* **2021**, *13*, 8551. [CrossRef]
8. Blasko, V.; Kaura, V. A new mathematical model and control of a three-phase AC-DC voltage source converter. *IEEE Trans. Power Electron.* **1997**, *12*, 116–123. [CrossRef]
9. Torquato, R.; Hax, G.R.T.; Freitas, W.; Nassif, A. Impact Assessment of High-Frequency Distortions Produced by PV Inverters. *IEEE Trans. Power Del.* **2021**, *36*, 2978–2987. [CrossRef]
10. Yazdani, A.; Iravani, R. A unified dynamic model and control for the voltage-sourced converter under unbalanced grid conditions. *IEEE Trans. Power Del.* **2006**, *21*, 1620–1629. [CrossRef]
11. Kamh, M.Z.; Iravani, R. Unbalanced Model and Power-Flow Analysis of Microgrids and Active Distribution Systems. *IEEE Trans. Power Del.* **2010**, *25*, 2851–2858. [CrossRef]
12. Karimi, H.; Karimi-Ghartemani, M.; Sheshyekani, K. Robust Control of Three-Phase Voltage Source Converters under Unbalanced Grid Conditions. *IEEE Trans. Power Electron.* **2019**, *34*, 11278–11289. [CrossRef]
13. Merritt, N.R.; Chakraborty, C.; Bajpai, P. New Voltage Control Strategies for VSC-Based DG Units in an Unbalanced Microgrid. *IEEE Trans. Sustain. Energy* **2017**, *8*, 1127–1139. [CrossRef]
14. Hwang, J.G.; Lehn, P.W.; Winkelnkemper, M. Control of grid connected AC-DC converters with minimized DC link capacitance under unbalanced grid voltage condition. In Proceedings of the 2007 European Conference on Power Electronics and Applications, Aalborg, Denmark, 2–5 September 2007; pp. 1–10.
15. Lo, Y.H.; Chen, Y.C.; Lian, K.L.; Karimi, H.; Wang, C.Z. An Iterative Control Method for Voltage Source Converters to Eliminate Uncharacteristic Harmonics Under Unbalanced Grid Voltages for High-Power Applications. *IEEE Trans. Sustain. Energy* **2019**, *10*, 1419–1429. [CrossRef]
16. Zhong, Q.; Lin, L.X.; Wang, G.; Zhang, Y.; Wu, Z.G. Harmonic analysis model for voltage source converter under unbalanced conditions. *IET Gener. Transm. Distrib.* **2015**, *9*, 12–21. [CrossRef]
17. Nascimento, C.F.; Watanabe, E.H.; Dietrich, A.B.; Dias, R.F.S.; Diene, O. Non-Characteristic Harmonics and DC Side Capacitor Calculation in VSC Connected to a Distribution System with Unbalanced Voltage. In Proceedings of the 2015 IEEE 24th International Symposium on Industrial Electronics (ISIE), Buzios, Brazil, 3–5 June 2015; pp. 349–354.
18. Nascimento, C.F.; Watanabe, E.H.; Diene, O.; Dietrich, A.B.; Goedtel, A.; Gyselinck, J.J.C.; Dias, R.F.S. Analysis of Noncharacteristic Harmonics Generated by Voltage-Source Converters Operating under Unbalanced Voltage. *IEEE Trans. Power Del.* **2017**, *32*, 951–961. [CrossRef]
19. Gao, B.; Wang, Y.; Xu, W. Modeling Voltage Source Converters for Harmonic Power Flow Studies. *IEEE Trans. Power Del.* **2021**, *36*, 3426–3437. [CrossRef]
20. Gao, B.; Wang, Y.; Xu, W. An Improved Model of Voltage Source Converters for Power System Harmonic Studies. *IEEE Trans. Power Del.* **2021**. *early access*. [CrossRef]
21. Khushalani, S.; Solanki, J.M.; Schulz, N.N. Development of three-phase unbalanced power flow using PV and PQ models for distributed generation and study of the impact of DG models. *IEEE Trans. Power Syst.* **2007**, *22*, 1019–1025. [CrossRef]
22. AlHajri, M.F.; El-Hawary, M.E. Exploiting the Radial Distribution Structure in Developing a Fast and Flexible Radial Power Flow for Unbalanced Three-Phase Networks. *IEEE Trans. Power Del.* **2010**, *25*, 378–389. [CrossRef]
23. Wang, Y.; Zhang, N.; Li, H.; Yang, J.W.; Kang, C.Q. Linear Three-Phase Power Flow for Unbalanced Active Distribution Networks with PV Nodes. *CSEE J. Power Energy Syst.* **2017**, *3*, 321–324. [CrossRef]
24. Yazdani, A.; Iravani, R. *Voltage-Sourced Converter in Power Systems: Modelling, Control, and Application*; Wiley: New York, NY, USA, 2010.
25. Liserre, M.; Blaabjerg, F.; Dell'Aquila, A. Step-by-step design procedure for a grid-connected three-phase PWM voltage source converter. *Int. J. Electron.* **2004**, *91*, 445–460. [CrossRef]

World Electric Vehicle Journal

MDPI

Article

Integration of Electric Vehicles into Microgrids: Policy Implication for the Industrial Application of Carbon Neutralisation in China

Yang (Andrew) Wu [1,*], Yui-Yip Lau [2] and Ju-Ai Wu [3]

[1] Division of Science, Engineering and Health Studies, College of Professional and Continuing Education, Hong Kong Polytechnic University, Hong Kong SAR, China
[2] Division of Business and Hospitality Management, College of Professional and Continuing Education, Hong Kong Polytechnic University, Hong Kong SAR, China; yuiyip.lau@cpce-polyu.edu.hk
[3] College of Automation & College of Artificial Intelligence, Nanjing University of Posts and Telecommunication, Nanjing 210023, China; wujuai@njupt.edu.cn
* Correspondence: andrew.wu@cpce-polyu.edu.hk

Abstract: With the dynamic development of renewable energies, energy storage devices, and electric vehicles, microgrids have been playing an increasingly vital role in smart power grids. Under the recent development of carbon neutralisation, microgrid systems containing multiple clean energy sources have become significant modules for energy conservation and emission reduction. Considering technological and environmental elements, we investigated the economic operation of microgrids with the integration of electric vehicles. In this paper, carbon trading mechanisms and operation scheduling strategies are analysed in the simulation models. Then, transaction costs and power balance are discussed. Industrial applications and policy implications are also presented.

Keywords: electric vehicle; carbon neutralisation; microgrid; energy storage; economic operation; carbon trading

Citation: Wu, Y.; Lau, Y.-Y.; Wu, J.-A. Integration of Electric Vehicles into Microgrids: Policy Implication for the Industrial Application of Carbon Neutralisation in China. *World Electr. Veh. J.* **2022**, *13*, 96. https://doi.org/10.3390/wevj13060096

Academic Editors: Wenbin Yu and Guang Zeng

Received: 4 April 2022
Accepted: 23 May 2022
Published: 26 May 2022

1. Introduction

The exploitation of fossil energy has shifted humanity's focus from agriculture to industrialisation. Industrialisation has driven the tremendous advancement of civilisation for over two centuries. The predicted total volume of CO_2 created by fossil fuels is estimated to be 2.2 trillion tons [1]. In recent years, fossil fuel-based energy utilisation models have resulted in serious carbon emissions and global warming. Greta Thunberg, a Swedish teenage girl and a world-renowned environmental activist, warned at the 2020 Davos Forum to "stop investing in fossil fuels immediately, or explain to your children why you did not protect them from the climate chaos you create" [2]. In the past decade, the power industry has actively promoted energy conservation and emission reduction. However, due to economic costs and load constraints, wind and solar curtailment persists [3–6]. With recent policy incentives, carbon neutralisation and the vigorous development of electric vehicles and renewable energies have become inevitable trends in the power industry [7–9]. As a comprehensive integrated technology combining multiple types of renewable energies, the microgrid can increase the utilisation of wind power and solar energy, reducing carbon emissions [10,11]. Renewable energy sources, such as wind power and solar energy, are unstable and intermittent due to unpredictable weather conditions (e.g., heatwaves, tropical cyclone, storms) during generation. Thus, applying these valuable electric energies continuously and stably is difficult [12]. Ultimately, it is imperative to contribute to energy conservation and carbon neutralisation.

Since 1906, the average global temperature has increased by 1.1 °C [1]. To effectively address the adverse effects of global warming on human society, the 2015 United Nations Framework Convention on Climate Change (UNFCCC) was held in Paris, France. The consensus of keeping global warming under 2 °C was reached. Countries worldwide have further clarified their scope of greenhouse gas emission reduction. Different from the 1997 Kyoto Protocol, the Paris Agreement allowed each party to clarify their submission on greenhouse gas emission reductions through a bottom-up approach based on their emission reduction responsibilities and capabilities, which were defined as Nationally Determined Contributions (NDCs) targets [13]. In China, President Xi Jinping has emphasised the importance of environmental protection and ecological civilisation since the 18th National Congress of the Communist Party of China [14]. This mechanism can avoid problems of selection and formulation of emission reduction responsibility allocation options among countries.

The carbon trading mechanism allows trading entities to adjust their carbon emission rights through trading and exchange within a certain range to achieve regional carbon reduction at the lowest cost. However, compared with the carbon tax, the policy setting and design process of the carbon trading mechanism are more sophisticated and complicated. Setting the appropriate amount of regionally total carbon emission rights and determining the coverage and allocation of appropriate initial carbon emission rights for various industrial sectors are extremely challenging for policy makers.

Carbon trading mechanisms have provided economic incentives for market players to reduce carbon emissions and promote the utilisation of renewable energies and electric vehicles [15]. The introduction of carbon trading mechanisms no longer defines carbon as emission costs, but additional economic gains through carbon trading [16]. Regulation has been deployed to encourage the power industry to transform its energy structure and to improve technological innovation, achieving an environmentally economic operation model. Ref. [17] considered the economic impact of carbon emission allowances on scheduling strategies. Ref. [18] established a mathematical model for calculating carbon transaction costs based on carbon emission intervals. Ref. [19] constructed an optimal dispatch model for large-scale solar farms accessing smart grids under carbon trading mechanisms, which examined the low-carbon performance of the proposed system, especially in the post-COVID-19 era. The microgrid provides an efficient approach to facilitate the high penetration of renewable energies and electric vehicles into the power grid [20–22], serving as an important link in energy saving and emission reduction. The rapid growth of electric vehicles has essentially affected the development of microgrids in the smart power system [23], which can further stimulate the achievement of carbon neutralisation [24]. The dynamic increase in carbon emissions in China can be neutralised with the promotion and penetration of electric vehicles [25].

We investigated the economic operation of microgrids with the integration of electric vehicles and renewable energies, including wind power and solar energy. In this paper, a microgrid optimal dispatch model is constructed. Carbon trading mechanisms and operation scheduling strategies are analysed through the simulation models. Transaction costs in economic dispatch and power balance are discussed. The related constraints of balance are designed to reduce the carbon emissions of the microgrid system, increase the utilisation of renewable energies, and promote carbon neutralisation in the power industry. Industrial and policy implications are provided in the conclusion.

The remainder of this paper is structured as follows. Section 2 presents the research methodology. Section 3 presents a robust low-carbon economic dispatch model. Sections 4 and 5 provide industrial applications and policy implications in carbon neutralisation, respectively. Section 6 summarises and discusses future research directions.

2. Research Methodology

Electric vehicles have been playing an increasingly vital role in the smart grid and power system technology, owing to updated charging technology, financial subsidies, and low cruise costs. With the expansion of the scale of using electric vehicles, the operational model of the microgrid must be modified accordingly. Electric vehicles have adjustable charging and discharging properties, and they have a dual role when they are connected to the grid: they can be used as a load to charge and as a power source to supply power to the microgrid. Therefore, research to incorporate electric vehicles into the microgrid to optimise the allocation of microgrid resources can improve the economic and stable operation of the microgrid, which will further enhance the dynamic performance of the main power grid. In this paper, we propose a microgrid economic dispatch model containing electric vehicles, wind power, solar energy, and energy storage devices. In addition, carbon neutralisation and carbon emission costs are added to the model as constraints. A standard structure of a microgrid connected to the main grid containing wind power, solar energy, electric vehicles, and energy storage devices is shown in Figure 1.

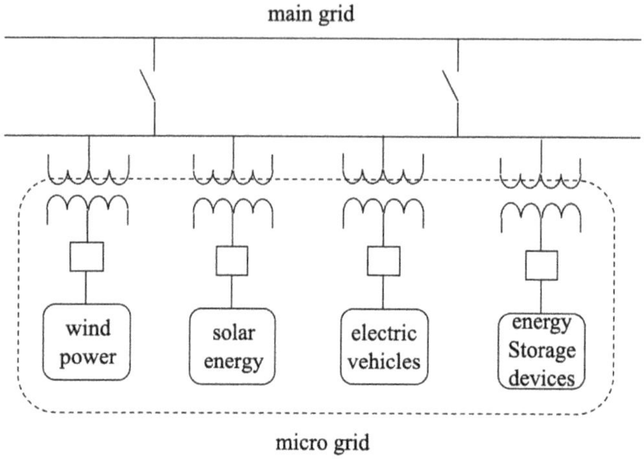

Figure 1. Sample structure of a microgrid connected to a main electric grid.

3. Low-Carbon Dispatch Model

The microgrid system is connected to the main grid and comprises electric vehicles, wind power, photovoltaic power generation, and storage batteries. If the electricity purchased by the microgrid from the main grid is generated by thermal power units, the microgrid's free carbon emission quota is:

$$C_m = \mu \sum_{t=1}^{T} \left(P_w \cdot t + P_s \cdot t + P_{ev} \cdot t - P_g \cdot t\right). \tag{1}$$

where:

C_m is the carbon emission quota of the microgrid;

μ is the co-efficiency as regulated by related authorities;

t is the time duration of the operation;

P_w is the electric power generated by the wind turbines;

P_s is the electric power generated from solar energy;

P_{ev} is the electric power charged by or used by electric vehicles connected to the microgrids, which can be positive or negative; and

P_g is the electric power purchased from the main grid.

Carbon emissions are lower than quotas, resulting in carbon emission benefits. Low carbon emissions can generate substantial carbon trading benefits and may even offset the increase in costs caused by gas turbine power generation from the main grid. This fully mobilises energy saving and emission reduction in the power generation industry.

To prevent excessive charging and discharging of the energy storage devices, charging constraints are introduced to fulfil upper and lower limits:

$$0 \leq P_b \leq P_{b,max}. \tag{2}$$

where:

P_b is the electric power saved in the energy storage devices; and

$P_{b,max}$ is the maximum charging capacity of the energy storage devices.

The trading data disclosed by the two leading carbon trading pilot centres in China, i.e., the Beijing Trading Centre and the Guangzhou Trading Centre, reveal a significant gap between the carbon trading patterns of northern and southern China, although the annual cumulative carbon emission rights and trading turnovers are close to each other, both being approximately 26 million tons at around USD 170 million [26,27], as shown in Figure 2.

Figure 2. Annual turnover of carbon trading pilot centres in China in 2021.

The peak trading season in Guangzhou is summer, mainly May to July; meanwhile, the peak trading season in Beijing is mainly November and December. The reason is due to different details on policy execution and settlement regimes. Market players in Guangzhou must face the mid-term inspection of carbon emissions, which is an important policy incentive for the market players with high carbon emission to increase their trading for sufficient emission right in the summer. Meanwhile, the market players in Beijing have only a year-end examination of their carbon emissions, and thus tend to wait until the year's end to complete their carbon trading. In addition to the different settlement regimes, another major reason is the different patterns of energy consumption between northern and southern China due to different weather conditions.

The above-mentioned pilot centres are only at the regional level, and can only provide a carbon trading platform for enterprises within their administrative regions. In July 2021, a national-level trading platform was launched in Shanghai under a trial mode. More information is to be released in the near future. With the official launch of the national carbon emission rights trading market, government authorities can utilise the market and pricing mechanism to regulate the use of carbon emission rights. When the carbon trading price is overly high, the carbon emission quota allocated to the industrial manufacturer may be reduced. Meanwhile, the pressure on the manufacturer to reduce emissions increases will drive the enterprise to spend more money to purchase carbon emission rights, which is not conducive to its production and operation. Therefore, an appropriate adjustment in the carbon trading price may not only reduce the emission reduction pressure of emission enterprises, but also effectively increase the manufactory output and broaden the eligible industrial sectors of the carbon quota mechanism.

4. Industrial Applications

By the end of 2020, the cumulative installed capacity of power generation from renewable energies reached 535,000 MW in China, a year-on-year increase of 29.4%, accounting for 24.3% of the total installed capacity. The newly installed capacity of renewable energy power generation in 2020 reached 120,000 MW, accounting for 63% of the total installed capacity of renewable energies in the whole nation [28]. Furthermore, the national power generation from renewable energies achieved 10% of the total power generation, highlighting a considerable improvement in renewable energy utilisation.

Under the guidance of the development goals of the electric power system, the future development of the new energy industry has shown a trend of deploying more renewable energies and electric vehicles. Distributed power generation, microgrids, and integration with electric vehicles have gained considerable importance in the industry. The widespread use of microgrids has intensified the demand for flexible adjustment resources, such as energy storage devices.

The shift from centralised generation to distributed generation has attracted the interest of academics, policy makers, and industrial engineers, thereby accelerating the development of microgrid systems. Energy storage devices are among the essential factors to facilitate such a trend. In addition to distributed renewable energy resources, such as roof-top wind turbines and solar panels, fuel cells are considered a clean and flexible option for promising energy resources in microgrid application [29]. Flow batteries have also been revived recently, especially for large-scale energy storage systems in microgrids, due to their low cost and eco-friendly materials, even with favourable design flexibility [30].

In recent years, electric vehicle sales have grown rapidly due to the incentive financial subsidy policy. The government's subsidy policy has greatly stimulated enthusiasm to invest in the electric vehicle industry. Domestic sales rose sharply with the implementation of the subsidy policy. Due to policy relaxation and tax benefits, electric vehicle manufacturers have also increased. The driving range and the safety of the power battery are constantly improving. Power battery technology is constantly advancing. Power semiconductor technology, intelligent drive technology, and power component integration technology have improved to increase power density, reduce losses, and ensure reliability. The main domestic charging method is AC slow charging, supplemented by DC fast charging. Convenience will be an important concern in the future. Microgrid power generation is in a low-carbon mode, interacting power transmission with the main grid. The sum of the total power generation in the microgrid is equal to the total load value in the microgrid, which satisfies the balance of supply and demand. When the load is low, the microgrid sells electricity to the main grid, the battery is charged, and the output of wind power and solar energy drops. When the load peaks, the output of wind power and solar energy increases, and the battery discharges to the main grid.

The electric vehicle industry policy aims to promote the development of the new automobile industry. Ref. [31] predicted electric vehicle sales in the next few years, as shown in Figure 3. In the implementation process, policy formulation has faced challenges from various aspects, including technical, financial, and environmental issues. Related policies and their implementation have encountered a lack of references. Therefore, policy formulation must be further improved. The competent authority shall formulate long-term policy plans considering standardisation and stability. The transition period of the policy should be well connected to ensure continuity and predictability, which will gradually form legal documents.

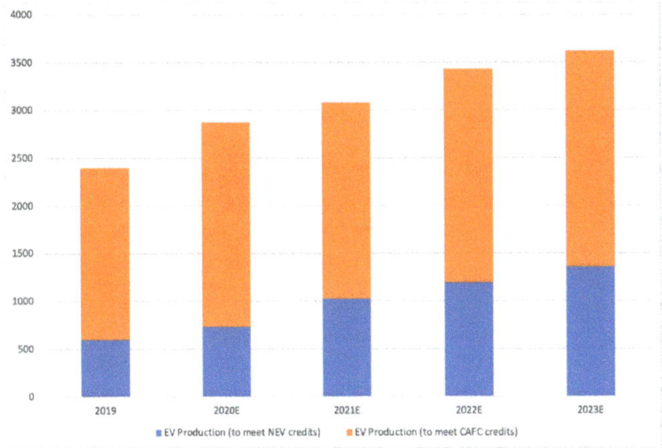

Figure 3. Predicted increase in EV production in China (unit: thousand).

To accommodate the flexibility of the microgrid system, energy storage devices of electric vehicles should be light with the potential to possess high energy density. The application of lithium batteries in recent years has been proven to be a feasible solution. The integration of lithium batteries in electric vehicles can significantly reduce carbon emissions and road-side air pollution [32].

Generation units of renewable energies are small in capacity, large in number, scattered in locations, and diverse in characteristics. Power electronic equipment adopts discrete control based on fast switching, which makes a fundamental change in the control mode of the distributed microgrid power system and has continuous adjustment and control capabilities.

The form of the power system has transformed from the traditional large power grid to the complementary symbiosis of the large power grid and the microgrids. The power system must not only satisfy the access and consumption of centralised new energy but also support the access and withdrawal of distributed power generation. Wind power and solar energy will become the installed capacity of electricity supply as well as bear the responsibility for the safe and stable operation of the power system.

China has launched carbon emission trading pilot schemes in eight regions, actively exploring the trading mechanism and future direction of the carbon trading market. In 2011, China first opened carbon emission rights in seven regions, Beijing, Tianjin, Shanghai, Chongqing, Hubei, Guangzhou, and Shenzhen. In 2017, Fujian joined as the eighth pilot region. Each pilot region followed a similar carbon emission measurement approach; however, the coverage for industrial sectors varied [33]. The development in those eight pilot regions built a solid foundation for the national carbon trading platform in the next stage.

The carbon trading pilot centre in Beijing is the most representative of the eight centres to reflect the carbon emission and trading situation in northern China. Meanwhile, the carbon trading pilot centre in Guangzhou is the most representative of the situation in southern China.

The essential purpose of running an enterprise is for production and sales. The reason why companies with emission pressure are willing to reduce emissions is that excess carbon emission rights exist that can be traded in the carbon trading market for profit. Therefore, related government authorities should encourage enterprises to invest in research and develop emission reduction technologies to reduce the operational and environmental costs of emission reduction. Furthermore, guidelines for consumers to buy low-carbon products will also boost market demand and increase production and sales revenue for enterprises.

5. Policy Implications

The power production structure has undergone profound changes. The primary energy supply for the new power system will shift from stable and controllable conventional energy sources, such as coal, gas, and water, to new energy, such as wind power and solar energy. The technical foundation of the power system has undergone significant changes. The synchronous operation of the AC power system of the new energy unit is driven by the physical characteristics of the steering control algorithm. This is a time to construct a carbon emission management system and a supervision system for low-carbon and green energy development [1]. Moreover, the policy support for industry–university research cooperation and policy support for new and in-depth collaboration between industry, academia, and research led by energy automobile companies must be strengthened. The support and incentives for technology innovation will provide a platform to encourage knowledge transfer and the transformation of scientific and technological achievements related to the automobile and power industry, which will encourage cooperation and integration amongst enterprises and research institutes, including collaborative technology development and industrialisation. The government may introduce various innovation and technology funds to support academic institutions and firms to purchase equipment and recruit research personnel.

Regarding subsidies and tax policies, direct subsidies should be gradually reduced, and low-interest loans and other financial policy support should be increased. The policy scope should consider the whole business chain of the electric vehicle industry. Subsidy policy can be returned to consumers to expand consumer demand. The requirements for products have driven automobile companies to invest in research and development. Subsidies and tax policies have also strengthened review and supervision, thereby increasing the execution power on regulations and laws. Other supporting policies, such as purchase discounts and exemption from traffic restrictions, will also improve the advantages and consumers' willingness to drive electric vehicles.

In addition to the global economic crisis and other factors, issues persist, such as the setting of the total amount of carbon emission rights, the determination of industry coverage, and the allocation of initial carbon emission rights in the design process of the carbon trading mechanism, which have become obstacles to the construction process of carbon trading markets in various countries and have affected the carbon trading mechanism. Government authorities can consider formulating the guidance price of carbon emission rights based on the emission reduction target and the industrial average emission reduction cost. The decision-making behaviours of automobile manufacturers under the carbon trading mechanism include energy intensity, carbon credit evaluation decisions, production decisions, automobile pricing decisions, and carbon credit accounting and trading considering energy intensity. In addition, car manufacturers can interact with car users when selling cars.

Given that information on China's carbon market has not been fully disclosed, investors' decisions based on known information are sophisticated, which exacerbates the randomness of investment decisions and expected yield from carbon trading in the market [34,35]. Therefore, the transparent information disclosure system in the carbon trading market must be improved. The current pricing mechanism in carbon trading pilot centres cannot fully reflect market information. Thus, market pricing efficiency must be improved. Furthermore, establishing appropriate mechanisms to prevent insider trading is essential for controlling market risks. In addition, an adjustment mechanism to effectively prevent abnormal price fluctuations and a risk prevention and control mechanism to prevent market manipulation should be established to form an effective price formation that reflects factors such as supply and demand, emission reduction costs, and so on.

To stimulate carbon trading activities in the pilot centres, government authorities can consider guiding relevant enterprises to increase their willingness to trade spontaneously in the carbon market by providing financial incentives and stimulation mechanisms. On the one hand, penalties for companies that fail to comply with the carbon emission targets should be implemented; on the other hand, more tax incentives and green credits to companies that actively participate in carbon trading and achieve the carbon emission targets should be provided. Such a stimulative mechanism can enhance the enthusiasm of companies to seek profit through active trading activities or to reduce carbon emissions.

Furthermore, the electric vehicle industry faces a challenge of long-term capital occupation and substantial initial financial investment. Based on the experiences of the conventional automobile industry, corporate investment and financing can appropriately reduce financing costs. Financing channels, such as corporate loans and bonds, must be broadened, and banks, financial intermediaries, and credit guarantee centres must be encouraged to provide priority services for electric automobile companies, moderately relax conditions, lower interest rates, and introduce foreign capital. Concurrently, a risk prevention and control mechanism should be implemented.

6. Conclusions and Future Work

In this paper, we proposed and analysed a multi-source microgrid economic dispatch model to reduce carbon emissions. The implemented carbon trading mechanism contributes to achieving carbon emissions control. Companies with surplus carbon emission quotas can increase their additional income by selling such quotas to reduce the total generation and operation cost of the power system. The low-carbon dispatch model enables wind power, solar energy, electric vehicles, and other distributed power generation units to benefit from carbon trading and to reduce the operation costs while expanding the microgrid system, which ultimately reduces the total cost of the power system.

The cost of distributed generation from renewable energies in microgrids is relatively higher than that of conventional generation. When the power generation exceeds a certain range, the system can sacrifice a certain economy to increase the penetration rate of new energy, thereby reducing the carbon emissions of the system and improving the environmental protection of the system.

The issuance and implementation of industrial policies on electric vehicles have become important factors that promote rapid development. With the advent of the post-subsidy era, the electric vehicle industry continues to face challenges related to core technologies and structures. Therefore, industrial policies must be continuously improved to promote the healthy and rapid development of the industry.

With the continuous introduction of new carbon trading-related financial products in China, the volatility of the carbon trading market may be affected by emerging factors. Future research must entail a comprehensive analysis and volatility assessment of the incentives for carbon trading volatility. Furthermore, automobile users are not included in the current emission reduction framework in China. To achieve carbon neutralisation, automobile users, such as the general public who drive EVs, should also be included.

Future study must incorporate related economic and microgrid management indicators into our mathematical model to enrich simulation scenarios with consideration of climate values. As such, it may contribute to the interdisciplinary research direction in the future. Moreover, future research must develop a climate-resilient infrastructure on how to minimise the vulnerabilities or risks influenced by climate change and improve climate resilience [36,37].

Author Contributions: Conceptualisation, methodology, and writing—original draft preparation, Y.W.; validation, formal analysis, and writing—review and editing, Y.-Y.L.; data curation and visualisation, J.-A.W.; funding and project administration, Y.W. All authors have read and agreed to the published version of the manuscript.

Funding: This research was funded by the Research Grant Council of Hong Kong SAR, project reference number UGC/IIDS24/E01/20; the PolyU CPCE Research Fund, project reference number SEHS-2019-169(J); and the PolyU CPCE Research Fund, project reference number SEHS-2021-228(J).

Institutional Review Board Statement: Not applicable.

Informed Consent Statement: Not applicable.

Data Availability Statement: Not applicable.

Conflicts of Interest: The authors declare no conflict of interest.

References

1. Li, Q. The view of technological innovation in coal industry under the vision of carbon neutralisation. *Int. J. Coal Sci. Technol.* **2021**, *8*, 1197–1207. [CrossRef]
2. Sengupta, S. Greta Thunberg's Message at Davos Forum: Our House Is Still on Fire. *The New York Times*, 21 January 2020. Available online: https://www.nytimes.com/2020/01/21/climate/greta-thunberg-davos.html(accessed on 30 March 2022).
3. Jiang, Y.; Deng, Z.; You, S. Size optimization and economic analysis of a coupled wind-hydrogen system with curtailment decisions. *Int. J. Hydrog. Energy* **2019**, *44*, 19658–19666. [CrossRef]
4. Davison-Kernan, R.; Liu, X.; McLoone, S.; Fox, B. Quantification of wind curtailment on a medium-sized power system and mitigation using municipal water pumping load. *Renew. Sustain. Energy Rev.* **2019**, *112*, 499–507. [CrossRef]
5. Canbulat, S.; Balci, K.; Canbulat, O.; Bayram, I.S. Techno-Economic Analysis of On-Site Energy Storage Units to Mitigate Wind Energy Curtailment: A Case Study in Scotland. *Energies* **2021**, *14*, 1691. [CrossRef]
6. Boddapati, V.; Nandikatti, A.S.R.; Daniel, S.A. Techno-economic performance assessment and the effect of power evacuation curtailment of a 50 MWp grid-interactive solar power park. *Energy Sustain. Dev.* **2021**, *62*, 16–28. [CrossRef]
7. Li, Y.; Mohammed, S.Q.; Nariman, G.S.; Aljojo, N.; Rezvani, A.; Dadfar, S. Energy Management of Microgrid Considering Renewable Energy Sources and Electric Vehicles Using the Backtracking Search Optimization Algorithm. *J. Energy Resour. Technol.* **2020**, *142*, 052103. [CrossRef]
8. Eseye, A.T.; Lehtonen, M.; Tukia, T.; Uimonen, S.; Millar, R.J. Optimal Energy Trading for Renewable Energy Integrated Building Microgrids Containing Electric Vehicles and Energy Storage Batteries. *IEEE Access* **2019**, *7*, 106092–106101. [CrossRef]
9. Abubakrab, H.; Mohamed, T.H.; Hussein, M.M.; Guerrero, J.M.; Agundis-Tinajero, G. Adaptive frequency regulation strategy in multi-area microgrids including renewable energy and electric vehicles supported by virtual inertia. *Int. J. Electr. Power Energy Syst.* **2021**, *129*, 106814. [CrossRef]
10. Iacobucci, R.; McLellan, B.; Tezuka, T. The Synergies of Shared Autonomous Electric Vehicles with Renewable Energy in a Virtual Power Plant and Microgrid. *Energies* **2018**, *11*, 2016. [CrossRef]
11. Elkadeem, M.R.; Wong, S.; Azmy, A.M.; Atiya, E.G.; Ullah, Z.; Sharshir, S.W. A systematic decision-making approach for planning and assessment of hybrid renewable energy-based microgrid with techno-economic optimization: A case study on an urban community in Egypt. *Sustain. Cities Soc.* **2020**, *54*, 102013. [CrossRef]
12. Sun, C.; Zhang, H. Review of the Development of First-Generation Redox Flow Batteries: Iron-Chromium System. *ChemSusChem* **2022**, *15*, e202101798. [CrossRef]
13. Van den Berg, N.J.; van Soest, H.L.; Hof, A.F.; den Elzen, M.G.J.; van Vuuen, D.P.; Chen, W.; Drouet, L.; Emmerling, J.; Fujimori, S.; Hohne, N.; et al. Implications of various effort-sharing approaches for national carbon budgets and emission pathways. *Clim. Chang.* **2020**, *162*, 1805–1822. [CrossRef]
14. Wang, Y.; Guo, C.; Chen, X.; Jia, L.; Guo, X.; Chen, R.; Zheng, M.; Chen, Z.; Wang, H. Carbon peak and carbon neutrality in China: Goals, implementation path and prospects. *China Geol.* **2021**, *4*, 720–746. [CrossRef]
15. Wu, C.; Guo, S.; Liu, Y.; Song, T.E.; Han, H. A model predictive control approach in microgrid considering multi-uncertainty of electric vehicles. *Renew. Energy* **2021**, *163*, 1385–1396. [CrossRef]
16. Shi, R.; Zhang, P.; Zhang, J.; Niu, L.; Han, X. Multidispatch for Microgrid including Renewable Energy and Electric Vehicles with Robust Optimization Algorithm. *Energies* **2020**, *13*, 2813. [CrossRef]
17. Liu, C.; Abdulkareem, S.S.; Rezvani, A.; Samad, S.; Aljojo, N.; Foong, L.K.; Nishihara, K. Stochastic scheduling of a renewable-based microgrid in the presence of electric vehicles using modified harmony search algorithm with control policies. *Sustain. Cities Soc.* **2020**, *59*, 102183. [CrossRef]
18. Shang, T.; Yangg, L.; Liu, P.; Shang, K.; Zhang, Y. Financing mode of energy performance contracting projects with carbon emissions reduction potential and carbon emissions ratings. *Energy Policy* **2020**, *144*, 111632. [CrossRef]
19. Tian, J.; Yu, L.; Xue, R.; Zhuang, S.; Shan, Y. Global low-carbon energy transition in the post-COVID-19 era. *Appl. Energy* **2022**, *307*, 118205. [CrossRef]
20. Noorollahi, Y.; Golshanfard, A.; Aligholian, A.; Mohammadi-ivatloo, B.; Nielsen, S.; Hajinezhad, A. Sustainable Energy System Planning for an Industrial Zone by Integrating Electric Vehicles as Energy Storage. *J. Energy Storage* **2020**, *30*, 101553. [CrossRef]
21. Sayed, K.; Abo-Khalil, A.G.; Alghamdi, A.S. Optimum Resilient Operation and Control DC Microgrid Based Electric Vehicles Charging Station Powered by Renewable Energy Sources. *Energies* **2019**, *12*, 4240. [CrossRef]
22. Hao, Y.; Dong, L.; Liang, J.; Liao, X.; Wang, L.; Shi, L. Power forecasting-based coordination dispatch of PV power generation and electric vehicles charging in microgrid. *Renew. Energy* **2020**, *155*, 1191–1210. [CrossRef]
23. Wang, M.; Wang, Y.; Chen, L.; Yang, Y.; Li, X. Carbon emission of energy consumption of the electric vehicle development scenario. *Environ. Sci. Pollut. Res.* **2021**, *28*, 42401–42413. [CrossRef] [PubMed]
24. Ghosh, A. Possibilities and Challenges for the Inclusion of the Electric Vehicle (EV) to Reduce the Carbon Footprint in the Transport Sector: A Review. *Energies* **2020**, *13*, 2602. [CrossRef]

25. Lin, B.; Wu, W. The impact of electric vehicle penetration: A recursive dynamic CGE analysis of China. *Energy Econ.* **2021**, *94*, 105086. [CrossRef]
26. Beijing Environment Exchange. Monthly Report of Carbon Trading at Beijing Environment Exchange. 2022. Available online: https://www.bjets.com.cn/ (accessed on 30 March 2022).
27. Guangzhou Emission Exchange. Weekly Report of Carbon Trading at Guangzhou Emission Exchange. 2022. Available online: http://www.cnemission.com/ (accessed on 30 March 2022).
28. Energy Bureau of Chinese Central Government. Continuous Optimization of Energy Structure in the First Three Quarters of 2021. In *Press Release of Chinese Central Government*; Energy Bureau of Chinese Central Government: Beijing, China, 2021.
29. Akinyele, D.; Olabode, E.; Amole, A. Review of Fuel Cell Technologies and Applications for Sustainable Microgrid Systems. *Inventions* **2020**, *5*, 42. [CrossRef]
30. Zhang, H.; Sun, C. Cost-effective iron-based aqueous redox flow batteries for large-scale energy storage application: A review. *J. Power Sources* **2021**, *493*, 229445. [CrossRef]
31. Wu, Y.A.; Ng, A.W.; Yu, Z.; Huang, J.; Meng, K.; Dong, Z.Y. A review of evolutionary policy incentives for sustainable development of electric vehicles in China: Strategic implications. *Energy Policy* **2021**, *148*, 111983. [CrossRef]
32. Speirs, J.; Contestabile, M.; Houari, Y.; Gross, R. The future of Lithium availability for electric vehicle batteries. *Renew. Sustain. Energy Rev.* **2014**, *35*, 183–193. [CrossRef]
33. Wu, Y.; Wang, P.; Liu, X.; Chen, J.; Song, M. Analysis of regional carbon allocation and carbon trading based on net primary productivity in China. *China Econ. Rev.* **2020**, *60*, 101401. [CrossRef]
34. Jia, Z.; Lin, B. Rethinking the choice of carbon tax and carbon trading in China. *Technol. Forecast. Soc. Change* **2020**, *159*, 120187. [CrossRef]
35. Lau, Y.Y.; Wu, A.W.; Mak, W.Y. A way forward for electric vehicle in Greater Bay Area: Challenges and opportunities for the 21st century. *Vehicles* **2022**, *4*, 420–432. [CrossRef]
36. Yang, Z.; Ng, A.K.Y.; Lee, P.T.W.; Wang, T.; Qu, Z.; Rodrigues, V.S.; Pettit, S.; Harris, I.; Zhang, D.; Lau, Y.Y. Risk and cost evaluation of port adaptation measures to climate change impacts. *Transp. Res. Part D Transp. Environ.* **2018**, *61*, 444–458. [CrossRef]
37. Ng, A.K.Y.; Zhang, H.; Afenyo, M.; Becker, A.; Cahoon, S.; Chen, S.L.; Esteban, M.; Ferrari, C.; Lau, Y.Y.; Lee, P.T.W.; et al. Port decision maker perceptions on the effectiveness of climate adaptation actions. *Coast. Manag.* **2018**, *46*, 148–175. [CrossRef]

energies

MDPI

Article

Slope Estimation Method of Electric Vehicles Based on Improved Sage–Husa Adaptive Kalman Filter

Jiawei Guo [1,2], Chao He [1,2], Jiaqiang Li [1,2,*] and Heng Wei [1,3]

1 School of Machinery and Transportation, Southwest Forestry University, Kunming 650224, China; williamgjw@swfu.edu.cn (J.G.); chao.he@swfu.edu.cn (C.H.); henrywei1@whut.edu.cn (H.W.)
2 Key Laboratory of Vehicle Emission and Safety on Plateau Mountain, Yunnan Provincial Department of Education, Kunming 650224, China
3 Hubei Key Laboratory of Advanced Technology for Automotive Components, Wuhan University of Technology, Wuhan 430070, China
* Correspondence: lijiaqiang@swfu.edu.cn

Abstract: In order to deal with many influence factors of electric vehicles in driving under complex conditions, this paper establishes the system state equation based on the longitudinal dynamics equation of vehicle. Combined with the improved Sage–Husa adaptive Kalman filter algorithm, the road slope estimation model is established. After the driving speed and rough slope observation are input into the slope estimation model, the accurate road slope estimation at the current time can be obtained. The road slope estimation method is compared with the original Sage–Husa adaptive Kalman filter road slope estimation method through three groups of road tests in different slope ranges, and the accuracy and stability advantages of the proposed algorithm in road conditions with large slopes are verified.

Keywords: road slope estimation; adaptive Kalman filter; electric car

Citation: Guo, J.; He, C.; Li, J.; Wei, H. Slope Estimation Method of Electric Vehicles Based on Improved Sage–Husa Adaptive Kalman Filter. *Energies* **2022**, *15*, 4126. https://doi.org/10.3390/en15114126

Academic Editors: Wenbin Yu, Guang Zeng and Tek Tjing Lie

Received: 10 April 2022
Accepted: 1 June 2022
Published: 3 June 2022

1. Introduction

Accurate road slope prediction method is of vital importance in the field of autonomous driving. The system collects parameters through algorithms, predicts the slope at the next moment in advance, and prepares to control the vehicle power system at the next moment, so as to improve the control ability of automatic transmission of vehicles, optimize the power and economy of vehicles, and reduce energy consumption while running smoothly. The parameters of the vehicle the system collects include torque, vehicle speed, acceleration, slope, etc. Among all the parameters, accurate slope information acquisition and prediction has a very special research significance for intelligent vehicle driving; details are shown below.

Figure 1 explains the basic principle of unmanned driving—all behaviors of human drivers in driving can be planned into three steps: environmental perception, decision and planning, control and execution. Environmental perception is the first and most important part. For example, human drivers collect the road condition information of the current vehicle through hearing and vision, and make decisions about the following driving behavior according to this information. Autonomous driving systems also need to collect road information before planning power systems, so high-precision collection algorithms are very important. In addition, the parameter of road slope is a very important part of the longitudinal dynamics equation of the vehicle. Both rolling friction and slope resistance are related to the road slope of the vehicle at the current moment, which to some extent determines the advantages and disadvantages of the adjustment of the dynamic performance of the vehicle in the automatic driving link.

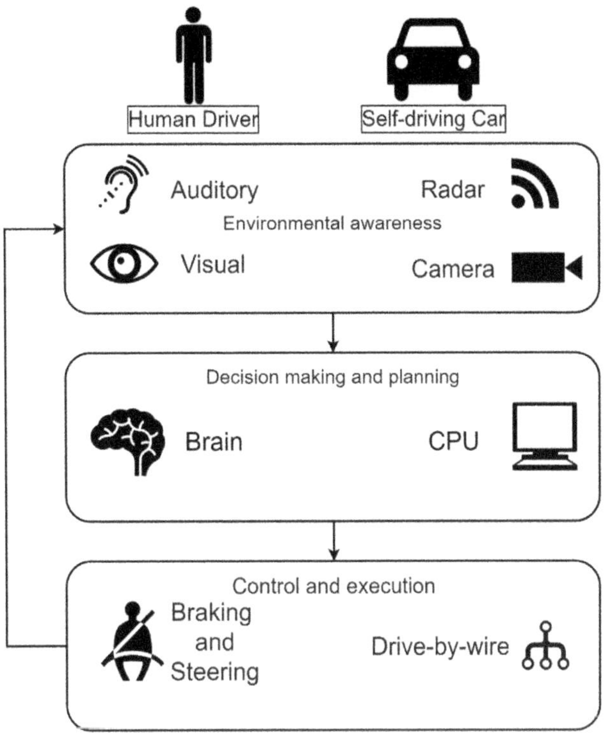

Figure 1. Concept of autonomous driving.

From the perspective of the long-term development of automatic driving, accurate slope prediction is the necessary guarantee for automatic driving technology to step into level 4 and level 5. Level 5 autonomous driving technology requires the vehicle to be able to navigate unknown roads and environments. Therefore, scholars all over the world have done a lot of research on slope prediction and have achieved some results.

In the past 5 years, most of the research on slope estimation algorithms has been based on longitudinal dynamics of vehicles. Some scholars designed a series of slope estimation methods using unscented Kalman filter, such as Dual Unscented Kalman Filter (DUKF) and Quaternion Unscented Kalman Filter (QUKF) [1–4]. A few scholars used extended Kalman filter to construct slope estimation methods and achieved some results [5,6]. The least square method was also used to construct slope estimation algorithms [7–9]. Jiang et al. proposed a two-stage estimation method for vehicle mass and road slope under longitudinal moving condition of mini fuel cell vehicle (FCV) [10]. Hu et al. established the longitudinal kinematics model of vehicles, using the recursive least squares method with adaptive forgetting factors and extended Kalman filter algorithm to estimate the vehicle mass and road grade, respectively [11]. Rodríguez et al. presented a novel accurate estimator based on errorEKF and UKF for vehicle dynamics [12]. Bian et al. presented a MPC based vehicular following control algorithm with road grade prediction [13]. Feng et al. proposed a slope estimation algorithm based on multi-model and multi-data fusion [14].

To sum up, the accuracy of slope prediction is closely related to autonomous driving technology. At the same time, slope estimation is an important cornerstone of the development of autonomous driving technology. Previous literatures have achieved certain results in slope prediction with good prediction accuracy, but there are still gaps in some aspects. The accuracy of the basic Kalman filter has struggled to meet the current requirements of automatic driving. The extended Kalman Filter (EKF) uses Taylor Expansion to construct approximate linear function by obtaining the slope of nonlinear function. The nonlinear system is approximated to linear system by means of ignoring the higher-order term, which inevitably introduces linear error and even leads to the divergence of filter. Thus, the accuracy of slope estimation is reduced. Unscented Kalman filter (UKF) is more accurate than EKF, and its accuracy is equivalent to second-order Taylor expansion, but the speed is slower. A large amount of calculation will require a longer calculation time, so whether it can be applied to the dynamic slope estimation of vehicles remains to be discussed. If the technical choice is to use hardware means to speed up the calculation, the cost of hardware is also higher.

In many Kalman filter algorithms, adaptive Kalman filter plays a very important role. Adaptive Kalman filter uses the measured data to filter and continuously judge whether the system dynamic changes by filtering itself. The model parameters and noise statistics are estimated and modified to improve the filter design and reduce the actual error of the filter. It automatically scales the system noise covariance matrix Q. It is considered that the algorithm has fewer tuning parameters and better robustness than the scaling state covariance matrix algorithm [15]. Therefore, scholars have done a lot of research on it. They combined adaptive steps with extended Kalman filter, untraced Kalman filter and other algorithms to derive a series of adaptive Kalman filters, and have made considerable achievements in various fields [16–23]. Due to the addition of adaptive steps, the accuracy of AEKF and AUKF algorithm is improved and the convergence speed is accelerated compared with the original algorithm. However, considering the complexity of calculation of UKF and EKF, on the premise of the same data sets and the same hardware devices, AKF's calculation speed is higher than AEKF's, and AEKF's calculation speed is higher than AUKF's. Adaptive Kalman filter has also made some contributions to slope estimation. Liao et al. proposed a road slope estimation method based on AEKF [24]. The method was based on the longitudinal dynamics equation of vehicle, and the state space system was discretized. Then, the innovation-based adaptive tuning part was designed to estimate time-varying process noise covariance and measurement noise covariance. Finally, the proposed method was verified by simulation on Carsim platform, and the result was better than the existing EKF algorithm. Sun et al. proposed a road slope estimation method based on AUKF [25]. By increasing the initial noise of the mass prediction model and designing the adaptive shrinkage coefficient to dynamically adjust the covariance matrix of the prediction error, the method realized the rapid and accurate joint estimation of vehicle mass and road slope under the condition of small acceleration. This adaptive step shortened the convergence time of UKF algorithm to some extent, but the convergence time was still about 10 s.

In this paper, an improved road grade estimation method based on Sage–Husa adaptive Kalman filter is proposed. In the method, the model is established based on the longitudinal dynamics of vehicles. The missing quantity in Sage–Husa algorithm is supplemented in the adaptive covariance matrix Q, and the updating of prediction noise q and observation noise r is cancelled. The effect of initial noise variance in Sage–Husa algorithm is retained to some extent, and the proportion of fixed noise variance can be controlled. At the same time, this method eliminates the deviation caused by approximate mathematical expectation in the process of recursive accumulation, thus further improving the accuracy of the algorithm, and maintaining the accuracy and stability of the method in a long time. In this paper, the test data measured by real vehicle is substituted into the slope algorithm proposed in this paper to calculate, and multiple groups of filtering results are obtained. By comparing the prediction results with the reference values, RMSE and MAE are used to evaluate the slope prediction effect, and the accuracy of the slope prediction algorithm is better. Compared with the traditional adaptive Kalman filtering algorithm, the accuracy of the improved algorithm is significantly improved, and the accuracy can be maintained for a long time. Compared with previous slope prediction algorithms, this algorithm has some advantages in accuracy and convergence time.

2. Modeling

2.1. Modeling Based on Electric Vehicle Longitudinal Dynamics

As shown in Figure 2 above, the longitudinal dynamics equation of an electric vehicle can be expressed as:

$$F_t = F_f + F_i + F_w + F_g + F_{err} \tag{1}$$

where F_t is the driving force. According to the characteristics of electric vehicles.

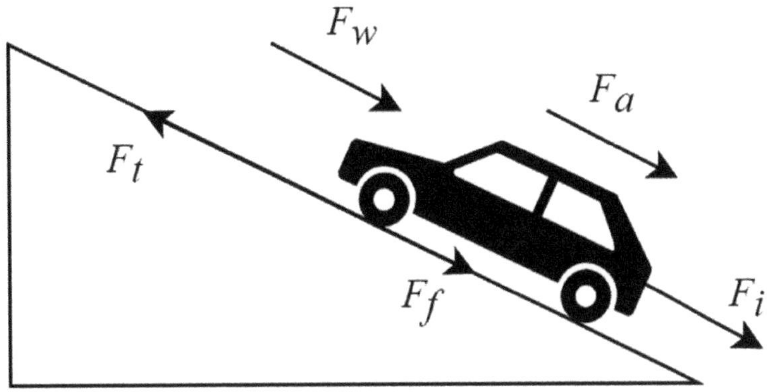

Figure 2. Diagram of electric vehicle longitudinal dynamics.

The driving force can be expressed as:

$$F_t = \frac{T_{tq} i_0 \eta}{r} \tag{2}$$

where T_{tq} is the motor torque, i_0 is the transmission ratio of the main reducer, η is the mechanical efficiency of the transmission system, and r is the rolling radius of the wheel, which is the default wheel radius when the tire pressure is sufficient

The rolling resistance F_f can be expressed as:

$$F_f = mgf \cos \alpha \tag{3}$$

where m is the vehicle mass, g is the gravitational acceleration, f is the rolling resistance coefficient, and α is the road slope angle where the vehicle is located. It is generally considered that the road slope angle is small, $\cos \alpha \approx 1$.

The acceleration resistance F_i can be expressed as:

$$F_i = \frac{\delta m du}{dt} \tag{4}$$

where δ is the conversion coefficient of vehicle rotating mass, which defaults to 1 in some common algorithms. However, in actual driving conditions, δ is usually distributed between 1.1 and 1.4, and the default value of 1 will affect the accuracy of the prediction equation to a certain extent.

The air resistance F_w can be expressed as:

$$F_w = \frac{C_D A \rho u^2}{2} \tag{5}$$

where C_D is the air resistance coefficient, A is the windward area, ρ is the air density, u is the vehicle speed.

The ramp resistance F_g, the component of gravity on the slope, can be expressed as:

$$F_g = mg \sin \alpha \tag{6}$$

The system error F_{err} is caused by uncertain environment disturbance in longitudinal dynamics [26].

To sum up, the longitudinal dynamics equation of an electric vehicle can be rewritten as [27]:

$$\frac{T_{tq} i_0 \eta}{r} = mgf + \frac{\delta m du}{dt} + \frac{C_D A \rho u^2}{2} + mgi \tag{7}$$

The windward area A of BAIC EX360 electric vehicle used in this paper is 2.77 square meters. The air resistance coefficient C_D generally takes an empirical value, and this paper sets C_D as 0.3.

2.2. Prediction Equation and Observation Equation

In this paper, appropriate state variables are selected, namely, speed u and road slope i, because they are easy to read. The state variable x can be expressed as:

$$x = \begin{bmatrix} u \\ i \end{bmatrix} \tag{8}$$

Generally, the slope of urban roads changes gently and the driving speed is low, so the differential equation of speed and slope can be obtained:

$$\begin{bmatrix} \dot{u} \\ \dot{i} \end{bmatrix} = \begin{bmatrix} \frac{T_{tq} i_0 \eta}{\delta m r} - \frac{fg}{\delta} - \frac{gi}{\delta} - \frac{C_D A \rho u^2}{2\delta m} \\ 0 \end{bmatrix} \tag{9}$$

Under urban road conditions, the vehicle speed is usually within 30 km/h, and the maximum is no more than 50 km/h. In the differential equation above, $\frac{T_{tq} i_0 \eta}{\delta m r}$ is usually small enough to be ignored.

Based on the arithmetic relation of velocity and acceleration and the longitudinal dynamic equation of electric vehicle above, the equation of state can be set as:

$$u_k = u_{k-1} + \dot{u}\Delta t \tag{10}$$

$$\begin{bmatrix} u_{k|k-1} \\ i_{k|k-1} \end{bmatrix} = \begin{bmatrix} 1 & g\Delta t \\ 0 & 1 \end{bmatrix} \begin{bmatrix} u_{k-1} \\ i_{k-1} \end{bmatrix} + \begin{bmatrix} \frac{\Delta t}{\delta m} \left(\frac{T_{tq} i_0 \eta}{r} - fmg \right) \\ 0 \end{bmatrix} + q \tag{11}$$

In the above formula, $u_{k|k-1}$ and $i_{k|k-1}$ represent the prior results, that is, the values at time k without Kalman filtering. Additionally, q is the noise vector of the prediction equation.

In the real vehicle test, the speed parameter u_k can be easily measured. Therefore, this paper first takes the speed u_k as the observed value and the observation equation can be expressed as:

$$Z_k = Hx_{k|k-1} + r \tag{12}$$

In the observation equation, $H = \begin{bmatrix} 1 & 0 \\ 0 & 1 \end{bmatrix}$, indicating that both velocity and slope have been observed. Because of the immeasurable noise in the actual experiment and the substitution of empirical values in some data of the prediction equation, the error of the prediction equation cannot be ignored. Therefore, a low-precision slope observation is added in this paper. r represents the measurement noise vector.

3. Adaptive Kalman Filtering

3.1. Flow Chart of Basic Kalman Filter

As shown in Figure 3, the application process of Kalman filter is divided into prediction and update. Kalman filter is a state optimal estimation algorithm. It calculates the priori estimate $x_{k|k-1}$ at time k by substituting the optimal estimate x_{k-1} at time $k-1$ into the prediction equation, compares $x_{k|k-1}$ with the measured value Z_k at time k, assigns weights to $x_{k|k-1}$ and Z_k by Kalman gain, and finally obtains the optimal estimate x_k at time k. After that, Kalman filter recursively computes the optimal estimate at the next time.

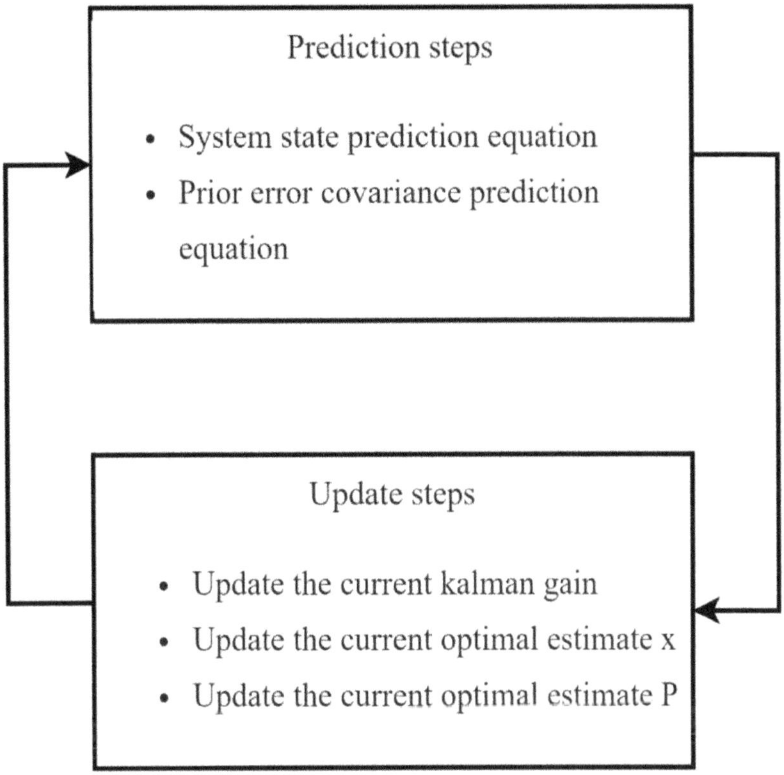

Figure 3. Flow chart of Kalman filter.

3.2. Prediction Equations

The prediction equation of the system can be expressed as:

$$x_{k|k-1} = Fx_{k-1} + B_{k-1} + q \tag{13}$$

Among them, $F = \begin{bmatrix} 1 & g\Delta t \\ 0 & 1 \end{bmatrix}$, $B_{k-1} = \begin{bmatrix} \frac{\Delta t}{\delta m}\left(\frac{T_{tq} i_0 \eta}{r} - fmg\right) \\ 0 \end{bmatrix}$.

The prediction equation of the prior error covariance can be expressed as:

$$P_{k|k-1} = FP_{k-1}F^T + Q \tag{14}$$

where P is the variance of the state estimate and represents the measure of the uncertainty in the predicted state, which comes from process error and the error of the estimate. Q is the covariance matrix of prediction noise q.

3.3. The Update Equation

The update of Kalman gain can be expressed as:

$$K_g = P_{k|k-1}H^T\left(HP_{k|k-1}H^T + R\right)^{-} \tag{15}$$

where K_g is Kalman gain and represents the weight relationship between predicted value and measured value. R is the covariance of the observed value and represents the uncertainty of the observed state. The smaller the value is, the more accurate the observation is.

The update of the optimal estimate of x_k can be expressed as:

$$x_k = x_{k|k-1} + K_g\left(Z_k - Hx_{k|k-1}\right) \tag{16}$$

The update of the optimal estimate P_k can be expressed as:

$$P_k = (I - K_g H)P_{k|k-1} \tag{17}$$

where I represents the identity matrix, whose order is equal to the number of elements of the state variable.

3.4. *Sage–Husa Adaptive Kalman Filter*

Adaptive Kalman filter updates q_k, Q_k, r_k and R_k on the basis of Kalman filter. A weighting coefficient d_k is given, d_k can be expressed as:

$$d_k = (1-b)\left(1-b^{k+1}\right)^{-} \tag{18}$$

where d_k is usually between 0.95 and 0.99. The weighting coefficient is used to enhance the effect of recent data and update noise.

The update formula is as follows:

$$\begin{aligned} \varepsilon_k &= Z_k - Hx_{k|k-1} - r_{k-1} \\ q_k &= (1-d_k)q_{k-1} + d_k(x_k - Fx_{k-1}) \\ Q_k &= (1-d_k)Q_{k-1} + d_k\left(K_g\varepsilon_k\varepsilon_k^T K_g^T + P_k - Fx_{k-1}F^T\right) \\ r_k &= (1-d_k)r_{k-1} + d_k\left(Z_k - Hx_{k|k-1}\right) \\ R_k &= (1-d_k)R_{k-1} + d_k\left(\varepsilon_k\varepsilon_k^T - HP_{k|k-1}H^T\right) \end{aligned} \tag{19}$$

3.5. *Improved Sage–Husa Adaptive Kalman Filter*

Wei et al. from Northwestern Polytechnical University studied Sage–Husa adaptive Kalman filter and found that the contribution of Q_0, the initial value of the covariance matrix, to Q_k would decrease sharply with the increase of k in the operation process, and would soon approach zero. When some systems need fixed noise with proportion, the applicability of traditional Sage–Husa adaptive Kalman filter will be reduced. The traditional Sage–Husa adaptive Kalman filter adopts the method of average information allocation for q_k, Q_k, r_k and R_k, and the proportion of initial value to contribution is only $1/k$. At the same time, Wei et al. found in their calculation that the deviation of q_k and r_k would affect the coordination relationship between R_k and Q_k, leading to the increase of subsequent deviation [15].

In view of the above situation, this paper improved the algorithm by canceling the calculation of q_k and r_k, and setting the forgetting factors b_1 and b_2 respectively. The traditional Sage–Husa Kalman filter formula is rewritten as follows:

$$\begin{aligned} d_{k1} &= (1-b_1)\left(1-b_1^{k+1}\right)^{-} \\ d_{k2} &= (1-b_2)\left(1-b_2^{k+1}\right)^{-} \\ R_k &= (1-d_{1k})R_{k-1} + d_{1k}\left(\varepsilon_k\varepsilon_k^T - HP_{k|k-1}H^T\right) \\ Q_k &= (1-d_{2k})Q_{k-1} + d_{2k}\Big(K_g\varepsilon_k\varepsilon_k^T K_g^T + P_k - Fx_{k-1}F^T - 2K_gR_kK_g^T \\ &\quad - 2K_gHP_{k|k-1}H^TK_g^T + P_{k|k-1}H^TK_g^T + K_gHP_{k|k-1}\Big) \end{aligned} \tag{20}$$

Combined with the above and Figure 4, it can be concluded that, compared with the previous algorithm, the calculation of q_k and r_k is cancelled, and the original forgetting factor b is replaced. Instead, two forgetting factors b_1 and b_2 are set for the update of R_k and Q_k respectively. Meanwhile the system noise Q_k is rederived by a series of approximate processing in the steps.

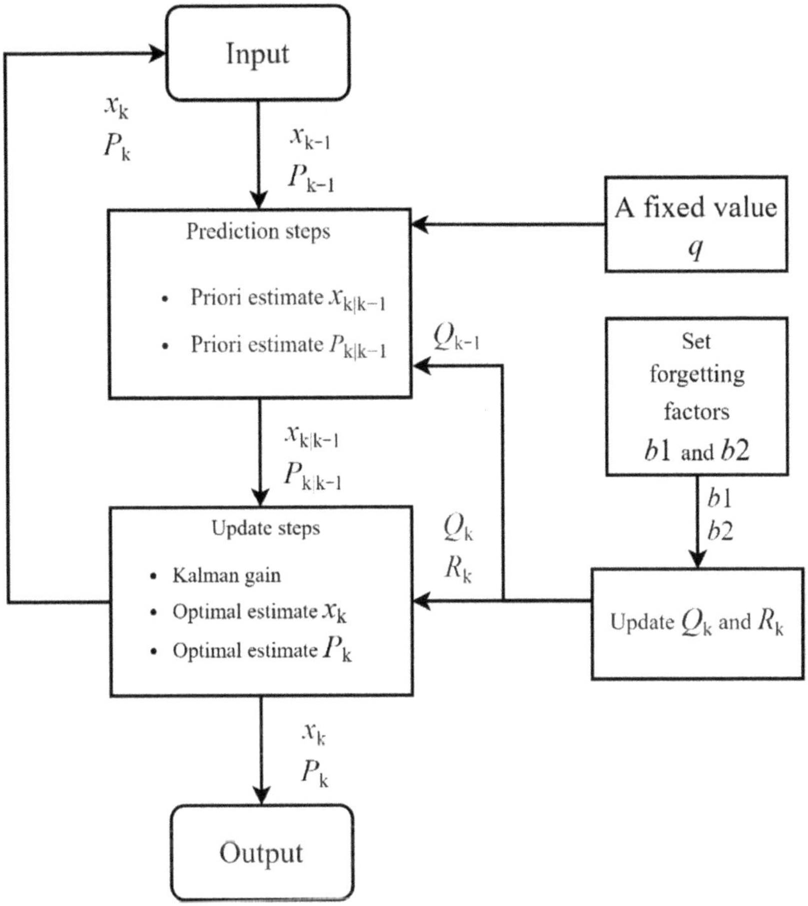

Figure 4. Flow chart of improved adaptive Kalman filter.

4. Experiment

4.1. Experiment Plan

In order to verify the effectiveness of the improved Sage–Husa adaptive Kalman filter for road slope estimation, a large number of vehicle tests were carried out. The driving data of multiple groups of roads with different slopes in and around southwest Forestry University in Kunming city, Yunnan Province were collected in the experiment, and the most representative groups of data were selected to verify the scheme.

4.2. Experiment Equipment and Parameters

The equipment used in the test includes BAIC New Energy EX-360 electric vehicle, on-board OBD, SD card, low-cost gyroscope, high-precision IMU and GPS. InVIEW is used as data processing and analysis software. Matlab is used to build algorithms model and produce results by inputting test data into the algorithm model.

Among them, the low-precision gyroscope is used to assist the algorithm proposed in this paper, that is, to provide rough observation value for the subsequent system to input the observation value into the algorithm. The high-precision IMU and GPS are used to provide an experimental slope value with a very small error, which can be used as a reference value for the real slope. The IMU used in this experiment is shown in Figure 5.

Figure 5. IMU of high accuracy.

Some vehicle parameters and model parameters are shown in Tables 1 and 2.

Table 1. Test vehicle parameters.

Parameter	VALUE
Vehicle type	BAIC EX360
Maximum motor power	80 kW
Maximum motor torque	230 N·m
Transmission type	fixed gear ratio
Curb weight	1480 kg
Size	4110 mm × 1750 mm × 1543 mm
Transmission efficiency	97%
Tire type	205/50 R16

Table 2. Prediction model parameters.

Parameter	VALUE
r	0.3 m
m	1715 kg
δ	1.1

The tire rolling radius r is the approximate tire rolling radius based on tire specification 205/50 R16 and taking into account the bearing time and sufficient tire pressure. Actual curb weight m includes the quality of the car itself, the instruments and the testers. The rotational mass conversion coefficient δ of the test vehicle is derived from the empirical value of the rotational mass of the car.

4.3. Experiment Section

According to the slope classification of geomorphic detail map application of geomorphic survey and Geomorphic Mapping Commission of international Geographical Union, the slope grade is defined as 0°~0.5° plain, 0.5°~2° micro slope, 2°~5° gentle slope and 5°~15° slope. In this paper, the angle value is converted into slope value, and the data of the three groups of sections in the test are classified according to the division basis, which are micro-slope model, gentle slope model R1 and gentle slope model R2. These sections are shown in Figure 6 below.

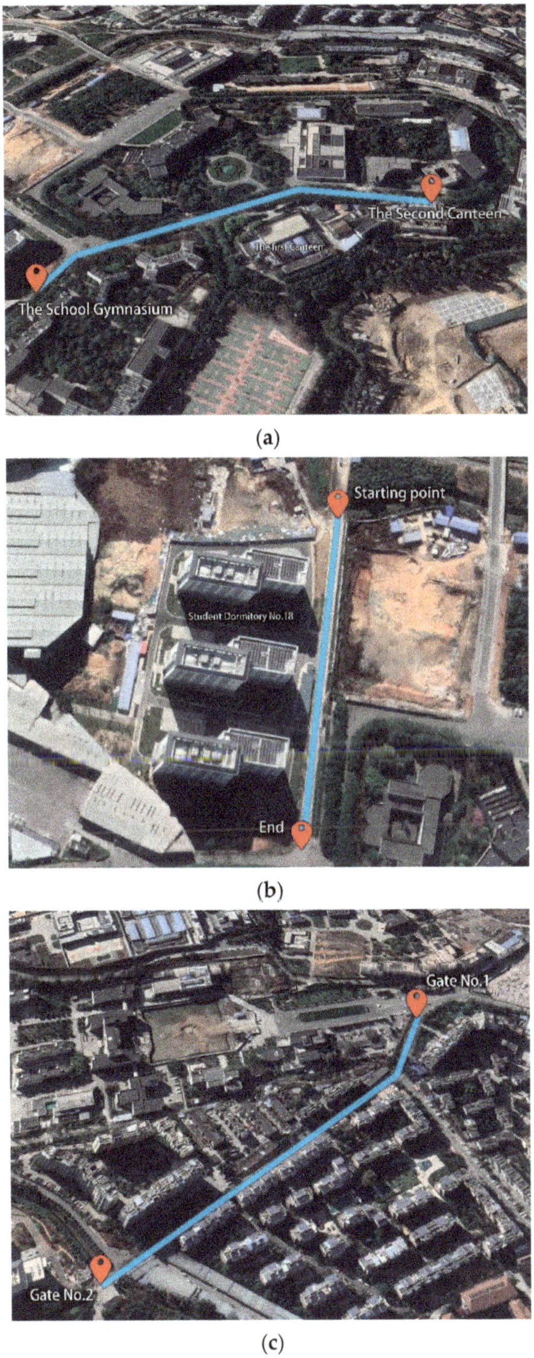

Figure 6. GPS images of the test sections. (**a**) Micro slope model, (**b**) Gentle slope model R1, (**c**) Gentle slope model R2.

Micro slope model: the road with a low slope, with a slope of 1 to 5%. The starting point of the road section is at the gate of the Second canteen of Southwest Forestry University, and it moves forward to the school gymnasium, then runs along the slope to the Engineering

Building, and stops in the middle of the slope. The slope of this section is mostly in the range of micro-slope, so this paper regards it as a micro-slope model.

Gentle slope model R1: the road with a moderate slope, with a slope of 2 to 6%, which is located on the west side of building 19 student dormitory of Southwest Forestry University. The slope of this section is mostly in the interval of gentle slope, so this paper regards it as a gentle slope model.

Gentle slope model R2: the road with a moderate slope, with a slope of 5 to 8%. This road is a long slope from gate No. 1 to gate No. 2 of Southwest Forestry University. The slope of this section is mostly in the interval of gentle slope, so this paper regards it as a gentle slope model.

4.4. Error Analysis

In order to evaluate the accuracy of this algorithm, Root Mean Square Error (RMSE) and Mean Absolute Error (MAE) are introduced. The errors of the data obtained by the common adaptive Kalman filter algorithm and the improved Sage–Husa adaptive Kalman filter algorithm are calculated with the real data respectively, and the size of the error index is analyzed.

RMSE and MAE are expressed as follows:

$$\text{RMSE} = \sqrt{\text{MSE}} = \sqrt{\frac{SSE}{n}} = \sqrt{\frac{1}{n}\sum_{i=1}^{n}(y_i - \hat{y}_i)^2} \tag{21}$$

$$\text{MAE} = \frac{1}{n}\sum_{i=1}^{n}|y_i - \hat{y}_i| \tag{22}$$

In this paper, AKF1 represents the initial adaptive Kalman filter algorithm, and AKF2 represents the improved adaptive Kalman filter algorithm. The black solid line is used to represent the reference measured road slope value, the blue solid line is used to draw the data results of the original adaptive Kalman filter algorithm, and the red solid line is used to draw the data results of the improved adaptive Kalman filter algorithm. In order to compare the two algorithms more clearly, some regions in the Figure 7 are enlarged and displayed on the right side of the original line chart. The RMSE and MAE of the algorithm results are drawn into tables, as shown in Figure 8.

As can be seen from Figures 7 and 8, the convergence speed of the two algorithms is fast. By longitudinal comparison of the two slope estimation methods under the three models, RMSE and MAE values of the original adaptive Kalman filter slope algorithm increase significantly with the increase of slope, RMSE value increases from 0.048 to 0.081%, MAE value increases from 0.035 to 0.064%, and the changes are particularly obvious. However, the RMSE and MAE values of the improved adaptive Kalman filter slope algorithm change only from 0.038 to 0.040%, and MAE from 0.029 to 0.030%. It can be concluded that the improved adaptive Kalman filter slope algorithm can better adapt to the road conditions in various slope ranges and has strong stability.

The two algorithms in the same slope range were compared horizontally. After the algorithm ran for a period of time, the accuracy of the original adaptive Kalman filter slope algorithm began to decline significantly, and errors that could not be ignored appeared. As shown in Figure 7, in the micro-slope model, the value of the original adaptive Kalman filter slope algorithm began to float significantly above the actual measured value near the moment of 24 s. In the gentle slope model R1, the error of the algorithm is more obvious when it starts near 26 s. In the gentle slope model R2, the error of the original adaptive Kalman filter slope algorithm cannot be ignored after 25 s, completely deviating from the real value. The improved adaptive Kalman filter slope algorithm can keep the data in accordance with the actual measured values in the three slope models, and will not lose the accuracy after running for a period of time.

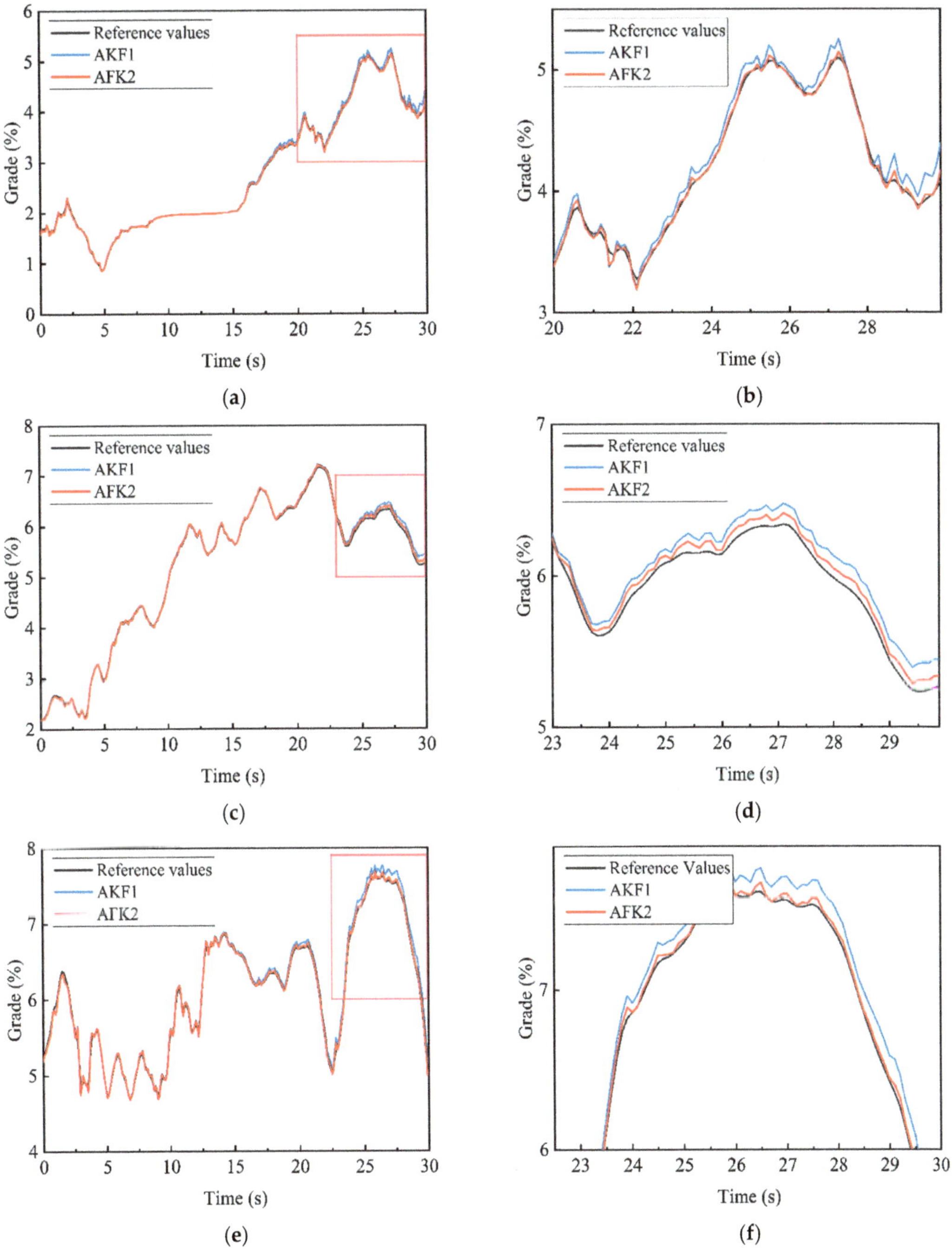

Figure 7. Contrast diagram of test effect. (**a**) Micro slope model, (**b**) part of micro slope model, (**c**) gentle slope model R1, (**d**) part of gentle slope model R1, (**e**) gentle slope model R2, (**f**) part of gentle slope model R2.

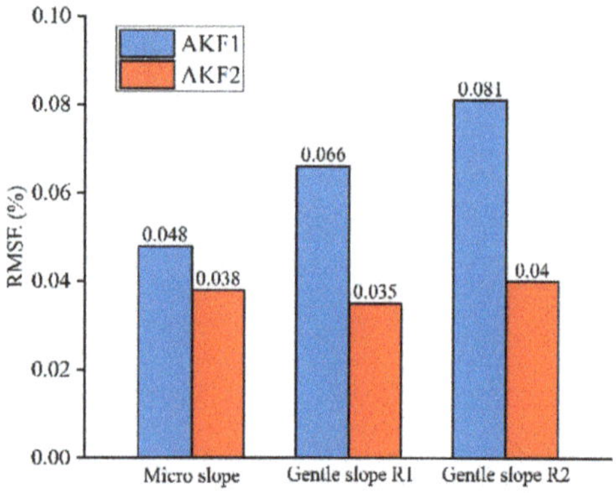

(a)

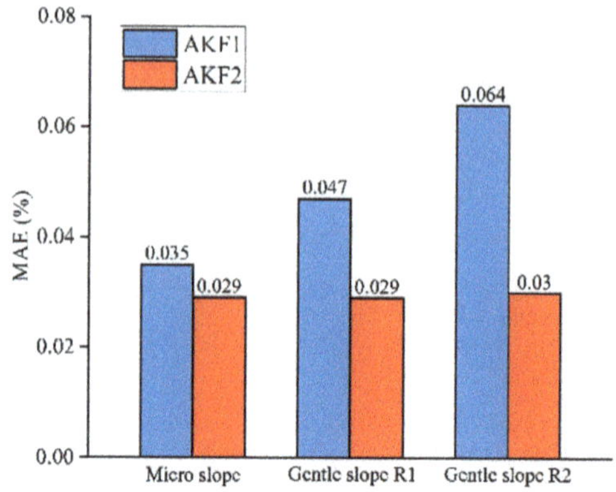

(b)

Figure 8. Comparison diagram of RMSE and MAE for two algorithms. (**a**) Comparison diagram of RMSE, (**b**) comparison diagram of MAE.

As shown in Figure 9, this paper also presents three groups of 100 s filtering results. In order to display the effect clearly, the original adaptive Kalman filter result is represented by the blue solid line, and the improved adaptive Kalman filter result is represented by the red dotted line. The filtering effect of the first few seconds is shown in the attached figure on the right. It can be seen that the error of the original adaptive Kalman filter gradually becomes obvious after a period of time. The results of the improved adaptive Kalman filter still fit the real value after a long time. From what has been discussed above, the gap between the two algorithms is greatly obvious. The RMSE value of the 100s filtering value of the slope prediction algorithm proposed in this paper increases to some extent compared with the RMSE value of the filtering value of the previous 30 s, but the result is still satisfactory.

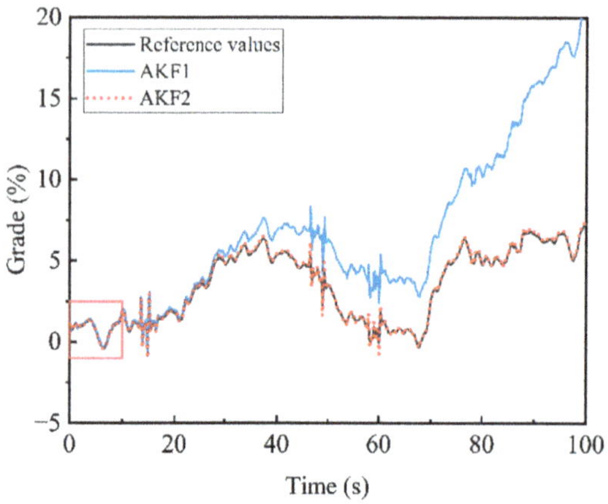

(a)

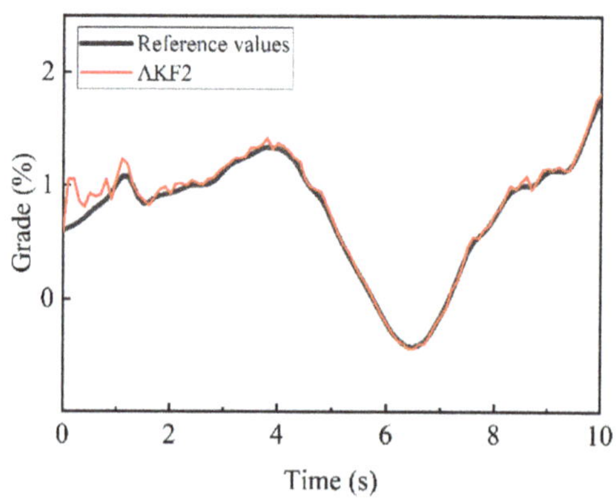

(b)

Figure 9. *Cont.*

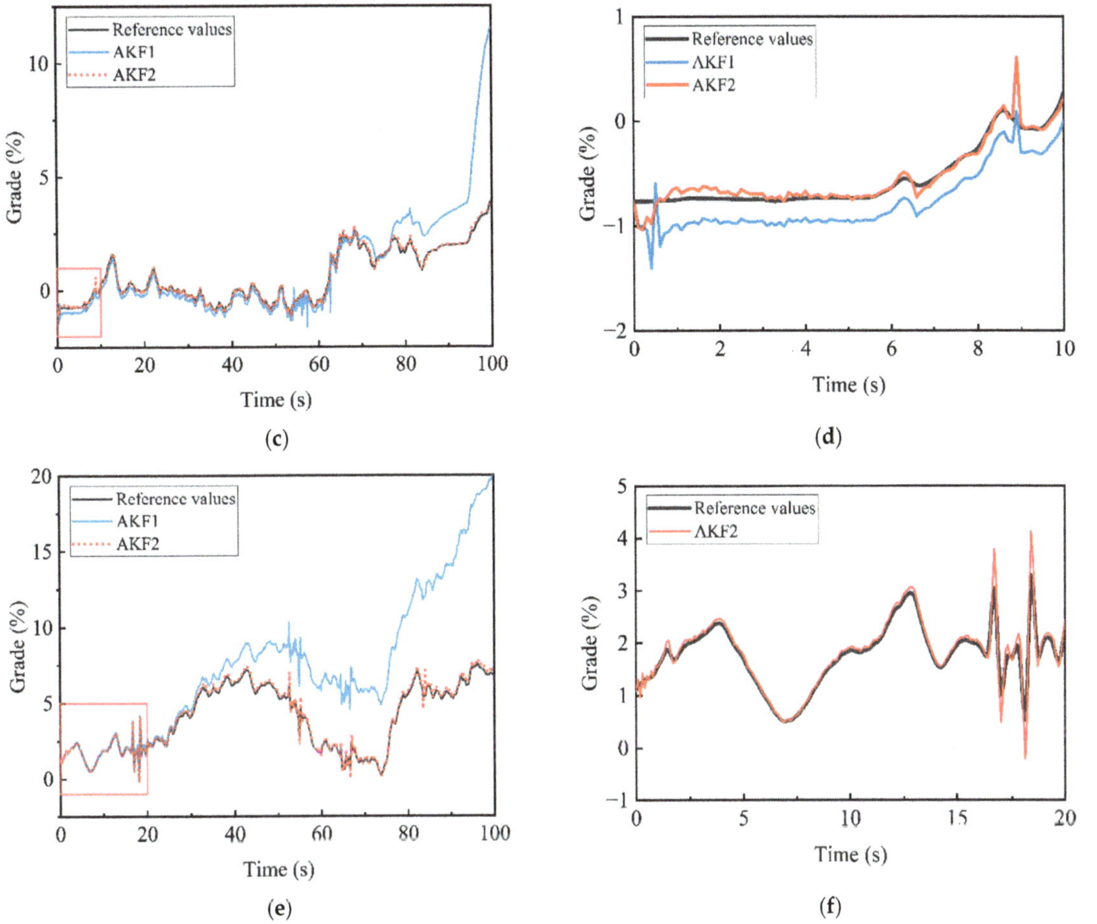

Figure 9. Contrast diagram of long-term test effect. (**a**) Micro slope model, (**b**) part of micro slope model, (**c**) gentle slope model R1, (**d**) part of gentle slope model R1, (**e**) gentle slope model R2, (**f**) part of gentle slope model R2.

Through the study and calculation of Wei et al., it is speculated that the reason for the increase of the time error may be that the recursive formula of q_k and r_k in the original Sage–Husa adaptive Kalman filter algorithm is the approximation of the system equation and the measurement equation. In the process of recursive accumulation, the deviation caused by the expectation of approximate mathematics may sometimes be large. However, the deviation of q_k and r_k will affect the coordination relationship between Q_k and R_k, leading to the increase of subsequent deviation, thus affecting the accuracy of estimation. Taking the gentle slope model R1 as an example, the prediction equation and observation equation of the two algorithms are consistent, so the initial set values of q_k, r_k, R_k and Q_k are consistent. In the subsequent update, the two algorithms adopt different processing methods. The improved adaptive Kalman filter algorithm eliminates the calculation of q_k and r_k, uses the variance P of state estimation to approximate Q_{k-1}, and uses correlation substitution, so less information is discarded.

As shown in the Table 3 below, compared with the QUKF algorithm proposed by He et al., according to RMSE standards, the algorithm proposed in this paper has higher accuracy under certain conditions and uses longer scene time [1].

Table 3. Algorithm RMSE comparison.

Algorithm		AKF2			QUKF
Experiment	Test 1	Test 2	Test 3	Test 1	Test 2
RMSE	1.3%	0.8%	1.9%	7.8%	7.8%

Compared with the DUKF algorithm proposed by Jin et al., the algorithm proposed in this paper has a faster convergence speed [2]. It can be seen from Figure 9 that the convergence time of the algorithm in this paper is less than 4 s, compared with the convergence time of the DUKF algorithm of about 7 s. Within the scope of adaptive Kalman filter, the convergence speed of the algorithm proposed in this paper is faster than the AEKF algorithm proposed by Liao et al. and the AUKF algorithm proposed by Sun et al. [24,25]. In the experiment of AEKF slope estimation method proposed by Liao et al., the distance was taken as abscissa and the slope was taken as ordinate to output filtered images. It can be seen from the literature that the algorithm converges when the driving distance is about 100 m, and the convergence speed is slow. The convergence time of the AUKF algorithm proposed by Sun et al. is about 10 s, which is longer than the convergence time of the slope estimation method proposed in this paper. Therefore, this algorithm has obvious advantages in the starting stage of electric vehicles. In the starting process of pure electric vehicle, the system directly controls the motor output torque to make the vehicle start normally. When electric vehicles start, if the output torque does not adapt to the starting slope, too much output torque may lead to vehicle running forward in small slope, while too little output torque may lead to vehicle sliding behind or insufficient starting power in large slope. Therefore, the slope algorithm proposed in this paper can improve the smoothness at the start time of EV autonomous driving.

5. Conclusions

(1) In this paper, the improved adaptive Kalman filtering algorithm draws on the valuable experience of predecessors and changes the traditional adaptive Kalman filtering algorithm. It removes the calculation of q_k and r_k, which may lead to a sharp increase in the subsequent deviation, and reasonably improves the update of Q_k and R_k by using double forgetting factors b_1 and b_2.

(2) The algorithm proposed in this paper has a wide range of application. Under the experimental data of multiple 30 s micro slope model and gentle slope model, RMSE can always be maintained within 0.04%, MAE can always be maintained within 0.03%, and short-term effect is relatively good. The RMSE and MAE can always be kept within 0.19% and 0.15%, respectively, under the demonstration of multiple groups of 100 s gentle slope model test data. Generally speaking, the algorithm is applicable to a wide range of slope and has a good general effect.

(3) After a comprehensive comparison of the results of the two algorithms above, it can be found that, compared with the results of the original adaptive Kalman filter slope estimation method, the RMSE and MAE of the improved algorithm are significantly reduced. The RMSE of the micro slope model is reduced by 0.01%, which is 20.8% lower than the original algorithm. The MAE of the micro-slope model is reduced by 0.006%, which is 17.1% lower than the original algorithm. The RMSE of the gentle slope model R1 is reduced by 0.031%, which is 47% lower than the original algorithm. The MAE of the gentle slope model R1 is reduced by 0.018%, which is 38.3% lower than the original algorithm. The RMSE of the gentle slope model R2 is reduced by 0.041%, which is 50.6% lower than the original algorithm. The MAE of the gentle slope model R2 is reduced by 0.034%, which is 53.1% lower than the original algorithm. The error of the 100 s filtering result of the algorithm proposed in this paper increases to some extent compared with the previous 30 s filtering result. However, the error is still reasonable. In conclusion, the improved adaptive Kalman filter slope estimation method is superior.

Author Contributions: Conceptualization, J.G.; methodology, J.G.; data curation, J.G.; funding acquisition, C.H.; software, H.W.; supervision, C.H. and J.L.; writing—review and editing, J.G. and C.H. All authors have read and agreed to the published version of the manuscript.

Funding: This research was funded by National Natural Science Foundation of China grant number 51968065 and Yunnan Provincial high level talent support project grant number YNWR-QNBJ-2018-066, YNQR-CYRC-2019-001.

Data Availability Statement: Not applicable.

Conflicts of Interest: The authors declare no conflict of interest.

References

1. He, W.; Xi, J. A Quaternion Unscented Kalman Filter for Road Grade Estimation. In Proceedings of the 2020 IEEE Intelligent Vehicles Symposium (IV), Las Vegas, NV, USA, 19 October–13 November 2020; pp. 1635–1640.
2. Jin, X.; Yang, J.; Li, Y.; Zhu, B.; Wang, J.; Yin, G. Online estimation of inertial parameter for lightweight electric vehicle using dual unscented Kalman filter approach. *IET Intell. Transp. Syst.* **2020**, *14*, 412–422. [CrossRef]
3. Boada, B.L.; Boada, M.J.L.; Zhang, H. Sensor fusion based on a dual kalman filter for estimation of road irregularities and vehicle mass under static and dynamic conditions. *IEEE/ASME Trans. Mechatron.* **2019**, *24*, 1075–1086. [CrossRef]
4. Pei, X.; Chen, Z.; Yang, B.; Chu, D. Estimation of states and parameters of multi-axle distributed electric vehicle based on dual unscented Kalman filter. *Sci. Prog.* **2020**, *103*, 0036850419880083. [CrossRef] [PubMed]
5. Sun, Y.; Li, L.; Yan, B.; Yang, C.; Tang, G. A hybrid algorithm combining EKF and RLS in synchronous estimation of road grade and vehicle' mass for a hybrid electric bus. *Mech. Syst. Signal Process.* **2016**, *68*, 416–430. [CrossRef]
6. Büyükköprü, M.; Uzunsoy, E. Reliability of Extended Kalman Filtering Technic on Vehicle Mass Estimation. *J. Innov. Sci. Eng.* **2021**, *5*, 1–11. [CrossRef]
7. Li, X.; Ma, J.; Zhao, X.; Wang, L. Intelligent Two-Step Estimation Approach for Vehicle Mass and Road Grade. *IEEE Access* **2020**, *8*, 218853–218862. [CrossRef]
8. Karoshi, P.; Ager, M.; Schabauer, M.; Lex, C. Robust and numerically efficient estimation of vehicle mass and road grade. In *Advanced Microsystems for Automotive Applications 2017*; Springer: Berlin/Heidelberg, Germany, 2018; pp. 87–100.
9. Zhang, Y.; Zhang, Y.; Ai, Z.; Feng, Y.; Hu, Z. A cross iteration estimator with base vector for estimation of electric mining haul truck's mass and road grade. *IEEE Trans. Ind. Inform.* **2018**, *14*, 4138–4148. [CrossRef]
10. Jiang, S.; Wang, C.; Zhang, C.; Bai, H.; Xu, L. Adaptive estimation of road slope and vehicle mass of fuel cell vehicle. *ETransportation* **2019**, *2*, 100023. [CrossRef]
11. Hu, M.; Gao, W.; Zeng, Y.; Li, H.; Yu, Z. Vehicle mass and road grade estimation based on adaptive forgetting factor RLS and EKF algorithm. In Proceedings of the 2020 5th International Conference on Power and Renewable Energy (ICPRE), Shanghai, China, 12–14 September 2020; pp. 342–346.
12. Rodríguez, A.J.; Sanjurjo, E.; Pastorino, R.; Naya, M.Á. State, parameter and input observers based on multibody models and Kalman filters for vehicle dynamics. *Mech. Syst. Signal Process.* **2021**, *155*, 107544. [CrossRef]
13. Bian, J.; Qiu, B.; Liu, Y.; Su, H. Adaptive Cruise Control for Electric Bus based on Model Predictive Control with Road Grade Prediction. In Proceedings of the VEHITS 2018, Funchal, Portugal, 16–18 May 2018; pp. 217–224.
14. Feng, J.; Qin, D.; Liu, Y.; You, Y. Real-time estimation of road slope based on multiple models and multiple data fusion. *Measurement* **2021**, *181*, 109609. [CrossRef]
15. Wei, W.; Qin, Y.Y.; Zhang, X.D.; Zhang, Y.C. Amelioration of the Sage-Husa algorithm. *J. Chin. Inert. Technol.* **2012**, *6*, 678–686. [CrossRef]
16. Narasimhappa, M.; Mahindrakar, A.D.; Guizilini, V.C.; Terra, M.H.; Sabat, S.L. MEMS-based IMU drift minimization: Sage Husa adaptive robust Kalman filtering. *IEEE Sens. J.* **2019**, *20*, 250–260. [CrossRef]
17. Liu, R.; Liu, F.; Liu, C.; Zhang, P. Modified sage-husa adaptive Kalman filter-based SINS/DVL integrated navigation system for AUV. *J. Sens.* **2021**, *2021*, 9992041. [CrossRef]
18. Song, K.; Wang, Y.; Hu, X.; Cao, J. Online Prediction of Vehicular Fuel Cell Residual Lifetime Based on Adaptive Extended Kalman Filter. *Energies* **2020**, *13*, 6244. [CrossRef]
19. Yan, W.; Ding, Q.; Chen, J.; Liu, Y.; Cheng, S.S. Needle Tip Tracking in 2D Ultrasound Based on Improved Compressive Tracking and Adaptive Kalman Filter. *IEEE Robot. Autom. Lett.* **2021**, *6*, 3224–3231. [CrossRef]
20. Huang, X.; Chen, G.; Liu, Z. Sage Husa Adaptive Integrated Navigation Algorithm Based on Variable Fading Factor. In Proceedings of the 2020 International Conference on Computer Network, Electronic and Automation (ICCNEA), Xi'an, China, 25–27 September 2020; pp. 378–383.
21. Zhang, F.; Yin, L.; Kang, J. Enhancing Stability and Robustness of State-of-Charge Estimation for Lithium-Ion Batteries by Using Improved Adaptive Kalman Filter Algorithms. *Energies* **2021**, *14*, 6284. [CrossRef]
22. Luo, Z.; Fu, Z.; Xu, Q. An Adaptive Multi-Dimensional Vehicle Driving State Observer Based on Modified Sage–Husa UKF Algorithm. *Sensors* **2020**, *20*, 6889. [CrossRef]

23. Xu, S.; Zhou, H.; Wang, J.; He, Z.; Wang, D. SINS/CNS/GNSS integrated navigation based on an improved federated Sage–Husa adaptive filter. *Sensors* **2019**, *19*, 3812. [CrossRef]
24. Liao, X.; Huang, Q.; Sun, D.; Liu, W.; Han, W. Real-time road slope estimation based on adaptive extended Kalman filter algorithm with in-vehicle data. In Proceedings of the 2017 29th Chinese Control and Decision Conference (CCDC), Chongqing, China, 28–30 May 2017; pp. 6889–6894.
25. Sun, E.; Yin, Y.; Xin, Z.; Li, S.; He, J.; Kong, Z.; Liu, X. Adaptive joint estimates of vehicle mass and road grades for small acceleration driving scenarios. *J. Tsinghua Univ. Sci. Technol.* **2022**, *62*, 125–132.
26. Lin, N.; Zong, C.; Shi, S. The method of mass estimation considering system error in vehicle longitudinal dynamics. *Energies* **2018**, *12*, 52. [CrossRef]
27. Wei, H.; He, C.; Li, J.; Zhao, L. Online estimation of driving range for battery electric vehicles based on SOC-segmented actual driving cycle. *J. Energy Storage* **2022**, *49*, 104091. [CrossRef]

energies

Article

Numerical Study on Aerodynamic Characteristics of Heavy-Duty Vehicles Platooning for Energy Savings and CO_2 Reduction

Junik Jo [1] and Chul-Ho Kim [2,*]

1 Department of Automotive Engineering, Graduate School, Seoul National University of Science & Technology, Seoul 01811, Korea; junik.jo@seoultech.ac.kr
2 Department of Mechanical & Automotive Engineering, Seoul National University of Science & Technology, Seoul 01811, Korea
* Correspondence: hokim@seoultech.ac.kr

Abstract: This study aims to analyze the aerodynamic interaction between moving vehicles platooning with the change in the platooning conditions on a freeway. The effect of the vortex generated by the forward vehicle reduces the value of the stagnation pressure generated at the front of the rear vehicle, which effectively reduces drag on the driving vehicle. To elucidate this, a total of four vehicles were applied to platooning at a speed of 100 km/h by altering the gap distance of heavy-duty vehicles (HDVs) such as 0.5 Length (L), 1 L, 1.5 L, 2.0 L, and 2.5 L under the conditions of 1 L equal to 13.16 m. The stagnation pressure at the front of the following vehicle (FV) was reduced, and quantitative analysis of drag force generated at each leading vehicle (LV) and following vehicle that is platooning exhibited a reduction of about 51%, 56%, and 52%, respectively, when compared to the single moving HDV. This is considered as a reduction in engine power for the driving vehicle. Taken together, these results are effective in improving fuel efficiency and reducing CO_2, a representative greenhouse gas, and predicting fuel and CO_2 reduction based on HDV annual mileage according to the highway conditions and logistics.

Keywords: aerodynamic drag; CFD (computational fluid dynamics); platooning; CO_2 reduction; GHG (greenhouse gas)

Citation: Jo, J.; Kim, C.-H. Numerical Study on Aerodynamic Characteristics of Heavy-Duty Vehicles Platooning for Energy Savings and CO_2 Reduction. *Energies* **2022**, *15*, 4390. https://doi.org/10.3390/en15124390

Academic Editors: Wenbin Yu and Guang Zeng

Received: 9 May 2022
Accepted: 13 June 2022
Published: 16 June 2022

1. Introduction

Greenhouse gas reduction has emerged as one of the biggest global problems. The six primary greenhouse gases are CO_2, CH_4, N_2O, HFCs, PFCs, and SF_6, which cause the greenhouse effect. Among them, CO_2 was increased by 80% in annual emissions from 1970 to 2004. To reduce greenhouse gas (GHG) emissions, the world government has established decarbonization policies, and automobiles are moving away from internal combustion engines in line with policies. However, due to logistics on the road it would take a long time to change internal combustion engine vehicles to eco-friendly vehicles. Previous studies have examined actively utilizing aerodynamic characteristics to improve the fuel efficiency of internal combustion engines. Representatively, recent studies were performed on mounting an airfoil-shaped spoiler at the tail end of a vehicle [1], mounting a vortex generator [2], attempting to change the rear design of the vehicle [3], mounting an air duct [4], and securing the seal of the side window of the vehicle [5]. In particular, research and development applying aerodynamic characteristics such as a roof fairing system is representative [6]. Recently, as autonomous driving such as V2X (Vehicle-to-Everything) became an issue, European countries tried to solve the problem of air resistance that occurs during driving rather than reducing air resistance by changing the vehicle's exterior. For example, in 2011, the SARTRE project [7], an empirical study conducted for GHG reduction using the aerodynamic characteristics, occurred during platooning of the freight vehicles. However, the technology for controlling the distance between vehicles was not developed at that time, so the project could only confirm the possibility of

GHG reduction through platooning. Therefore, the SARTRE project, an empirical study, researched controlling the distance between vehicles by lowering the freight vehicle to 6 m, thereby confirming the possibility of improving fuel economy and reducing GHG. Those effects lead to the reduction in logistics costs [8]. In addition, the Korean government leads the Truck Platooning Project of Korea (TROOP) [9], an empirical study for platooning large freight vehicles. This project has been conducted to control the distance between vehicles at about 15 m. Based on V2X for the safety of autonomous vehicles, diversified platooning safety research is underway, such as distance control of ACC (Adaptive Cruise Control) [10], platooning for CACC (Cooperative Adaptive Cruise Control) [11,12], analysis of acceleration driving of lead vehicles during platooning [13], and emergency braking control [14]. Previous studies demonstrated that platooning driving might contribute to an eco-friendly road environment by reducing air resistance. Therefore, the objective of this study is to understand the variation in aerodynamic characteristics of each model vehicle in platooning by analyzing the change in pressure and drag force of each vehicle in platooning.

2. Description of the Model Vehicle with Its Numerical Grid

In this study, we chose the SCANIA R-series heavy-duty vehicle with a 13.0 liter diesel engine with its maximum brake power, 500 hp. This model vehicle has been used for on-road tests for aerodynamic performance analysis of vehicle platooning in previous empirical studies, due to occupying a large position in the transportation and logistics market worldwide [15,16]. The simplified model vehicle was used for this numerical study, and the solid model with the geometric dimensions 13.16 m × 2.495 m × 3.725 m (Length × Width × Heigth) is given in Figure 1.

Figure 1. A perspective view of the simplified model vehicle with its dimensions.

2.1. Numerical Domain and Its Conditions

The numerical domain size was defined as shown in Figure 2. Especially, 8 L was given at the rear side of the last model vehicle for stable convergence on the simulation result.

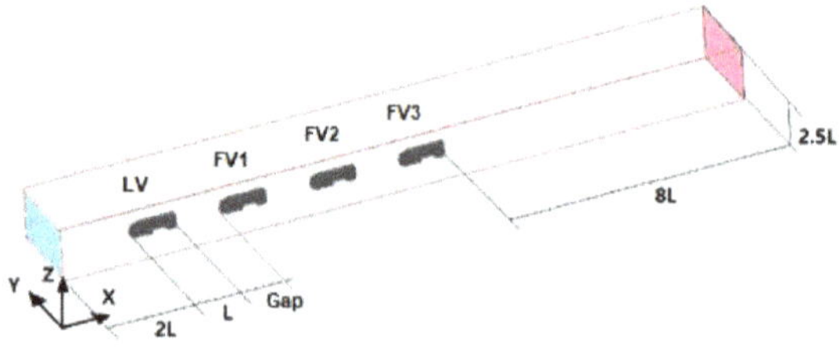

Figure 2. Numerical domain with the model heavy-duty vehicles platooning.

The airflow field was assumed as a 3-dimensional steady incompressible and turbulent flow field for the numerical calculation. The details of the initial and boundary conditions are given in Table 1. The inlet was defined as the velocity boundary with the plugged flow condition, and the outlet was assumed as fully developed and defined as the pressure boundary with the ambient condition. The moving boundary was set to the ground with the same speed as the inlets.

Table 1. Boundary and initial conditions on the control volume.

Boundary Surface	Boundary and Initial Conditions
Inlet	Velocity boundary: 100 km/h
Outlet	Pressure boundary: ambient pressure, 1 atm
Sides and top	Open boundary: symmetric conditions
Ground	Moving boundary condition: 100 km/h
Model surface	No-slip wall

2.2. Numerical Grid of the Physical Model

The CAD-to-CFD method [17] in conjunction with the orthogonal grid was used for the discretization of the physical domain in this study. First, the 3D CAD program, Pro-Engineers, was used as solid model of the vehicle and it was transferred into the numerical domain to generate the hexagonal meshes in the Cartesian coordinate system.

Figure 3 shows a typical grid in the numerical domain with four model vehicles keeping their distance (1 L) in between. For the optimum grid in the domain, the grid convergence test was conducted by thorough grid dependence evaluation with different iterations and very nicely converged with the cut-off error of the residual fraction of the main properties less than 0.01%. The optimum grid size was decided to be (658 × 99 × 87), and the total number of the grid was 5,667,354.

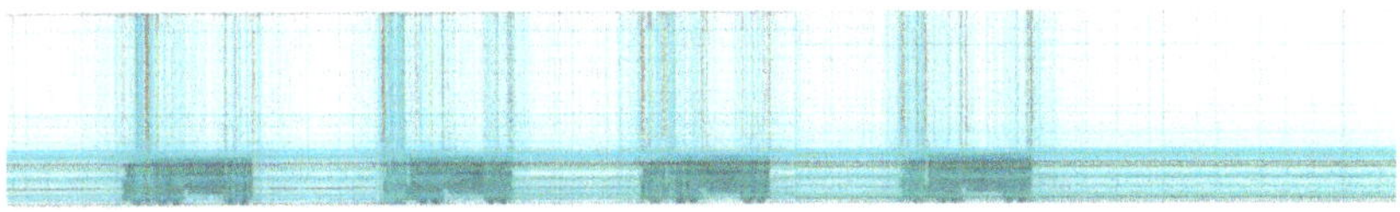

Figure 3. A typical numerical grid with the four model vehicles in platooning (658 × 99 × 87).

3. Numerical Scheme and Its Condition

The general-purpose FVM (finite volume method) CFD code, PHOENICS (ver.2020), was used for a numerical calculation of the turbulent incompressible flow field. The 3-dimensional Naiver–Stokes equations were solved with the KECHEN turbulent model (Chen–Kim κ-ε model), a modified standard κ-ε turbulent model that had good agreement with experimental data [18]. The turbulent no-slip condition near the solid boundary was modeled by logarithmic law.

The airflow field for the analysis was defined as follows:

- Quasi-3D flow.
- Turbulent flow.
- Incompressible flow.
- Steady flow.

3.1. Governing Equations

The governing equation for the steady, incompressible, and turbulent flow fields is given below:

(1) Continuity equation:

$$\frac{\partial U_i}{\partial x_i} + \frac{\partial U_j}{\partial y_i} + \frac{\partial U_k}{\partial z_i} = 0 \tag{1}$$

(2) Momentum equation:

$$\frac{\partial U_i}{\partial t} + \frac{\partial}{\partial x_j}(U_i U_i) = -\frac{1}{\rho}\frac{\partial P}{\partial x_i} + \frac{\partial U}{\partial x_j}\left[v\left(\frac{\partial U_i}{\partial x_j} + \frac{\partial U_j}{\partial x_i}\right) - \overline{u_i u_j}\right] - g_i \tag{2}$$

where $-\overline{u_i u_j} = v_t\left(\frac{\partial U_i}{\partial x_j} + \frac{\partial U_j}{\partial x_i}\right) - \frac{2}{3}k\delta_{ij}$.

(3) κ-ε turbulent energy model (KECHEN);

- Turbulent kinetic energy equation:

$$\frac{\partial}{\partial x_i}(U_j k) = \frac{\partial}{\partial x_i}\left[\left(v + \frac{v_t}{\sigma_k}\right)\frac{\partial \varepsilon}{\partial x_j}\right] + G - \varepsilon \tag{3}$$

where $G = -\overline{u_i u_j}\frac{\partial U_i}{\partial x_j}$, $v_t = C_\mu \frac{k^2}{\varepsilon}$.

- Energy dissipation equation:

$$\frac{\partial}{\partial x_i}(U_j \varepsilon) = \frac{\partial}{\partial x_i}\left[\left(v + \frac{v_t}{\sigma_\varepsilon}\right)\frac{\partial \varepsilon}{\partial x_j}\right] + \frac{\varepsilon}{k}(C_{\varepsilon 1} G - C_{\varepsilon 2}\varepsilon) \tag{4}$$

where $(C_\mu = 0.09,\ C_{\varepsilon 1} = 1.15,\ C_{\varepsilon 2} = 1.92,\ \sigma_k = 1.0,\ \sigma_\varepsilon = 1.0)$.

v_t is turbulent kinematic viscosities, σ_k, σ_ϵ Prandtl number connected to the diffusivity of K and ε to eddy viscosity [18].

3.2. *Aerodynamic Pressure Drag*

The drag force (F_D) and drag coefficient (C_D) of the model vehicle are calculated using the following equations.

- Drag force (F_D)

$$F_D = C_D \cdot \frac{1}{2} \cdot \rho_{air} \cdot A_y \cdot V_{HDV}^2 \tag{5}$$

where A_y is the projected area of the model vehicle in the longitudinal direction.

- Drag coefficient (C_D)

$$C_D = \frac{2F_D}{\rho_{air} \cdot A_y \cdot V_{HDV}^2} \tag{6}$$

3.3. *Fuel Savings and CO_2 Reduction*

3.3.1. Traction Power Saved on Each Model Vehicle Platooning

The reduction in the tractive power of each model vehicle platooning compared to the single moving vehicle (SV) can be calculated by Equation (7) [19].

- Tractive power saved,

$$PW_{sav} = \left(F_{D_{Single}} - F_{D_{platoon}}\right) \times V_{HDV} \tag{7}$$

where:

PW_{sav}: tractive power saved [kW].
$F_{D_{Single}}$: drag force of a single vehicle moving (SV) [kN].
$F_{D_{platoon}}$: drag force of each vehicle platooning (or FV1, 2, 3) [kN].
V_{HDV}: velocity of the model vehicle [m/s].

3.3.2. Fuel Saved on Each Model Vehicle Platooning Compared to SV

The mass flow rate of the fuel saved by the aerodynamic reduction can be calculated with the power saved as follows:

- Fuel mass and volume flow rate,

$$\dot{m}_{fuel} = PW_{sav} / \left(Q_{LHV} \times \eta_{engine}\right) \tag{8}$$

$$\dot{Q}_{fuel} = \dot{m}_{fuel} / \rho_{fuel} \tag{9}$$

- Fuel consumption (fc, km/liter) of a vehicle by aerodynamic resistance,

$$fc = \frac{V_{HDV}}{\dot{Q}_{fuel}}(km/h) \tag{10}$$

where, Q_{LHV}, ρ_{fuel} are the lower heating value and density of diesel fuel, η_{engine} is the brake thermal efficiency of the diesel engine. The values are given in Table 2.

Table 2. Property of the diesel fuel and IC engine thermal efficiency [20–22].

Q_{LHV} MJ/kg	ρ_{fuel} kg/m^3	η_{engine} [%]
42	815	0.40

3.3.3. Reduction in Carbon Dioxide

The reduction in CO_2 emission is calculated by the equation given below [23]

$$CO_2 \text{ emissions} = \sum(AL \cdot CL \cdot OF)_i \cdot 44/12 \tag{11}$$

where,

CO_2 emissions: incineration of fossil liquid waste, [Gg].
AL_i: the amount of incinerated fossil liquid waste type i, [Gg].
CL_i: carbon content of fossil liquid waste type i, (fraction).
OF_i: oxidation factor for fossil liquid waste type i, (fraction).
44/12: conversion factor from C to CO_2.

4. Results and Discussion

4.1. Aerodynamics Characteristics of the Model Vehicles in Platooning

The C_D of the leading vehicle (LV) is 0.6 at 0.5 L condition, which is 15% lower than the single vehicle moving (SV) (Table 3). It is due to the rear pressure recovery because of the higher stagnation pressure formed on the frontal surface of the following vehicle (FV1) However, the C_D of LV does not change much under the different distance conditions. However, FV1, FV2, and FV3 have a serious reduction in $C_{D,}$ and the average reduction compared to SV is about 44% which is similar to the results reported in the previous study [24].

Figure 4 shows the variation in C_D of each model vehicle in platooning with the change in gap. The LV in platooning has the highest drag coefficient in all gaps but the averaged C_D (Figure 4) is still lower than the SVs (0.71). The drag decreases continuously as the distance between the model vehicles is shortened.

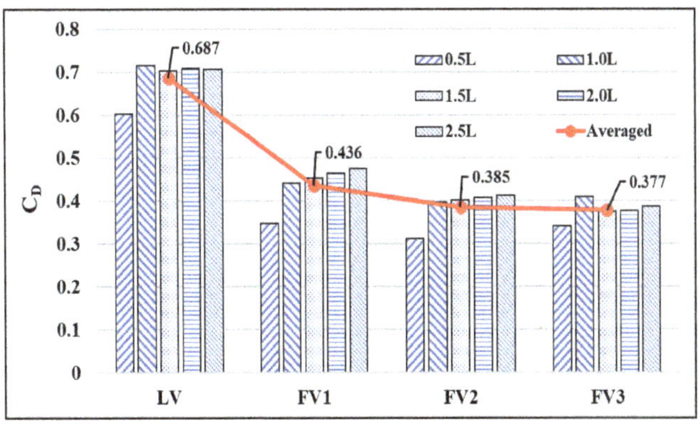

Figure 4. Variation in drag coefficient (C_D) of each model vehicle in platooning with the change in the gap.

Table 3. Variation in drag coefficient of the model vehicle with the gap and the position.

Gap		C_D	C_D Reduction Rate [%]
Single Moving Vehicle (SV)		0.710	-
0.5 L	LV	0.604	15%
	FV1	0.347	51%
	FV2	0.311	56%
	FV3	0.340	52%
1.0 L	LV	0.710	0%
	FV1	0.440	38%
	FV2	0.396	44%
	FV3	0.409	42%
1.5 L	LV	0.704	1%
	FV1	0.453	36%
	FV2	0.402	43%
	FV3	0.374	47%
2.0 L	LV	0.709	0%
	FV1	0.464	35%
	FV2	0.406	43%
	FV3	0.376	47%
2.5 L	LV	0.707	0%
	FV1	0.474	33%
	FV2	0.411	42%
	FV3	0.386	46%

Figure 5 shows the averaged reduction in C_D of each vehicle platooning compared to the SV. As shown, the front leading vehicle has only 3.3% less drag than the SV and the drag drops significantly from the 2nd vehicle (FV1) and the reduction reaches 46.9% on the last-placed vehicle (FV3) (Figure 5).

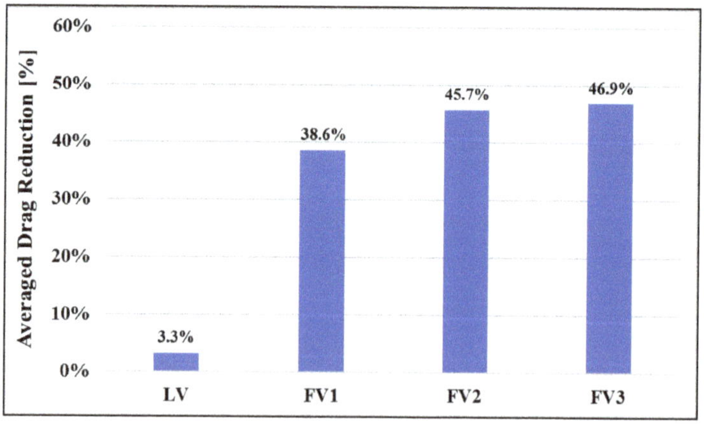

Figure 5. Reduction in the averaged C_D of each vehicle in platooning compared to SV.

Figure 6 shows the static pressure distribution around the vehicles platooning and it indicates that the stagnation pressure is highest at the front surface of the LV and gradually decreases for the following vehicles because the incoming air velocity decreases due to the circulation flow at the front region of each following vehicle. Table 3 shows that LV has 15% lower drag than the SV under the 0.5 L condition. It is due to the higher station pressure formed on the frontal region of FV1 which recovers the rear pressure of the LV. Additionally, the effect disappears as the gap distance increases.

Figure 6. Comparison of the static pressure distribution with the change in the gap at 100 km/h.

The drag force on a moving vehicle generates air resistance due to the pressure difference between the front and rear sides of the vehicle. When two vehicles are moving in parallel, a serious vortex is generated at the rear side of the front vehicle as shown in Figure 7. This complicated flow phenomenon decreases the stagnation pressure on the front side of the following vehicle and decreases air resistance on it.

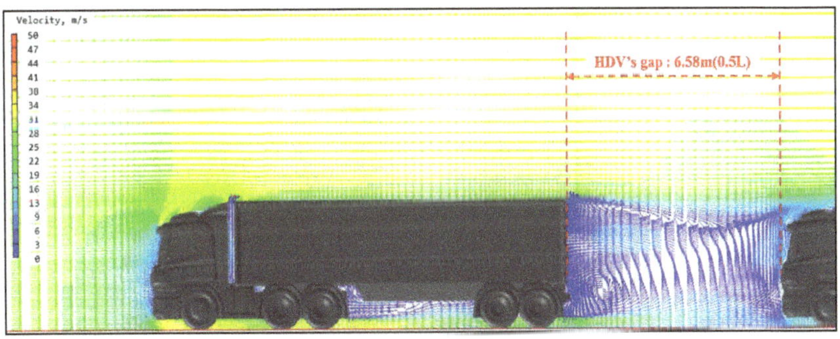

Figure 7. Vortex generated in between LV and FV1 at 0.5 L and 100 km/h in speed.

It is caused by the lower stagnation pressure formed on the frontal face of each model vehicle in platooning (Figure 6) and the lower turbulent kinetic energy formed at the rear side of each model vehicle (Figure 7).

Figure 8 shows the turbulent kinetic energy distribution between the model vehicles with the different gaps. As the gap is shorter, the kinetic energy formed between the vehicles is higher and it is the reason for the lower static pressure formed in the gap, and it contributes to the lower stagnation pressure formed on the frontal face of the following vehicles.

4.2. Fuel Efficiency and GHG Emission Improvement in the Model Vehicles

According to the research reports from the EU FT7 (SARTRE project, 2016), it was found that close distance between the vehicles in platooning improves the fuel efficiency for all vehicles [24]. In the case of a 6 m gap, which is half-length of the model HDV, the LV saved 8% of fuel compared to SV and up to 16% for the following FV [25]. We also predicted fuel consumption and GHG reduction in this study.

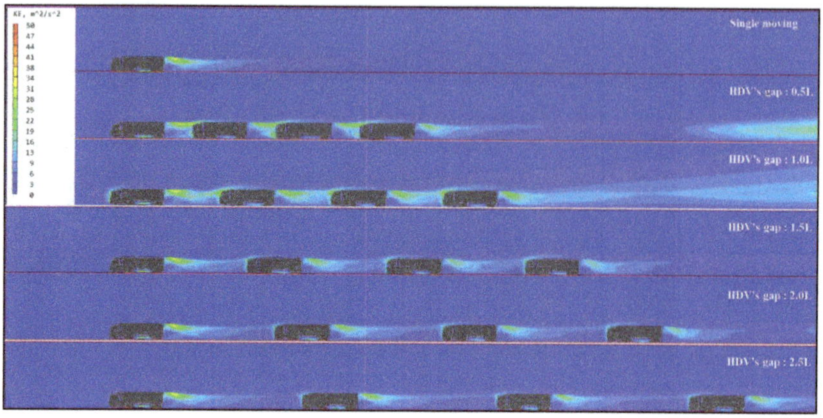

Figure 8. Comparison of the turbulent kinetic energy distribution with the change in the gap at 100 km/h.

Figure 9 shows the tractive power saved by the vehicles platooning compared to the single moving vehicle (SV). For the leading vehicle (LV), the power energy was not saved much compared to SV but from the 2nd vehicle (FV1, FV2, FV3) the tractive power distinctively decreased. The power saved on the model vehicles platooning versus SV was calculated with Equation (7).

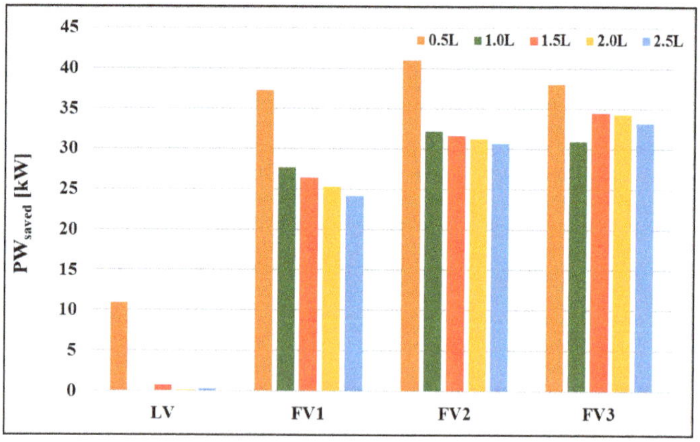

Figure 9. Tractive power (PW_{saved}) saved on the vehicles platooning compared to SV at 100 km/h.

LV saved approximately 10 kW of power at the 0.5 L gap condition. Under 1.0 L~2.5 L gap conditions, a relatively narrower gap does result in a better traction power reduction than SV, and the power saved increases as the location of the model vehicle is far from the leading position. When the HDV's gap is 0.5 L, the averaged traction power saved for FV1, FV2, and FV3 is 38 kW compared to the SV. Additionally, the FVs kept an average of 30 kW of power under the condition of the 1.0 L~2.5 L gap.

The fuel savings for each model vehicle platooning compared to SV was calculated with Equations (8)–(10).

The fuel consumption of SV by aerodynamic resistance was calculated as 5.2 km/liter at 100 km/h from Equation (10). When a single vehicle was moving, the fuel efficiency of FV1, FV2, and FV3 can be predicted as shown in Table 4. As a result, among simulation cases, the possibility of increasing the fuel efficiency of FV 1, 2, and 3 to 10 km/L is also confirmed under the 0.5 L conditions with maximum aerodynamic effects (100 km/h cruising driving condition).

Table 4. Fuel mileage of each model vehicle platooning compared to the single moving vehicle (SV).

Vehicle's Gap	Fuel Mileage [km/Liter]				
	Single Moving	LV	FV1	FV2	FV3
0.5 L	5.2	6.1	10.7	11.9	10.9
1.0 L		5.2	8.4	9.3	9.0
1.5 L		5.2	8.2	9.2	9.9
2.0 L		5.2	8.0	9.1	9.8
2.5 L		5.2	7.8	9.0	9.6

According to KOTI (The Korea Transport Institute) report. 2021, Vol.62, the average daily driving distance of a loaded HDV was about 300 km/day in Korea for 2020 [26]. The averaged cruising speed of the HDV on a highway is 100 km/h; thus, the daily drive time is 3 h.

Based on HDV's gap of 0.5 L, LV kept 2100 liter per year compared to SV. The saved result of an average of 7435 liter of FVs is shown in Figure 10. CO_2 reduction is calculated through Equation (11), resulting in LV saved by 5.3 tons per year and FVs saved by an average about 18.8 tons per year compared to the SV (Figure 11).

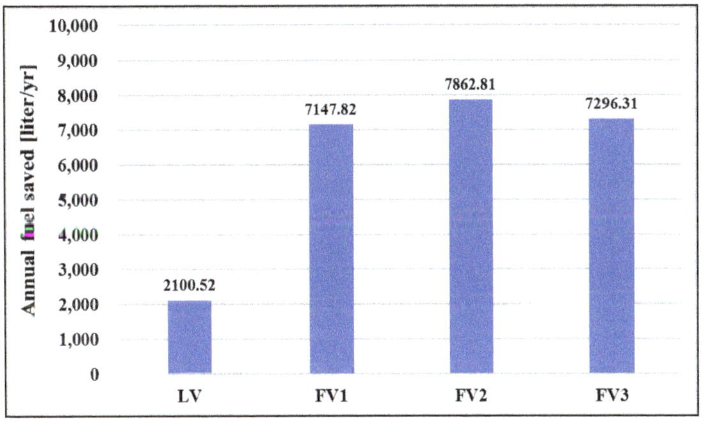

Figure 10. Annual fuel saving on the model vehicles platooning compared to the single moving vehicle (SV) at a 0.5 L gap.

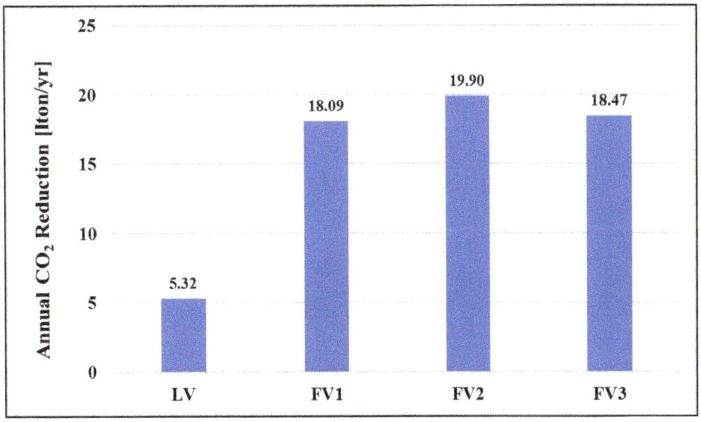

Figure 11. Annual CO_2 reduction in platooning vehicle driving compared to the single moving vehicle (SV) at a 0.5 L gap.

Thus, reducing drag force from the platoon can reduce the power required by the engine, resulting in reduced fuel consumption and GHG. Furthermore, aerodynamic drag is proportional to the square of vehicle speed, so high speed platooning could be more effective in saving energy.

5. Conclusions

This numerical study aimed to understand the aerodynamic effect of the heavy-duty vehicles platooning with the change in the platooning conditions; the gap between the model vehicles and the platooning position of the model vehicle. With the platooning condition change, aerodynamic resistance was analyzed to have the possibility of GHG emission reductions and fuel efficiency improvement. This study only considered steady driving conditions on a level road with no side wind. In addition, the aerodynamic characteristics of the vehicle were theoretically analyzed to determine the energy saving of the vehicle. Further study is needed to analyze the real driving state. From the research, the following were found:

i. The stagnation pressure of FV is reduced due to the rear vortex generated by the front leading vehicle. This is the main cause of the drag reduction on the FV. In this study, the drag of FVs was reduced over 50% compared to SV at the gap (0.5 L) (Table 3).
ii. The shorter the gap between the model vehicles platooning, the smaller C_D of FV, which indicates the shorter the gap distance, the more significant the influence of vortex on the FV. As the gap between the model vehicles widens, the rear vortex of the LV gradually decreases and the static pressure recovers to the ambient pressure. It is the reason for increasing the stagnation pressure of FV causing to C_D increase.
iii. Platoon driving has very positive effect not only to the fuel savings but also on the GHG reduction. Thus, the platoon driving mode of heavy-duty vehicles would seriously contribute to the logistics industry, economically and environmentally.
iv. This study hypothesized the driving conditions of autonomous vehicles based on V2X and suggested that the aerodynamic effect can be maximized when an appropriate platoon gap is set in vehicle-to-vehicle distance control. However, this study aimed to confirm the possibility of reducing GHG according to the platooning concept.
v. This study performed a theoretical analysis using the FVM numerical simulation method. Platooning simulations were performed under the steady driving condition on a level road with no side wind to determine the fuel economy effect so this study might be used for reference. In future research, we will undertake the aerodynamic driving stability research on vehicle platooning. In the study, the weight of the vehicle with the road condition such as the friction coefficient and slippery factor of the road and the side wind, the gyration radius of the road with the tilted angle, etc., should be the important parameters to evaluate the road-load power for the study.

Author Contributions: Conceptualization, J.J. and C.-H.K.; methodology, J.J. and C.-H.K.; software, J.J.; validation, C.-H.K.; writing—original draft preparation, J.J. and C.-H.K.; writing—review and editing, J.J.; visualization, J.J.; supervision, C.-H.K.; project administration, C.-H.K. All authors have read and agreed to the published version of the manuscript.

Funding: This research received no external funding.

Institutional Review Board Statement: Not applicable.

Informed Consent Statement: Not applicable.

Data Availability Statement: The data presented in this study are available on request from the corresponding author.

Acknowledgments: This research is sponsored by the foundation of research and industry of Seoul National University of Science and Technology.

Conflicts of Interest: The authors declare no conflict of interest.

References

1. Buljac, A.; Džijan, I.; Korade, I.; Krizmanić, S.; Kozmar, H. Automobile aerodynamics influenced by airfoil-shaped rear wing. *Int. J. Automot. Technol.* **2016**, *17*, 377–385. [CrossRef]
2. Cihan, B. Numerical drag reduction of a ground vehicle by NACA2415 airfoil structured vortex generator and spoiler. *Int. J. Automot. Technol.* **2019**, *20*, 943–948. [CrossRef]
3. Song, K.S.; Kang, S.O.; Jun, S.O.; Park, H.I.; Kee, J.D.; Kim, K.H.; Lee, D.H. Aerodynamic design optimization of rear body shapes of a sedan for drag reduction. *Int. J. Automot. Technol.* **2012**, *13*, 905–914. [CrossRef]
4. Huluka, A.W.; Kim, C.H. A Numerical Analysis on Ducted Ahmed Model as a New Approach to Improve Aerodynamic Performance of Electric Vehicle. *Int. J. Automot. Technol.* **2021**, *22*, 291–299. [CrossRef]
5. Zhu, W.F.; Jiang, X.H.; Chen, X.; Lin, P.J. Automotive window seal design considering external aerodynamic load and surrogate constraint modeling. *Int. J. Automot. Technol.* **2016**, *17*, 853–864. [CrossRef]
6. Banks, J. Turbulence Modeling in PHOENICS. 2007. Available online: http://www.cham.co.uk/phoenics/d_polis/d_enc/turmod/enc_t342.htm (accessed on 13 June 2021).
7. Davila, A.; Nombela, M. *Sartre-Safe Road Trains for the Environment Reducing Fuel Consumption through Lower Aerodynamic Drag Coefficient*; SAE Technical Paper 2011-36-0060; SAE International: Warrendale, PA, USA, 2011.
8. Sun, X.; Yin, Y. Behaviorally stable vehicle platooning for energy savings. *Trans. Res. Part C Emerg. Technol.* **2019**, *99*, 37–52. [CrossRef]
9. Ministry of Land, Infrastructure and Transport. *Development of Operation Technology for V2X Truck Platooning. Land Infrastructure and Transport R&D Report*; 20TLRP-B147674-03; Ministry of Land, Infrastructure and Transport: Sejong, Korea, 2018.
10. Shin, K.; Choi, J.; Huh, K. Adaptive cruise controller design without the transitional strategy. *Int. J. Automot. Technol.* **2020**, *21*, 675–683. [CrossRef]
11. Emirler, M.T.; Guvenc, L.; Guvenc, B.A. Design and evaluation of robust cooperative adaptive cruise control systems in parameter space. *Int. J. Automot. Technol.* **2018**, *19*, 359–367. [CrossRef]
12. Saleem, A.; Al Maashri, A.; Al-Rahbi, M.A.; Awadallah, M.; Bourdoucen, H. Cooperative cruise controller for homogeneous and heterogeneous vehicle platoon system. *Int. J. Automot. Technol.* **2019**, *20*, 1131–1143. [CrossRef]
13. Wi, H.; Park, H.; Hong, D. Model predictive longitudinal control for heavy-duty vehicle platoon using lead vehicle pedal information. *Int. J. Automot. Technol.* **2019**, *21*, 563–569. [CrossRef]
14. Kang, Y.; Hedrick, J. Emergency braking control of a platoon using string stable controller. *Int. J. Automot. Technol.* **2004**, *5*, 89–94.
15. Alam, A.; Gattami, A.; Johansson, K.H. An experimental study on the fuel reduction potential of heavy duty vehicle platooning. In Proceedings of the 13th International IEEE Conference on Intelligent Transportation Systems, Funchal, Portugal, 19–22 September 2010; pp. 19–22.
16. Liang, K.-Y.; Martensson, J.; Johansson, K.H. Heavy-Duty Vehicle Platoon Formation for Fuel Efficiency. *IEEE Trans. Intell. Transp. Syst.* **2015**, *17*, 1051–1061. [CrossRef]
17. Ludwig, J. *Concentration Heat and Momentum Limited*; CHAM Technical Report TR326; POLIS, Phoenics On-line Information System: London, UK, 2008.
18. Chen, Y.-S.; Kim, S.-W. *Computation of Turbulent Flows Using an Extended K-Epsilon Turbulence Closure Interim Report*; No. NAS 1.26: 179204; NASA: Huntsville, AL, USA, 1987.
19. Kim, C.H. A streamline design of a high-speed coach for fuel savings and reduction of carbon dioxide. *Int. J. Automot. Eng.* **2011**, *2*, 101–107. [CrossRef]
20. World Nuclear Association. Heat Values of Various Fuels. Available online: http://www.world-nuclear.org/information-library/facts-and-figures/heat-values-of-various-fuels.aspx (accessed on 15 August 2021).
21. K-Petro (Korea Petroleum Quality & Distribution Authority). Quality Standard Chart according to Laws. Available online: https://www.kpetro.or.kr/eng/lay1/S210T239C379/contents.do (accessed on 15 August 2021).
22. Suppes, G.J.; Storvick, T.S. *Sustainable Power Technologies and Infrastructure: Energy Sustainability and Prosperity in a Time of Climate Change*; Academic Press: Cambridge, MA, USA, 2015; Chapter 5.
23. Guendehou, S.; Koch, M.; Hockstad, L.; Pipatti, R.; Yamada, M. *Draft 2006 IPCC Guidelines for National Greenhouse Gas Inventories*; Institute for Global Environmental Strategies: Hayama, Japan, 2006; Chapter 5.
24. Davila, A.; del Pozo, E.; Aramburu, E.; Freixax, A. *Environmental Benefits of Vehicle Platooning*; 2013-26-0142; SAE International: Warrendale, PA, USA, 2013.
25. Chan, E. SARTRE automated platooning vehicles. In *Towards Innovative Freight and Logistics*; Wiley: Hoboken, NJ, USA, 2016; Volume 2, pp. 137–150.
26. KOTI (The Korea Transport Institute). Report 62. 2021. Available online: https://www.koti.re.kr/user/bbs/BD_selectBbs.do?q_bbsCode=1093&q_bbscttSn=20210726174721314 (accessed on 20 August 2021).

energies

MDPI

Article

Evaluation Method of Naturalistic Driving Behaviour for Shared-Electrical Car

Shaobo Ji [1,*], Ke Zhang [1], Guohong Tian [2], Zeting Yu [1], Xin Lan [1], Shibin Su [3] and Yong Cheng [1]

1 College of Energy and Power Engineering, Shandong University, Jinan 250061, China; 202134527@mail.sdu.edu.cn (K.Z.); yuzt1979@126.com (Z.Y.); lampxane@163.com (X.L.); cysgdb@126.com (Y.C.)
2 Department of Mechanical Engineering Sciences, University of Surrey, Guildford GU2 7XH, UK; g.tian@surrey.ac.uk
3 Research Management Department, Hisense TransTech Co., Ltd., Qingdao 266071, China; sushibin@hisense.com
* Correspondence: ytjobo@126.com

Abstract: Evaluation of driving behaviour is helpful for policy development, and for designing infrastructure and an intelligent safety system for a car. This study focused on a quantitative evaluation method of driving behaviour based on the shared-electrical car. The data were obtained from the OBD interface via CAN bus and transferred to a server by 4G network. Eleven types of NDS data were selected as the indexes for driving behaviour evaluation. Kullback–Leibler divergence was calculated to confirm the minimum data quantity and ensure the effectiveness of the analysis. The distribution of the main driving behaviour parameters was compared and the change trend of the parameters was analysed in conjunction with car speed to identify the threshold for recognition of aberrant driving behaviour. The weights of indexes were confirmed by combining the analytic hierarchy process and entropy weight method. The scoring rule was confirmed according to the distribution of the indexes. A score-based evaluation method was proposed and verified by the driving behaviour data collected from randomly chosen drivers.

Keywords: driving behaviour evaluation; naturalistic driving study; shared-electrical car; Kullback–Leibler divergence; analytic hierarchy process; entropy weight method

Citation: Ji, S.; Zhang, K.; Tian, G.; Yu, Z.; Lan, X.; Su, S.; Cheng, Y. Evaluation Method of Naturalistic Driving Behaviour for Shared-Electrical Car. *Energies* **2022**, *15*, 4625. https://doi.org/10.3390/en15134625

Academic Editor: Giovanni Lutzemberger

Received: 10 May 2022
Accepted: 22 June 2022
Published: 24 June 2022

1. Introduction

Road accidents kill approximately 1.24 million people every year and they are the eighth leading cause of death globally [1]. Aberrant driving and violation of traffic rules cause 74% of traffic accidents [2]. In addition, these behaviours also lead to excessive fuel consumption and vehicle emissions [3]. Therefore, it is necessary to understand the influence of driving behaviour on road risks and vehicle performance. Additionally, quantitative methods of the aberrant driving behaviour should be proposed to improve the driving behaviour.

In former studies, driving behaviour data have been mainly obtained by the following methods: driver self-reported survey [4], driver behaviour questionnaires [5], driver simulators [6], field tests and Naturalistic Driving Studies (NDS) [7]. The first two methods are subjective evaluation methods. For the driver simulators and field tests, the driving behaviour may be different due to the pre-arranged test environment and procedure. Naturalistic driving data were obtained with an unobtrusive data acquisition system during everyday driving. NDS can observe the driving behaviour in a natural driving condition without experimental control. Therefore, the method can accurately reflect driving habits [8]. NDS can be categorized further into on-site study and individual driver study [9]. For the on-site study, naturalistic driving data can be collected with video cameras or microwave and other equipment at a particular site. The individual driver study can obtain various driving behaviour data, such as vehicle speed, acceleration, location and vehicle clearance and so on. For the individual driver study, abundant driving behaviour data can provide multi-dimensional analysis for driving behaviour.

NDS data from the study of individual drivers can be obtained by Global Position System (GPS), video cameras, independent sensors, mobile phones and On-Board Diagnostics (OBD). GPS can provide vehicle location directly and vehicle speed can be estimated based on the location. The acceleration and deceleration behaviour of drivers were studied using GPS in [10,11]. Video cameras were used to capture the drivers' facial expressions and eye movements to analyse the driving behaviour [12]. Video cameras were also used to record drivers' body movements, such as feet movements, to evaluate the reaction time of the drivers [13]. Independent sensors such as accelerometer [14], Radar sensor [15] and Inertial Measurement Unit (IMU) [16] were also used to obtain driving behaviour data. Nowadays, most mobile phones are equipped with GPS and an accelerometer. Driving behaviour such as acceleration, braking and speeding were studied using mobile phones in [17]. OBD II interface was mandatory for vehicles and the system can collect variable vehicle information from the Electronic Control Units (ECU) of the vehicles via the Controller Area Network (CAN) bus. OBD II interface can collect vehicle status and operation information, including vehicle speed, acceleration, position of accelerator/ brake pedal, angle of steering wheel and fuel consumption and so on [18]. Considering the abundant information and reliability of the OBD II, the data collected from OBD was used to analyse the driving behaviour [19].

Driving behaviour can be influenced by driver's driving ability, driver characteristics, driving duration and driver's distraction [20]. Driving ability was influenced by the driver's experience, skill and knowledge. Compared with experienced drivers, young drivers had a higher rate of traffic accidents [21]. Driver characteristics include age, gender, and education level and etc. Drivers' behaviour was influenced by the drivers' age. Young drivers had a higher probability of accidents than middle-aged and older drivers due to their higher tendency of speeding [22]. Most studies showed that the drive behaviour of males was more aggressive that of than females and that males had higher risk of having an accident [23]. The behaviour of drivers with a high education level was more compliant, such as less lane-changing [24]. Driving duration had a significant influence on driving behaviour and there was more speeding for the drivers who had a longer driving distance [25]. A driver can be distracted by the driving environment, their mobile phone, their co-passenger and so on [26]. The distractions had adverse impacts on road safety, especially when the driver glanced away from the road.

Driving behaviour has a great influence on the road safety and vehicle performance. The evaluation of driving behaviour is helpful in providing positive feedback to drivers so they reduce the dangerous driving, thus avoiding traffic accidents and enhancing vehicle performance. According to the conclusion of the previous paragraph, driving behaviour is influenced by multiple factors. Considering the stochastic feature of the driving behaviour, it is better to extract the driving behaviour feature from a large sample, including the driver selection and driving route. Nowadays, shared-electrical cars are used in several cities in China, the driver and driving route are random for each trip, and the driving behaviour is purely naturalistic. The driving behaviour data contains drivers of different ages, genders and driving experiences. This makes it suitable for the evaluation of the driving behaviour. While there have been a few studies that have conducted an evaluation of the driving behaviour using data from shared-electrical cars, the statistical characteristics of NDS data with a large and stochastic sample is still unclear, and should be studied in more depth.

To bridge the knowledge gap, this paper was conducted using the data from shared-electrical cars used in Shanghai and Tianjin cities of China. On this basis, a quantitative evaluation method of driving behaviour was proposed and verified. This paper was organized as follows. Firstly, the NDS data obtained from the OBD-II interface of the shared-electrical car was pre-treated to improve the data quality. Secondly, an estimation method was employed to confirm the appropriate amount of NDS data. Thirdly, the relationship between different driving behaviour parameters was studied. Finally, an evaluation of driving behaviour was proposed and verified in practical application. The study flow chart is shown in Figure 1.

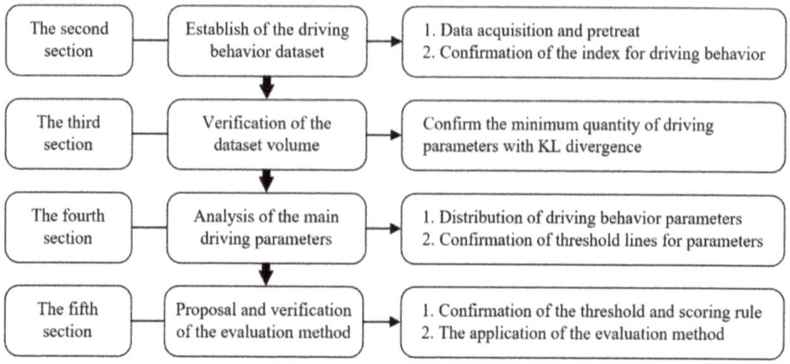

Figure 1. The main study flow chart.

2. Data Acquisition and Treatment

2.1. Data Acquisition

The NDS data were collected from 400 electrical cars, in which 305 cars were shared-electrical cars and others were online car-hiring cars. A data acquisition system was used to obtain the NDS data from the OBD-II interface via CAN bus. All the sampled data were sent to a server by a 4G network. The data were saved and treated thereafter for driving behaviour analysis. The data acquisition process is shown in Figure 2. The data were collected during a period of about four months. The total car mileage was about 1.2 million kilometres, with the total car travelling hours adding up to about 45,000, and the sample points totaling about 1.28×10^9.

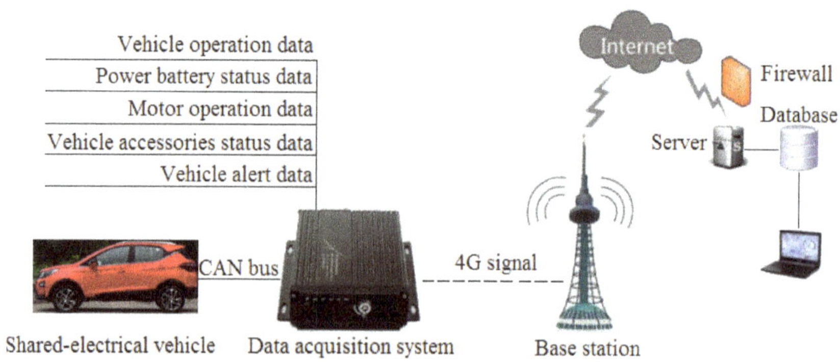

Figure 2. The acquisition process of the NDS data.

The data acquisition system can obtain more than 50 driving behaviour parameters, including vehicle operation, power battery status, motor operation, vehicle accessories status and vehicle alert. The primary data were shown in Table 1. All the data were collected at a rate of 10 Hz, which was adequate for transient process analysis, such as acceleration or deceleration.

Table 1. The primary parameters collected by the data acquisition system.

Signal Type	The Primary Parameters
Vehicle operation	Vehicle speed, mileage, position of the acceleration and brake pedal, steering wheel angle.
Power battery status	Total voltage and current of the battery pack, insulation resistance and temperature.
Motor operation	Voltage, current, motor speed, motor torque and temperature.
Vehicle accessories status	Voltage and current of the air conditioner, Voltage and current of the DC-DC.
Vehicle alert	Alert signal of power battery, motor and thermal management system.

It should be noted that the collected data mentioned above were obtained from the OBD-II interface, and do not include any personal information, such as GPS data.

2.2. Data Treatment

The vehicle data were collected in 'Charge', 'Standby' and 'Operation' phases. For the analysis of driving behaviour, the data collected in the 'Charge' and 'Standby' phases were neglected. A valid driving event was recognised by both the power-on status and the vehicle speed being greater than zero. When the vehicle speed was zero for less than 10 min with power-on status, it was considered in the same driving event. Figure 3 showed an example of the vehicle speed versus time in a complete driving event.

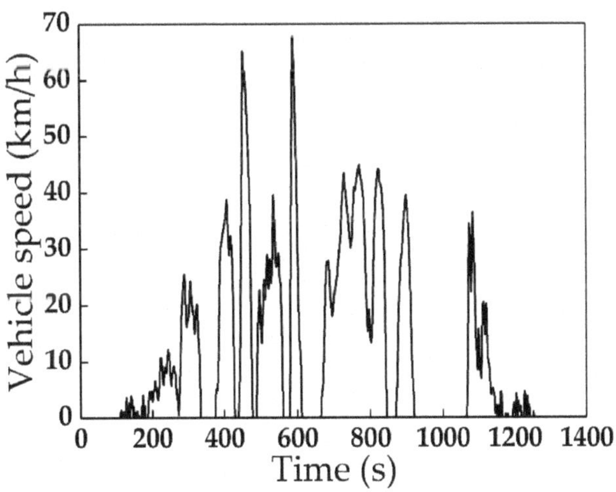

Figure 3. The vehicle speed versus time in a complete driving event.

The data acquisition and transmission can be affected by occasional interferences, such as the voltage fluctuation in the vehicle start and stop phase, abnormal power failure of the data acquisition system, the interruption of the communication signal caused by tunnels or tall buildings and so on [27]. The data received by the server include occasional flaws; therefore, a data quality control method must be used before processing. The sliding-window averaging filter was implemented to reduce the random noise. The filter can be expressed as follows:

$$x[k] = \frac{1}{M}\sum_{i=0}^{M-1} x[k-i] \tag{1}$$

where i is the number of the original data, $x[k-i]$ is the original data, M is the length of the sliding-window averaging filter, which is set to five, and $x[k]$ is the treated data.

The vehicle data were obtained from the ECUs of the vehicles via CAN bus. Most of the data were deemed normal. To eliminate the occasional abnormal data, the box diagram method was used, which was introduced in the previous paper [28]. The sliding-window averaging filter and box diagram method were used to treat the original data. Figure 4a showed the treatment effect of the filter on vehicle velocity. It can be seen that the signal fluctuation was eliminated but the trend remained the same. Figure 4b showed the treatment effect of the box diagram method for the acceleration and the abnormal value can be eliminated effectively.

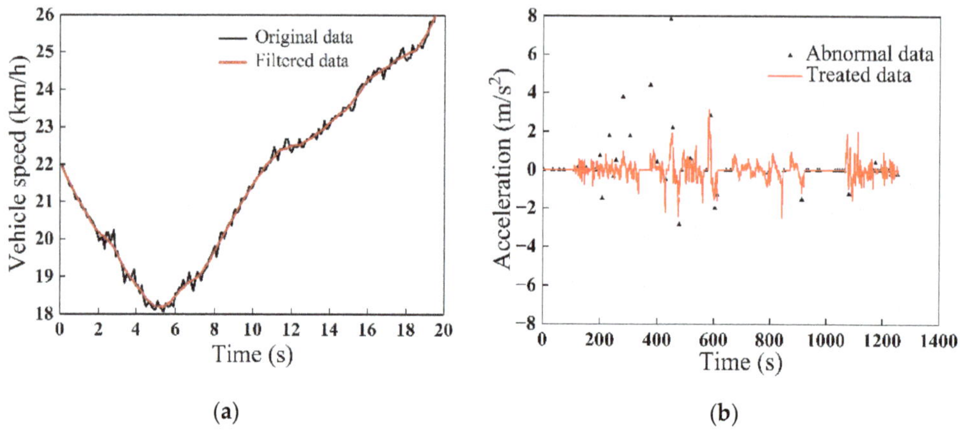

Figure 4. The comparison of the signal with different treat method: (**a**) Sliding-window averaging filter; (**b**) The box diagram method.

2.3. The Indexes for Driving Behaviour

There were multiple indexes to evaluate the driving behaviour. This study mainly focused on the indexes related to driving safety, including frequent adjustment of vehicle speed and steering wheel, rapid acceleration, sudden braking, rapid turning, fatigue driving and power consumption. The proposed evaluation method of driving behaviour was based on 11 indexes. The index, symbol, definition and unit are listed in Table 2. All values were calculated based on a complete driving event.

Table 2. The indexes used to evaluate the driving behaviour.

Number	Index	Symbol	Definition	Unit
1	Standard deviation of vehicle speed	v_{std}	$v_{std}=\sqrt{\frac{\sum_{i=1}^{n}(v_i-v_m)^2}{n}}$	km/h
2	Average value of acceleration	$a_m{}^+$	$a_m{}^+=\frac{\sum_{i=1}^{n}a_i{}^+}{n}$	m/s^2
3	Average value of deceleration	$a_m{}^-$	$a_m{}^-=\frac{\sum_{i=1}^{n}a_i{}^-}{n}$	m/s^2
4	Standard deviation of acceleration	$a_{std}{}^+$	$a_{std}{}^+=\sqrt{\frac{\sum_{i=1}^{n}(a_i{}^+-a_m{}^+)^2}{n}}$	m/s^2
5	Standard deviation of deceleration	$a_{std}{}^-$	$a_{std}{}^-=\sqrt{\frac{\sum_{i=1}^{n}(a_i{}^--a_m{}^-)^2}{n}}$	m/s^2

Table 2. *Cont.*

Number	Index	Symbol	Definition	Unit
6	The number of rapid acceleration per 100 km	F_a	$F_a = \frac{N_a}{D} \times 100$	times/100 km
7	The number of sudden braking per 100 km	F_d	$F_d = \frac{N_d}{D} \times 100$	times/100 km
8	The number of rapid turning per 100 km	F_t	$F_t = \frac{N_t}{D} \times 100$	times/100 km
9	The number of speeding during steering per 100 km	F_{st}	$F_{st} = \frac{N_{st}}{D} \times 100$	times/100 km
10	Driving time per trip	T	-	hour
11	Power consumption per 100 km	P_m	$P_m = \frac{W}{D} \times 100$	kW·h/100 km

where i is the number of sample point, n is the total number of sample point in a complete driving event, v_i is the vehicle speed of each sample point, v_m is the average speed in a complete driving event, a_i^+ is the acceleration of each sample point, a_i^- is the deceleration of each sample point, N_a, N_d, N_t and N_{st} are the number of occurrences of rapid acceleration, sudden braking, rapid turning and speeding during steering, respectively, in a complete driving event, D is the mileage in a complete driving event, and W is the power consumption in a complete driving event.

3. Quantity Estimation of Driving Behaviour Data

For an individual driver, the driving style is relatively stable after a certain period of driving experience and the statistical characteristics will be convergent [29]. The evaluation accuracy of driving behaviour depends on the volume of the NDS dataset, which should cover as many of the driving behaviour characteristics as possible. From the statistical perspective, if the NDS data are adequate, the distribution density of the data will remain the same with additional sample data. In this case, the amount of NDS data can be considered suitable for driving behaviour analysis. To ensure the appropriate quantity of dataset, the amount of NDS data was estimated by Kullback–Leibler (*KL*) divergence [30]. The quantity estimation of a given dataset $x = \{x_i\}_{i=1}^{n}$ included the following two steps.

Firstly, the distribution density ($f_n(x)$) of the dataset (x) was calculated with the kernel density method, which was a non-parametric method, and the equations are as follows.

$$f_n(x) = \frac{1}{n}\sum_{i=1}^{n}\frac{1}{h}K(\frac{x - x_i}{h}) \tag{2}$$

$$h = 1.06{\cdot}\sigma{\cdot}n^{-1/5} \tag{3}$$

$$K(u) = \frac{1}{2\pi}\exp(-\frac{u^2}{2}) \tag{4}$$

where n is the length of the dataset, σ is standard deviation of the dataset.

Secondly, the *KL* divergence was calculated to assess the difference in kernel density between the two datasets with the data length of n and $n+m$, respectively. Assuming the kernel densities of the two datasets are expressed as $f_n(x)$ and $f_{m+n}(x)$, the *KL* divergence of the two datasets can be expressed as Equation (5). If *KL* remains as a small value with the increase in m, it indicates that the kernel density is almost unchanged with the adjustment of the dataset and the quantity of the dataset is adequate. The variation value of *KL* can affect the data quantity, which was set as $KL < 1 \times 10^{-5}$ in this study.

$$KL(f_{n+m}(x)||f_n(x)) = \int f_{n+m}(x) \times \log\frac{f_{n+m}(x)}{f_n(x)} \tag{5}$$

The *KL* divergence of different driving behaviour parameters was calculated and Figure 5 showed the changing trend of vehicle speed, acceleration, steering wheel angle and steering wheel speed. It can be seen that the *KL* of different parameters decreased with the increase in data quantity. Although the convergence rate of the parameters was different, the *KL* convergence of the four parameters all satisfied the requirement when the data quantity exceeded 231×10^5. The similar calculation of *KL* was applied to other driving behaviour parameters. Based on this analysis, the quantity of the sampled data in this study was considered adequate.

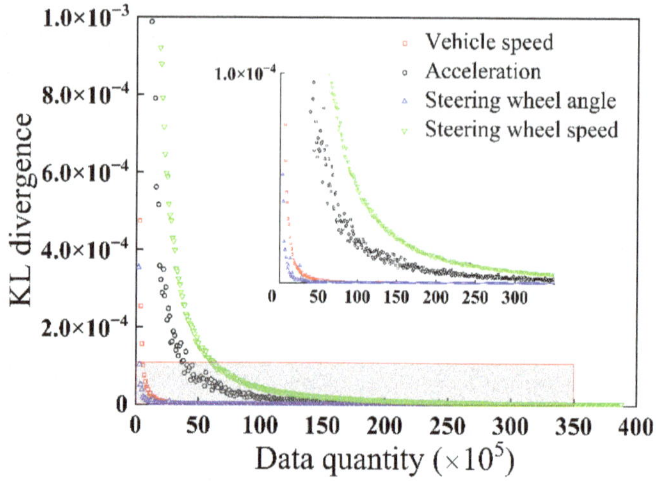

Figure 5. The *KL* of different parameters versus data quantity.

4. Study on the Relationship between Different Driving Behaviour Parameters

4.1. Statistical Characteristics of the Main Driving Behaviour Parameters

Vehicle speed, acceleration, deceleration, steering wheel angle and steering wheel speed contain abundant information reflecting driving behaviour. The statistics of these parameters were firstly calculated to understand the general characteristics of the driving behaviour.

According to Figure 5, the *KL* of vehicle speed converged when the data quantity exceeded 20×10^5. The vehicle speed was selected randomly from the total dataset of shared-electrical vehicles to the quantity of 20×10^5. The distribution characteristic of the vehicle speed is shown in Figure 6. Most of the vehicles' speeds were lower than 80 km/h and the average speed was 33.4 km/h. The relatively low speed was mainly attributed to the heavy urban traffic. The largest proportion of the speed was lower than 5 km/h, which reflected the frequent starting or parking mode due to red lights or traffic jams in urban traffic.

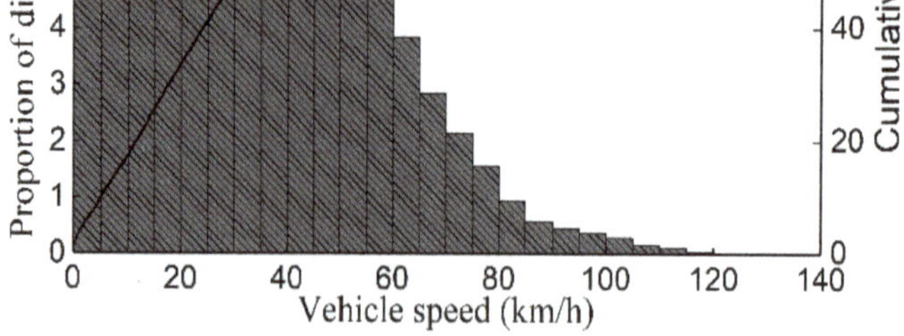

Figure 6. The distribution characteristic of the vehicle speed.

According to Figure 5, the *KL* of acceleration and deceleration converged when the data quantity exceeded 163×10^5. The corresponding data quantity of acceleration and deceleration were selected randomly from the total dataset of shared-electrical vehicle. The distribution characteristic was compared and shown in Figure 7. Over 90% of the acceleration and deceleration were lower than 1.5 m/s^2. The average values of acceleration and deceleration were 2.05 and 2.72 m/s^2, respectively. Compared to deceleration, acceleration was mainly located within the small value sections, which was mainly attributed to the heavy traffic.

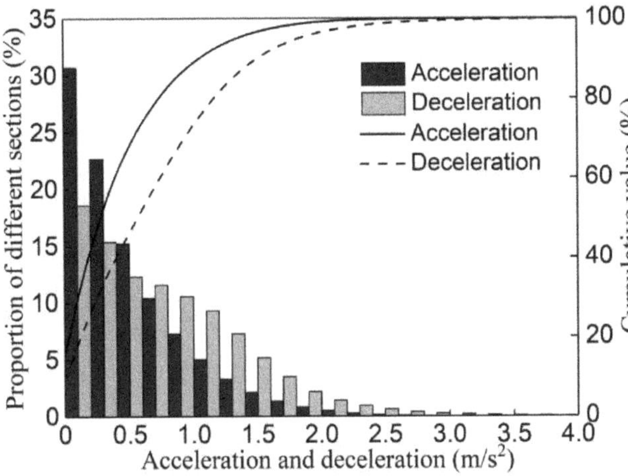

Figure 7. The distribution characteristic of acceleration and deceleration.

The *KL* of steering wheel angle converged when the data quantity was more than 10×10^5, and the corresponding quantity of steering wheel angle data was selected randomly. The distribution characteristic of steering wheel angle is shown in Figure 8. Over 60% of the steering wheel angle was lower than 25°, which meant that most of the angle change was attributed to the slight adjustment in driving direction. The percentage decreased with the increase in steering wheel angle until 300°, and these steering actions were mainly due to lane-changing or vehicle turning. There was also angle distribution around 550°, and this was mainly caused by vehicle U-turn or parking.

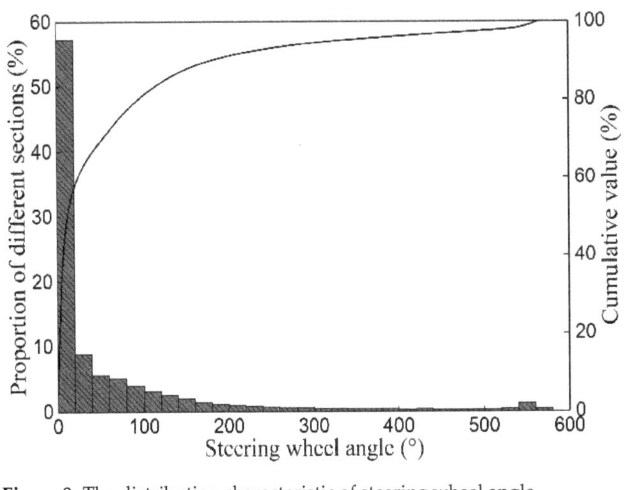

Figure 8. The distribution characteristic of steering wheel angle.

The *KL* of steering wheel speed converged when the data quantity was more than 231×10^5. The same amount of steering wheel speed data was selected randomly and compared in Figure 9. According to the steering characteristic, the steering wheel speed can be classified to steering speed (positive values) and return speed (negative values). Comparison showed that the two speeds had a similar changing pattern. Most of the steering speeds and return speeds were lower than 25°/s, which meant that the majority of the steering actions were careful. The averages of the steering speed and return speed were calculated, which were 38.11°/s and 30.35°/s, respectively. The steering speed was larger than the return speed, which corresponded to the normal driving habit.

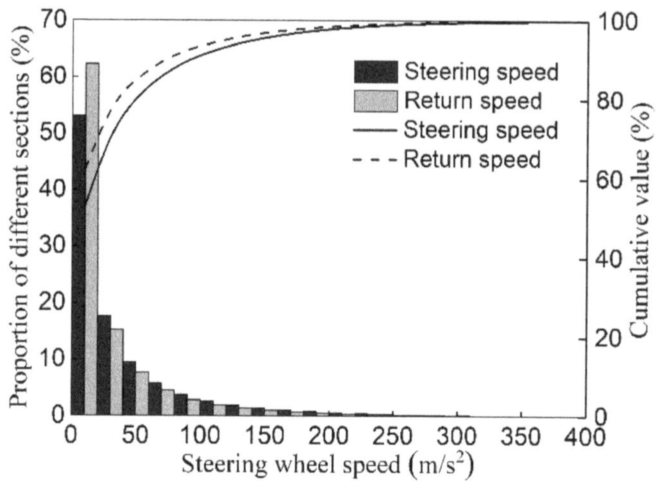

Figure 9. The distribution characteristic of steering wheel speed.

4.2. Statistical Characteristics of Parameters at Different Vehicle Speed

The acceleration, deceleration, steering wheel angle and steering wheel speed are affected by vehicle speed. To reveal the influence of speed on different driving parameters, the statistical characteristics of the parameters at different vehicle speeds were analysed. The analysis can provide guidance for threshold settings of driving behaviour evaluation.

Figure 10a showed the distribution characteristic of acceleration and deceleration versus vehicle speed. The acceleration decreased, along with the increase in vehicle speed. The deceleration firstly increased and then decreased with the increase in vehicle speed, and the inflection point was approximately 25 km/h. An explanation of this scenario is introduced in the following section. The overall change trend suggested that the acceleration and deceleration approached zero with the increase in vehicle speed. This can be explained by the fact that the higher vehicle speeds normally appeared during smooth traffic and the vehicle speed change will reduce accordingly in this situation. Additionally, the drivers tended to keep the speed stable to ensure safety at high vehicle speeds. Figure 10b showed the distribution comparison of acceleration and deceleration at different vehicle speeds. The acceleration distribution was relatively stable in a wide range of vehicle speeds from 0 to 60 km/h and the largest proportion of acceleration appeared around the speed of 20 km/h, which was the vehicle starting condition. The deceleration was mainly distributed in the lower speed condition and the largest proportion of deceleration appeared at speeds ranging from 0 to 5 km/h.

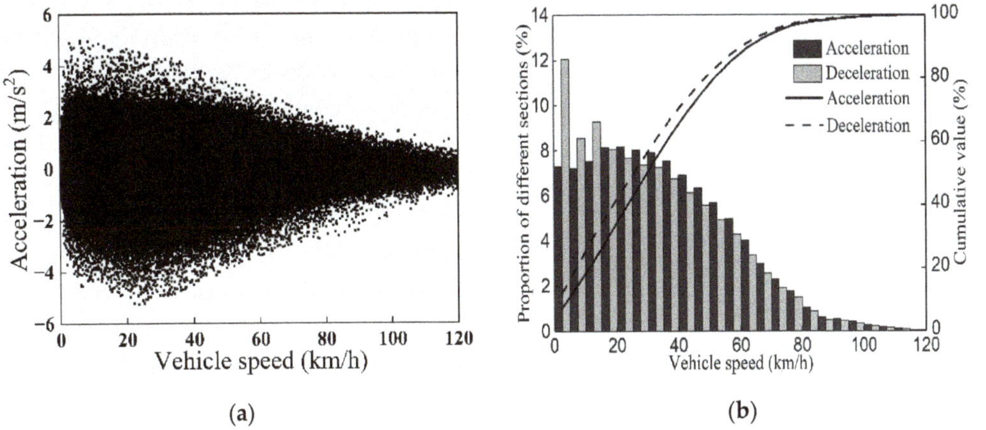

Figure 10. The distribution characteristic of acceleration and deceleration versus vehicle speed: (**a**) Overall trend; (**b**) Comparison of distribution characteristic.

Figure 11 showed the different quantile values of acceleration and deceleration at different vehicle speeds. It can be seen that the acceleration increased significantly during the speed from 0 to 5 km/h, which was attributed to the electric motors' high torque output at low speed. The quantile of the acceleration decreased with the increase in speed. This can be explained by the fact that the drivers tended to accelerate the vehicle more rapidly at lower speed conditions. The quantile of the deceleration firstly increased and then decreased. The reason for this is that the shared-electrical cars have braking energy recovery and the maximum braking torque appears at around 25 km/h, which led to this inflection point. The braking energy recovery in the deceleration phase was the main difference between the electrical vehicles and vehicles powered by Internal Combustion Engine (ICE).

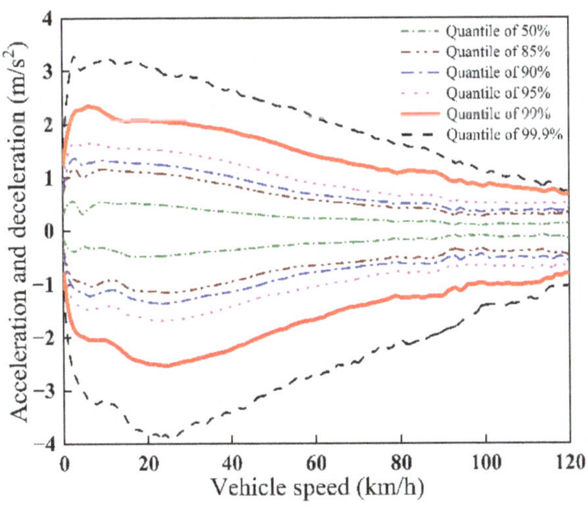

Figure 11. The comparison of different quantile value for acceleration and deceleration.

To evaluate the abnormal acceleration or deceleration, some researchers employed the fixed threshold value. According to Figure 11, both the acceleration and deceleration tended to decrease with the increase in vehicle speed in general. Therefore, it was reasonable that the threshold of abnormal acceleration or deceleration changed with speed. To confirm the threshold, a linear fit was applied to the quantile of acceleration and deceleration during the speed range from 10 to 120 km/h. The linear fit of the acceleration is expressed as Equation (6).

$$f(x) = \beta_1 x + \beta_2 \tag{6}$$

where x is vehicle speed and β_1 and β_2 are the coefficients of linear fit. The linear fit result is shown in Table 3, in which R^2 is the coefficient of determination.

Table 3. The linear fit result of the acceleration.

Coefficient	50%	85%	90%	95%	99%	99.9%
β_1	−0.0042	−0.0090	−0.0102	−0.0118	−0.0150	−0.0238
β_2	0.5312	1.2056	1.4002	1.7076	2.5000	3.5534
R^2	0.9123	0.9351	0.9386	0.9477	0.9708	0.9930

Similarly, the quantile of deceleration also had a linear fitted. Considering the piece-wise characteristic of the deceleration, the piece-wise linear fit was employed, which is shown in Equation (7).

$$f(x) = \begin{cases} \beta_1 x + \beta_2 x < 25\text{km/h} \\ \beta_3 x + \beta_4 x \geq 25\text{km/h} \end{cases} \tag{7}$$

where x is vehicle speed and β_1, β_2, β_3 and β_4 are the coefficients of linear fit. The linear fit result is shown in Table 4.

Table 4. The linear fit result of the deceleration.

Coefficient	50%	85%	90%	95%	99%	99.9%
β_1	−0.0087	−0.0128	−0.0146	−0.0182	−0.030	−0.0565
β_2	−0.2903	−0.8529	−1.0147	−1.2648	−1.8500	−2.6433
β_3	0.0047	0.0105	0.012	0.0144	0.0220	0.0301
β_4	−0.5456	−1.3544	−1.5801	−1.9741	−3.150	−4.57
R^2	0.8266	0.8263	0.8402	0.8838	0.9494	0.9220

The linear fit result of the acceleration and deceleration were compared in Figure 12. Greater acceleration and deceleration can cause a feeling of discomfort and risk of traffic accidents. According to the quantile and distribution of acceleration and deceleration, the quantile value of 99% was arbitrarily employed to distinguish the abnormal acceleration and deceleration in the following evaluation of driving behaviour.

Vehicle speed has a great influence on driving safety and it is dangerous to steer heavily at high speeds. Figure 13 showed the comparison of distribution and quantile of steering wheel angle at different speeds. Most of the steering action occurred at the speed of 20 km/h, and the steering wheel angle decreased with the increase in speed. The trend corresponded to the driving habits that the steering demand was decreased with the increasing speed to ensure driving safety. The quantile value of the steering wheel angle was almost unchanged when the steering wheel angle was larger than 300°. This indicated that drivers tended to decrease vehicle speed at high steering angle conditions to avoid traffic accidents.

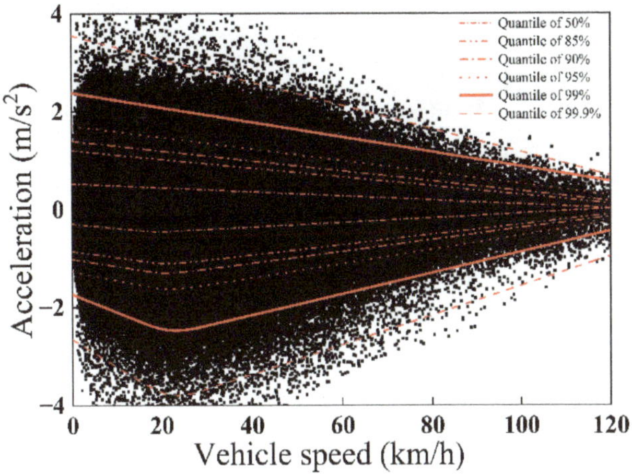

Figure 12. The linear result of quantile for acceleration and deceleration.

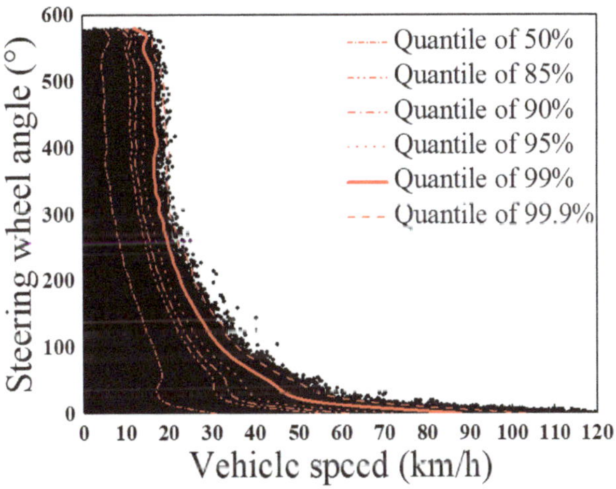

Figure 13. The comparison of distribution and quantile for steering wheel angle at different speeds.

Rapid steering can reduce the lateral stability of vehicles. Steering wheel speed reflects the operation speed of the drivers on the steering wheels. A faster speed is usually associated with a higher risk of rollover and skid of the vehicle. Figure 14 showed the distribution of steering wheel speed versus vehicle speed. The maximum steering wheel speed can reach approximately 600°/s. The high steering wheel speeds mainly occurred at vehicle speeds lower than 30 km/h, which suggested that drivers tended to turn the steering wheel rapidly at safe speeds. The trends of steering speed and return speed were similar. They both decreased with the increase in vehicle speed. Considering the similar trend, the quantiles of the steering speed was obtained. The comparison of the quantile showed that the steering speed was almost unchanged when the vehicle speed was higher than 60 km/h, which meant that most of the steering actions were gentle at high vehicle speeds to ensure driving safety.

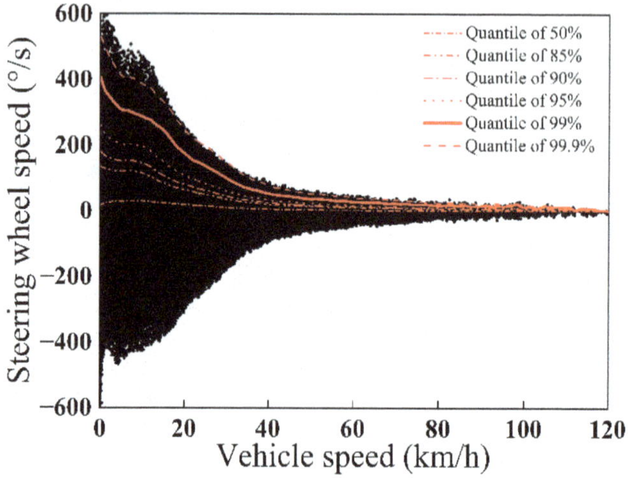

Figure 14. The comparison of distribution and quantile for steering wheel speed at different speeds.

The changing trend of the steering wheel speed was also exponentially fitted with vehicle speed to obtain the threshold of rapid turning. The exponential fitting equation is expressed as Equation (8).

$$f(x) = a + b \times x^c \tag{8}$$

where x is vehicle speed and a, b and c are the coefficients of exponential fit. The exponential fit result is shown in Table 5.

Table 5. The exponential fit result of the steering wheel speed.

Coefficient	50%	85%	90%	95%	99%	99.9%
a	0.97	5.82	7.64	9.83	19.56	3.80
b	−53.27	−209.30	−270.77	−373.06	−680.00	−692.02
c	0.948	0.940	0.936	0.936	0.935	0.949
R^2	0.9932	0.9982	0.9985	0.9985	0.9942	0.9935

Considering the influence of the speed on the safety of steering action, the evaluation of the rapid turning should consider the speed. According to the distribution and quantile of rapid turning, the quantile of 99% was selected as the threshold to distinguish the abnormal steering action, which is shown in Figure 15. The threshold of rapid turning decreased with the increase in vehicle and the setting method of the threshold can distinguish the rapid turning more effectively.

The previous analysis investigated the influence of vehicle speed on driving behaviours, such as acceleration, deceleration and steering action. According to the analysis, the threshold of aberrant driving behaviours were established to identify rapid acceleration, sudden braking, and rapid turning. In the next section, these thresholds were employed to calculate the number of occurrences for aberrant driving behaviour, which can be used to evaluate the driving behaviours.

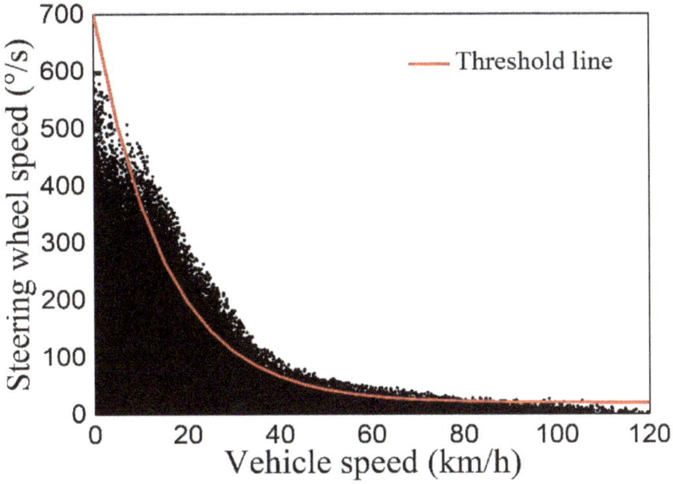

Figure 15. The distribution of steering wheel speed and threshold of abnormal steering action.

5. Evaluation Method of Driving Behaviour

5.1. Confirmation of Weight and Scoring Rule

According to the previous analysis, 11 indexes related to driving safety were utilized to evaluate driving behaviour. The evaluation result is influenced by the weight of indexes which can be confirmed by either the subjective or objective evaluation method. Subjective evaluation confirms the weight according to experience. Objective evaluation calculates the weight based on the relationship between the indexes. This study confirmed the weight of indexes by combining the subjective and objective evaluation method.

The comprehensive multi-index evaluation system was composed of two hierarchies. The first layer was the criterion layer, which consisted of three indexes: driving action, vehicle operation and fatigue driving/power consumption. The second layer was the index layer that consisted of the 11 indexes. The detailed index structure is shown in Table 6.

Table 6. The detailed index structure of driving behaviour evaluation.

Criterion Layer	Index Layer	Symbol of the Index
vehicle operation (u_1)	The number of rapid accelerations per 100 km	u_{11}
	The number of sudden brakings per 100 km	u_{12}
	The number of rapid turns per 100 km	u_{13}
	The number of speeding occurrences during steering per 100 km	u_{14}
driving action (u_2)	Standard deviation of vehicle speed	u_{21}
	Average value of acceleration	u_{22}
	Average value of deceleration	u_{23}
	Standard deviation of acceleration	u_{24}
	Standard deviation of deceleration	u_{25}
fatigue driving/power consumption (u_3)	Time of a driving event	u_{31}
	Power consumption per 100 km	u_{32}

Firstly, the weights of the indexes were preliminarily determined with the Analytic Hierarchy Process (AHP) method, which was a subjective evaluation. According to the experience, the judgement matrix was confirmed and used to indicate the relative importance degree of the indexes in the same layer. Assuming the index number of criterion layer m and the number of index layer n, the weight of indexes for the criterion layer ($AB_i(i = 1, 2, \ldots, m)$) and index layer ($AS_j(j = 1, 2, \ldots, n)$) can be calculated by Equation (9).

$$AB_i = \frac{\sqrt[m]{M_i}}{\sum\limits_{i=1}^{m} \sqrt[m]{M_i}} \quad AS_j = \frac{\sqrt[n]{N_j}}{\sum\limits_{j=1}^{n} \sqrt[n]{N_j}} \tag{9}$$

where M_i and N_j are the product of value in every row of the judgement matrix.

Secondly, the weight was further confirmed by the Entropy Weight Method (EWM), which was an objective evaluation. The EWM confirmed the weight of indexes according to the difference in the samples. Assuming the evaluation objectives are expressed as $A_i(i = 1, 2, \ldots, m)$, the indexes are as follows: $X_j(i = 1, 2, \ldots, n)$, AX'_{ij} is the original value of the jth index in the ith sample and AX_{ij} is the treated value by the Equation (10). The weight of index (EW_j) can be calculated by EWM method using Equation (11).

$$AX_{ij} = \frac{AX'_{ij} - \min(AX'_{ij})}{\max(AX'_{ij}) - \min(AX'_{ij})} \tag{10}$$

$$EW_j = \frac{1 + (1/\ln m) \sum\limits_{i=1}^{m} P_{ij} \ln P_{ij}}{\sum\limits_{j=1}^{n} (1 + (1/\ln m) \sum\limits_{i=1}^{m} P_{ij} \ln P_{ij})} \tag{11}$$

where P_{ij} is the weight of the j_{th} index in the i_{th} sample and this can be expressed as Equation (12).

$$P_{ij} = \frac{AX_{ij}}{\sum\limits_{i=1}^{m} AX_{ij}} \tag{12}$$

Finally, the comprehensive index weight (τ_j) can be calculated from AS_j and EW_j using Equation (13). Then, the normalisation can be applied to index layer, which is expressed as Ω_{ij}. The final weight (W_j) can be obtained from the product of Ω_{ij} and AB_i, as shown in Equation (14).

$$\tau_j = \frac{AS_j \cdot EW_j}{\sum\limits_{j=1}^{n} (AS_j \cdot EW_j)} \tag{13}$$

$$W_j = \frac{AB_i \cdot \Omega_{ij}}{\sum\limits_{i=1}^{m} (AB_i \cdot \Omega_{ij})} (i = 1, \ldots, m; j = 1, \ldots, n) \tag{14}$$

The judgement matrix was established based on the questionnaire introduced in previous studies [31–33], as shown in Table 7. The values in the matrix indicated the relative importance degree of the indexes. The final weights (W_j) were calculated by combining the AHP and EWM, which were shown in Table 8. It can be seen that the number of speeding occurrences during steering per 100 km had the maximum value of the weight and score. This suggests that the speeding during steering was the most dangerous behaviour among the selected indexes. Other indexes with higher weight or score were also risky in terms of causing traffic accidents.

Table 7. The judgement matrix for evaluation of the driving behaviour.

Criterion Layer	Index Layer
	$\begin{array}{ccccc} \mu_1 & \mu_{11} & \mu_{12} & \mu_{13} & \mu_{14} \\ \mu_{11} & 1 & 2 & 1/2 & 1/2 \\ \mu_{12} & 1/2 & 1 & 1/3 & 1/3 \\ \mu_{13} & 2 & 3 & 1 & 1 \\ \mu_{14} & 2 & 3 & 1 & 1 \end{array}$
$\begin{array}{cccc} \mu & \mu_1 & \mu_2 & \mu_3 \\ \mu_1 & 1 & 3 & 2 \\ \mu_2 & 1/3 & 1 & 1/2 \\ \mu_3 & 1/2 & 2 & 1 \end{array}$	$\begin{array}{cccccc} \mu_2 & \mu_{21} & \mu_{22} & \mu_{23} & \mu_{24} & \mu_{25} \\ \mu_{21} & 1 & 2 & 1 & 2 & 1 \\ \mu_{22} & 1/2 & 1 & 1/2 & 1 & 1/2 \\ \mu_{23} & 1 & 2 & 1 & 2 & 1 \\ \mu_{24} & 1/2 & 1 & 1/2 & 1 & 1/2 \\ \mu_{25} & 1 & 2 & 1 & 2 & 1 \end{array}$
	$\begin{array}{ccc} \mu_3 & \mu_{31} & \mu_{32} \\ \mu_{31} & 1 & 2 \\ \mu_{32} & 1/2 & 1 \end{array}$

Table 8. The final weight for different indexes.

Index	Weight (W_j)	Value (Percentage)
The number of rapid accelerations per 100 km	0.0858	9
The number of sudden brakings per 100 km	0.0677	7
The number of rapid turns per 100 km	0.1580	16
The number of speeding occurrences during steering per 100 km	0.2280	22
Standard deviation of vehicle speed	0.0428	4
Average value of acceleration	0.0233	2
Average value of deceleration	0.0206	2
Standard deviation of acceleration	0.0434	4
Standard deviation of deceleration	0.0333	4
Driving time per trip	0.1525	15
Power consumption per 100 km	0.1445	15

After the confirmation of the index weight, the scoring rule for different indexes was set to accurately evaluate the driving behaviour. The distribution of the 11 indexes was analysed and compared in Figure 16. Figure 16a shows the distribution of the number of rapid accelerations per 100 km. It can be seen that the number gradually decreased and in over 90% of cases, it was lower than 50 times. According to the distribution and the weight of the index, the scoring rule was set with a cut-off point of 50. The other scoring rules of the indexes were confirmed with the similar treatment by combining the distribution and the weight, and the results are shown in Table 9.

5.2. Proposed Driving Behaviour Evaluation Method

Based on the previous analysis, a driving behaviour evaluation method was proposed. This method included driving behaviour data acquisition, data pre−processing, recognition of the aberrant driving behaviour by statistics, index value calculation and output of the score. The flow chart of the evaluation process is shown in Figure 17.

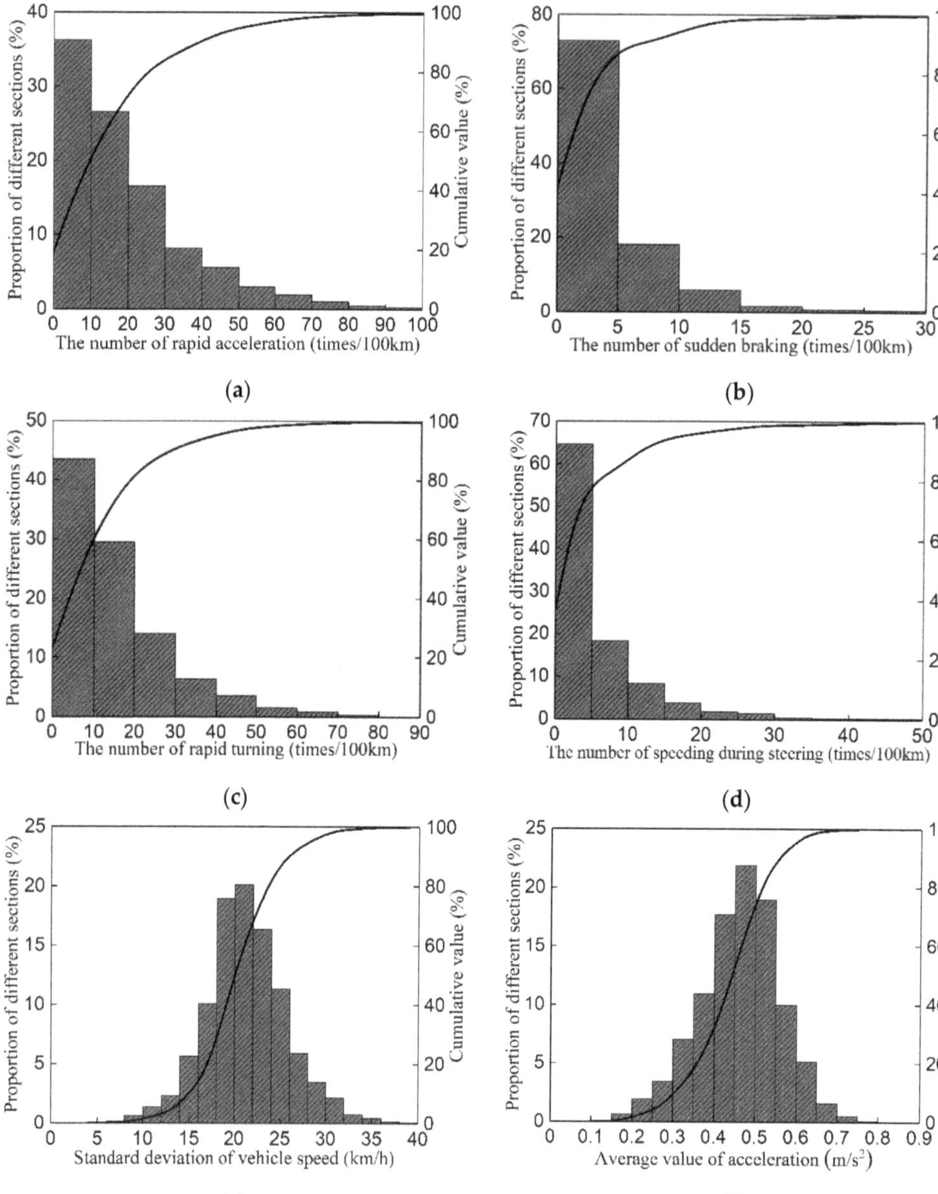

Figure 16. *Cont.*

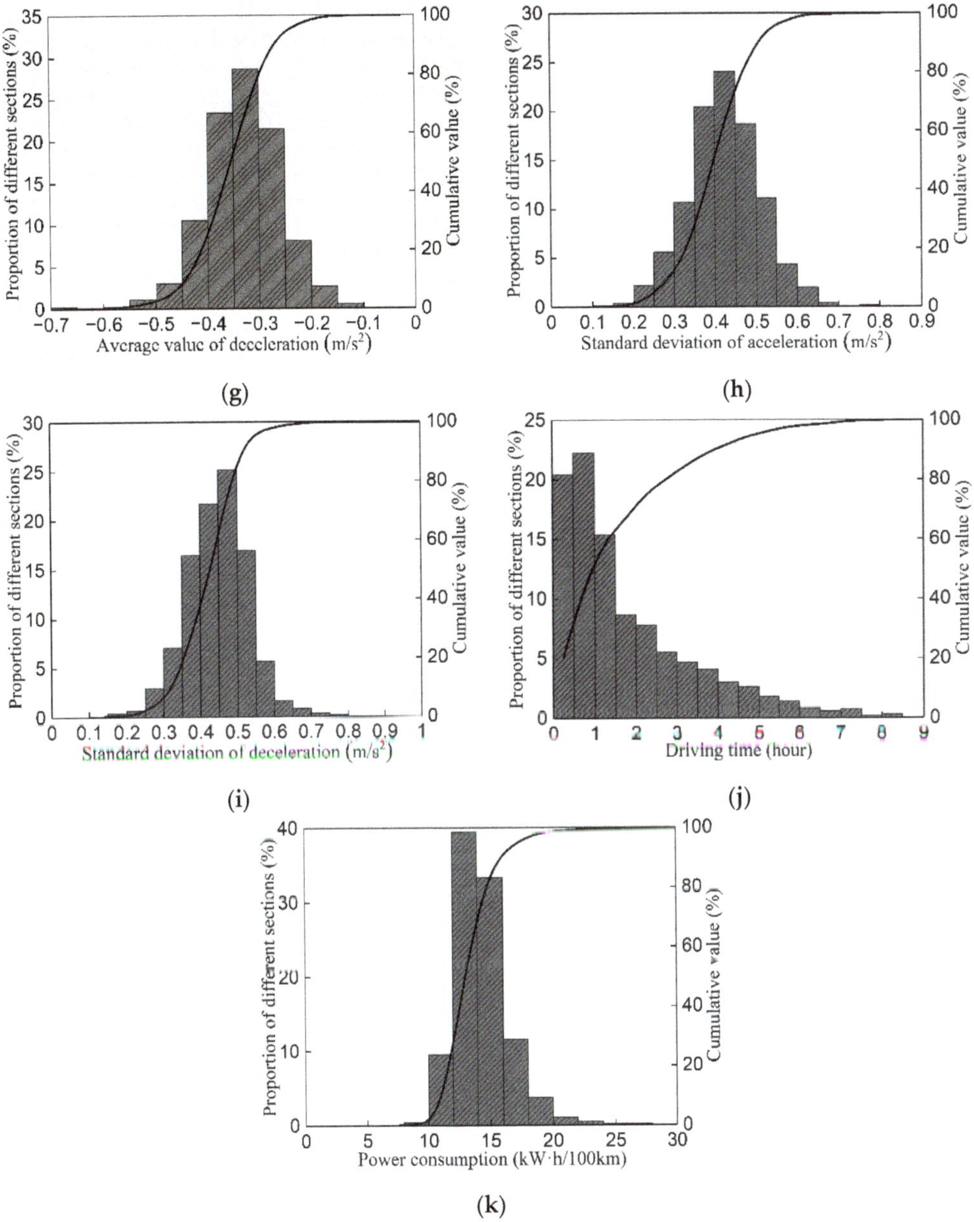

Figure 16. The distribution of different indexes: (**a**) The number of rapid accelerations per 100 km; (**b**) The number of sudden brakings per 100 km; (**c**) The number of rapid turns per 100 km; (**d**) The number of speeding occurences during steering per 100 km; (**e**) Standard deviation of vehicle speed; (**f**) Average value of acceleration; (**g**) Standard deviation of acceleration; (**h**) Average value of deceleration; (**i**) Standard deviation of deceleration; (**j**) Driving time per trip; (**k**) Power consumption.

Table 9. The scoring rule of different indexes.

Index	Unit	Score	Score Rule
The number of rapid acceleration per 100 km	times/100 km	9	$y = \begin{cases} 9 - 0.18x; 0 \leq x \leq 50 \\ 0; x > 50 \end{cases}$
The number of sudden braking per 100 km	times/100 km	7	$y = \begin{cases} 7 - 0.28x; 0 \leq x \leq 25 \\ 0; x > 25 \end{cases}$
The number of rapid turning per 100 km	times/100 km	16	$y = \begin{cases} 16 - 0.32x; 0 \leq x \leq 50 \\ 0; x > 50 \end{cases}$
The number of speeding during steering per 100 km	times/100 km	22	$y = \begin{cases} 22 - 0.88x; 0 \leq x \leq 25 \\ 0; x > 25 \end{cases}$
Standard deviation of vehicle speed	km/h	4	$y = \begin{cases} 4; 0 \leq x \leq 20 \\ 2; 20 < x \leq 35 \\ 0; x > 35 \end{cases}$
Average value of acceleration	m/s^2	2	$y = \begin{cases} 2; 0 \leq x \leq 0.5 \\ 1; 0.5 < x \leq 0.7 \\ 0; x > 0.7 \end{cases}$
Standard deviation of acceleration	m/s^2	4	$y = \begin{cases} 4; 0 \leq x \leq 0.4 \\ 2; 0.4 < x \leq 0.8 \\ 0; x > 0.8 \end{cases}$
Average value of deceleration	m/s^2	2	$y = \begin{cases} 2; -0.35 \leq x \leq 0 \\ 1; -0.5 \leq x < 0.35 \\ 0; x < -0.5 \end{cases}$
Standard deviation of deceleration	m/s^2	4	$y = \begin{cases} 4; 0 \leq x \leq 0.4 \\ 2; 0.4 < x \leq 0.8 \\ 0; x > 0.8 \end{cases}$
Driving time per trip	hour	15	$y = \begin{cases} 15; t \leq 4 \\ 5; 4 < t < 8 \\ 0; t \geq 8 \end{cases}$
Power consumption	kW·h/100 km	15	$y = \begin{cases} 15; 0 < x \leq 15 \\ 15 - 3(x - 15); 15 < x \leq 20 \\ 0; x > 20 \end{cases}$

To verify the effectiveness of the evaluation method, thirty drivers were selected from the online car–hiring. The driving behaviour data were selected in one month with 2227 driving events. The scoring result for each driving event was calculated and compared in Figure 18. Most of the scoring results were located from 70 to 100, which indicates that most of the driver behaviours were cautious. There were also some scoring results lower than 60, which can be considered as unsafe driving behaviours. To quantitatively evaluate the driving behaviour of a driver, the average score for every driver in a month was obtained by averaging the score of each driving event, which is shown in Figure 18b. The comparison result showed that the score difference was considerable for each driver. The evaluation result suggested that the scores of the driver number 2 and 14 were relatively low.

In order to investigate the reasons for the low scores of drivers 2 and 14, the score of the four indexes were compared in Figure 19. The four indexes were the number of rapid accelerations per 100 km, the number of sudden brakings per 100 km, the number of rapid turns per 100 km and the number of speeding occurrences during steering per 100 km, which were closely related with aberrant driving behaviour. The scores of drivers 2 and 14 were also compared with the average scores of the thirty drivers. It can be seen that the scores of the selected four indexes for drivers 2 and 14 were obviously lower than the average scores. Additionally, the score of driver number 14 was lower than that of driver number 2. A lower score meant more aberrant driving behaviour, which showed the effectiveness of the evaluation method. The scoring result can be an important reference for the passenger transport company to distinguish the driver with abnormal or dangerous driving behaviour. The driving behaviour scores can also be an important basis for evaluation for the insurance company to collect the premium.

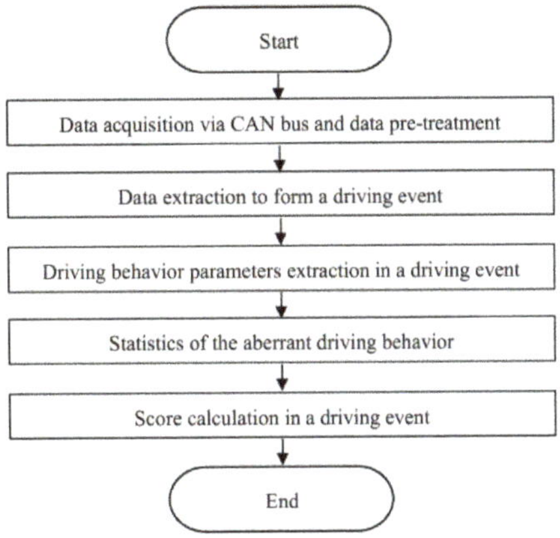

Figure 17. The evaluation flow chart of driving behaviour.

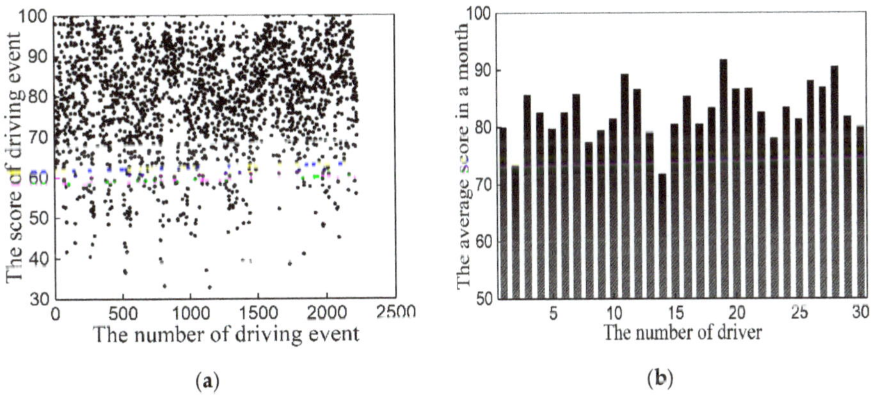

Figure 18. The evaluation result of driving behaviour: (**a**) The score distribution of driving events; (**b**) The score for every driver.

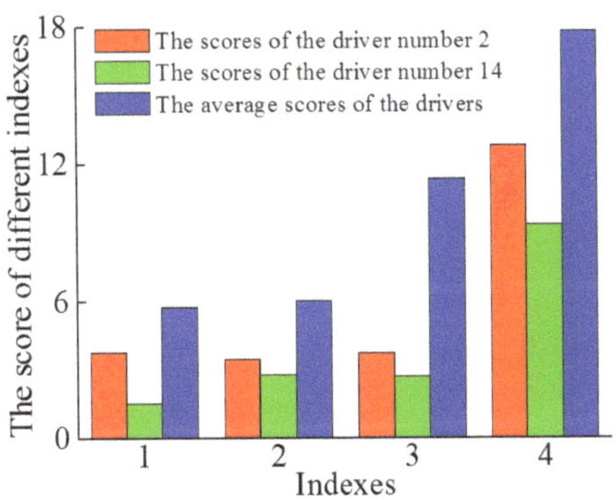

Figure 19. The comparison of scores for four indexes (numbers from 1 to 4 represent the number of rapid accelerations per 100 km, the number of sudden brakings per 100 km, the number of rapid turns per 100 km and the number of speeding occurrences during steering per 100 km, respectively).

6. Conclusions

This study proposed a quantitative evaluation method of driving behaviour based on NDS data collected from shared-electrical cars and online car-hiring services. Data acquisition, treatment method and data volume verification were analysed to ensure the effectiveness of the dataset. The distribution characteristics of the main driving behaviour parameters were studied. On this basis, the evaluation method was proposed and verified. The main conclusions were shown as follows. The main differences between the electrical vehicles and vehicles powered by ICE were in the deceleration phase. There was braking energy recovery for the electrical vehicles. Therefore, the conclusion can be basically applied to the ICE vehicles.

1. The NDS data were collected from the OBD−II interface via CAN bus with the rate of 10 Hz. This sampling frequency satisfies the requirement of transient process analysis. The sliding-window averaging filter and the box diagram method were used to improve the data quality. Eleven indexes were selected to evaluate the driving behaviour, including vehicle running data, driver operation data and power consumption of the vehicles.
2. *KL* divergence was applied to confirm the appropriate data quantity for the driving behaviour analysis. The result showed that the minimum data quantity for vehicle speed, acceleration, steering wheel angle and steering wheel speed were 20×10^5, 63×10^5, 10×10^5, 231×10^5, respectively, with the variation value of *KL* lower than 1×10^5.
3. The changing trend of acceleration and deceleration, steering wheel angle and steering wheel speed versus vehicle speed were compared. Based on the distribution characteristics, the thresholds of aberrant driving were determined in correlation with vehicle speed to enhance the recognition accuracy of the aberrant driving behaviour. The thresholds can be used to evaluate the aberrant driving behaviour.
4. The weights for the 11 indexes were obtained by combining the AHP and EWM methods. The scoring rules of the 11 indexes were confirmed based on the distribution of the indexes. An evaluation method of driving behaviour was proposed and verified according to the driving behaviour data of the car-hiring driver.

In future research, more parameters will be considered including, road slope, weather, driving experience and so on. Additionally, other useful sensors will be used to obtain more driving behaviour parameters. On this basis, the scoring rule will be further optimized and the accuracy of the evaluation method can be subsequently improved.

Author Contributions: Formal analysis, X.L.; methodology, Y.C.; software, K.Z. and S.S.; validation, Z.Y.; writing—original draft, S.J.; writing—review and editing, G.T. All authors have read and agreed to the published version of the manuscript.

Funding: This research was funded by National Key Research and Development Program, grant number 2020YFB1600501; Major Technology Innovation Projects of Shandong Province, grant number 2019TSLH0203; Natural Science Foundation of Shandong Province, grant number ZR2020ME180; National Natural Science Foundation of China, grant number 51976107.

Institutional Review Board Statement: Ethical review and approval were waived for this study due to the reason that the driving behaviour data in this study was obtained from the vehicle operation data and did not involve any personal information, such as GPS data and so on.

Informed Consent Statement: Informed consent was obtained from all subjects involved in the study.

Data Availability Statement: The data were obtained from the OBD−II interface, which only contained the vehicle and components operation data. The driving behaviour data were obtained from the vehicle operation data.

Conflicts of Interest: The authors declare that they have no known competing financial interest or personal relationships that could have appeared to influence the work reported in this paper.

Abbreviations

AHP	Analytic Hierarchy Process
CAN	Controller Area Network
ECU	Electronic Control Units
EWM	Entropy Weight Method
GPS	Global Positioning System
ICE	Internal Combustion Engine
IMU	Inertial Measurement Unit
KL	Kullback–Leibler
NDS	Naturalistic Driving Study
OBD	On-Board Diagnostics

References

1. Singh, H.; Kathuria, A. Analyzing driver behavior under naturalistic driving conditions: A review. *Accid. Anal. Prev.* **2021**, *150*, 105908. [CrossRef] [PubMed]
2. Singh, S. *Critical Reasons for Crashes Investigated in the National Motor Vehicle Crash Causation Survey*; National Center for Statistics and Analysis: Washington, DC, USA, 2015.
3. Tanvir, S.; Chase, R.T.; Roupahil, N.M. Development and analysis of eco-driving metrics for naturalistic instrumented vehicles. *J. Intell. Transp. Syst.* **2021**, *25*, 235–248. [CrossRef]
4. Ersan, Ö.; Üzümcüoğlu, Y.; Azık, D.; Fındık, G.; Kaçan, B.; Solmazer, G.; Özkan, T.; Lajunen, T.; Öz, B.; Pashkevich, A.; et al. The relationship between self and other in aggressive driving and driver behaviors across countries. *Transp. Res. Part F Traffic Psychol. Behav.* **2019**, *66*, 122–138. [CrossRef]
5. Deng, Z.; Chu, D.; Wu, C.; He, Y.; Cui, J. Curve safe speed model considering driving style based on driver behaviour questionnaire. *Transp. Res. Part F Traffic Psychol. Behav.* **2019**, *65*, 536–547. [CrossRef]
6. Han, H.; Kim, S.; Choi, J.; Park, H.; Yang, J.H.; Kim, J. Driver's avoidance characteristics to hazardous situations: A driving simulator study. *Transp. Res. Part F Traffic Psychol. Behav.* **2021**, *81*, 522–539. [CrossRef]
7. Papazikou, E.; Thomas, P.; Quddus, M. Developing personalised braking and steering thresholds for driver support systems from SHRP2 NDS data. *Accid. Anal. Prev.* **2021**, *160*, 106310. [CrossRef]
8. Akamatsu, M.; Green, P.; Bengler, K. Automotive technology and human factors research: Past, present, and future. *Int. J. Veh. Technol.* **2013**, *2013*, 1–27. [CrossRef]
9. Precht, L.; Keinath, A.; Krems, J.F. Effects of driving anger on driver behavior—Results from naturalistic driving data. *Transp. Res. Part F Traffic Psychol. Behav.* **2017**, *45*, 75–92. [CrossRef]
10. Ellison, A.B.; Greaves, S.P.; Bliemer, M.C.J. Driver behaviour profiles for road safety analysis. *Accid. Anal. Prev.* **2015**, *76*, 118–132. [CrossRef]
11. Shridhar Bokare, P.; Kumar Maurya, A. Study of effect of speed, acceleration and deceleration of small petrol car on its tail pipe emission. *Int. J. Traffic Transp. Eng.* **2013**, *3*, 465–478. [CrossRef]
12. Sun, Q.; Xia, J.; Nadarajah, N.; Falkmer, T.; Foster, J.; Lee, H. Assessing drivers' visual-motor coordination using eye tracking, GNSS and GIS: A spatial turn in driving psychology. *J. Spat. Sci.* **2016**, *61*, 299–316. [CrossRef]
13. Sheykhfard, A.; Haghighi, F. Performance analysis of urban drivers encountering pedestrian. *Transp. Res. Part F Traffic Psychol. Behav.* **2019**, *62*, 160–174. [CrossRef]
14. Bokare, P.S.; Maurya, A.K. Acceleration-deceleration behaviour of various vehicle types. *Transp. Res. Procedia* **2017**, *25*, 4733–4749. [CrossRef]
15. Chen, R.; Kusano, K.D.; Gabler, H.C. Driver behavior during overtaking maneuvers from the 100-car naturalistic driving study. *Traffic Inj. Prev.* **2015**, *16*, S176–S181. [CrossRef]
16. Mahapatra, G.; Maurya, A.K. Study of vehicles lateral movement in non-lane discipline traffic stream on a straight road. *Procedia Soc. Behav. Sci.* **2013**, *104*, 352–359. [CrossRef]
17. Papadimitriou, E.; Argyropoulou, A.; Tselentis, D.I.; Yannis, G. Analysis of driver behaviour through smartphone data: The case of mobile phone use while driving. *Saf. Sci.* **2019**, *119*, 91–97. [CrossRef]
18. Saleh, K.; Hossny, M.; Nahavandi, S. Driving behavior classification based on sensor data fusion using LSTM recurrent neural networks. In Proceedings of the 2017 IEEE 20th International Conference on Intelligent Transportation Systems (ITSC), Yokohama, Japan, 16–19 October 2017.
19. Sangster, J.; Rakha, H.; Du, J. Application of naturalistic driving data to modeling of driver car-following behavior. *Transp. Res. Rec.* **2013**, *2390*, 20–33. [CrossRef]
20. Jachimczyk, B.; Dziak, D.; Czapla, J.; Damps, P.; Kulesza, W. IoT on-board system for driving style assessment. *Sensors* **2018**, *18*, 1233. [CrossRef]
21. Mayhew, D.R.; Simpson, H.M. The safety value of driver education and training. *Inj. Prev. J. Int. Soc. Child Adolesc. Inj. Prev.* **2002**, *8* (Suppl. 2), ii3–ii8.

22. Zhao, Y.; Yamamoto, T.; Morikawa, T. An analysis on older driver's driving behavior by GPS tracking data: Road selection, left/right turn, and driving speed. *J. Traffic Transp. Eng.* **2018**, *5*, 56–65. [CrossRef]
23. Feng, F.; Bao, S.; Sayer, J.R.; Flannagan, C.; Manser, M.; Wunderlich, R. Can vehicle longitudinal jerk be used to identify aggressive drivers? An examination using naturalistic driving data. *Accid. Anal. Prev.* **2017**, *104*, 125–136. [CrossRef] [PubMed]
24. Das, A.; Ghasemzadeh, A.; Ahmed, M.M. Analyzing the effect of fog weather conditions on driver lane-keeping performance using the SHRP2 naturalistic driving study data. *J. Saf. Res.* **2019**, *68*, 71–80. [CrossRef] [PubMed]
25. Kong, X.; Das, S.; Jha, K.; Zhang, Y. Understanding speeding behavior from naturalistic driving data: Applying classification based association rule mining. *Accid. Anal. Prev.* **2020**, *144*, 105620. [CrossRef] [PubMed]
26. Morgenstern, T.; Schott, L.; Krems, J.F. Do drivers reduce their speed when texting on highways? A replication study using European naturalistic driving data. *Saf. Sci.* **2020**, *128*, 104740. [CrossRef]
27. Hallmark, S.L.; Tyner, S.; Oneyear, N.; Carney, C.; McGehee, D. Evaluation of driving behavior on rural 2-lane curves using the SHRP 2 naturalistic driving study data. *J. Saf. Res.* **2015**, *54*, 17.e1–27. [CrossRef]
28. Ji, S.; Chen, Q.; Shu, M.; Tian, G.; Liao, B.; Lv, C.; Li, M.; Lan, X.; Cheng, Y. Influence of operation management on fuel consumption of coach fleet. *Energy* **2020**, *203*, 117853. [CrossRef]
29. Sagberg, F.; Selpi; Bianchi Piccinini, G.F.; Engström, J. A review of research on driving styles and road safety. *Hum. Factors* **2015**, *57*, 1248–1275. [CrossRef]
30. Wang, W.; Liu, C.; Zhao, D. How much data is enough? A statistical approach with case study on longitudinal driving behavior. *IEEE Trans. Intell. Veh.* **2017**, *2*, 85–98. [CrossRef]
31. Wang, Z. Research on Vehicle Insurance Pricing Based on UBI Driving Behaviour Score. Master's Thesis, Hunan University, Hunan, China, 2019.
32. Peng, J. Research on Usage-Based Insurance Premiums and Driving Behaviour Scoring Based on GID. Master's Thesis, Nanjing University of Posts and Telecommunications, Nanjing, China, 2016.
33. Tselentis, D.I.; Yannis, G.; Vlahogianni, E.I. Innovative motor insurance schemes: A review of current practices and emerging challenges. *Accid. Anal. Prev.* **2017**, *98*, 139–148. [CrossRef]

sustainability

MDPI

Article

A High-Gain DC Side Converter with a Ripple-Free Input Current for Offshore Wind Energy Systems

Ran Tao [1], Jingpeng Yue [1], Zhenlin Huang [1], Ranran An [1], Zou Li [2] and Junfeng Liu [2,*]

1 Guangdong Key Laboratory of New Technology for Smart Grid, Guangzhou 510062, China
2 School of Automation Science and Engineering, South China University of Technology, Guangzhou 510641, China
* Correspondence: jf.liu@connect.polyu.hk

Abstract: Considering that the distance between offshore wind farms and onshore converters is getting farther and farther, dc transmission becomes increasingly more applicable than conventional ac transmission. To reduce the transmission loss, a feasible solution is using a high-gain dc/dc converter to boost the rectified output voltage to thousands of volts. Thus, a novel single-switch high-gain dc/dc converter with a ripple-free input current is presented in this paper. The structure consists of two cells—a coupled-inductor cell and a switched-capacitor cell. The coupled-inductor cell in the proposed converter provides a ripple-free input current. The switched-capacitor cell provides a high voltage gain. The converter has a simple control strategy due to the use of a single switch. Moreover, the output capacitor is charged and discharged continuously by a 180° phase shift to eliminate the output voltage ripple. A steady-state analysis of the converter is proposed to determine the parameters of the devices. In addition, a 240 W, 40/308 V laboratory prototype at 35 kHz switching frequency has been developed, in which the input current ripple is only 1.1% and a peak efficiency of 94.5% is reached. The experimental results verify the validity and feasibility of the proposed topology.

Keywords: dc/dc converter; high voltage gain; ripple-free input current; offshore wind farms

Citation: Tao, R.; Yue, J.; Huang, Z.; An, R.; Li, Z.; Liu, J. A High-Gain DC Side Converter with a Ripple-Free Input Current for Offshore Wind Energy Systems. *Sustainability* **2022**, *14*, 11574. https://doi.org/10.3390/su141811574

Academic Editors: Wenbin Yu and Guang Zeng

Received: 19 August 2022
Accepted: 8 September 2022
Published: 15 September 2022

1. Introduction

With the beginning of global carbon neutrality, offshore wind energy has become one of the main and growing sources of renewable energy worldwide [1–3]. The European Commission stated that the offshore wind power capacity in Europe would reach 450 GW by 2050, making it a key part of renewable energy [4]. Compared with its onshore counterpart, an offshore wind farm has the merits of less land occupation, higher wind speeds, and more stable wind conditions [5–7]. However, there are some problems that need to be solved, such as the difficulties of installation and maintenance [8,9]. Once an accident occurs, the long time for fault correction will have an adverse impact on the continuous power supply. Moreover, with the increase in offshore distance, conventional high voltage ac (HVAC) transmission is no longer suitable for long-distance offshore wind farms, as it brings higher power loss and significant power fluctuation [10–12].

Considering the above problems, high voltage dc (HVDC) transmission appears to be a more promising solution for long-distance and large-scale offshore wind farms [13–15]. The traditional HVAC transmission system of an offshore wind farm consists of a medium voltage ac (MVAC) collection grid as shown in Figure 1. Each wind turbine is connected to a transformer to boost the turbine's output voltage. To avoid the use of large volume transformers, a HVDC transmission system uses a medium voltage dc (MVDC) collection grid as shown in Figure 2 [16–18], where the traditional MV transformers are replaced by MV step-up dc/dc converters. The use of MV dc/dc converters can significantly reduce the volume and weight of the offshore platforms which leads to lower installation costs. Meanwhile, due to the low output voltage generated by wind turbines, high-gain dc/dc converters become one of the key levels of MV dc collection grids [19–21]. In the existing research, there are bidirectional and unidirectional high-gain dc/dc converters. However, there is no need for bidirectional power flow capacity due to the inherent characteristics of offshore wind farms, so a simpler unidirectional dc/dc converter is more applicable for offshore wind energy systems [22]. In addition to a high voltage gain, there are some other challenges such as low input current ripple, high conversion efficiency, and high-power density. To overcome these challenges, a large amount of relevant research has been done.

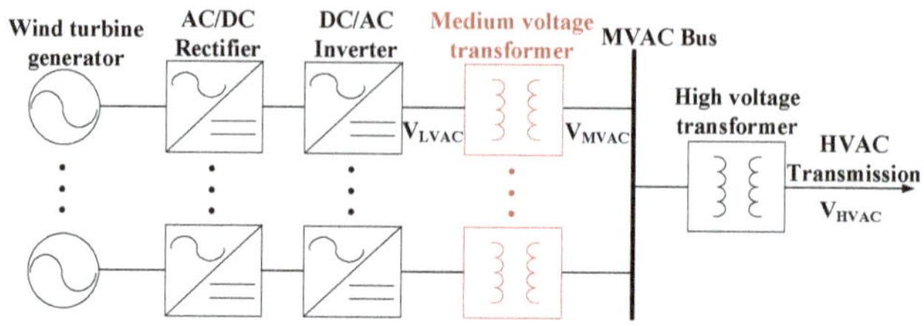

Figure 1. HVAC transmission system with MVAC grid.

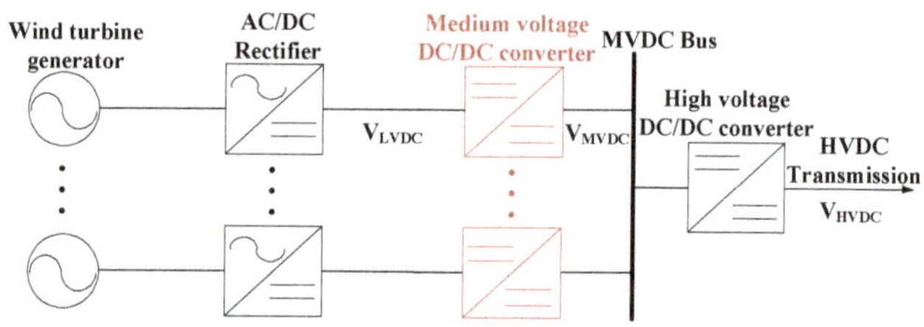

Figure 2. HVDC transmission system with MVDC grid.

To reduce the power loss and maintain high efficiency, researchers have generated much interest in high-gain dc/dc converters for offshore wind farms. In [23], a high-gain resonant switched-capacitor (RSC) dc/dc converter was introduced, which provided low switching losses and high efficiency by the resonant switching transitions. In addition, the voltage gain was increased through the series-modular configuration. However, to reach a high voltage gain, lots of switching devices and passive components were used in the topology. Meanwhile, the voltage stress increment of the switches and diodes blocked its application in offshore wind energy systems. To reduce the number of power switches, [24,25] both presented step-up dc/dc converters which worked only by one switch. The control of the converters was easy, and the conduction loss of the switch was decreased due to the use of a single-switch structure. But the voltage gain was not high enough. In [26], a high-power multilevel step-up dc/dc converter was studied with the merits of outstanding dynamic performance and low voltage stress. Although there was a filter inductor in this converter that could reduce the input current ripple, the step-up ratio was not high enough for offshore wind farms. Moreover, the concept of modularization used in offshore wind farms has attracted considerable interests recently due to its high reliability and excellent expandability [27–30]. Nevertheless, a complex switching scheme is usually required in the modular structure. Furthermore, a large number of power devices connected in series increase the volume and weight of the offshore platforms and may raise the overall costs.

In order to ensure the long-distance and stable transmission of electricity, a ripple-free input current is necessary for a high-gain dc/dc converter. Generally, bulky and huge electrolytic capacitors are used at the input stage of the dc/dc converters to decrease the large input current ripple. To avoid the use of bulky electrolytic capacitors, a filter inductor is placed at the input stage to reduce the current ripple [31]. However, the current ripple still exists, and the effect is not ideal. By utilizing the interleaved structure of switches and inductors on the input side, the input current ripple was significantly reduced in [32]. Nevertheless, the switching scheme is relatively complex and the input direct current still consists of a little ripple. An improved dc/dc converter with a ripple-free input current was proposed in [33]. The input current ripple was reduced to zero with the use of the coupled inductor. Moreover, the converter can reach a high voltage gain through the use of a transformer. However, the use of too many magnetic components also limits its application in offshore wind farms due to the increase in volume and weight. Nonetheless, the coupled inductor is a promising component to achieve a ripple-free input current.

Considering the above problems, this paper presents a single-switch high-gain dc/dc converter with a ripple-free input current. The proposed converter combines the coupled inductor with the switch-capacitor structure and has the following features by comparing with the existing converters: (1) high voltage gain; (2) ripple-free input current; (3) simplicity of control strategy; (4) low voltage stress across the components; (5) high efficiency. Given all of the advantages, the converter is very suitable for offshore wind energy systems.

The rest of this paper is organized as follows. The proposed topology and operating principle are discussed in Section 2. The detailed steady-state analysis is presented in Section 3. The performance comparisons of the converters are provided in Section 4. Section 5 illustrates the parameters design and the selection of the components. Experimental results are shown in Section 6. Finally, conclusions are drawn in Section 7.

2. Topology and Operating Principle

2.1. Topology

Figure 3a shows the proposed single-switch dc/dc converter with a high voltage gain and a ripple-free input current. A coupled inductor L_C is inserted at the input stage to eliminate the input current ripple. Figure 3b shows the equivalent circuit of the converter where the coupled inductor L_C is described by the magnetic inductor L_m, the leakage inductor L_k, and the ideal transformer (turns ratio $n = N_s/N_p$). The switched-capacitor cell uses only one switch with a new arrangement of the diodes and capacitors to raise the voltage gain significantly. The control strategy is very simple due to the use of one switch which can reduce the incidence of failure in offshore wind energy systems. The theoretical waveforms of the main devices are shown in Figure 4. The voltage at both ends of the primary side and the secondary side of the coupled inductor are defined as v_p and v_s, respectively. The analysis of the converter during a switching period T_S can be divided into two operating modes, and they are shown in Figure 5.

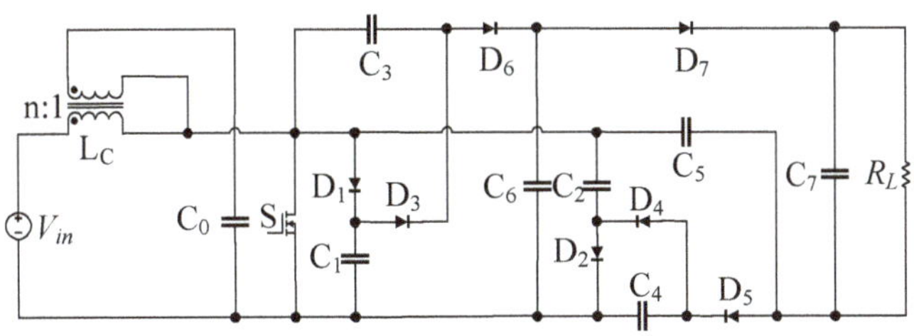

(a)

Figure 3. *Cont.*

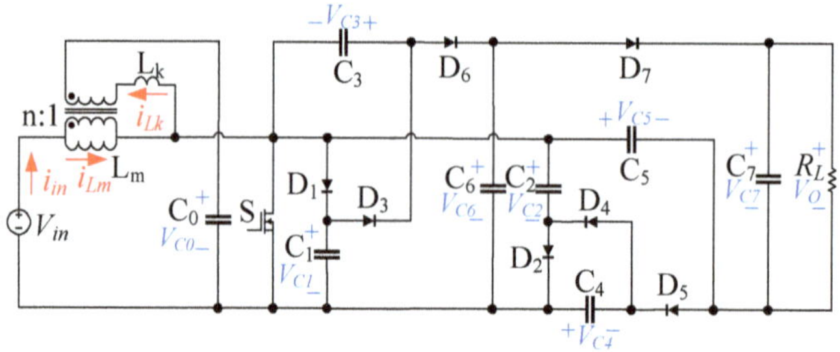

(b)

Figure 3. Proposed dc/dc converter: (**a**) Topology; (**b**) Equivalent circuit.

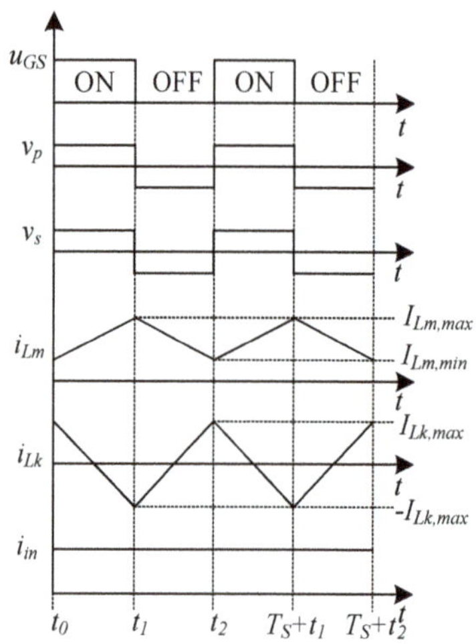

Figure 4. Theoretical waveforms of the proposed converter.

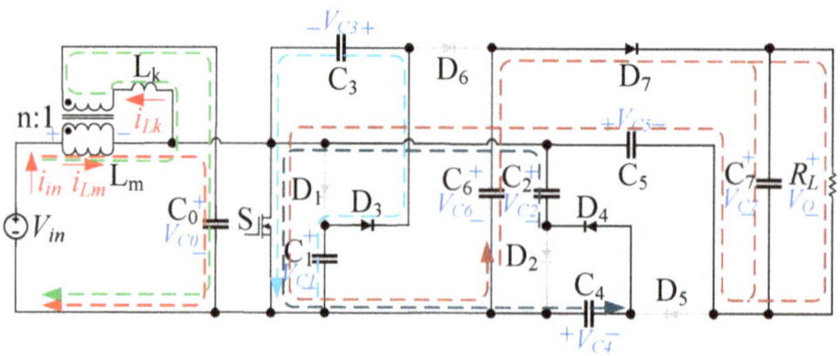

(a)

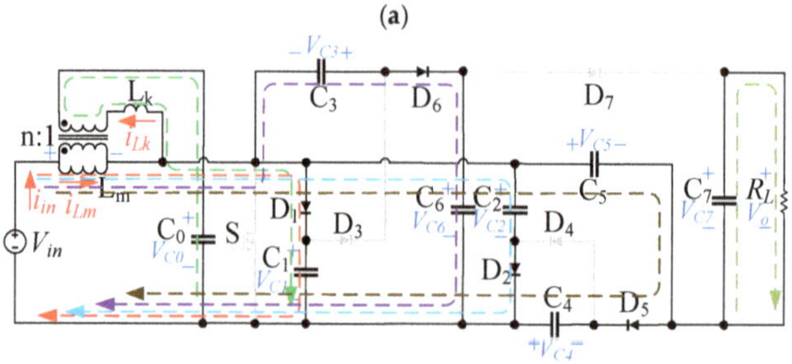

(b)

Figure 5. Operating modes of the proposed converter: (**a**) Mode 1; (**b**) Mode 2.

2.2. Operating Principle

Since the circuit is controlled by one switch, there are only two operating modes during a switching period. The following assumptions are made before the analysis.

(1) All the switches, capacitors, diodes, and inductors used in the circuit are assumed to be ideal components;
(2) All the capacitors are large enough to maintain output voltage constant;
(3) V_{in} is an ideal dc voltage source, and the load is modeled by a pure resistor R_L.

Mode 1 $[t_0,t_1]$ in Figure 5a: In this mode, the switch S begins to conduct under the action of the gate driving signal. The current i_{Lm} increases linearly from its minimum value due to the positive voltage. The diodes D_1, D_2, D_5, and D_6 are reverse biased. The magnetic inductor L_m starts to be charged by the input voltage source V_{in}. C_1 is discharged to C_3 through diode D_3 and C_2 is discharged to C_4 through diode D_4. Meanwhile, C_5 and C_6 are discharged in series to C_7 and R_L through diode D_7. Figure 6 shows the simplified equivalent circuits in this mode.

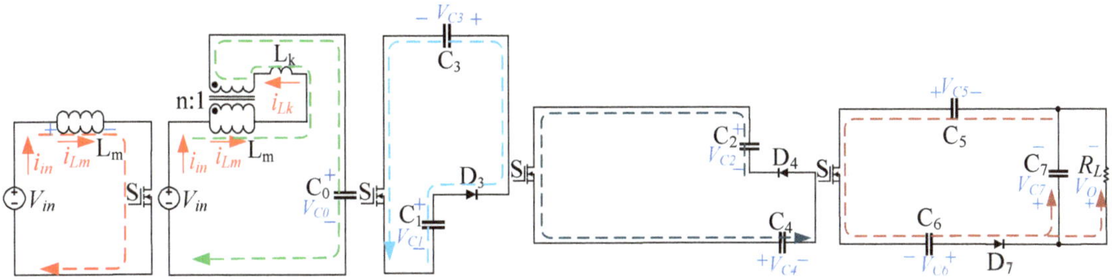

Figure 6. Simplified equivalent circuits of Mode 1.

Here, the voltages across the capacitors, diodes, inductor, and load are defined as V_{C0}–V_{C7}, V_{D1}–V_{D7}, v_{Lm}, v_{Lk}, and V_O, respectively. Similarly, the currents flowing through the input voltage source, magnetic inductor, and leakage inductor are defined as i_{in}, i_{Lm}, and i_{Lk}, respectively.

Mode 2 $[t_1,t_2]$ in Figure 5b: At t_1, the switch S is turned off. The diodes D_3, D_4, and D_7 are reverse biased. The input voltage source V_{in} and magnetic inductor L_m are discharged in series to C_1 and C_2, respectively. Meanwhile, the input voltage source V_{in}, magnetic inductor L_m, and C_4 are discharged in series to C_5 through diode D_5. The input voltage source V_{in}, magnetic inductor L_m, and C_3 are discharged in series to C_6 through diode D_6. Therefore, the current i_{Lm} starts to decrease linearly from its maximum value, and the output capacitor C_7 is discharged to the load R_L. This mode ends when the driving signal of the switch S comes in the next period. Figure 7 shows the simplified equivalent circuits in this mode.

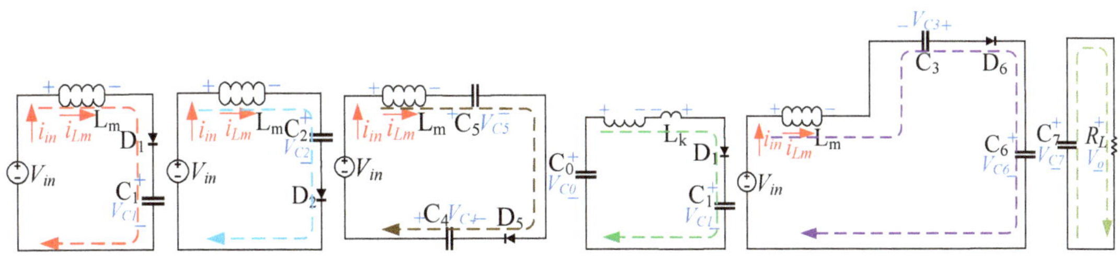

Figure 7. Simplified equivalent circuits of Mode 2.

3. Steady-State Analysis

3.1. Voltage Gain

Referring to Figures 5a and 6, when the switch is turned on, the following equations can be obtained according to Kirchhoff's voltage law, where $v_{Lm(on)}$ is the voltage across the magnetic inductor when the switch is turned on.

$$\begin{cases} v_{\mathrm{Lm}(on)} = V_{in} \\ V_{C3} = V_{C1} \\ V_{C4} = V_{C2} \\ V_{C5} + V_{C6} = V_{C7} = V_O \end{cases} \tag{1}$$

Similarly, from Figures 5b and 7, the voltage relationship of each loop can be expressed as follows, where $v_{Lm(off)}$ is the voltage across the magnetic inductor when the switch is turned off.

$$\begin{cases} V_{in} - v_{\mathrm{Lm}(off)} = V_{C1} = V_{C2} \\ V_{in} - v_{\mathrm{Lm}(off)} + V_{C4} = V_{C5} \\ V_{in} - v_{\mathrm{Lm}(off)} + V_{C3} = V_{C6} \\ V_O = V_{C7} \end{cases} \tag{2}$$

By applying the volt-second balance principle to the magnetic inductor L_m under steady-state conditions, the equation is given below:

$$\int_{t_0}^{t_0+T_S} v_{Lm} dt = 0 \tag{3}$$

where v_{Lm} is the voltage across the magnetic inductor.

From (1)–(3), the steady-state voltage expressions of the capacitors and the load are given below, where D is the duty cycle.

$$\begin{cases} V_{C1} = V_{C2} = V_{C3} = V_{C4} = \frac{V_{in}}{1-D} \\ V_{C5} = V_{C6} = \frac{2V_{in}}{1-D} \\ V_O = V_{C7} = \frac{4V_{in}}{1-D} \end{cases} \tag{4}$$

Therefore, the voltage gain M of the proposed converter can be derived as follows:

$$M = \frac{V_O}{V_{in}} = \frac{4}{1-D} \tag{5}$$

3.2. Ripple-Free Condition

As shown in Figure 5, the direction of the current can be obtained. From Figure 5a, when the switch S is turned on, the voltage v_{Lm} across L_m is V_{in}. Hence, the current i_{Lm} increases linearly from its minimum value $I_{Lm,min}$ as follows:

$$i_{Lm}(t) = I_{Lm,\min} + \int_{t_0}^{t} \frac{V_{in}}{L_m} dt \tag{6}$$

The voltage v_{Lk} across L_k is $-(V_{C0} - nV_{in})$. Therefore, the current i_{Lk} decreases linearly from its maximum value $I_{Lk,max}$ as follows:

$$i_{Lk}(t) = I_{Lk,\max} - \int_{t_0}^{t} \frac{V_{C0} - nV_{in}}{L_k} dt \tag{7}$$

Since the average inductor voltage must be zero under the steady-state conditions, the voltage V_{C0} can be shown as follows:

$$V_{C0} = V_{in} \tag{8}$$

It is apparent that the input current i_{in} is the sum of i_{Lm} and ni_{Lk}. Therefore, combining (6)–(8), i_{in} can be derived as follows:

$$i_{in}(t) = I_{Lm,\min} + nI_{Lk,\max} + \int_{t_0}^{t} \left(\frac{1}{L_m} - \frac{n(1-n)}{L_k}\right)V_{in}dt \tag{9}$$

To achieve a ripple-free condition, the variation of input current must be zero during this stage as follows:

$$\frac{d}{dt}i_{in}(t) = 0 \tag{10}$$

From (9) and (10), since ($I_{Lm,min}$ + $nI_{Lk,max}$) is a constant value, the input current ripple is eliminated with the following condition:

$$L_k = n(1-n)L_m \tag{11}$$

Consequently, the input current i_{in} in Mode 1 is only determined by

$$i_{in}(t) = I_{Lm,\min} + nI_{Lk,\max} \tag{12}$$

Similarly, with the turn-off of the switch S shown in Figure 5b, the voltage v_{Lm} across L_m is $-(V_{C1} - V_{in})$. The current i_{Lm} decreases linearly from its maximum value $I_{Lm,max}$ as follows:

$$i_{Lm}(t) = I_{Lm,\max} - \int_{t_1}^{t} \frac{V_{C1} - V_{in}}{L_m}dt \tag{13}$$

The voltage v_{Lk} across L_k is ($V_{C1} - V_{CO} + nv_{Lm}$), which can also be written as $(1 - n)$ $(V_{C1} - V_{in})$. Therefore, the current i_{Lk} increases linearly from its minimum value $-I_{Lk,max}$ as follows:

$$i_{Lk}(t) = -I_{Lk,\max} + \int_{t_1}^{t} \frac{(1-n)(V_{C1} - V_{in})}{L_k}dt \tag{14}$$

Combining (13) and (14), the input current i_{in} can be derived as follows:

$$i_{in}(t) = I_{Lm,\max} - nI_{Lk,\max} + \int_{t_1}^{t} \left(\frac{n(1-n)}{L_k} - \frac{1}{L_m}\right)\frac{DV_{in}}{1-D}dt \tag{15}$$

Since ($I_{Lm,max} - nI_{Lk,max}$) is a constant value, the input current ripple is also eliminated with the condition of (11). Therefore, the input current i_{in} in Mode 2 is only determined by

$$i_{in}(t) = I_{Lm,\max} - nI_{Lk,\max} \tag{16}$$

From Figures 4 and 5a, the value of Δi_{Lm} can be obtained as follows:

$$\Delta i_{Lm} = I_{Lm,\max} - I_{Lm,\min} = \frac{V_{in}}{L_m}DT_S \tag{17}$$

From (7), (8), and Figure 4, the maximum value $I_{Lk,max}$ can be derived as follows:

$$I_{Lk,\max} = \frac{(1-n)V_{in}}{2L_k}DT_S \tag{18}$$

Combining (11), (17), and (18), $I_{Lk,max}$ can be further expressed by

$$I_{Lk,\max} = \frac{1}{2n}\frac{V_{in}}{L_m}DT_S = \frac{1}{2n}(I_{Lm,\max} - I_{Lm,\min}) \tag{19}$$

From (19), it can be concluded that

$$i_{in}(t) = I_{Lm,\min} + nI_{Lk,\max} = I_{Lm,\max} - nI_{Lk,\max} \tag{20}$$

Consequently, the input current i_{in} is a constant value during the whole switching period with the condition of (11).

3.3. Voltage Stress Analysis

Referring to the circuit diagrams shown in Figure 5 and according to Kirchhoff's voltage law, the following voltage relationships can be obtained:

$$\begin{cases} V_{DS} = V_{D1} = V_{C1} \\ V_{D2} = V_{D4} = V_{C4} \\ V_{D3} = V_{C3} \\ V_{D5} = V_{C5} - V_{C2} \\ V_{D6} = V_{C6} - V_{C1} \\ V_{D7} = V_{C7} - V_{C4} - V_{C6} \end{cases} \tag{21}$$

where V_{DS} and V_{D1}~V_{D7} are the voltage stresses of the switch S and the diodes D_1~D_7, respectively.

From (4), the simplified voltage stress relationship is given as follows:

$$V_{DS} = V_{D1} = V_{D2} = V_{D3} = V_{D4} = V_{D5} = V_{D6} = V_{D7} = \frac{V_{in}}{1-D} \tag{22}$$

Therefore, all the diodes and the switch have the same voltage stress which means the same type of diodes can be selected. Due to the relatively low voltage stress, the components with a lower-rated voltage and a lower on-resistance can be used to further reduce the power losses and costs.

3.4. Real-Gain Analysis

In fact, all the active and passive components contain some non-idealities in practice that influence the voltage gain and efficiency of high-gain dc/dc converters. Figure 8 shows the equivalent circuits of the proposed topology considering all the non-idealities in the two operating modes. In Figure 8, R_{Lp} and R_{Ls} are the equivalent series resistance (ESR) of the primary and secondary sides of the coupled inductor, respectively, R_{C0}–R_{C7} are the ESRs of capacitors, R_{D1}–R_{D7} are the forward diode resistances of diodes, V_{d1}–V_{d7} are the forward voltage drops of diodes, and R_S is the on-state resistance of the switch.

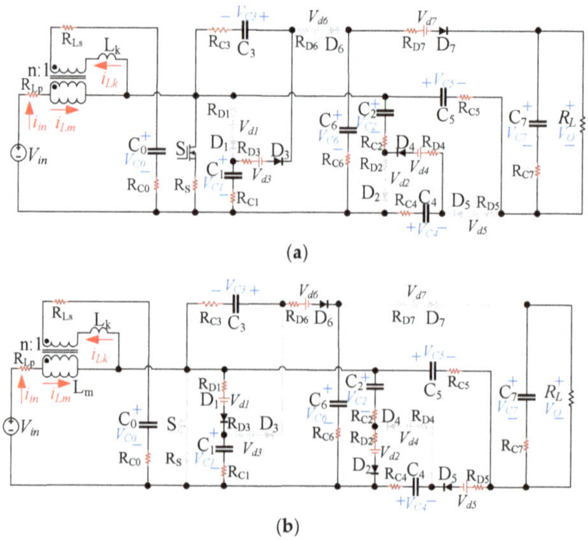

Figure 8. The equivalent circuits considering all the non-idealities: (**a**) Mode 1; (**b**) Mode 2.

Referring to Figure 8a, when the switch is turned on, the following equations can be obtained according to Kirchhoff's voltage law.

$$\begin{cases} v_{Lm(on)} + i_{in}R_{Lp} = V_{in} \\ V_{C3} = V_{C1} - i_S R_S - i_{D3}(R_{C1} + R_{C3} + R_{D3}) - V_{d3} \\ V_{C4} = V_{C2} - i_S R_S - i_{D4}(R_{C2} + R_{C4} + R_{D4}) - V_{d4} \\ V_{C7} = V_{C5} + V_{C6} - i_S R_S - i_{D7}(R_{C5} + R_{C6} + R_{C7} + R_{D7}) - V_{d7} \end{cases} \tag{23}$$

Similarly, from Figure 8b, the voltage relationship of each loop can be expressed as follows when the switch is turned off.

$$\begin{cases} V_{C1} = V_{in} - v_{Lm(off)} - i_{in}R_{Lp} - i_{D1}(R_{C1} + R_{D1}) - V_{d1} \\ V_{C2} = V_{in} - v_{Lm(off)} - i_{in}R_{Lp} - i_{D2}(R_{C2} + R_{D2}) - V_{d2} \\ V_{C5} = V_{in} - v_{Lm(off)} + V_{C4} - i_{in}R_{Lp} - i_{D5}(R_{C4} + R_{C5} + R_{D5}) - V_{d5} \\ V_{C6} = V_{in} - v_{Lm(off)} + V_{C3} - i_{in}R_{Lp} - i_{D6}(R_{C3} + R_{C6} + R_{D6}) - V_{d6} \\ V_O = V_{C7} - i_O R_{C7} \end{cases} \tag{24}$$

By applying the volt-second balance principle to the magnetic inductor L_m, the real-gain of the proposed converter can be deduced and is given below after simplification.

$$V_O = \frac{1}{H} \cdot \left[V_{in} - \frac{1-D}{4}(V_{d1} + V_{d2} + V_{d3} + V_{d4} + V_{d5} + V_{d6} + V_{d7}) \right] \tag{25}$$

where the parameter H is defined as follows:

$$H = \frac{1-D}{4} + \frac{4R_{Lp}}{(1-D)R_L} + \frac{(3+D)R_S}{4R_L} + \frac{1-D}{4R_L} \cdot \begin{bmatrix} (R_{D1} + R_{D2} + R_{D3} + R_{D4} + R_{D5} + R_{D6} + R_{D7}) + \\ 2(R_{C1} + R_{C2} + R_{C3} + R_{C4} + R_{C5} + R_{C6} + R_{C7}) \end{bmatrix} \tag{26}$$

3.5. Losses Analysis

The power losses of the proposed converter are caused by diodes, capacitors, the switch, and the coupled inductor.

In the diodes D_1-D_7, the forward voltage drop and forward resistance are the reasons for the power loss $P_{D,}$ and it can be derived as follows:

$$P_D = V_d I_D + R_D {I_D}^2 \tag{27}$$

where V_d, R_D, and I_D are the forward voltage drop, the forward resistance, and the average current of the diodes, respectively.

As for capacitors C_0–C_7, the power loss P_C caused by the ESR can be calculated by

$$P_C = \frac{f_S \cdot C \cdot \Delta U^2}{2} \tag{28}$$

where C and ΔU represent the capacitance and voltage ripple of the capacitor, respectively.

As for switch S, the power losses comprise conduction loss $P_{S\text{-}C}$ and switching loss $P_{S\text{-}S}$. The on-resistance is the reason for the conduction loss of a switch. By defining the on-resistance and rms current of the switch as R_{DSon} and I_S, respectively, the conduction loss $P_{S\text{-}C}$ can be obtained as follows:

$$P_{S\_C} = {I_S}^2 R_{DSon} \tag{29}$$

The switching loss $P_{S\text{-}S}$ can be estimated by linearizing the voltage and current of the switch during the turn-on and turn-off processes as follows:

$$\begin{cases} P_{S\_SON} = V_{DS} \cdot I_{on} \cdot t_{ondelay} \cdot f_S/6 \\ P_{S\_SOFF} = V_{DS} \cdot I_{off} \cdot t_{offdelay} \cdot f_S/6 \end{cases} \tag{30}$$

where I_{on} and I_{off} are the turn-on and turn-off currents, and $t_{ondelay}$ and $t_{offdelay}$ are the turn-on and turn-off time delays.

As for the coupled inductor, the power losses are mainly composed of copper loss $P_{L\text{-}copper}$ and core loss $P_{L\text{-}core}$. According to [34], the theoretical estimation formula of copper loss can be obtained as follows:

$$P_{L\_copper} = {I_L}^2 r_L \tag{31}$$

where I_L and r_L represent the rms current and the ESR of the coupled inductor, respectively.

The core loss can be calculated by

$$P_{L\_core} = K_{Fe} \cdot V_e \cdot f_S \cdot (\frac{\Delta B}{2})^{\alpha} \tag{32}$$

where K_{Fe} and α are constants determined by the core material, V_e is the volume of the core, and ΔB is decided by the current ripple of the coupled inductor.

The total power loss of the proposed converter can be obtained as follows:

$$P_{total} = P_D + P_C + P_{S\_C} + P_{S\_S_{ON}} + P_{S\_S_{OFF}} + P_{\text{L\_copper}} + P_{\text{L\_core}} \tag{33}$$

In order to exhibit the losses distribution of the proposed converter intuitively, the losses of each component at 240 W are calculated through (27)–(32) and shown graphically in Figure 9. It can be seen that most of the total power loss occurs in the diodes, which is mainly caused by the large output current. However, the conduction loss of the switch is significantly reduced due to the use of a single switch compared with other multi-switch high-gain converters.

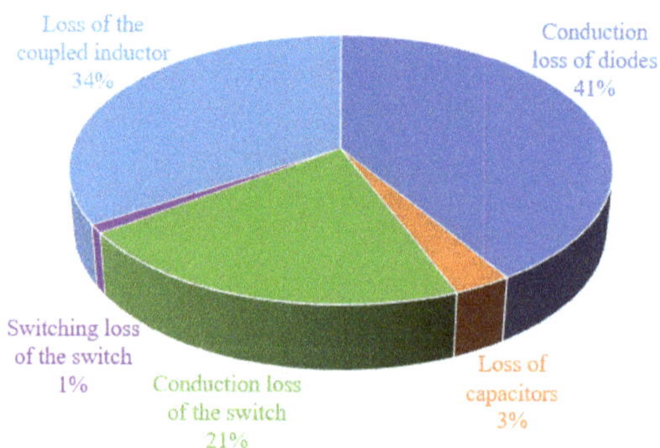

Figure 9. Loss distribution of the proposed converter.

4. Performance Comparisons

The performance indexes of relevant high-gain dc/dc converters are summarized in Table 1, including the number of switches, voltage gain, voltage stress of switches and diodes, total standing voltage (TSV), and input current ripple. According to [35], the total voltage rating of switching power devices can be reflected by TSV which is defined as

$$TSV = \frac{\sum_{i=1}^{n} V_{Sn} + \sum_{j=1}^{m} V_{Dj}}{V_o} \tag{34}$$

where V_{Sn} and V_{Dj} represent the voltage stress of each switch and diode, respectively.

Table 1. Comparisons among different converters.

Parameters	[24]	[25]	[26]	[32]	Proposed
Number of switches	1	1	3	2	1
Voltage gain	$\frac{3D}{1-D}$	$\frac{3+D}{2(1-D)}$	$\frac{1+D}{1-D}$	$\frac{D(1+D)}{(1-D)^2}$	$\frac{4}{1-D}$
Voltage stress of switches	$\frac{V_O}{3D}$	$\frac{2V_O}{3+D}$	$\frac{V_O}{3(1+D)}$	$\frac{V_O}{1+D}, \frac{(1-D)V_O}{D(1+D)}$	$\frac{V_O}{4}$
Voltage stress of diodes	$\frac{V_O}{3D}$	$\frac{V_O}{3+D}, \frac{2V_O}{3+D}$	$\frac{V_O}{3(1+D)}$	$\frac{V_O}{1+D}, \frac{(1-D)V_O}{D(1+D)}$	$\frac{V_O}{4}$
TSV	$\frac{4}{3D}$	$\frac{8}{3+D}$	$\frac{3}{1+D}$	$\frac{2+D}{D(1+D)}$	2
Input current ripple	Low	High	High	Low	Zero

Figure 10 gives the comparison curves of different converters in Table 1. From Figure 10a, the proposed converter has the highest voltage-boosting capability compared to other converters in the optimal duty cycle range. From Table 1, the switches and diodes of all the converters have the same maximum voltage stress. Thus, the maximum voltage stress curve of switches and diodes is plotted in Figure 10b. The voltage stress in the proposed converter is lower than other converters except for the converter in [26]. Although the voltage stress in [26] is lower when the duty cycle is greater than 1/3, its voltage gain is much lower than that in the proposed converter. Similarly, Figure 10c shows that the proposed converter has the lowest TSV when the duty cycle is smaller than 0.5. The TSV in [26] is lower than that in this paper when the duty cycle is greater than 0.5; however, its voltage gain is also much lower. It can be deduced from Table 1 and Figure 10 that the proposed converter has a high voltage gain and a low voltage stress. That is to say, the active power devices with low withstand voltage can be selected.

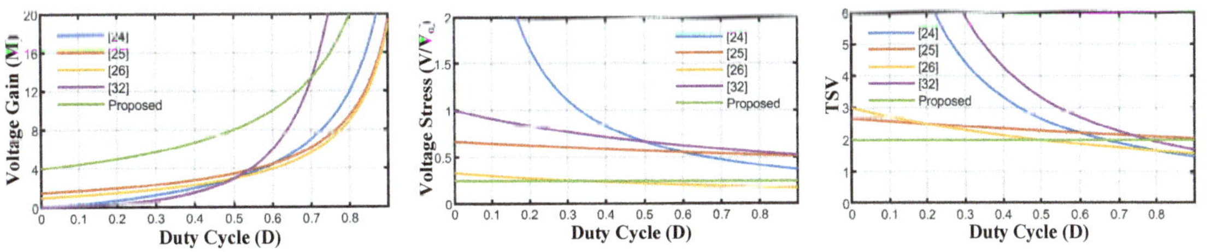

(a) (b) (c)

Figure 10. Comparative results of the converters versus the duty cycle D: (**a**) Voltage gain; (**b**) Voltage stress of switches and diodes; (**c**) TSV.

Moreover, from Table 1, the number of switches used in [24,25] and the proposed converter is the smallest. The proposed converter uses only one switch which can significantly simplify the control strategy. Meanwhile, the proposed converter has the lowest input current ripple, and it achieves a ripple-free input current condition which is of great importance in offshore wind energy systems. Owing to the ripple-free input current, the HVDC transmission will be more stable. Consequently, the proposed converter is well suited for offshore wind farms due to the above-mentioned superiorities.

5. Design Guideline

5.1. Design of the Coupled Inductor

To achieve ripple-free conditions, the converter should operate in CCM mode which means the current i_L must be continuous. Thus, combining the current waveform in Figure 4 and (13), the minimum value $I_{Lm,min}$ should be greater than zero. Equation (13) can also be written as follows:

$$I_{Lm,\min} = I_{Lm}(1+\frac{\lambda}{2}) - \frac{V_{C1}-V_{in}}{L_m}(1-D)T_S \tag{35}$$

where λ is the ripple rate of the current i_{Lm}.

Since the average current I_{Lk} is zero, the average current on the primary side of the coupled inductor is also zero. The average current I_{Lm} can be expressed by

$$I_{Lm} = I_{in} \tag{36}$$

Combining (4), (35), and (36), the minimum value of L_m can be obtained below:

$$L_{m,\min} = \frac{2V_{in}T_S D}{(2+\lambda)I_{in}} \tag{37}$$

Therefore, to ensure that the circuit operates under CCM mode, L_m should be selected with the following condition:

$$L_m > L_{m,\min} \tag{38}$$

When the value of L_m is determined, L_k can be determined by (11).

5.2. Design of Capacitors

In mode 1, the current flowing through C_0 is i_{Lk}. The currents flowing through C_3, C_4, and C_7 are i_{D3}, i_{D4}, and i_{D7}, respectively. Assuming that the capacitor voltage ripple rate is $x_c\%$, the minimum values of the capacitors C_3, C_4, and C_7 can be obtained from the capacitor's characteristic equation as follows:

$$C_{n\min} = \int_0^{DT_S} \frac{i_{Dn}}{x_c\%V_{Cn}}dt \tag{39}$$

and C_{0min} can be expressed by

$$C_{0\min} = \int_0^{DT_S} \frac{i_{Lk}}{x_c\%V_{C0}}dt \tag{40}$$

Similarly, in mode 2, the currents flowing through C_1, C_2, C_5, and C_6 are i_{D1}, i_{D2}, i_{D5}, and i_{D6}, respectively. Thus, the minimum values of the capacitors C_1, C_2, C_5, and C_6 can be obtained as follows:

$$C_{n\min} = \int_{DT_S}^{T_S} \frac{i_{Dn}}{x_c\%V_{Cn}}dt \tag{41}$$

5.3. Selection of Switch and Diodes

For switch S and diodes D_1-D_7, according to the voltage stress relationship obtained from (22) and considering an appropriate margin, the maximum withstand voltage value is given below:

$$V_{DS} = V_{D1} = V_{D2} = V_{D3} = V_{D4} = V_{D5} = V_{D6} = V_{D7} = \frac{k_V V_{in}}{1-D} \tag{42}$$

where k_V represents the voltage margin factor.

6. Experimental Results

To verify the validity and feasibility of the proposed topology, a 240 W laboratory prototype converter at 35 kHz switching frequency was designed. Detailed parameters and selected components are given in Table 2. Since the experimental leakage inductance of the coupled inductor was about 2 uH, an auxiliary inductor was connected in series with the secondary side of the coupled inductor to meet the requirements. The main voltage and current experimental waveforms of the converter are shown in Figure 11.

Table 2. Parameters of the converter.

Parameters	Value/Model
Input voltage V_{in}/V	40
Output voltage V_O/V	308
Switching frequency f_S/kHz	35
Magnetic inductor L_m/uH	250
Leakage inductance L_k/uH	62.5
Turn ratio n	1:2
Capacitor C_0/uF	330
Diodes D_1–D_7	MBR10200CT
Switch	IRFP260NPBF

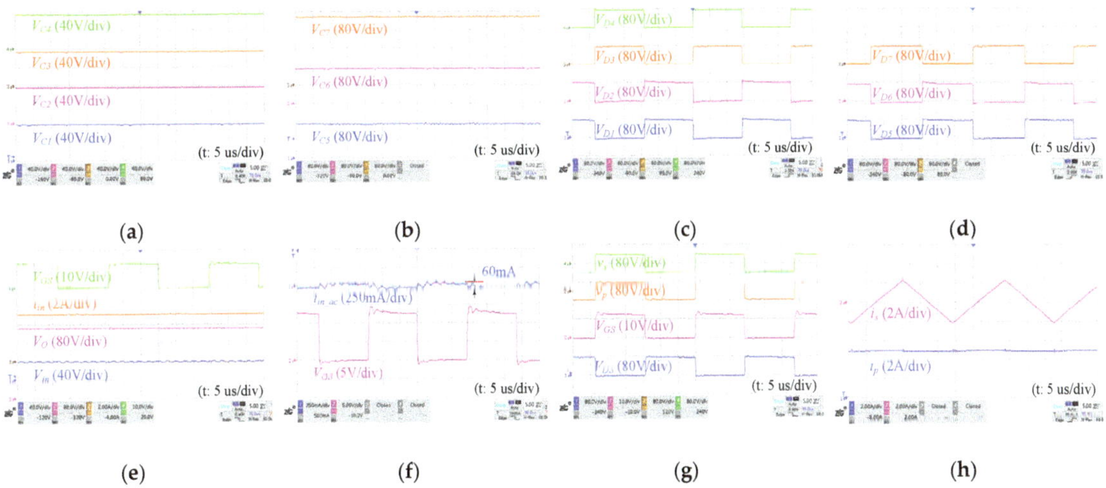

Figure 11. Experimental waveforms of the prototype: (**a**) V_{C1}, V_{C2}, V_{C3}, and V_{C4}; (**b**) V_{C5}, V_{C6}, and V_{C7}; (**c**) V_{D1}, V_{D2}, V_{D3}, and V_{D4}; (**d**) V_{D5}, V_{D6}, and V_{D7}; (**e**) V_{in}, V_O, i_{in}, and V_{GS}; (**f**) $i_{in\text{-}ac}$, and V_{GS}; (**g**) V_S, V_{GS}, v_p, and v_s; (**h**) i_p and i_s.

From Figure 11a, V_{C1}–V_{C4} reach nearly 80 V. Due to the R_{DSon} of the switch and the forward voltage drop of the diodes, V_{C1}–V_{C4} are slightly lower than the theoretical values. From Figure 11b, V_{C5} and V_{C6} reach nearly 160 V, and V_{C7} reaches nearly 320 V. Due to the loss of devices on different loops, V_{C5}–V_{C7} are also slightly lower than the theoretical values.

From Figure 11c,d, the voltage stresses on the diodes D_1–D_7 are about 80 V which is consistent with the theoretical analysis. Figure 11e shows that the output voltage reaches up to 308 V under 40 V input voltage. Hence, the experimental results verify that the proposed converter has the characteristic of high voltage gain. According to the theoretical calculation, the output voltage should have been 320 V under ideal conditions. The difference between experimental results and ideal conditions is caused by the non-idealities in the circuit. As can be seen from the waveform of i_{in} in Figure 11e, the input current i_{in} is constant. The ac component of i_{in} shown in Figure 11f is about 60 mA. The ripple rate of i_{in} is only 1.1%, thus the proposed converter provides a ripple-free input current through the aforementioned parameter design. From Figure 11g, the voltages at both ends of the primary and secondary sides of the coupled inductor change at the same time and have the same value. Figure 11h shows the currents of the coupled inductor, where i_p and i_s represent the current on the primary and secondary sides of the coupled inductor, respectively. Also, the voltage and current waveforms of the coupled inductor in Figure 11g,h are consistent with the theoretical waveforms in Figure 4.

In summary, the experimental results verify the validity and feasibility of the proposed converter. Some deviations from the theoretical analysis are inevitable. The proposed converter exhibits an efficiency of 93.7% at a 240 W load. Figure 12 shows the measured efficiency curve of the proposed converter under different loads and the maximum efficiency is 94.5%. Figure 13 shows the photograph of the experimental prototype.

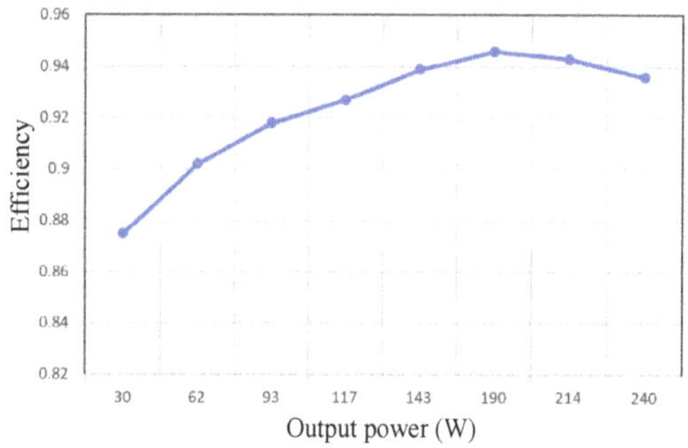

Figure 12. Measured efficiency of the proposed converter.

Figure 13. Photograph of the experimental prototype.

7. Conclusions

Aiming at the MVDC system in offshore wind farms, a novel single-switch high-gain dc/dc converter with a ripple-free input current is proposed in this paper. Since the converter uses only one switch, the control strategy is not complicated which is beneficial for the stability of offshore wind energy systems. The converter provides a high voltage gain through a switched-capacitor structure. Additionally, the converter provides a ripple-free input current by utilizing a coupled inductor which can avoid the use of a large electrolytic capacitor. Hence, the volume and weight of the converter are reduced. Moreover, the output capacitor is charged and discharged continuously by a 180° phase shift to eliminate output voltage ripple which can further improve the stability of the systems. The steady-state characteristic under CCM of the converter is analyzed. Comparisons of the proposed converter with its counterparts show various beneficial characteristics as follows: (1) high voltage gain; (2) ripple-free input current; (3) simple control strategy; (4) low voltage stress on devices; and (5) high efficiency. Finally, to verify the validity and feasibility of the proposed converter, a laboratory prototype has been built for a power of 240 W, input and output voltages of 40 and 308 V, respectively, and a switching frequency of 35 kHz. The input current ripple is only 1.1% and the maximum efficiency is measured to be 94.5%.

Experimental results confirm that the proposed converter is well suited for high-gain offshore wind energy applications.

Author Contributions: Conceptualization, R.T. and J.Y.; methodology, R.T. and Z.L.; software, Z.H.; validation, Z.L. and J.L.; formal analysis, Z.L.; investigation, R.A.; resources, R.A.; data curation, Z.H.; writing—original draft preparation, R.T. and Z.L.; writing—review and editing, J.L. and J.Y.; project administration, J.L. All authors have read and agreed to the published version of the manuscript.

Funding: This research was supported by the Open Fund of Guangdong Key Laboratory of New Technology for Smart Grid under GDDKY2021KF02.

Institutional Review Board Statement: Not applicable.

Informed Consent Statement: Not applicable.

Data Availability Statement: Not applicable.

Conflicts of Interest: The authors declare no conflict of interest.

References

1. Fjellstedt, C.; Ullah, M.I.; Forslund, J.; Jonasson, E.; Temiz, I.; Thomas, K. A Review of AC and DC Collection Grids for Offshore Renewable Energy with a Qualitative Evaluation for Marine Energy Resources. *Energies* **2022**, *15*, 5816. [CrossRef]
2. Lam, J.; Jain, P.K. A High Efficient Medium Voltage Step-up DC/DC Converter with Zero Voltage Switching (ZVS) and Low Voltage Stress for Offshore Wind Energy Systems. In Proceedings of the 2014 16th European Conference on Power Electronics and Applications, Lappeenranta, Finland, 26–28 August 2014; IEEE: Piscataway, NJ, USA, 2014; pp. 1–10.
3. Lee, Y.; Vakil, G.; Feldman, R.; Watson, A.J.; Wheeler, P.W. A High-Power DC-DC Converter Based Dual Active Bridge for MVDC Grids on Offshore Wind Farms. In Proceedings of the 2016 18th European Conference on Power Electronics and Applications (EPE'16 ECCE Europe), Karlsruhe, Germany, 5–9 September 2016; IEEE: Piscataway, NJ, USA, 2016; pp. 1–10.
4. Abeynayake, G.; Van Acker, T.; Hertem, D.V.; Liang, J. Analytical Model for Availability Assessment of Large-Scale Offshore Wind Farms Including Their Collector System. *IEEE Trans. Sustain. Energy* **2021**, *12*, 1974–1983. [CrossRef]
5. Basbas, H.; Liu, Y.-C.; Laghrouche, S.; Hilairet, M.; Plestan, F. Review on Floating Offshore Wind Turbine Models for Nonlinear Control Design. *Energies* **2022**, *15*, 5477. [CrossRef]
6. Wang, B.; Tian, M.; Lin, T.; Hu, Y. Distributed Complementary Control Research of Wind Turbines in Two Offshore Wind Farms. *Sustainability* **2018**, *10*, 553. [CrossRef]
7. Vagiona, D.; Kamilakis, M. Sustainable Site Selection for Offshore Wind Farms in the South Aegean—Greece. *Sustainability* **2018**, *10*, 749. [CrossRef]
8. Torres, J.P.N.; De Jesus, A.S.; Marques Lameirinhas, R.A. How to Improve an Offshore Wind Station. *Energies* **2022**, *15*, 4873. [CrossRef]
9. Bernal-Camacho, D.F.; Fontes, J.V.H.; Mendoza, E. A Technical Assessment of Offshore Wind Energy in Mexico: A Case Study in Tehuantepec Gulf. *Energies* **2022**, *15*, 4367. [CrossRef]
10. Rong, F.; Wu, G.; Li, X.; Huang, S.; Zhou, B. ALL-DC Offshore Wind Farm With Series-Connected Wind Turbines to Overcome Unequal Wind Speeds. *IEEE Trans. Power Electron.* **2019**, *34*, 1370–1381. [CrossRef]
11. Bahmani, M.A.; Thiringer, T.; Rabiei, A.; Abdulahovic, T. Comparative Study of a Multi-MW High-Power Density DC Transformer with an Optimized High-Frequency Magnetics in All-DC Offshore Wind Farm. *IEEE Trans. Power Delivery* **2016**, *31*, 857–866. [CrossRef]
12. Hu, P.; Yin, R.; Wei, B.; Luo, Y.; Blaabjerg, F. Modular Isolated LLC DC/DC Conversion System for Offshore Wind Farm Collection and Integration. *IEEE J. Emerg. Sel. Top. Power Electron.* **2021**, *9*, 6713–6725. [CrossRef]
13. Herrera, D.; Tricarico, T.; Oliveira, D.; Aredes, M.; Galván-Díez, E.; Carrasco, J.M. Advanced Local Grid Control System for Offshore Wind Turbines with the Diode-Based Rectifier HVDC Link Implemented in a True Scalable Test Bench. *Energies* **2022**, *15*, 5826. [CrossRef]
14. Xie, L.; Cheng, F.; Wu, J. Control Strategy for Offshore Wind Farms with DC Collection System Based on Series-Connected Diode Rectifier. *Sustainability* **2022**, *14*, 7860. [CrossRef]
15. Liu, H.; Dahidah, M.; Naayagi, R.T.; Armstrong, M.; Yu, J. Unidirectional DC/DC Modular Multilevel Converter for Offshore Windfarm with the Control Strategy Based on Stationary Frame. *J. Eng.* **2019**, *2019*, 4309–4314. [CrossRef]
16. Liu, J.; Li, B.; Li, L.; Liu, X.; Wu, J.; Xu, D.; Wei, T.; Li, W. A High Step-up Ratio DC-DC Converter with Fault Blocking Capability for Offshore Wind Farms. In Proceedings of the 2021 IEEE 12th Energy Conversion Congress & Exposition-Asia (ECCE-Asia), Singapore, 24 May 2021; IEEE: Singapore, 2021; pp. 702–707.
17. Dincan, C.; Kjaer, P.; Chen, Y.; Munk-Nielsen, S.; Bak, C.L. Analysis of a High-Power, Resonant DC–DC Converter for DC Wind Turbines. *IEEE Trans. Power Electron.* **2018**, *33*, 7438–7454. [CrossRef]
18. Zhao, X.; Li, B.; Zhang, B.; Xu, D. A High-Power Step-Up DC/DC Converter Dedicated to DC Offshore Wind Farms. *IEEE Trans. Power Electron.* **2022**, *37*, 65–69. [CrossRef]

19. Deng, F.; Chen, Z. Control of Improved Full-Bridge Three-Level DC/DC Converter for Wind Turbines in a DC Grid. *IEEE Trans. Power Electron.* **2013**, *28*, 314–324. [CrossRef]
20. Chen, W.; Huang, A.; Lukic, S.; Svensson, J.; Li, J.; Wang, Z. A Comparison of Medium Voltage High Power DC/DC Converters with High Step-up Conversion Ratio for Offshore Wind Energy Systems. In Proceedings of the 2011 IEEE Energy Conversion Congress and Exposition, Phoenix, AZ, USA, 17–22 September 2011; IEEE: Piscataway, NJ, USA, 2011; pp. 584–589.
21. Liu, H.; Dahidah, M.S.A.; Yu, J.; Naayagi, R.T.; Armstrong, M. Design and Control of Unidirectional DC–DC Modular Multilevel Converter for Offshore DC Collection Point: Theoretical Analysis and Experimental Validation. *IEEE Trans. Power Electron.* **2019**, *34*, 5191–5208. [CrossRef]
22. Denniston, N.; Massoud, A.M.; Ahmed, S.; Enjeti, P.N. Multiple-Module High-Gain High-Voltage DC–DC Transformers for Offshore Wind Energy Systems. *IEEE Trans. Ind. Electron.* **2011**, *58*, 1877–1886. [CrossRef]
23. Parastar, A.; Seok, J.-K. High-Gain Resonant Switched-Capacitor Cell-Based DC/DC Converter for Offshore Wind Energy Systems. *IEEE Trans. Power Electron.* **2015**, *30*, 644–656. [CrossRef]
24. Banaei, M.R.; Sani, S.G. Analysis and Implementation of a New SEPIC-Based Single-Switch Buck–Boost DC–DC Converter With Continuous Input Current. *IEEE Trans. Power Electron.* **2018**, *33*, 10317–10325. [CrossRef]
25. Saravanan, S.; Babu, N.R. Design and Development of Single Switch High Step-Up DC–DC Converter. *IEEE J. Emerg. Sel. Topics Power Electron.* **2018**, *6*, 855–863. [CrossRef]
26. Gandomkar, A.; Parastar, A.; Seok, J.-K. High-Power Multilevel Step-Up DC/DC Converter for Offshore Wind Energy Systems. *IEEE Trans. Ind. Electron.* **2016**, *63*, 7574–7585. [CrossRef]
27. Hu, Y.; Zeng, R.; Cao, W.; Zhang, J.; Finney, S.J. Design of a Modular, High Step-Up Ratio DC–DC Converter for HVDC Applications Integrating Offshore Wind Power. *IEEE Trans. Ind. Electron.* **2016**, *63*, 2190–2202. [CrossRef]
28. Abbasi, M.; Lam, J. A Step-Up Transformerless, ZV–ZCS High-Gain DC/DC Converter With Output Voltage Regulation Using Modular Step-Up Resonant Cells for DC Grid in Wind Systems. *IEEE J. Emerg. Sel. Topics Power Electron.* **2017**, *5*, 1102–1121. [CrossRef]
29. Parastar, A.; Kang, Y.C.; Seok, J.-K. Multilevel Modular DC/DC Power Converter for High-Voltage DC-Connected Offshore Wind Energy Applications. *IEEE Trans. Ind. Electron.* **2015**, *62*, 2879–2890. [CrossRef]
30. Yin, R.; Shi, M.; Hu, W.; Guo, J.; Hu, P.; Wang, Y. An Accelerated Model of Modular Isolated DC/DC Converter Used in Offshore DC Wind Farm. *IEEE Trans. Power Electron.* **2019**, *34*, 3150–3163. [CrossRef]
31. Zhang, J.; Sha, D.; Ma, P. A Dual Active Bridge DC–DC-Based Single Stage AC–DC Converter With Seamless Mode Transition and High Power Factor. *IEEE Trans. Ind. Electron.* **2022**, *69*, 1411–1421. [CrossRef]
32. García-Vite, P.M.; Rosas-Caro, J.C.; Martínez-Salazar, A.L.; Chavez, J.d.J.; Valderrábano-González, A.; Sánchez-Huerta, V.M. Quadratic Buck–Boost Converter with Reduced Input Current Ripple and Wide Conversion Range. *IET Power Electron.* **2019**, *12*, 3977–3986. [CrossRef]
33. Do, H.-L. Improved ZVS DC-DC Converter With a High Voltage Gain and a Ripple-Free Input Current. *IEEE Trans. Circuits Syst. I* **2012**, *59*, 846–853. [CrossRef]
34. Yang, N.; Zeng, J.; Hu, R.; Liu, J. Analysis and Design of an Isolated High Step-Up Converter Without Voltage-Drop. *IEEE Trans. Power Electron.* **2022**, *37*, 6939–6950. [CrossRef]
35. Liu, J.; Lin, W.; Wu, J.; Zeng, J. A Novel Nine-Level Quadruple Boost Inverter With Inductive-Load Ability. *IEEE Trans. Power Electron.* **2019**, *34*, 4014–4018. [CrossRef]

World Electric Vehicle Journal

MDPI

Article

How to Cross the Chasm for the Electric Vehicle World's Laggards—A Case Study in Kuwait

Andri Ottesen [1,*], Sumayya Banna [2] and Basil Alzougool [2]

1 LSE Middle East Centre (MEC)—Sustainability Research and Consultancy (CSRC), Australian University (AU), West Mishref P.O. Box 1411, Kuwait
2 LSE Middle East Centre (MEC), Arab Open University (AOU), Ardiya P.O. Box 830, Kuwait
* Correspondence: a.ottesen@au.edu.kw

Abstract: Ever since the discovery of oil in 1938, the State of Kuwait has increasingly sought out international brands in the car market due to the high purchasing power of Kuwaiti nationals. However, the makers of electric vehicles (EVs) have not been able to penetrate this market, with the exception of innovators and early adopters. The phenomenon in disruptive innovation theory—called "Crossing the Chasm"—regarding a mass market appeal has not yet occurred in Kuwait. Through deep interviews with 12 Kuwaiti owners of EVs and automotive dealers who sold either EVs or Hybrid Electric Vehicles (HEVs), 10 key reasons for this phenomenon have been previously revealed, which were used to develop an extensive questionnaire. A total of 472 car drivers aged from 18 to 30, identified as the "early majority", completed the questionnaire to achieve the objective of identifying the factors required to create a mass market for EVs in Kuwait. The results demonstrated that potential customers highly preferred three different types of attributes of EVs: environmental, financial, and technological. There were significant differences in the identified attributes preferred by Kuwaiti individuals for EVs in terms of the number of cars owned and the sector of employment. Moreover, the results of our study indicate that potential customers are very willing to buy EVs in the future, considering both their financial and infrastructure attributes. There were further significant differences in the identified necessary conditions to buy EVs in terms of educational level and monthly income. This study discusses a variety of valuable promotional tactics, which may be implemented in conjunction with public incentives and policy changes in the State of Kuwait. This information is considered useful for marketers and designers who wish to tap into this lucrative market, which is significantly different from that in the global North.

Keywords: electric vehicles; disruptive innovation; customer preferences; emerging market; Kuwait; EV infrastructure

Citation: Ottesen, A.; Banna, S.; Alzougool, B. How to Cross the Chasm for the Electric Vehicle World's Laggards—A Case Study in Kuwait. *World Electr. Veh. J.* **2023**, *14*, 45. https://doi.org/10.3390/wevj14020045

Academic Editors: Wenbin Yu and Guang Zeng

Received: 23 December 2022
Revised: 30 January 2023
Accepted: 6 February 2023
Published: 8 February 2023

1. Introduction

This paper is the fifth published paper stemming from the project called "Breaking the Internal Combustion Engine Reign: A Mixed-Methods Study of Attitudes Towards Using and Purchasing Electric Vehicles in Kuwait," funded by the Kuwait Foundation for Scientific Advancement (KFAS) and managed by the Middle East Center of the London School of Economics. The project is complementary to the research of the KFAS on infrastructure and top-down policy, as well as research conducted by the National Laboratory of the Kuwaiti Institute for Scientific Research (KISR), mainly focused on EV batteries and their charging capacity under extreme heat conditions. This study, on the other hand, follows a bottom-up approach to capture customer preferences and attitudes towards EVs through the use of information gained from deep interviews with existing owners and automotive dealers in order to develop an extensive questionnaire, which we issued to potential consumers. The final outcome of this study is a series of suggestions for national and local governments, service providers, and automobile dealers that wish to break into this market, of which policies, practices, procedures, and infrastructural and automobile design are likely to yield the biggest impact, in terms of making EVs appealing to mainstream automobile consumers [1–4].

The concept of sustainable mobility is changing for the better. EV sales are expected to increase dramatically in the coming years; in particular, EV sales have been predicted to rise from 2.7% at present to 58% by 2040 [5]. Accordingly, European Economic Area (EEA) countries are seeking to ban the sales of all fossil-fueled cars by 2040 in order to achieve net-zero emissions by 2050 and have stated their apparent intentions to halt investing in new ICE platforms and models to end ICE vehicle production [6,7]. Hence, major car manufacturers have agreed to stop producing new ICE cars by 2040. For instance, some developed countries—namely, Norway, Iceland, and the Netherlands—are at the forefront of this transition, with up to 20% of their car fleet being fully or partially electric [8]. On the other hand, some developing countries—particularly those such as Kuwait—are considered to be the world's laggards regarding EV adoption, with less than 1/1000th of the world's average [3]. Consumer mindsets have also shifted toward sustainable mobility [9]. Consequently, the EV landscape in Kuwait needs to be examined and inspected, particularly from the perspective of consumer behaviors.

The concepts of consumer behavior and awareness are changing as more consumers come to accept alternative and sustainable mobility modes [10]. Customer preferences have been studied by numerous researchers and marketers in order to support the strategic planning and decision-making of business managers [11,12]. The acknowledgment of customer preferences provides useful insights for serving and executing EV marketing campaigns, which might lead them to better design features that might be appealing to new EV customers in the target market. Consequently, this might enhance customer satisfaction and, hence, create an EV community beyond early adopters or early users of EVs.

Although researchers have begun to examine the EV landscape worldwide, very little research has examined this landscape in emerging markets such as that of Kuwait. Furthermore, limited research has considered conventional car drivers in this context. Therefore, the aim of this study was twofold:

(i) To identify the preferences of EV early majority/pragmatic consumers in Kuwait, as perceived by conventional ICE car drivers;
(ii) To examine the influence of several selected demographic variables on preferences of early majority/pragmatic consumers, which might sway them to switch over from ICE vehicles to EVs, thus creating a mass market for EVs in Kuwait.

The present study hence contributes to various existing bodies of literature: (i) Five types of attributes—financial/economic, technological, infrastructure, brand, and social/environmental attributes—are considered collectively and comprehensively within a single choice set for EVs. It is important to note that, while the attributes have been examined separately among existing studies, they have not been collectively considered previously; (ii) emphasis is placed on conventional car drivers, which has not been explored previously in the literature; and (iii) this study fills a gap in the existing literature by adding more significant knowledge on the EV phenomenon, in terms of preferences, along with consumer demographic background data obtained from a large-scale survey in Kuwait.

The burgeoning sustainable development needs to be accomplished through the use of green energy to protect human health and the environment, ensuring the target of zero CO_2 by 2050 for all countries [13–15]. Regarding sustainable consumption, the automotive industry has undergone a major transformation and is endeavoring to lessen the environmental damage (e.g., air pollution, carbon footprints, greenhouse gas emissions) caused by excessive oil consumption, which is expected to reach 100,600,000 barrels per day by the end of 2023, compared with 3,100,000 barrels in 2021 [16]. The industry is trying to rapidly increase the production investment into EVs, as well as shifting the consumer response in this regard.

2. Problem Statement and Significance: Highest EV Adopters vs. the Lowest Ones

Even though geographically different, Kuwait and Norway have many similarities in terms of their wealth accumulation, economy, and population. As of the end of 2022, Norway had a population of 5.5 million, whereas Kuwait had close to 4.3 million residents [17]. Both nations are heavily reliant on oil production for wealth generation, with Kuwait in 10th and Norway in 13th place, in terms of wealth derived from such production. Kuwait's crude oil reserves account for 6.1% of the world's reserves—the 6th highest in the world. Petroleum production accounts for over half of Kuwait's GDP, 92% of its export revenues, and 90% of its governmental income. In comparison, Norwegian oil reserves only account for 0.3% of world reserves, or only 18th in the world [18]. However, Norwegians have invested their proceeds from oil more prudently, as they have accumulated a 1.1 trillion-USD sovereign wealth fund, compared with the 0.8 trillion-USD sovereign fund in Kuwait. [19]. In terms of GDP per capita (adjusted for purchasing power), these countries are also similar, with Norway having USD 78,128 vs. USD 51,528 GDP per capita for Kuwait, according to IMF estimates for 2022 [20].

One of the reasons that Kuwait has a lower GDP per capita than Norway is because of the high number of relatively low-paid ex-pats living in Kuwait. Expats account for 70% of the population of Kuwait, of which one-fourth are low-paid domestic workers, which equates to one domestic worker for every two Kuwaiti citizens. Norway, on the other hand, has a very homogeneous population, with only 20% of the population being immigrants or first-generation. In Norway, half of the immigrants are now citizens—something that would be next to impossible in Kuwait [17]. When it comes to car ownership, the rate is 2.3 million in Kuwait vs. 3.4 million in Norway. The reason for this difference can mainly be explained in terms of restrictions on acquiring driving licenses for ex-pats, who only own about 500,000 cars [3]. Kuwaiti citizens are about six times more likely to have a car than the ex-pats living there. Surprisingly, the lower number of cars in Kuwait has not led to lower CO_2 emissions than Norway. Kuwait emits 21 tCO_2e/person, with transport being the third-highest greenhouse gas (GHG)-emitting sector. Ground transportation only accounts for 12% of the total release of GHGs. Although this percentage might appear low, in comparison to other sectors, one should keep in mind that the emissions per capita in Kuwait are ranked the second highest in the world (after Qatar) and are about three times higher than the average in the European Union. This ratio also holds with respect to oil-producing Norway, which emits approximately one-fifth of the per capita emissions of Kuwait (6.7 tCO_2e/person) [21].

One of the reasons for this stark difference is that Norway has already replaced 20% of its ground transportation fleet with Electric Vehicles (about 800,000 EVs), with plans for 100% replacement. As a result, Norway achieved a 3% permanent reduction in the country's release of GHGs [22]. As this transition has yet to take place in Kuwait and EVs on Kuwaiti streets, only account for a fraction of a percent (around 300 at the end of 2021 and 600 at the end of 2022), mass EV adoption offers a tremendous opportunity to lower GHG emissions in Kuwait. EVs, as a replacement for Internal Combustion Engine (ICE) cars, offer a viable solution to Kuwait's pledge to lower GHG emissions. This would allow progress to be made towards both the United Nations, through their Nationally Determined Contributions (NDCs), and for their own national vision for the State of Kuwait by 2035, especially towards achieving its sustainability goal number 13, which states that Kuwait must "Take urgent action to combat climate change and its impacts" [23].

In a recently published article by the authors, entitled "How to sell zero-emission vehicles when petrol is almost for free" [3], as well as in research conducted by the Kuwait Institute for Scientific Research (KISR) on EV battery performance in high-temperature environments [24,25], we can state the ten main reasons for the ultra-low EV adoption rate in Kuwait:

1. Absence of fast-charging and powerful EV public charging stations that rely on 300 kW Direct Current (DC to DC), which could charge the most popular large-battery EVs in Kuwait to an 80% charge in about 20 min [26].

2. The reluctance of Kuwaiti landlords (as ex-pats are not permitted to own real estate by law in the State of Kuwait) to allow ex-pat-owners of EVs to install fast-charging 11 kW EV amplifier wall-boxes in or around their rented apartments, which could reduce the charging-time from up to 48 h for the biggest batteries down to only 5–10 h.
3. The State of Kuwait subsidizes petrol for its residents and, as a result, has one of the world's lowest retail gasoline prices (at USD 0.34 per liter). In comparison, one liter of retail petrol costs just over USD 2 (almost six times more) in Norway [27].
4. Neither the State of Kuwait nor its municipalities offer financial incentives to buy or own EVs instead of ICE cars. In comparison, Norway offers a long list of incentives, including
    a. Import and value-added tax exceptions from the purchasing price;
    b. Road tax exceptions;
    c. Ferry and toll-road fee exceptions;
    d. Permission to drive EVs on designated fast lanes for buses;
    e. Free municipal parking.

These incentives make it both cheaper to buy and own EVs over ICE cars in Norway, whereas the opposite is true for purchasing EVs in Kuwait. Furthermore, when considering the potential depreciation in value when re-selling, maintenance costs, as well as low fuel costs due to subsidies, drivers in Kuwait still simply do not see any economic benefit to converting to EVs, mainly as the purchasing cost is about 20% higher than equivalent ICE cars [3].

5. The lack of an EV community and exposure, stemming from the low number of EVs on the streets, indicate that EVs have not yet "crossed the chasm" in Kuwait—a term used for a disruptive innovation in which a certain type of technology eventually takes over the existing one [28]. The main hurdle is when the market is dominated by an early niche market made up of "innovators and techies" on the one hand, along with "visionaries and early adopters," while the market has not yet reached the "early-majority or the pragmatists." The reasons for this inability to reach the mass market can be explained by the Technology Acceptance Model (TAM), which provides two explanations: "the lack of perceived usefulness" along with "the unease of use" [29]. Our study supports this theory, as the most commonly sold EV in 2021 in Kuwait was the Porsche Taycan EV for approximately USD 200,000. This was bought primarily as the third or fourth car by affluent Kuwaiti males in their fifties and sixties as a status symbol rather than as a primary mode of transport. According to a dealership interview, the benefit of ownership for buyers was not primarily environmental but rather to be significantly faster than supercars such as those of Ferrari, Lamborghini, and so on, as the gearless EV powertrain allows for acceleration which no ICE car can compete with [3].
6. Potential EV drivers have apprehension as to how many years the battery will last in the extreme heat of Kuwait, as many have witnessed their cell phones automatically shutting down outside or inside of a car due to heat exposure. Generally, there is an 8-year guarantee on the battery (or about 150,000 km driven). The average life of an ICE car is 12–13 years, about 5 years longer than the EV battery warranty lasts. As a new EV car battery in the ninth year might cost more than the market value of the car at that time, replacing the battery might not be deemed worth the money. Thus, the life of the battery might dictate the life of the car. With potentially 30–40% less lifetime, rapid depreciation might represent the highest cost of ownership to EV owners. Luckily, KISR—the National Laboratory of Kuwait—has researched this phenomenon and has and will continue to publish data that will hopefully appease the concerns of consumers regarding this issue.
7. The almost total lack of maintenance, due to EVs only having 20 moving parts vs. up to 2000 in ICE cars, can actually pose a problem. EV owners complain that because EVs do not need as much maintenance, dealerships are reluctant to build up technical capacity or parts inventories. For example, Tesla does not even have a dealership in

Kuwait, as all updates and inspections are conducted online. Such a lack of facilities has proven problematic in the case of accidents or other mishaps [3].

8. EV owners have pointed out that the ground clearance of the car is especially important for EVs in Kuwait (i.e., the distance from the lowest point of the car to the ground). High speedbumps in residential areas aiming to keep out low-riding power cars are a problem, as they may damage the battery at the bottom of the EV.
9. Some efforts have been made by municipalities, shopping centers, and transportation authorities to have designated parking spaces with or without charging facilities. However, as no penalty is typically levied on ICE car drivers—in contrast to those who park in handicapped parking spaces—virtually all EV owners we talked to complained that the designated parking was not respected as exclusive to EVs.
10. Although farfetched, there is a moderate to strong correlation when comparing the percentage of women in national congresses worldwide, and the percentage of EVs sold that year. For example, the national parliament in Norway is represented by 46% women, with 9 out of 19 ministers being women. The Kuwait parliament, on the other hand, only has 2 women out of 50 seats and 1 woman out of 12 ministers. Several studies conducted by the Organization for Economic Cooperation and Development (OECD), among others, have demonstrated that women in power demonstrate more environmental concern than their male counterparts. This ratio might explain, in part, why the State of Kuwait has been slower than Norway to provide a support system for EV adoption [4].

Our research did not indicate that the perceived safety of EVs was a problem for current EV owners or potential EV drivers in Kuwait, as EVs are generally viewed as safer than ICE cars. This notion corresponds to American safety data from 2021, which states that EVs are about 60 times less likely to catch fire than a combustion engine car [30]; or, as Ian Must worded it, "What part of combustion in an Internal combustion engine do you not understand" [31]. Neither was the price of electricity for charging the EV a problem, as this is also subsidized and only costs about KWD 0.009 (or about USD 0.03) per kWh for a home, amounting to about USD 15 per month for average EV use [32]. Finally, our interviewees were largely aware that, due to about 3-fold higher efficiency of EV engines over ICEs regarding the transformation of the energy from the tank/battery to the wheels, EVs are still significantly environmentally friendlier than ICE vehicles, even though electricity is made using natural gas in Kuwait and more CO_2 is emitted in the construction of an EV due to the size of its battery [33].

Whether one of these ten reasons for the low EV adoption rate is more prevalent than the other is not the focal point of this paper. Experience from the oil-producing country Norway clearly indicates that mass adoption of EVs is a viable option for lowering greenhouse gas emissions and is a worthwhile tool for the State of Kuwait to meet its national and international commitments to lowering such emissions. Additional benefits derived from mass EV adoption potentially include improved air quality (as EVs are emission-free) and reduced sound pollution (as EV engines are also soundless).

3. Literature Review

In the recent literature, several reviews related to barriers and opportunities that affect the economics and development of public charging infrastructures have been conducted [34,35]. Reviews have comprehensively studied aspects such as charging solutions and optimization techniques, charging scheduling, data mining, load forecasting, wireless charging technologies, cybersecurity of onboarding charging systems, power system quality in smart microgrids and multi-microgrid networks, the application of green energy to supply EV loads, and prediction-based mechanisms for dynamic response schemes compatible with smart prosumer behaviors [36–49].

This study is a part of wider research, starting with a pilot study with 50 participants [2], that attempted to provide initial insight into new conventional car purchasing behaviors among consumers in Kuwait. Interestingly, the study concluded that there are three potential new conventional car buyer segments. The first was identified as the 'Value Seeker' group, as early majority pragmatists which are not likely to become early adopters of EVs in the GCC region. The second segment is the 'Performance Seeker' group, mainly including early-adopter younger men who prefer EVs due to their high torque (0 to 100 km in a few seconds). Finally, the third segment was the 'Safety Seekers,' a niche market mainly consisting of environmentally conscious younger women who may prefer to drive EVs, with low maintenance as a predictive determining factor for EV adoption in the GCC region in the future. It is clear that, as the study was only a pilot study delivering various predictions, there was a lack of empirical evidence. These predictions were certainly not based on collective factual data but, instead, based on the assumptions and subjective interpretations of the authors. This study was followed by qualitative deep interviews with automotive dealers in Kuwait, as well as several EV (Tesla) owners [3]. This study revealed that two dealerships dominated the market: Porsche and Mercedes. As of 2022, there was no dealership for Tesla, and these cars were imported from neighboring countries (e.g., Saudi Arabia, Bahrain, and the United Emirates). The interviews confirmed that were market niches for EVs, as per the pilot study. The majority of women, according to Mercedes car dealers, preferred to buy EVs due to environmental concerns, their soundlessness, and lack of maintenance. However, contradictory to the pilot study, the high torque and high performance of EVs did not affect their purchase by younger men but, instead, men in their 50 s to 60 s who were not environmentally concerned at all but were attracted by technology, luxury, and the faster pick-up to 100 km than high-performance ICE sports cars. A large, extended survey superseded these two studies. The purpose of this study was two-fold, including studying how to facilitate mass EV adaptation to lower the CO_2 emissions in Kuwait, which is the second-highest in the world per capita, by examining attitudes towards EVs and the environment [1]. The present study, on the other hand, explores the marketing potential of EVs. Hence, we conducted an extensive literature review to identify the attributes of EVs that have been studied around the world (i.e., developed, developing countries, and the MENA region) as perceived by potential customers of EVs (see Table 1 and Figure 1).

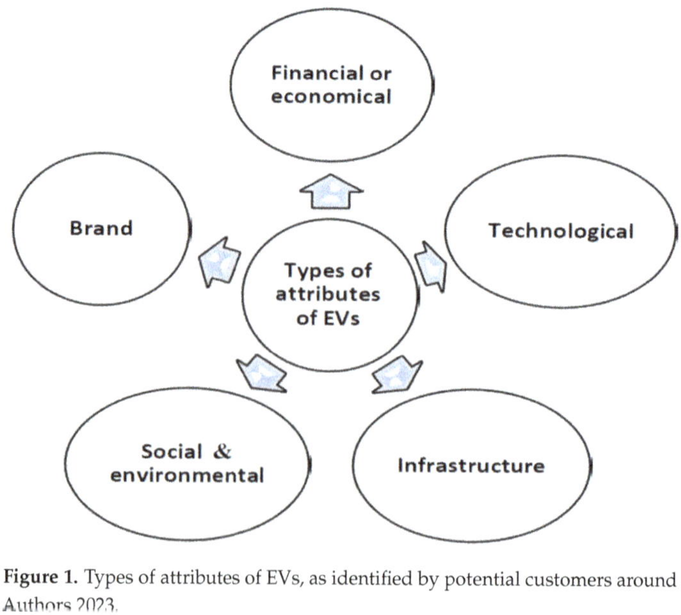

Figure 1. Types of attributes of EVs, as identified by potential customers around the world. Source: Authors 2023.

Table 1. Types of attributes of EVs as identified by potential customers around the world.

Types of Attributes	Industrialized Countries (Global North)	Developing Countries (Global South Excluding MENA Region)	MENA Region (Arab World)
Financial or economical attributes	Archsmith et al. (2021) [50], Mandys (2021) [51], Guerra and Daziano (2020) [52] Miranada and Delgado (2020) [53], Higueras-Castillo et al. (2019) [54], Rietmann and Lieven (2019) [55], Kowalska-Pyzalska et al. (2021–22) [56,57]	De Oliveira et al. (2022) [58], Lashari et al. (2021) [59], Dasharathraj et al. (2020) [60], Colak and Kaya (2020) [61], Li et al. (2020) [62], Bhalla et al. (2018) [63], Zhang et al. (2018) [64]	Eneizan (2019, Jordan) [65]
Technological attributes	Mandys (2021) [51], Archsmith et al. (2021), [50], Higueras-Castillo et al. (2019) [54]	De Oliveira et al. (2022) [58], Kim et al. (2022) [66] Ho and Huang (2022) [67], Kowalska-Pyzalska et al. (2021–22) [56,57], Kongklaew et al. (2021) [68], Khurana et al. (2020) [69], Colak and Kaya (2020) [61], Khurana et al. (2020) [70], Dasharathraj et al. (2020) [60], Haider et al. (2019) [70]	Hamwi (2022, Kuwait) [24,25]
Infrastructure attributes	Archsmith et al. (2021) [50], Miranada and Delgado (2020) [53], Guerra and Daziano (2020) [52], Rietmann and Lieven (2019) [55], Ottesen and Banna (2018) [71], Hardman et al. (2018) [72]	Kim et al. (2022) [66], De Oliveira et al. (2022) [58], Kongklaew et al. (2021) [68], Khurana et al. (2020) [70], Bhaskar et al. (2020) [73], Haider et al. (2019) [70], Bhalla et al. (2018) [63]	Shareeda et al. (2021, Bahrain) [74], Jreige et al. (2021, Lebanon) [75]
Social and environmental attributes	Archsmith et al. (2021) [50], Higueras-Castillo et al. (2020) [54], Miranada and Delgado (2020) [53], Vilchez et al. (2019) [76], Rietmann and Lieven (2019) [55]	Fan and Chen (2022) [77], Kim et al. (2022) [66], De Oliveira et al. (2022) [58], Dasharathraj et al. (2020) [60], Colak and Kaya (2020) [61], Haider et al. (2019) [70], Lin and Wu (2018) [78], Zhang et al. (2018) [35]	Al-Buenain et al. (2021, Qatar) [79], Shareeda et al. (2021, Bahrain), [74], Eneizan (2019, Jordan), [65]
Brand attributes	Kowalska-Pyzalska et al. (2021–22) [56,57]	Vongurai (2020) [80], Dasharathraj et al. [60]	

As shown in Table 1, five types of attributes have been investigated in the previous literature, including financial/economic, technological, infrastructure, brand, and social/environmental attributes. However, most studies have investigated one, two, and/or three types of these attributes, while no study has examined these five attributes collectively. Specifically, financial/economic attributes include features relating to the cost of EVs, such as purchase price, repair cost, maintenance costs, insurance costs, charging costs, warranty, and guarantee costs. Technological attributes encompass features relating to the technology used in EVs, such as battery life, battery performance, driving range, maximum speed, recharging time, acceleration, technology advancement, operating condition (i.e., heater and air condition), safety, trust, reliability, and engine performance (i.e., low noise). Infrastructure attributes include features relating to the infrastructure supporting EVs, such as charging stations and networks, commercial and public recharging infrastructure, home-based charging infrastructure, road and public infrastructure. Social and environmental attributes include features related to the governmental policies that serve social and environmental purposes, such as free parking spots, reduction in sales price, governmental subsidies and incentive policies, health and safety, tax reduction policy, and penalizing policies for conventional cars, as well as environmental friendliness. These four attribute types have been extensively investigated in the previous literature from both developed and developing countries. Brand attributes, including features such as design, brand reputation, credibility, and comfort, have been investigated to a lesser extent in the previous literature in both developed and developing countries. However, very little research has explored all five attribute types in the MENA region, particularly in Kuwait. Furthermore, little research has examined these types from the perspectives of conventional car drivers as potential customers.

It is necessary to point out that the previous literature investigating consumers' preferences for EVs typically considered a single layer of the five attributes, while other valuable factors for prediction were ignored and neglected. Interestingly, very few studies examined more than the two attributes in combination. Therefore, for the present study, we aimed to investigate all five types of attributes together in order to provide a comprehensive and up-to-date holistic picture of EV preferences. As such, the present paper helps to address this fundamental gap in the literature. Another advantage is that, by considering all five attribute types altogether, we hope to lay a solid foundation for upcoming studies to consider future re-occurring issues, as well as assist in the prediction of possible solutions for promoting the adoption of EVs, based on the present findings of this study.

In light of the above, none of the studies in the existing literature have specifically explored consumer preferences regarding EVs in Kuwait, as a critical part of the MENA region. Moreover, few studies have considered conventional car drivers in a similar context. Therefore, there is a need to explore consumer preferences for EVs from the perspective of conventional car drivers in order to identify which of these five types of attributes are considered most important by potential consumers in Kuwait, as this region constitutes a very promising market. Hence, we aimed to fill this gap in the literature by adding more significant knowledge, as stated in the introduction.

4. Data Collection and Methodology

4.1. Research Instrument

We employed a survey method by issuing a questionnaire to 472 conventional car drivers in order to achieve the objective of this study. This is part of a wider study, called "Breaking the ICE reign: mixed method study of attitudes towards buying and using EVs in Kuwait". A large-scale online questionnaire was used to provide a broader picture of the EV landscape in Kuwait and to confirm the prior survey results. A pilot test evaluation of the questionnaire has been described by Ottesen et al. (2022) [2], following which some changes and revisions were made to the format and overall design of the questionnaire. The first part focuses on data related to the demographic characteristics of the respondents, covering gender, age, marital status, education, nationality, employment, field of employment, number of cars owned by the household, job role, and household income. This first part consisted of 10 items concerning the demographic characteristics of the respondents. The second part of the questionnaire consisted of 18 closed-ended questions, which served to enrich our understanding regarding the preferences and viewpoints of conventional car drivers in Kuwait.

4.2. Sampling Procedures and Size

As this study aimed to collect data about EVs, we focused on the drivers and owners of conventional cars, who could be considered potential buyers of EVs. We decided to use the random sampling technique for data collection. For this study, the population is composed of all people who drive and/or own conventional cars in Kuwait. As the population of this study was over 100,000, the minimum required sample size for survey research was 384, as suggested by Krejcie and Morgan [81], and any sample size over 500 was considered very good, as recommended by Comrey and Lee [82]. A large sample size enables the collection of meaningful demographic data and allows one to reach actionable conclusions regarding the population. We stopped collecting data after collecting questionnaires from 604 participants. The final sample included 472 (78.1% response rate) questionnaires, which were used in the subsequent analysis.

4.3. Data Collection Procedure

Two web links were formulated, one for the Arabic version and another for the English version of the questionnaire. The authors formulated the questionnaire using Google Forms, which is a validated tool used for data collection. The researchers distributed the questionnaires among several groups in Kuwait who usually drive and/or own conventional cars, including students, the general public, faculty members, and tutors, in order to collect their feedback and comments using the two web-based links. The respondents to the questionnaire had to be at least 18 years old and conventional car drivers/owners to participate in the study. The data collection stage started in February 2022 and ran until May 2022. The purposes of the study were explained to participants, and they were asked to complete the questionnaire. The instructions for completing the questionnaire were given on the cover page in order to avoid any misunderstanding about the issue. To ensure the objectivity of the study, the respondents were asked one qualifying question to ensure whether they drive and/or own a conventional car. Only after this were they allowed to answer the rest of the questionnaire. A total of 604 persons participated in this study. After removing 132 questionnaires (i.e., those who did not have or drive a car), a total of 472 questionnaires were analyzed. As shown in Table 2, there was an approximate gender balance within the data (238 males and 234 females). Additionally, 64% of the sample (n = 304) were in the age category of 26 to 60 years, reflecting the car ownership status in Kuwait.

Table 2. Summary of the demographic characteristics of the respondents.

Variable	Categories	N = 472	%
Gender	Male	238	50.4%
	Female	234	49.6%
Age Range	18–25 years	168	35.6%
	26–39 years	222	47.0%
	40–49 years	61	12.9%
	50–60 years	21	4.4%
Marital Status	Single	272	57.6%
	Married without kids	37	7.8%
	Married with 1 kid	35	7.4%
	Married with 2 kids	42	8.9%
	Married with 3 kids or more	86	18.2%
Ethnicity	Kuwaiti	287	60.8%
	Arab Non-Kuwaiti	144	30.5%
	Asian Non-Arab	38	8.1%
	American, European, or Australian	2	0.4%
	African Non-Arab	1	0.2%
Number of Cars in household	One car	62	13.1%
	Two cars	139	29.4%
	Three cars	70	14.8%
	Four cars	66	14.0%
	Five cars or more	135	28.6%

Table 2. *Cont.*

Variable	Categories	N = 472	%
Educational Level	Less than high school	8	1.7%
	High School diploma	108	22.9%
	Trade/Commerce degree	55	11.7%
	Bachelor's degree	259	54.9%
	Master's degree	31	6.6%
	PhD	11	2.3%
Employment	Private sector	176	37.3%
	Public sector	152	32.2%
	Unemployed	83	17.6%
	Self-employed	35	7.4%
	Family-owned business	26	5.5%
Field of employment	Other private services	139	29.4%
	Government and Ministries	125	26.5%
	Family business	61	12.9%
	Education—government or private	46	9.7%
	Oil and Gas sector	32	6.8%
	Large Kuwaiti corporation	29	6.1%
	Health Care—government or private	26	5.5%
	Military or police	14	3.0%
Which of the following best describes your role in industry?	Middle Management	102	21.6%
	Administrative Staff	80	16.9%
	Upper Management	59	12.5%
	Student—Not working	54	11.4%
	Lower Management	38	8.1%
	Support Staff	34	7.2%
	Temporary Employee	28	5.9%
	Self-employed/Business Partner	27	5.7%
	Trained Professional expert	22	4.7%
	Researcher	12	2.5%
	Consultant	8	1.7%
	Skilled Laborer	8	1.7%
Monthly Income	Less than KWD 500 (USD 1650)	149	31.6%
	KWD 500–999	104	22.0%
	KWD 1000–1499	111	23.5%
	KWD 1500–1999	59	12.5%
	KWD 2000 and above (USD 6600)	49	10.4%

4.4. Statistical Analysis

We conducted statistical analysis using IBM SPSS 19 software. Descriptive statistics (i.e., frequencies and percentages) were computed to analyze the data relating to the demographic characteristics and closed-ended questions. Therefore, no hypothesis was formulated, as the study was conducted to explore the prevalence of attributes of EVs among potential consumers. ANOVA and *t*-tests were also conducted in order to determine significant differences. The *t*-test was used to compare the mean scores of two different groups of one independent variable that had only two distinct categories and one continuous dependent variable [82], while ANOVA was used to determine whether statistically significant differences in means occurred between more than two groups [83]; that is, ANOVA was used when considering one independent variable with more than two distinct or continuous categories and one continuous dependent variable.

4.5. Data Analysis and Findings

A summary of the demographic characteristics of the respondents is presented in Table 2. An approximately equal number of males (50.4%) and females (49.6%) completed the questionnaire. Approximately half of the participants (47%) were in the age range of 26 to 39 years, while more than a third of them (35.6%) were in the age range of 18 to 25 years. More than half of the respondents were single (57.6%), while approximately a fifth of them (18.2%) were married with three or more children. Approximately two-thirds (60.8%) of the participants were Kuwaiti. More than half of the respondents (54.9%) had a bachelor's degree. More than a quarter of the participants owned two cars (29.4%) or five or more cars (28.6%). More than a third of participants (37.3%) were employed in the private sector, while approximately another third of them (32.2%) were employed in the public sector. Approximately a third of participants (29.4%) worked in other private services, more than a quarter of them (26.5%) worked in the government and ministries sector, and more than a fifth of them (21.6%) were in middle management. The monthly income of more than half of the participants (53.6%) was less than KWD 1000 (USD 3300).

The following scale is used to facilitate reporting the rest of the results:

- High agreement: Calculated mean ($M \geq 3.5$);
- Medium agreement: Calculated mean ($2.5 \geq M < 3.5$);
- Low agreement: Calculated mean ($M < 2.5$)

With regard to the most favorable features of EVs, as perceived by participants (see Table 3 and Figure 2), the participants highly agreed on three features: namely, environmental friendliness, with lower CO_2 leading to better air quality ($M = 3.74$), much lower fuel price than gasoline ($M = 3.54$), and soundless engine ($M = 3.52$). Other features had a medium agreement, such as increased safety in terms of fire and crash tests ($M = 3.49$), faster and more powerful air conditioning ($M = 3.46$), much faster acceleration from 0 to 100 km ($M = 3.32$), and much lower maintenance and associated costs ($M = 3.24$).

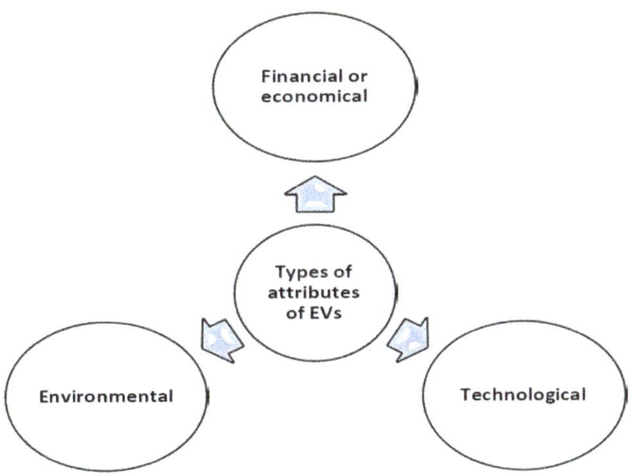

Figure 2. Attributes of EVs highly preferred by consumers in Kuwait. Source: Authors 2023.

Table 3. The most favorable features of EVs as perceived by participants (N = 472).

Type of Attribute (Features)	To What Extent You Agree/Disagree about the Most Favorable Features of EV?	Mean	SD
Social attributes—pro-environmental	Environmental friendliness, less CO_2 that leads to better air quality	3.74	1.308
Financial/economic attributes	Much lower fuel price than gasoline	3.54	1.279
Technological attributes	Soundless engine	3.52	1.305
Technological attributes	Increased safety in terms of fire and crash tests	3.49	1.305
Technological attributes	Faster and more powerful air conditioning	3.46	1.246
Technological attributes	Much faster acceleration (from 0 to 100 km)	3.32	1.236
Financial/economic attributes	Much lower maintenance and associated cost	3.24	1.311

With regard to the requirements to buy EVs, as perceived by participants (see Table 4 and Figure 3), participants highly agreed on two requirements; namely, if the guarantee of the battery would last at least 10 years or 150,000 km (M = 3.54), and if there was a fast-charging station within 5 km from almost every place in Kuwait (M = 3.51). Other requirements had a medium agreement, such as if they start to see a noticeable change in air quality due to people driving EVs (M = 3.47) if the range (how long they can drive) per full charge would be at least 400 km (M = 3.47), if they had a cool and unique design (M = 3.43), if the price was the same or lower than an equivalent gasoline car (M = 3.39) if the re-selling value would be equivalent or higher than that of a gasoline car (M = 3.39), and if the price of gasoline would increase three-fold (M = 3.33).

Table 4. The requirements to buy EVs as perceived by participants (N = 472).

Types of Attributes (Features)	I Would Buy an EV if ...	Mean	SD
Financial/economic attributes	If the guarantee of the battery lasted at least 10 years or 150,000 km	3.54	1.189
Infrastructure attributes	If there was a fast-charging station within 5 km from almost every place in Kuwait	3.51	1.193
Social attributes—pro-environmental	If I start to see noticeable change in air quality because people are driving EVs	3.47	1.162
Technological attributes	If the range (how far you can drive) per full charge would be at least 400 km	3.47	1.160
Brand attribute	If they were cool and unique design	3.43	1.217
Financial/economic attributes	The price was same or lower than equivalent gasoline car	3.39	1.287
Financial/economic attributes	If the reselling value was equivalent or higher than gasoline car	3.39	1.164
Financial/economic attributes	If gasoline prices increased three-fold	3.33	1.220
Infrastructure attributes	There was a special EV lane on major highways such as highway 30 and 40	3.27	1.165
Social attributes—social acceptance	If most of my friends or family bought an EV	2.97	1.152

The *t*-test and ANOVA were conducted to determine significant differences between the most favorable features of EVs identified in the study in terms of four background variables having two groups (i.e., gender, marital status, ethnicity, and the number of cars owned) and four background variables with more than two groups (i.e., age range, education, employment, and monthly income), respectively. The results are summarized in Table 5. When the mean score of features was compared across the eight background variables, there were strong significant differences in the mean scores of the features regarding the number of cars owned and employment. Participants who had one or two cars highly agreed on the environmental friendliness (M = 3.91, $p = 0.02$) and soundless engine (M = 3.68, $p = 0.024$) attributes of EVs, while participants who had three cars or more moderately agreed on the soundless engine feature, and highly agreed on the environmental friendliness of EVs. Furthermore, participants employed in the public sector highly agreed on the environmental friendliness (M = 3.91, $p = 0.001$), much lower fuel price (M = 3.69, $p = 0.004$), and soundless engine (M = 3.68, $p = 0.011$) of EVs, while participants who were employed in a family-owned business moderately agreed on these features. On the other hand, no significant differences in the most favorable features of EVs identified in the study could be related to the other six background variables of the participants.

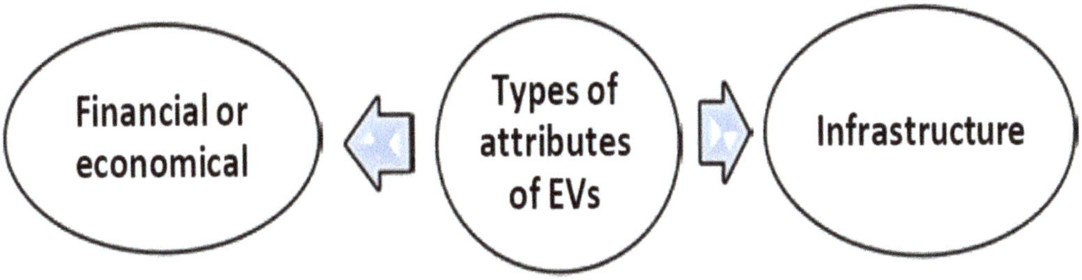

Figure 3. Types of attributes of EVs in Kuwait as identified by customers as highly necessary to buy EVs in the near future. Source: Authors 2023.

Table 5. Comparison of mean scores of the highly preferred favorable features of EVs in terms of eight background variables of the participants.

Variables	Categories (N)	The Highly Preferred Favorable Features of EVs		
		Environmental Friendliness, Less CO_2 and Sod That Lead to Better Air Quality	Much Lower Fuel Price than Gasoline	Soundless Engine
		Mean	Mean	Mean
Gender	Male (N = 238)	3.79	3.63	3.61
	Female (N = 234)	3.70	3.46	3.42
	Sig. (2-tailed)	0.482	0.163	0.113
Marital Status	Single (N = 272)	3.69	3.57	3.43
	Married (N = 200)	3.82	3.51	3.64
	Sig. (2-tailed)	0.277	0.668	0.084
Ethnicity	Kuwaiti (N = 287)	3.73	3.56	3.57
	Non-Kuwaiti (N = 185)	3.76	3.52	3.43
	Sig. (2-tailed)	0.805	0.728	0.247
Number of cars	One to Two cars (N = 201)	3.91	3.65	3.68
	Three cars or more (N = 271)	3.62	3.46	3.40
	Sig. (2-tailed)	0.020	0.117	0.024
Age Range	18–25 years (N = 168)	3.67	3.49	3.40
	26–39 years (N = 222)	3.75	3.55	3.55
	40–60 years (N = 82)	3.89	3.65	3.67
	Sig. (2-tailed)	0.447	0.655	0.260

Table 5. *Cont.*

Variables	Categories (N)	The Highly Preferred Favorable Features of EVs		
		Environmental Friendliness, Less CO_2 and Sod That Lead to Better Air Quality	Much Lower Fuel Price than Gasoline	Soundless Engine
		Mean	Mean	Mean
Education	Less than high school (N = 8)	3.00	3.00	3.00
	High School diploma (N = 108)	3.83	3.67	3.44
	Trade/Commerce degree (N = 55)	3.38	3.31	3.15
	Bachelor's degree (N = 259)	3.78	3.56	3.65
	Master's degree (N = 31)	4.00	3.55	3.55
	Ph.D. (N = 11)	3.64	3.64	3.36
	Sig. (2-tailed)	0.118	0.499	0.104
Monthly Income	Less than KD 500 (N = 149)	3.61	3.45	3.34
	KD 500–999 (N = 104)	3.75	3.53	3.52
	KD 1000–1499 (N = 111)	3.89	3.77	3.74
	KD 1500 and above (N = 108)	3.77	3.46	3.55
	Sig. (2-tailed)	0.392	0.204	0.105
Employment	Self-employed (N = 35)	3.77	3.69	3.46
	Family-owned business (N = 26)	2.81	2.81	2.77
	Private sector (N = 176)	3.81	3.63	3.58
	Public sector (N = 152)	3.91	3.69	3.68
	Unemployed (N = 83)	3.58	3.28	3.35
	Sig. (2-tailed)	0.001	0.004	0.011

Furthermore, the *t*-test and ANOVA were conducted to determine significant differences in the necessary requirements to buy EVs identified in the study in terms of the four background variables having two groups (i.e., gender, marital status, ethnicity, and the number of cars owned) and the four background variables with more than two groups (i.e., age range, education, employment, and monthly income), respectively. The results are summarized in Table 6. When the mean score of features was compared across the eight background variables, there were strong significant differences in the mean scores of the two requirements with respect to the education level and monthly income of participants. Participants who had a master's degree or Ph.D. highly agreed on if there was a fast-charging station within 5 km from almost every place in Kuwait (Ph.D. mean = 4.18, Master's mean = 3.81, $p = 0.002$), while participants who had a bachelor's degree moderately agreed on this condition. Furthermore, participants who had a monthly income between KD 1000 and 1499 or KD 1500 and above highly agreed on if the guarantee of the battery would last at least 10 years or 150,000 km (M = 3.81, M = 3.60, $p = 0.014$), and if there was a fast-charging station within 5 km from almost every place in Kuwait (M = 3.82, M = 3.52, $p = 0.009$), while participants who had a monthly income less than KD 500 moderately agreed on these features. On the other hand, no significant differences in the necessary requirements to buy EVs identified in the study were observed with respect to the other six background variables of the participants.

Table 6. Comparison of mean scores of the identified necessary requirements to buy EVs in terms of eight background variables of the participants.

Variables	Categories (N)	I Would Buy an EV if . . .	
		If the Guarantee of the Battery Would Last as Least 10 Years or 150.000 km	**If There Was a Fast-Charging Station Within 5 km from Almost Every Place in Kuwait**
		Mean	**Mean**
Gender	Male (N = 238)	3.57	3.51
	Female (N = 234)	3.51	3.51
	Sig. (2-tailed)	0.592	0.998
Marital Status	Single (N = 272)	3.55	3.53
	Married (N = 200)	3.53	3.48
	Sig. (2-tailed)	0.837	0.666
Ethnicity	Kuwaiti (N = 287)	3.60	3.58
	Non-Kuwaiti (N = 185)	3.45	3.41
	Sig. (2-tailed)	0.189	0.136
Number of cars	One to Two cars (N = 201)	3.56	3.56
	Three cars or more (N = 271)	3.52	3.48
	Sig. (2-tailed)	0.705	0.486
Age Range	18–25 years (N = 168)	3.53	3.55
	26–39 years (N = 222)	3.48	3.42
	40–60 years (N = 82)	3.72	3.68
	Sig. (2-tailed)	0.288	0.198
Education	Less than high school (N = 8)	2.63	2.13
	High School diploma (N = 108)	3.41	3.64
	Trade/Commerce degree (N = 55)	3.55	3.36
	Bachelor's degree (N = 259)	3.60	3.47
	Master's degree (N = 31)	3.68	3.81
	PhD (N = 11)	3.64	4.18
	Sig. (2-tailed)	0.206	0.002
Monthly Income	Less than KD 500 (N = 149)	3.44	3.43
	KD 500–999 (N = 104)	3.33	3.30
	KD 1000–1499 (N = 111)	3.81	3.82
	KD 1500 and above (N = 108)	3.60	3.52
	Sig. (2-tailed)	0.014	0.009
Employment	Self-employed (N = 35)	3.40	3.34
	Family-owned business (N = 26)	3.00	3.12
	Private sector (N = 176)	3.56	3.52
	Public sector (N = 152)	3.64	3.66
	Unemployed (N = 83)	3.52	3.42
	Sig. (2-tailed)	0.129	0.162

5. Discussion

We aimed to examine the consumer preferences of what we labeled as the 'early majority or pragmatists' (83% of the sample was under 40 years old and 75% had at least some level of higher education) regarding the EV market in Kuwait, as well as to explore any differences in these identified preferences with respect to their demographic background. According to the findings, drivers highly preferred three different types of attributes for EVs: Environmental attributes (e.g., in terms of leading to better air quality), financial attributes (e.g., in terms of lower fuel price), and technological attributes (e.g., in terms of their soundless engines). This means that drivers in Kuwait can be expected to prefer EVs over gasoline cars in the future due to their environmental, economic, and technological values. This result is consistent with previous studies in other countries [35,61,62,78]. One plausible explanation for this result is that drivers in Kuwait care about the triple bottom line of sustainability. The government could, therefore, implement awareness plans and agendas to better inform consumers about environmental issues and promote sustainability. This finding implies that the driving force for the adoption of EVs could be the promotion of sustainability. We believe that this form of promotion should be applied by car manufacturers in Kuwait, targeting a wider segment of the population in order to increase the rate of adoption of EVs.

There was a strongly significant difference in the identified preferred attributes for EVs in terms of the employment sector of and the number of cars owned by participants. Participants who were employed in the public sector highly preferred the three types of attributes for EVs more than those who worked in a family-owned business. This means that individuals in Kuwait who are employed in the public sector may prefer EVs over gasoline cars in the future due to their environmental, financial, and technological values. Furthermore, participants who owned one or two cars highly preferred the environmental and technological attributes of EVs, more than those who owned three or more cars. No significant differences in financial attributes could be related to the number of cars owned by participants. This means that individuals in Kuwait who own one or two cars can be expected to prefer EVs over gasoline cars in the future due to their environmental and technological values only, and they do not care as much about the price of fuel. These results characterize individuals who own a maximum of two cars and are employed in the public sector as a potential segment of EV consumers in Kuwait who place value on environmental and technological attributes, particularly regarding better air quality and having a soundless engine. This implies that manufacturers and dealerships should consider targeting these potential buyers and should continually create or maintain a competitive advantage over their competitors in terms of innovating technologies that respond to the evolving nature of environmental concerns. On the other hand, no significant differences in the three preferred attributes for EVs identified in the study could be related to the other six background variables of the participants.

Additionally, drivers were found to be highly willing to buy EVs in the future when considering their financial and infrastructure attributes. This means that drivers in Kuwait can be expected to buy EVs in the future on two conditions: Economy in battery life and the availability of nearby fast-charging station infrastructure. This result is consistent with previous studies in other countries [35,78]. One possible explanation for this result is that drivers in Kuwait are aware of both the importance of the battery life in EVs, as they commonly witness their cell phone turning off due to heat exposure and are aware of the poor EV infrastructure EVs in Kuwait. As such, the manufacturers of EVs should update consumers on the development of battery technology in a timely manner, as well as encourage the government to invest in infrastructure that supports EVs in Kuwait. This implies that EV recharging infrastructure should be accessible and available in all residential areas of Kuwait. This finding implies that the Ministry of Transport should adhere to the suggestions proposed by prior studies concerning the recharging scheduling and optimization of recharging networking in order to prevent any potential infrastructure issues [36–38]. Hence, we recommend that policymakers and government regulators start constructing infrastructure-related facilities to promote of the adoption of EVs in Kuwait in a secure manner.

We observed a strongly significant difference in the identified necessary conditions to buy EVs when considering the educational level and monthly income of participants. Participants who had a master's degree or Ph.D. were more willing to buy EVs in the future for their infrastructure attributes compared with those who had a bachelor's degree. No significant differences in financial attributes could be related to the educational level of participants. This means that people in Kuwait who had a master's degree or Ph.D. would buy EVs in the future on one condition only; namely, the availability of infrastructure (i.e., nearby fast-charging stations), and they care less about the battery life of EVs. This result characterized highly educated individuals as a potential segment of EVs in Kuwait who favor infrastructure (i.e., recharging networks). Surprisingly, this segment was unique and distinct from other groups, as they did not present favorable attributes highly related to the product itself. This signifies the important role of manufacturers in providing recharging stations and making them accessible across the country of Kuwait. Further, participants who had a monthly income of KWD 1000 (USD 3300) and above were highly willing to buy EVs in the future due to their financial and infrastructure attributes. This means that individuals in Kuwait who have a monthly income of KD 1000 and above can be expected to buy EVs in the future on two conditions: Their economic battery life and the availability of infrastructure (in terms of nearby fast-charging stations). This result characterized individuals who earned KD 1000 and above as a potential segment of EVs in Kuwait who place value on battery warranties and the required infrastructure (i.e., recharging stations) being accessible and widespread. This requires marketers to tailor sustainable products and place emphasis on warranties and guarantees associated with EVs. On the other hand, no significant differences in the two necessary conditions to buy EVs identified in the study could be related to the other six background variables of the participants.

One interesting finding of this study was that drivers were not highly willing to buy EVs in the future due to their brand attributes. This result contradicts the existing literature [56,57], in which brand attributes were considered important by consumers. One plausible explanation for this result is that drivers in Kuwait are already aware of the quality design, brand reputation, credibility, and comfort of EVs, such that there is no need to consider the brand as an important condition to buy EVs in the future. This result deserves further investigation in future research.

6. Limitations and Future Studies

First, although the present study utilized a quantitative questionnaire, it remained essentially focused on a qualitative approach using descriptive statistics due to the lack of a focus on hypothesis testing and empirical testing. For this reason, a range of reliability tests was not conducted for this study. Therefore, future studies should involve empirical investigations of various hypotheses and variables, which may yield a more reliable conclusion regarding the relevant population in Kuwait. Second, the generalizability of the results was limited by the lack of information regarding non-participants. Finally, future studies should place emphasis on both EV owners and managers of car dealership companies, collecting their viewpoints and obtaining more insight into EV-related phenomena in Kuwait. Such a proposed further study should emphasize the 'late majority' EV market, composed of drivers over the age of 40 with a lower level of education, which might reveal some perceptions based on outdated safety records (e.g., perceived safety in terms of fire) or even loyalty to gasoline (as it is the main source of wealth in Kuwait).

7. Conclusions and Implications

This is the first known study exploring consumer preferences for EVs in Kuwait in terms of which factors are influential for the 'early majority' (i.e., a part of the general market vs. a niche one) that could influence their purchasing behavior in favor of EVs. In this way, we provided a comprehensive and up-to-date picture of the preferences regarding this market. Potential consumers highly preferred three different types of attributes for EVs: environmental, financial, and technological. There were strongly significant differences in the identified preferred attributes of individuals in Kuwait for EVs with respect to the number of cars owned and sector of employment. Drivers who valued EVs most typically owned a maximum of two cars and were employed in the public sector, as well as being environmentally and technologically sensitive. Moreover, potential customers were highly willing to buy EVs in the future due to both their financial and infrastructure attributes. There were also strong significant differences in the identified necessary conditions to buy EVs in terms of educational level and monthly income. Drivers who expressed early interest in adopting EVs were typically highly educated, of mid-range income, and concerned regarding the availability of nearby fast-charging stations. This implies that the upfront purchasing price of EVs might not be important to this segment of the market in Kuwait.

The theoretical and practical implications of this study are significant. On a theoretical level, this study contributes to the limited literature on EVs in developing countries in general and Kuwait in particular, which can help researchers to compare the adoption of EVs between developing and developed countries. On a practical level, the findings suggest that drivers in Kuwait prefer EVs over gasoline cars due to their environmental, economic, and technological benefits. Therefore, marketing campaigns should highlight these values when targeting segments, including car drivers and owners. The number of cars owned by people in Kuwait was found to play an important role in preferring EVs over gasoline cars. People in Kuwait who owned one or two cars preferred EVs due to their environmental and technological benefits alone. Furthermore, the employment sector also plays an important role in the preference for EVs over gasoline cars. People in Kuwait who were employed in the public sector preferred EVs due to their environmental, economic, and technological values. Therefore, there is a need for the customization of marketing campaigns in order to address the preferences of each group.

Moreover, the results of this study suggested that drivers in Kuwait can be expected to use EVs in the future if fast-charging station infrastructure was available in a widespread manner. Therefore, policymakers and government agencies should be encouraged to start building and providing these stations in order to promote the adoption of EVs. Furthermore, this study suggested that drivers in Kuwait can be expected to use EVs in the future for economic purposes, particularly in terms of a long-lasting battery. As such, the manufacturers of EVs should try to develop batteries that last as long as possible.

The availability of nearby fast-charging station infrastructure for EVs was found to be a very important pre-condition for people in Kuwait who have a master's degree or Ph.D. to buy EVs. Therefore, it is important to raise awareness among this group of people about available charging stations in Kuwait. Monthly income was also found to play an important role in encouraging the purchase of EVs in Kuwait. Therefore, marketing campaigns should target this group of people, especially those who earn a monthly salary of more than KD 1000 (KWD 3300). These campaigns should highlight the economic value of EVs and the available EV charging infrastructure. We also suggest that the government develops subsidy programs and financial support for EV buyers in order to combat the elevated EV purchase prices.

Author Contributions: A.O. is the principal investor of the project and is responsible for the conceptualization, methodology, and editing, as well as funding acquisition and administration. S.B. and B.A. were responsible for the literature review, conceptualization, synthesis, methodology, validation, formal analysis, data curation, writing—original draft preparation, and editing. All authors have read and agreed to the published version of the manuscript.

Funding: This paper is a part of a wider study called "Breaking the ICE reign: mixed method study of attitudes towards buying and using EVs in Kuwait." The study was funded by the Kuwait Foundation for the Advancement of Sciences and administrated by the Middle East Center of the London School of Economics and Political Science and the LSE Research Ethics Committee approval.

Institutional Review Board Statement: This study was approved by the London School of Economics and Political Science Ethics Committee (00558000004KJE9AAO, dated 24 November 2021).

Informed Consent Statement: Informed statement about the usage and purpose of the study was included in the questionnaire, as directed by the LSE Ethics Committee.

Data Availability Statement: Not applicable.

Conflicts of Interest: The authors declare no conflict of interest.

References

1. Ottesen, A.; Banna, S.; Alzougool, B. Attitudes of Drivers towards Electric Vehicles in Kuwait. *Sustainability* **2022**, *14*, 12163. [CrossRef]
2. Ottesen, A.; Banna, S. Why so few EVs are in Kuwait and how to amend it. *Int. J. Eng. Technol.* **2021**, *10*, 181–189. Available online: https://www.sciencepubco.com/index.php/ijet/issue/view/559 (accessed on 30 January 2023). [CrossRef]
3. Ottesen, A.; Toglaw, S.; AlQuaoud, F.; Simovic, V. How to Sell Zero Emission Vehicles when the Petrol is almost for Free: Case of Kuwait. *J. Manag. Sci.* **2022**, *9*, 1–20. [CrossRef]
4. Ottesen, A.; Banna, S.; Alzougool, B.; Simovic, V. Driving factors for women's switch to electric vehicles in conservative Kuwait. *J. Women Entrep. Educ.* **2022**, *3*, 1–21.
5. Rosellon, M.A.D. *Clean Energy Technology in the Philippines: Case of the Electric Vehicle Industry (No. 2021-15)*; PIDS Discussion Paper Series; Public Institute for Development Study Manila: Makati, Philippines, 2021.
6. Pardi, T. *Heavier, Faster and Less Affordable Cars: The Consequence of EU Regulations for Car Emissions*; ETUI Research Paper-Report 7; European Trade Union Institute: Brussels, Belgium, 2022.
7. Kalghatgi, G. Is it really the end of internal combustion engines and petroleum in transport? *Appl. Energy* **2018**, *225*, 965–974. [CrossRef]
8. Bloomberg, N.E.F. Electric Vehicle Outlook. 2022. Available online: https://about.bnef.com/electric-vehicle-outlook/ (accessed on 30 January 2023).
9. Tirachini, A. Ride-hailing, travel behavior and sustainable mobility: An international review. *Transportation* **2020**, *47*, 2011–2047. [CrossRef]
10. Yegin, T.; Ikram, M. Analysis of Consumers' Electric Vehicle Purchase Intentions: An Expansion of the Theory of Planned Behavior. *Sustainability* **2022**, *14*, 12091. [CrossRef]
11. Cid-López, A.; Hornos, M.J.; Carrasco, R.A.; Herrera-Viedma, E. Decision-making model for designing telecom products/services based on customer preferences and non-preferences. *Technol. Econ. Dev. Econ.* **2022**, *28*, 1818–1853. [CrossRef]
12. Musumali, B. An Analysis why customers are so important and how marketers go about in understanding their decisions. *Bus. Mark. Res. J.* **2019**, *23*, 230–246.
13. Crespo, B.; Míguez-Álvarez, C.; Arce, M.E.; Cuevas-Alonso, M.; Míguez, J.L. The Sustainable Development Goals: An Experience on Higher Education. *Sustainability* **2017**, *9*, 1353. [CrossRef]
14. Ikram, M.; Sroufe, R.; Awan, U.; Abid, N. Enabling Progress in Developing Economies: A Novel Hybrid Decision-Making Model for Green Technology Planning. *Sustainability* **2021**, *14*, 258. [CrossRef]
15. Ikram, M.; Ferasso, M.; Sroufe, R.; Zhang, Q. Assessing Green Technology Indicators for Cleaner Production and Sustainable Investments in a Developing Country Context. *J. Clean. Prod.* **2021**, *322*, 129090. [CrossRef]
16. IEA. Oil Market Report. 2022. Available online: https://www.iea.org/reports/oil-market-report-august-2022 (accessed on 28 January 2023).
17. Worldometer. Norway and Kuwait Population. 2023. Available online: https://www.worldometers.info/world-population (accessed on 28 January 2023).
18. World Population Review—Reserve by Country. 2023. Available online: https://worldpopulationreview.com/country-rankings/oil-reserves-by-country (accessed on 28 January 2023).
19. Sovern Investor Leadership Conference. 2023. Available online: https://www.swfinstitute.org/fund-rankings/sovereign-wealth-fund (accessed on 28 January 2023).
20. IMF. GDP Per Capita. 2023. Available online: https://www.imf.org/external/datamapper/NGDPDPC@WEO/OEMDC/ADVEC/WEOWORLD (accessed on 28 January 2023).
21. World Bank (2023) CO2 Emission (Metric Ton Per Capita). Available online: https://data.worldbank.org/indicator/EN.ATM.CO2E.PC (accessed on 28 January 2023).
22. Alvik & Bakken. How Norway's EVs Have Cut Emission Globally. 2021. Available online: https://eto.dnv.com/2019/how-norways-evs-have-cut-emissions-globally (accessed on 20 January 2023).
23. UN Sustainable Development Goals Report Kuwait. 2022. Available online: https://www.un.org/sustainabledevelopment/climate-change/ (accessed on 28 January 2023).
24. Hamwi, H.; Rushby, T.; Mahdy, M.; Bahaj, A.S. Effect of High Ambient Temperature on Electric Vehicle Efficiency and Range: Case Study of Kuwait. *Energies* **2022**, *15*, 3178. [CrossRef]

25. Hamwi, H.; Alasseri, R.; Aldei, S.; Al-Kandari, M. A Pilot Study of Electric Vehicle Performance, Efficiency, and Limitation in Kuwait's Harsh Weather and Environment. *Energies* **2022**, *15*, 7466. [CrossRef]
26. Charged Kuwait. Charging Station in Kuwait. 2023. Available online: https://www.chargedkw.com/where-to-charge (accessed on 28 January 2023).
27. Global Petrol Prices. Gasoline Prices, Liter, 19th of December 2022. Available online: https://www.globalpetrolprices.com/gasoline_prices/ (accessed on 22 December 2022).
28. Christensen, C.; Horn, M.B.; Johnson, C. *Disrupting Class: How Disruptive Innovation Will Change the Way the World Learns*, 2nd ed.; MacGraw Hill: New York, NY, USA, 2011.
29. Muller, J.M. Comparing Technology Acceptance for Autonomous Vehicles, Battery Electric Vehicles, and Car Sharing—A study across Europe, China and North America. *Sustainability* **2019**, *11*, 4333. [CrossRef]
30. National Transportation Safety Board. 2021. Available online: https://data.ntsb.gov/Docket/?NTSBNumber=HWY19SP002 (accessed on 28 January 2023).
31. Insider. Elon Musk Calls Gas-Power Cars a Passing Fad. 2019. Available online: https://www.businessinsider.com/teslas-elon-musk-calls-gas-powered-cars-a-passing-fad-2019-8 (accessed on 28 January 2023).
32. Global Petrol Prices. Kuwait Electricity Prices. 2022. Available online: https://www.globalpetrolprices.com/Kuwait/electricity_prices/#:~{}:text=Kuwait%2C%20June%202022%3A%20The%20price,of%20power%2C%20distribution%20and%20taxes (accessed on 28 January 2023).
33. United State Environmental Agency. Electric Vehicle Myths. 2022. Available online: https://www.epa.gov/greenvehicles/electric-vehicle-myths#Myth1 (accessed on 28 January 2023).
34. Nour, M.; Chaves-Ávila, J.P.; Magdy, G.; Sánchez-Miralles, Á. Review of Positive and Negative Impacts of Electric Vehicles Charging on Electric Power Systems. *Energies* **2020**, *13*, 4675. [CrossRef]
35. Zhang, Q.; Li, H.; Zhu, L.; Campana, P.E.; Lu, H.; Wallin, F.; Sun, Q. Factors influencing the economics of public charging infrastructures for EV—A review. *Renew. Sustain. Energy Rev.* **2018**, *94*, 500–509. [CrossRef]
36. Mansouri, S.A.; Nematbakhsh, E.; Jordehi, A.R.; Tostado-Véliz, M.; Jurado, F.; Leonowicz, Z. A Risk-Based Bi-Level Bidding System to Manage Day-Ahead Electricity Market and Scheduling of Interconnected Microgrids in the presence of Smart Homes. In Proceedings of the IEEE International Conference on Environment and Electrical Engineering and 2022 IEEE Industrial and Commercial Power Systems Europe (EEEIC/I&CPS Europe), Prague, Czech Republic, 28 June–1 July 2022; pp. 1–6. [CrossRef]
37. Mansouri, S.A.; Nematbakhsh, E.; Ahmarinejad, A.; Jordehi, A.R.; Javadi, M.S.; Marzband, M. A hierarchical scheduling framework for resilience enhancement of decentralized renewable-based microgrids considering proactive actions and mobile units. *Renew. Sustain. Energy Rev.* **2022**, *168*, 112854. [CrossRef]
38. Mansouri, S.A.; Jordehi, A.R.; Marzband, M.; Tostado-Véliz, M.; Jurado, F.; Aguado, J.A. An IoT-enabled hierarchical decentralized framework for multi-energy microgrids market management in the presence of smart prosumers using a deep learning-based forecaster. *Appl. Energy* **2023**, *333*, 120560. [CrossRef]
39. Aziz, H.; Tabrizian, M.; Ansarian, M.; Ahmarinejad, A. A three-stage multi-objective optimization framework for day-ahead interaction between microgrids in active distribution networks considering flexible loads and energy storage systems. *J. Energy Storage* **2022**, *52*, 104739. [CrossRef]
40. Ravi, S.S.; Aziz, M. Utilization of electric vehicles for vehicle-to-grid services: Progress and perspectives. *Energies* **2022**, *15*, 589. [CrossRef]
41. Ben Slama, S. Prosumer in smart grids based on intelligent edge computing: A review on Artificial Intelligence Scheduling Techniques. *Ain Shams Eng. J.* **2021**, *13*, 101504. [CrossRef]
42. Solanke, T.U.; Ramachandaramurthy, V.K.; Yong, J.Y.; Pasupuleti, J.; Kasinathan, P.; Rajagopalan, A. A review of strategic charging–discharging control of grid connected electric vehicles. *J. Energy Storage* **2020**, *28*, 101193. [CrossRef]
43. Ahmad, A.; Khan, Z.A.; Saad Alam, M.; Khateeb, S. A review of the electric vehicle charging techniques, standards, progression and evolution of EV technologies in Germany. *Smart Sci.* **2018**, *6*, 36–53. [CrossRef]
44. Ahmadi, A.; Tavakoli, A.; Jamborsalamati, P.; Rezaei, N.; Miveh, M.R.; Gandoman, F.H.; Heidari, A.; Nezhad, A.E. Power quality improvement in smart grids using electric vehicles: A review. *IET Electr. Syst. Transp.* **2019**, *9*, 53–64. [CrossRef]
45. Ahmad, A.; Alam, M.S.; Chabaan, R. A Comprehensive Review of Wireless Charging Technologies for Electric Vehicles. *IEEE Trans. Transp. Electrif.* **2017**, *4*, 38–63. [CrossRef]
46. Amjad, M.; Ahmad, A.; Rehmani, M.H.; Umer, T. A review of EVs charging: From the perspective of energy optimization, optimization approaches, and charging techniques. *Transp. Res. Part D Transp. Environ.* **2018**, *62*, 386–417. [CrossRef]
47. Knez, M.; Zevnik, G.K.; Obrecht, M. A review of available chargers for electric vehicles: United States of America, European Union, and Asia. *Renew. Sustain. Energy Rev.* **2019**, *109*, 284–293. [CrossRef]
48. Al-Ogaili, A.S.; Hashim Tengku, T.J.; Rahmat, N.A.; Ramasamy, A.K.; Marsadek, M.B.; Faisal, M.; Hannan, M.A. Review on scheduling, clustering, and forecasting strategies for controlling electric vehicle charging: Challenges and recommendations. *IEEE Access* **2019**, *7*, 128353–128371. [CrossRef]
49. Adnan, N.; Nordin, S.M.; Rahman, I.; Amini, M.H. A market modeling review study on predicting Malaysian consumer behavior towards widespread adoption of PHEV/EV. *Environ. Sci. Pollut. Res.* **2017**, *24*, 17955–17975. [CrossRef]

50. Archsmith, J.; Muehlegger, E.; Rapson, D. Future Paths of Electric Vehicle Adoption in the United States: Predictable Determinants, Obstacles and Opportunities. NBER Working Paper No. w28933. 2021. Available online: https://ssrn.com (accessed on 30 January 2023).
51. Mandys, F. Electric vehicles and consumer choices. *Renew. Sustain. Energy Rev.* **2021**, *142*, 110874. [CrossRef]
52. Guerra, E.; Daziano, R. Electric vehicles and residential parking in an urban environment: Results from a stated preference experiment. *Transp. Res. Transp. Environ. Part D Transp. Environ.* **2020**, *79*, 10222. [CrossRef]
53. Miranda, J.L.; Delgado, C.J.M. Determinants of Electric Car Purchase Intention in Portugal. In *Governance and Sustainability (Developments in Corporate Governance and Responsibility)*; Crowther, D., Seifi, S., Eds.; Emerald Publishing Limited: Bingley, UK, 2020; Volume 15, pp. 161–172.
54. Higueras-Castillo, E.; Molinillo, S.; Coca-Stefaniak, J.; Liébana-Cabanillas, F. Potential Early Adopters of Hybrid and Electric Vehicles in Spain—Towards a Customer Profile. *Sustainability* **2020**, *12*, 4345. [CrossRef]
55. Rietmann, N.; Lieven, T. How policy measures succeeded to promote electric mobility—Worldwide review and outlook. *J. Clean. Prod.* **2019**, *206*, 66–75. [CrossRef]
56. Kowalska-Pyzalska, A.; Kott, M.; Kott, J. How Much Polish Consumers Know about Alternative Fuel Vehicles? Impact of Knowledge on the Willingness to Buy. *Energies* **2021**, *14*, 1438. [CrossRef]
57. Kowalska-Pyzalska, A.; Michalski, R.; Kott, M.; Skowrońska-Szmer, A. Consumer preferences towards alternative fuel vehicles. Results from the conjoint analysis. *Renew. Sustain. Energy Rev.* **2022**, *155*, 111776. [CrossRef]
58. De Oliveira, B.; Ribeiro da Silva, M.; Jugend, H.; De Camargo Fiorini, D.; Carlos Eduardo, P. Factors influencing the intention to use electric cars in Brazil. *Transp. Res. Part A Policy Pract.* **2022**, *155*, 418–433. [CrossRef]
59. Lashari, Z.A.; Ko, J.; Jang, J. Consumers' Intention to Purchase Electric Vehicles: Influences of User Attitude and Perception. *Sustainability* **2021**, *13*, 6778. [CrossRef]
60. Dasharathraj, S.; Smaran, S.; Lewlyn, R.; Nithesh, N.; Chetana, M.; Namesh, M.; Nilakshman, S. Barriers to widespread adoption of plug-in electric vehicles in emerging Asian markets: An analysis of consumer behavioral attitudes and perceptions. *Cogent Eng.* **2020**, *7*, 1796198. [CrossRef]
61. Colak, M.; Kaya, I. Providing the spark: Impact of financial incentives on battery electric vehicle adoption. *J. Environ. Econ. Manag.* **2020**, *98*, 102295.
62. Li, L.; Wang, Z.; Chen, L.; Wang, Z. Consumer preferences for battery electric vehicles: A choice experimental survey in China. *Transp. Res. Part D Transp. Environ.* **2020**, *78*, 102185. [CrossRef]
63. Bhalla, P.; Ali, I.; Nazneen, A. A Study of consumer perception and purchase intention of electric vehicles. *Eur. J. Sci. Res.* **2018**, *149*, 362–368.
64. Zhang, X.; Bai, X.; Shang, J. Is subsidized electric vehicles adoption sustainable: Consumers' perceptions and motivation toward incentive policies, environmental benefits, and risks. *J. Clean. Prod.* **2018**, *192*, 71–79. [CrossRef]
65. Eneizan, B. The adoption of electrics vehicles in Jordan based on theory of planned behavior. *Am. J. Econ. Bus. Manag.* **2019**, *2*, 1–14. [CrossRef]
66. Kim, S.; Choi, J.; Yi, Y.; Kim, H. Analysis of Influencing Factors in Purchasing Electric Vehicles Using a Structural Equation Model: Focused on Suwon City. *Sustainability* **2022**, *14*, 4744. [CrossRef]
67. Ho, J.C.; Huang, Y.H.S. Evaluation of electric vehicle power technologies: Integration of technological performance and market preference. *Clean. Responsible Consum.* **2022**, *5*, 100063. [CrossRef]
68. Kongklaew, C.; Phoungthong, K.; Prabpayak, C.; Chowdhury, S.; Khan, I.; Yuangyai, N.; Yuangyai, C.; Techato, K. Barriers to Electric Vehicle Adoption in Thailand. *Sustainability* **2021**, *13*, 12839. [CrossRef]
69. Khurana, A.; Kumar, R.; Sidhpuria, M. A Study on the Adoption of Electric Vehicles in India: The Mediating Role of Attitude. *Vision* **2020**, *24*, 23–34. [CrossRef]
70. Haider, S.W.; Zhuang, G.; Ali, S. Identifying and bridging the attitude-behavior gap in sustainable transportation adoption. *J. Ambient. Intell. Humaniz. Comput.* **2019**, *10*, 3723–3738. [CrossRef]
71. Ottesen, A.; Banna, S. Early Adopter Nation for Electric Vehicles: The Case of Iceland. In *Gulf Conference on Sustainable Built Environment*; Bumajdad, A., Bouhamra, W., Alsayegh, O., Kamal, H., Alhajraf, S., Eds.; Springer: Cham, Switzerland, 2020. [CrossRef]
72. Hardman, S.; Jenn, A.; Tal, G.; Axsen, J.; Beard, G.; Daina, N.; Figenbaum, E.; Jakobsson, N.; Jochem, P.; Kinnear, N.; et al. A review of consumer preferences of and interactions with electric vehicle charging infrastructure. *Transp. Res. Part D Transp. Environ.* **2018**, *62*, 508–523. [CrossRef]
73. Bhaskar, M.G.; Narahari, N.S.; Guptha, C.K.G. Consumer Preferences for mid-Segment electric cars—An Indian Perspective. *Int. J. Eng. Res. Technol.* **2020**, *9*, 1303–1309. [CrossRef]
74. Shareeda, A.; Al-Hashimi, M.; Hamdan, A. Smart cities and electric vehicles adoption in Bahrain. *J. Decis. Syst.* **2021**, *30*, 321–343. [CrossRef]
75. Jreige, M.; Abou-Zeid, M.; Kaysi, I. Consumer preferences for hybrid and electric vehicles and deployment of the charging infrastructure: A case study of Lebanon. *Case Stud. Transp. Policy* **2021**, *9*, 466–476. [CrossRef]
76. Vilchez, J.J.G.; Smyth, A.; Kelleher, K.; Lu, H.; Rohr, C.; Harrison, G.; Thiel, C. Electric Car Purchase Price as a Factor Determining Consumers' Choice and their Views on Incentives in Europe. *Sustainability* **2019**, *11*, 6357. [CrossRef]

77. Fan, R.; Chen, R. Promotion policies for electric vehicle diffusion in China considering dynamic consumer preferences: A network-based evolutionary analysis. *Int. J. Environ. Res. Public Health* **2022**, *19*, 5290. [CrossRef]
78. Lin, B.; Wu, W. Why people want to buy electric vehicle: An empirical study in first-tier cities of China. *Energy Policy* **2018**, *112*, 233–241. [CrossRef]
79. Al-Buenain, A.; Al-Muhannadi, S.; Falamarzi, M.; Kutty, A.A.; Kucukvar, M.; Onat, N.C. The Adoption of Electric Vehicles in Qatar Can Con-tribute to Net Carbon Emission Reduction but Requires Strong Government Incentives. *Vehicles* **2021**, *3*, 618–635. [CrossRef]
80. Vongurai, R. Factors Affecting Customer Brand Preference toward Electric Vehicle in Bangkok, Thailand. *J. Asian Financ. Econ. Bus.* **2020**, *7*, 383–393. [CrossRef]
81. Krejcie, R.V.; Morgan, D.W. Determining sample size for research activities. *Educ. Psychol. Meas.* **1970**, *30*, 607–610. [CrossRef]
82. Comrey, A.L.; Lee, H.B. *A First Course in Factor Analysis*; Erlbaum: Hillsdale, NJ, USA, 1992.
83. Pallant, J. *SPSS Survival Manual: A Step-by-Step Guide to Data Analysis Using SPSS for Windows (Version 15)*, 7th ed.; Allen & Unwin: London, UK, 2022.

designs

MDPI

Article

Optimal Hybridization with Minimum Fuel Consumption of the Hybrid Fuel Cell Train

Mario Mišić [1,*], Marinko Stojkov [1], Rudolf Tomić [2] and Mario Lovrić [1]

[1] Mechanical Engineering Faculty, University of Slavonski Brod, 35000 Slavonski Brod, Croatia
[2] Faculty of Mechanical Engineering and Naval Architecture, University of Zagreb, 10000 Zagreb, Croatia
* Correspondence: mario.misic@yahoo.com

Abstract: This paper describes a numerical study of the optimal distribution of energy between fuel cells and auxiliary energy storages in the hybrid train. Internal combustion engines (ICEs) are currently under pressure from environmental agencies due to their harmful gas emissions, and pure battery vehicles have a short range; a hybrid train powered by fuel cells, batteries, and supercapacitors can provide a viable propulsion solution. In this study, special energy management on the mountain railway with optimal power distribution and minimum hydrogen consumption is proposed. Considering the characteristics of the mountain railway, the vehicle uses recuperation of regenerative braking energy and thus charges additional power devices, and hybridization optimization gives favorable power to each power source device with a minimum consumption of hydrogen in the fuel cell. In this study, a simulation model was created in a Matlab/Simulink environment for the optimization of hybridized power systems on trains, and it can be easily modified for the hybridization of any type of train. Optimization was performed by using Sequential quadratic programming (SQP). The results show that this hybrid train topology has the ability to recover battery and supercapacitor state of charge (SOC) while meeting vehicle speed and propulsion power requirements. The effect of battery and supercapacitor parameters on power distribution and fuel consumption was also simulated.

Keywords: optimization; hybridization; train; fuel cells; power management; SQP

Citation: Mišić, M.; Stojkov, M.; Tomić, R.; Lovrić, M. Optimal Hybridization with Minimum Fuel Consumption of the Hybrid Fuel Cell Train. *Designs* **2023**, *7*, 45. https://doi.org/10.3390/designs7020045

Academic Editors: Wenbin Yu and Guang Zeng

Received: 3 February 2023
Revised: 10 March 2023
Accepted: 13 March 2023
Published: 16 March 2023

1. Introduction

Since greenhouse gases have a great impact on the environment, various measures are being taken to reduce or perhaps even eliminate harmful emissions of exhaust gases. Trains are not big polluters in themselves [1], but limits for harmful gas emissions are still prescribed for diesel trains by Leaflet UIC624 (Table 1).

In order to decrease or completely eliminate the harmful emission of exhaust gases, hybrid vehicles powered by a fuel cell can be introduced [2,3]. Such vehicles do not pollute the environment, unlike diesel vehicles, and compared to pure battery vehicles, they have a much longer range.

The advantages of introducing hybrid fuel cell trains are:

- Zero harmful emissions of exhaust gases;
- Energy savings by recuperation of regenerative braking;
- Reduced mass of all power sources;
- Optimal power distribution on each auxiliary power source.

Hybrid vehicles can have numerous advantages over each individual component. A supercapacitor could enable instantaneous cold-start operation of auxiliary devices while the fuel cell warms up. If the temperature is not too low, the battery could run the train from the depot to the departure station during a cold start. A hybrid system can allow all drive components to be smaller in size and operate with greater efficiency since none would have to provide full load and capacity.

Table 1. UIC624 ICE Emission Standards [4].

Stage	Date	Power, P [kW]	Engine Speed, r [min^{-1}]	CO	HC	NOx	PM	Smoke
				g/kWh				
UIC I	up to 31 December 2002			3	0.8	12	-	1.6–2.5 [a]
UIC II	1 January 2003	$P \leq 560$		2.5	0.6	6.0	0.25	
		$P > 560$	$r > 1000$	3	0.8	9.5	0.25 [b]	
			$r \leq 1000$	3	0.8	9.9	0.25 [b]	

[a]—Bosch smoke number (BSN) = 1.6 for engines with an air throughput of above 1 kg/s; BSN = 2.5 for engines below 0.2 kg/s; linear BSN interpolation applies between these 2 values. [b]—For engines above 2200 kW, a PM emission of 0.5 g/kWh is accepted on an exceptional basis until 31 December 2004.

Fuel cells could have zero harmful emissions of exhaust gases only if the fuel is hydrogen. In general, fuel cells can also work with methanol, natural gas, and gasoline. Since such fuels, in addition to hydrogen, contain carbon, it is impossible to avoid the exhaust gas carbon dioxide. Since the aim of this paper is to achieve zero harmful emissions of exhaust gases, only hydrogen is considered as a fuel. When using hydrogen as fuel, a chemical process produces water and energy.

Software and algorithms for simulations and optimization can contribute significantly to research. Matlab software has already been used to develop train motion simulations with Object Oriented Programming (OOP). Research in [5] shows the movement of a train with optimal driving times, which was compared with the actual railway. It showed that the simulation is very effective and applicable for the research and application of train operations.

Likewise, ADVISOR software is also a good modeling and cost-effectiveness analysis tool that was used to introduce a fuel cell hybrid locomotive. According to the traction characteristics of the Indian WDM-7 locomotive and the fuel cell as the main power source, in a certain driving cycle, the dynamic performance of the hybrid system was verified using an advanced vehicle simulator [6].

Fuel cells, batteries, and supercapacitors are very expensive devices, and it is necessary to efficiently determine their hybridization ratios. The aim is to reduce the power of the fuel cell to the lowest value and to design the battery and supercapacitor with the lowest losses [7]. In ref. [8], the numerically verified methods of hybridization are presented, and thus the optimal values of the power source are selected.

The possibility of installing fuel cells in trains has already been researched [9]. In the mentioned paper, the operation of the shunting locomotive according to the actual working cycle is presented. Energy management and power distribution between power sources were developed but without optimization. Since the shunting locomotive has demanding work, it still works on a straight railway in shorter time intervals. This is a good template for simulating a train on a mountain railway and can still show better system performance.

Matlab-Simulink™ (M/S™) can also be used to develop tools such as TrEnO, which provides tools for optimizing total energy consumption [10]. The tool is also able to estimate the efficiency, power dissipation, and thermal behavior of the traction system components. It is used during the conceptual faze of train prototypes to optimize the overall traction and braking of trains on a given railway.

The SQP optimization tool is incorporated into the M/S™ environment [11]. One of the first steps is to model the train speed trajectory. The minimum energy consumption can be set as an optimization goal, and a trade-off between energy consumption and accuracy can be made using the train trajectory model [12].

In this study, the SQP algorithm is used to optimize the hybridization of power sources of the train [13]. The model uses SQP to find the optimal hybridization ratio between the fuel cell, the battery, and the supercapacitor according to the trajectory of the train movement on the mountain railway between Knin and Perković in Croatia. The simulation itself calculates the power of the fuel cell, battery, and supercapacitor according to the given load currents of each power source [14]. The aim is the lowest consumption of hydrogen in fuel cells [15].

The energy management developed in this paper is based on SOC control. It is designed to maintain the charge of the battery and supercapacitor with optimal values, and the design was developed based on the prototype of a passenger train and a mountain railway.

2. Materials and Methods

2.1. Model and Topology of the Train

The train model in this work is based on the HŽ7022 prototype train of Croatian Railways. The train's propulsion system is modeled as a hybrid of a fuel cell (FC), battery, and supercapacitor.

A passenger train has a uniform demand for maximum power during acceleration and cruising for reduced power that must overcome the force of movement resistance. In order to analyze the electricity demand of the train, a typical timetable with a railway profile was set according to the real railway in Croatia between Knin and Perković [16].

The simulation model in M/S™ calculates the traction power at the wheels according to the railway trajectory and calculates the distribution of total train power between the power sources. The traction system topology used in trains can be connected in series or in parallel. The traction hybrid system for this train is connected in series [17] (Figure 1). Induction motors are connected to an IGBT (insulated gate bipolar transistor) converter that is connected to DC/DC converters that are directly connected to power sources [18]. The battery and supercapacitor are connected to bidirectional DC/DC converters for charging/discharging, while FC is connected to a chopper, a converter in one direction [19].

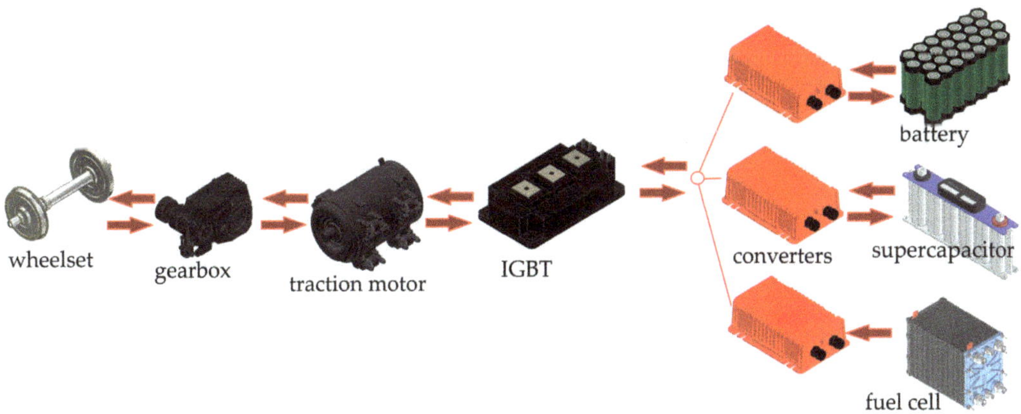

Figure 1. Power flow according to traction system topology.

According to the existing DMU (diesel multiple unit) train that will be converted into a hybrid fuel cell train, the maximum traction power is 1255 kW, and the maximum traction force is 125 kN. In a hybrid fuel cell train, the total power will be obtained from the fuel cell and auxiliary energy storage devices and taken over by the IGBT. The voltage on the IGBT is 2.4 kV, which is provided by the converters of power devices that receive a voltage of 0.8 kV. Such a high voltage generally corresponds well to the requirements of the drive system and enables lower losses in the power system.

The advantage of such a parallel system is the different voltages of the power sources. Each power source can work at its nominal voltage because the converter will raise the voltage to 2.4 kV, which is necessary for IGBT. However, it is desirable that the voltages of all power sources are approximately the same and as high as possible because the losses will be the lowest.

2.2. Energy Management System

The power flow passes with losses, and the energy management coordinates the operation of power devices and distributes the power to devices according to the hybridization ratio (Figure 2). When accelerating, energy management uses a fuel cell, battery, and supercapacitor for traction. For cruising, the supercapacitor performs traction until a state of charge SOC_{sc} = 0.01, and then the battery is used. If, during acceleration and cruising, the climb is greater than 15%, all devices are switched on. If SOC_b = 0.2, traction is performed by supercapacitor, and fuel cell will charge the battery. The battery drives the traction only after the supercapacitor is discharged (SOC_{sc} = 0.01). Regenerative braking will first charge the supercapacitor to SOC_{sc} =1, and the battery only will be charged after supercapacitor is fully charged. Energy management does not allow SOC_b = 0.2 and SOC_{sc} = 0.01 at the same time.

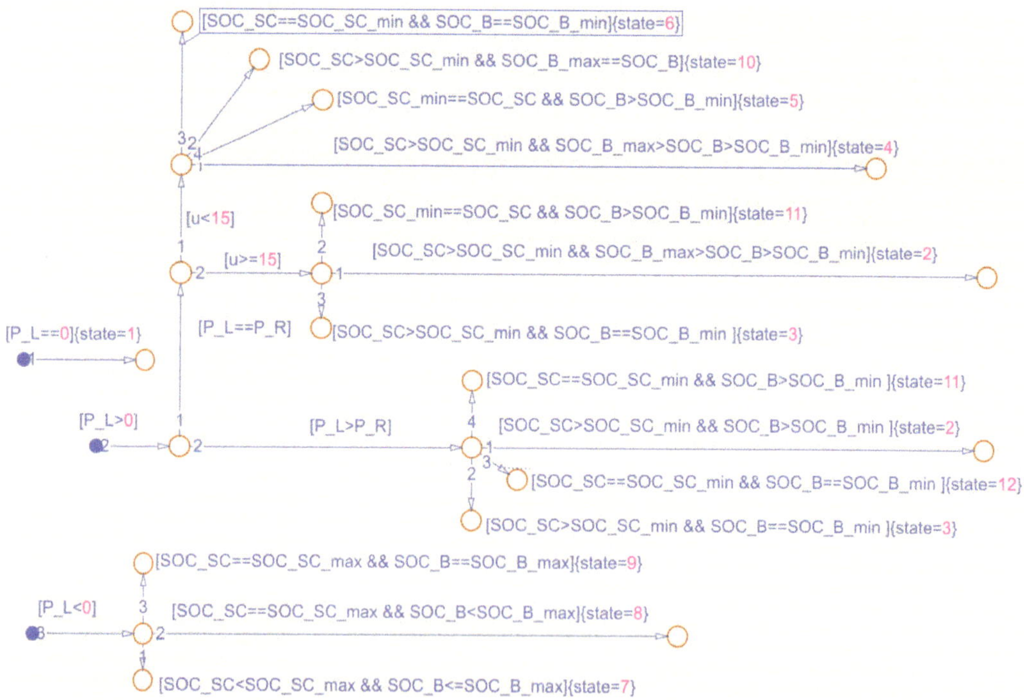

Figure 2. Energy management.

The hybridization ratio is affected by the total power, the properties of the railway, and the strength of the discharge currents of the fuel cell, battery, and supercapacitor.

2.3. Traction Force

Total power of the train is calculated as

$$P_{tot} = F_{tot} \cdot v_{tr} = (P_{fc} \cdot \eta_{fc} \cdot \eta_{dc} + P_b \cdot \eta_{dc} + P_{sc} \cdot \eta_{dc}) \cdot \eta_{ti} \cdot \eta_{tm} \cdot \eta_{gb}, \quad (1)$$

where F_{tot} is the total force for movement in N, v_{tr} is the train speed in m/s, P_{fc} is the fuel cell power in W, η_{fc} is the fuel cell efficiency, η_{dc} is the DC/DC converter efficiency, P_b is the battery power in W, P_{sc} is the supercapacitor power in W, η_{ti} is the traction inverter efficiency, η_{tm} is the traction motor efficiency, and η_{ti} is the gear box efficiency. For regenerative braking, each efficiency is calculated reciprocally ($1/\eta$), except for the fuel cell.

The total power (the total force) must be sufficient to overcome all resistances of the train:

$$F_{tot} \cdot a = F_{tr,\max} - F_{dr} - F_{gr} - F_{cu} \tag{2}$$

$$a = \pm\frac{1}{m_{tot}}(F_{tr,\max} - (r_{rr} + r_{pm} \cdot v_{tr} + r_{ar} \cdot v_{tr}{}^2)) - g \cdot (g_{rt} + \frac{0.5g}{R-30}), \tag{3}$$

$$v_{tr} = \int a\mathrm{d}t, \tag{4}$$

where F_{tr} is the traction force on wheels in N (positive for traction, negative for braking); m_{tot} is the total mass of the train in kg, increased by 6% due to the inertia of the rotating masses [20]; r_{rr} is the rolling resistance force in N; r_{pm} is the resistance coefficient of parasitic movements in N/kmh^{-1}; r_{ar} is the air resistance coefficient in N/(kmh^{-1})^2; v_{tr} is the train speed in km/h; g is the gravitational acceleration; g_{rt} is the gradient of the railway in ‰; and R is the radius of curvature of the railway in m.

The traction force of the train is achieved up to the critical speed through torque, and after that, through the power of the traction motor. The torque is constant, and therefore the traction force is also constant. The manufacturer always provides the highest constant traction force on the wheels resulting from the torque. When a critical speed is reached, the traction force decreases with increasing speed. Accordingly, the traction force is:

$$F_{tr,\max} = \begin{cases} F_{tr,\max};\ \forall 0 \leq v_{tr} \leq v_{cr} \\ F_{tr,\max} = \frac{P_{tot}}{v_{tr}},\ \forall v_{cr} < v_{tr} \leq v_{\max} \end{cases} \tag{5}$$

For acceleration, the train will use the maximum traction force; for cruising, the traction force will be equal to the total movement resistance forces; and for regenerative braking, it will use the maximum braking force (Figure 3).

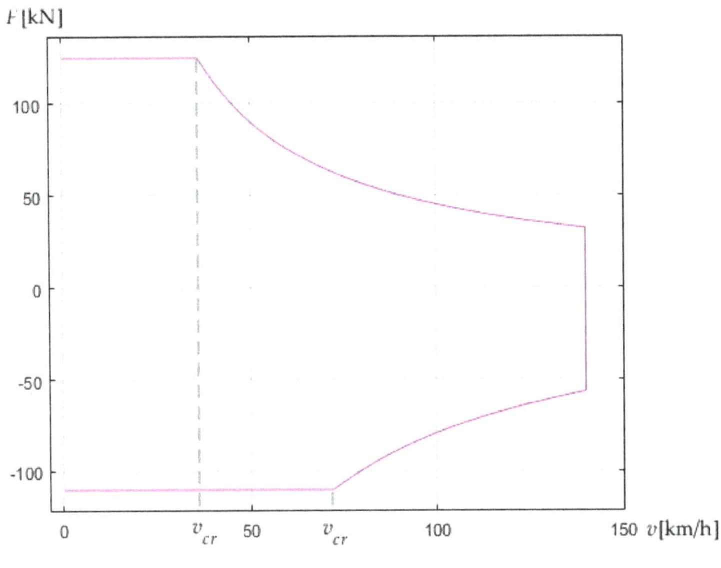

Figure 3. Traction properties of the train.

In Figure 4, the railway profile is depicted. The yellow line represents the gradient of the railway in ‰, the red line represents the speed limit, and the blue line represents train speed.

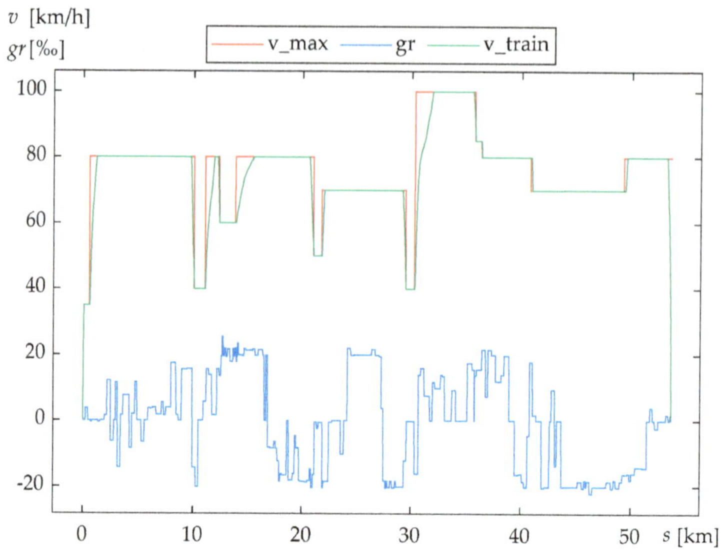

Figure 4. The railway profile.

In Figure 5, red line represents power demand, and green line represents resistance force of the railway.

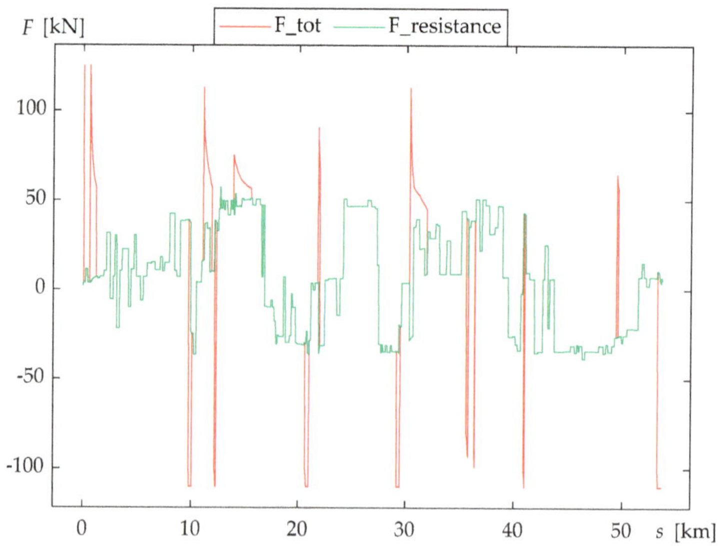

Figure 5. Demand force and the resistance force of the railway.

Based on the railway profile data depicted in Figure 5 and train parameters, resistance power and demand power were calculated. Figure 4 shows the calculated power distribution on the railway. Green line represents calculated resistance power, and red line represents the necessary traction power for the simulated train.

The railway is extremely hilly and full of radiuses, which thoroughly tests the endurance of all power sources and shows whether regenerative braking can store enough energy to make the system sustainable.

2.4. Fuel Cell

Fuel cells are devices that convert chemical energy into electrical energy. The fuel cell system contains a fuel supply system, an air supply system, a water management system, and a fuel cooling system. They include AFC (alkaline fuel cell), PEMFC (Polymer Electrolyte Membrane), DMFC (Direct Methanol), PAFC (Phosphoric Acid), MCFC (Molten Carbonate), and SOFC (Solid Oxide). Only PEMFC has zero harmful emissions of exhaust gases.

PEMFC operates at relatively low temperatures and has a high power density. It cannot change the output power very fast, so PEMFC has slow dynamics. This can be compensated by faster dynamics from storage devices. The voltage of one cell is about 1 V with a current density of 0.5 A/cm^2 to 1 A/cm^2. To obtain higher power, the cells are joined in stacks. The fuel is hydrogen, and pure oxygen or oxygen from the air can be used as an oxidant.

PEMFC can reach 1.3 kW/L power density, 0.6 kW/L system power density, and 0.6 kW/kg mass-specific power density. It operates at low temperatures and can be started and operated in sub-zero temperatures, although normal operating temperatures are 20–90 °C [21]. For the above reasons, PEMFC was chosen for this study, even though other cells exist.

The load power of the fuel cell stack is:

$$P_{fc} = U_{fc} \cdot I_{Load,fc} \tag{6}$$

$$P_{fc} = (E_{oc} - U_{act} - U_{ohm}) I_{Load,fc} \tag{7}$$

where U_{fc} is the fuel cell stack voltage in V, $I_{Load,fc}$ is the fuel cell current load in A, E_{oc} is the open circuit voltage in V, U_{ohm} is the ohmic voltage drop in V, and U_{act} is the activation voltage drop in V [22].

2.5. Battery

Lithium iron phosphate ($LiFePO_4$) is the safest cathode material used for the high-power modules required in hybrid vehicles. The advantages of this type of battery are the theoretical specific capacity of 170 Ah/kg and greater thermal stability against the release of oxygen, which makes it safer and more tolerant under extreme operating conditions [23].

It can change the output power much faster than a fuel cell, so it can compensate for power in the fuel cell. In the nominal operating range of the battery, during discharge, the voltage changes slightly. When the rated capacity is discharged, the battery enters the operating range, where the battery voltage decreases rapidly.

The load power of the battery cell for discharging is:

$$P_b = U_b \cdot I_{Load,b} \tag{8}$$

$$P_b = \left(E_0 - K\frac{Q_b}{Q_b - I_{load,b} \cdot t} I_{load,b} \cdot t - K\frac{Q_b}{Q_b - I_{load,b} \cdot t} I_{load,b} - R_b \cdot I_{load,b} + A \cdot e^{(-B \cdot I_{load,b} \cdot t)}\right) I_{load,b} \tag{9}$$

where U_b is the battery cell voltage in V, E_0 is the constant voltage in V, K is the polarization constant or the polarization resistance in Ω, Q_b is the standard battery capacity in Ah, $I_{Load,b}$ is the battery current load in A, R_b is the battery internal resistance in Ω, A is the voltage drop during the exponential zone in V, and B is the exponential time inverse constant in Ah^{-1}.

The load power of battery for charging is:

$$P_b = (E_0 - K\frac{Q_b}{Q_b - I_{load,b} \cdot t} I_{load,b} \cdot t - K\frac{Q_b}{I_{load,b} \cdot t - 0.1Q_b} I_{load,b} - R_b \cdot I_{load,b} + A \cdot e^{(-B \cdot I_{load,b} \cdot t)}) I_{load,b} \tag{10}$$

State of charge is:

$$SOC_b = 1 - \frac{1}{Q_b} \int I_{Load,b} \mathrm{d}t \tag{11}$$

2.6. Supercapacitor

Supercapacitor (EDCL, Electrochemical Double Layer Capacitors) is a device used for energy storage, and it has high energy and power densities, high efficiency (almost 95%), and long lifetime. The main property of supercapacitor is the possibility of rapid charging and discharging without loss of efficiency (>95%) during thousands of cycles [24].

Supercapacitors can be recharged in a very short time and have an excellent ability to change the output power very fast, faster than battery, and to operate with frequent peak power demands. All these reasons improve the efficiency of the vehicle and save energy, so they are suitable for installation. They especially show their excellent properties during regenerative braking.

The load power of the supercapacitor cell is:

$$P_{sc} = U_{sc} \cdot I_{Load,sc} \tag{12}$$

$$P_{sc} = (U_1 + R_1 \cdot I_{Load,sc}) \cdot I_{Load,sc} \tag{13}$$

$$P_{sc} = (\frac{-C_0 + \sqrt{C_0{}^2 + 2C_v \cdot Q_1}}{C_v} + R_1 \cdot I_{Load,sc}) \cdot I_{Load,sc} \tag{14}$$

where U_{sc} is the voltage of the supercapacitor cell in V, R_1 is the resistance of the supercapacitor's main cell in Ω, $I_{Load,sc}$ is the supercapacitor's load current in A, U_1 is the voltage of the supercapacitor's main cell in V, C_0 is the constant capacitance in F, C_v is the constant parameter in F/V, and Q_1 is the instantaneous charge in the supercapacitor's main cell in As.

State of charge is:

$$SOC_{sc} = 1 - \frac{1}{Q_{sc}} \int I_{Load,sc} \mathrm{d}t \tag{15}$$

2.7. Method of Sequential Quadratic Programming

To solve the nonlinearity and non-convexity of the resulting optimal control problem, sequential quadratic programming is used [25]. SQP is used to optimally control the management of power sources with the aim of the lowest amount of fuel consumption.

The main objective function for minimization is:

$$\min f(x) = \int_{t_0}^{t_f} P_{tot}(v(t), s(t), u(t)) \mathrm{d}t \tag{16}$$

where $v(t)$ is the speed of the train; $s(t)$ is the distance; and $u(t)$ is the set of all variables that affect driving, as shown by Equation (3).

To obtain the minimum fuel consumption, which is directly a function of energy, it is necessary to determine the discrete-time formulation of the problem. The problem of energy consumption is defined by hybridization ratios [26]. Minimization is achieved by composing a Quadratic Programming (QP) subproblem based on the quadratic approximation of the Lagrange function [27].

$$L(x, \lambda) = f(x) + \sum_{i=1}^{m} \lambda_i g_i(x) \tag{17}$$

where $f(x)$ is the main objective function, $g_i(x)$ are the inequality constraints, λ_i are Lagrange multipliers under the non-negativity constraint, and m is the total number of restrictions.

QP subproblem is given by

$$\begin{aligned} &\min_{d \in \Re^n} \tfrac{1}{2} d^T H_k d + \nabla f(x_k)^T d \\ &\nabla g_i(x_k)^T d + g_i(x_k) = 0, \ i = 1, \ldots, m_e \\ &\nabla g_i(x_k)^T d + g_i(x_k) \leq 0, \ i = m_e + 1, \ldots, m \end{aligned} \tag{18}$$

where d is the number of iterations, H_k is the Hessian Matrix, and m_e is the equality constraints number. Updated, the Hessian Matrix is:

$$H_{k+1} = H_k + \frac{q_k q_k^T}{q_k^T s_k} - \frac{H_k s_k s_k^T H_k^T}{s_k^T H_k s_k} \tag{19}$$

where

$$\begin{aligned} &s_{k+1} = x_{k+1} - x_k \\ &q_k = (\nabla f(x_{k+1}) + \sum_{i=1}^{m} \lambda_i \nabla g_i(x_{k+1})) - (\nabla f(x_k) + \sum_{i=1}^{m} \lambda_i \nabla g_i(x_k)) \end{aligned} \tag{20}$$

Duration of the simulation in time, boundary conditions, and time-dependent constraints and control variables are:

$$g_i(x) = \begin{cases} I_{sc,\min} \leq I_{Load,sc} \leq I_{sc,\max} \\ I_{b,\min} \leq I_{Load,b} \leq I_{b,\max} \\ SOC_{sc,\min} \leq SOC_{sc} \leq SOC_{sc,\max} \\ SOC_{b,\min} \leq SOC_b \leq SOC_{b,\max} \\ hr_{sc,\min} < hr_{sc} \leq hr_{sc,\max} \\ hr_{b,\min} \leq hr_b \leq hr_{b,\max} \end{cases} \tag{21}$$

where $I_{Load,sc}$ is the discharge current of the supercapacitor, $I_{Load,b}$ is the discharge current of the battery, SOC_{sc} is the state of charge of the supercapacitor, SOC_b is the state of charge of the battery, hr_{sc} is the supercapacitor hybridization ratio, and hr_b is the battery hybridization ratio.

2.8. Model Parameters

In the simulation model, the PEMFC modules "Ballard" Fcvelocity-HD6, rechargeable LFP battery "Lithium Werks" 26650, and "Maxwell" supercapacitor BCAP3000 were used (Table 2). According to the proportion of hybridization, the power used to discharge the battery and the supercapacitor was determined. In the simulation model, it was set that the algorithm optimizes the discharge current and thus obtains the optimal hybridization ratio for both the battery and the supercapacitor. As a result, the optimal discharge current of both the battery and the supercapacitor was obtained. The load currents are:

$$I_{Load,sc} = \frac{hr_{sc} \cdot P_{tot}}{U_{sc} \cdot n_{par,sc} \cdot n_{ser,sc}} \tag{22}$$

$$I_{Load,b} = \frac{hr_b \cdot P_{tot}}{U_b \cdot n_{par,b} \cdot n_{ser,b}} \tag{23}$$

where $n_{par,sc}$ is the number of supercapacitor cells in parallel, $n_{ser,sc}$ is the number of supercapacitor cells in series, $n_{par,b}$ is the number of battery cells in parallel, $n_{ser,b}$ is the number of battery cells in series, and P_{tot} is demand power.

Table 2. Power sources properties.

Fuel Cell Stack			Battery Cell		
Rated power	P_{fc}	150 kW	Rated voltage	U_0	3.3 V
Idle power	$P_{fc,\min}$	6 kW	Rated capacity	Q_0	2.56 Ah
Maximum load current	$I_{fc,\max}$	320 A	Maximum charging current	$I_{ch,\max}$	10 A
Mass	m_{fc}	404 kg	Maximum discharging current	$I_{dis,\max}$	20 A
Average hydrogen consumption	m_{H2}	2.5 g/s	Mass	m_B	76 g
Supercapacitor Cell			**Train**		
Nominal capacitance	C_{sc}	3000 F	Tractive power	$P_{tr,\max}$	1255 kW
Rated voltage	U_{sc}	2.7 V	Tractive force	$P_{tr,\max}$	125 kN
Maximum discharging current	$I_{sc,\max}$	160 A	Braking power	$P_{br,\max}$	2200 kW
Mass	m_{sc}	506.7 g	Braking force	$P_{br,\max}$	110 kN
			Mass	m_{tot}	191 t

The fuel cell stack load current is:

$$I_{Load,fc} = \frac{hr_{fc} \cdot P_{tot}}{U_{fc} \cdot n_{par,sc}} \tag{24}$$

where hr_{fc} is the fuel cell stack hybridization ratio, and $n_{par,fc}$ is the number of fuel cell stacks in parallel.

First, the optimization parameters with boundaries in the simulation and the objective function with the final desired value are selected (Figure 6).

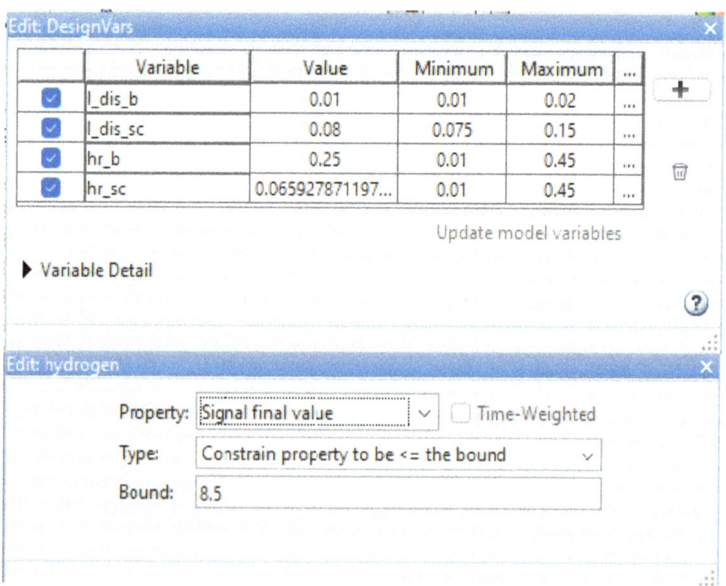

Figure 6. Setting of the main objective function, constraints, and boundaries (current in kA).

Boundaries for the hybridization ratio have been proven, and it has already been shown that with the ratio hr_b/hr_{fc} = 0.33, the vehicle can travel the most kilometers per kilogram of hydrogen. Since the supercapacitor also participates in traction, the upper boundary is 0.45 for both energy stores, as there is space for optimization. If the system does not converge, the values will expand [28].

Energy storages discharge current constraints are set according to the manufacturer's datasheet.

3. Results and Discussion

In the simulation, randomly selected parameters were taken and then optimized. Simulation results are values for the battery hybridization ratio, supercapacitor hybridization ratio, battery discharge current, and supercapacitor discharge current. Optimal parameter values for the presented study were obtained through iteration (Table 3, Figure 7).

Table 3. Optimized parameters.

Parameter		Start	Optimized
Battery hybridization ratio	hr_b	0.25	0.39
Supercapacitor hybridization ratio	hr_{sc}	0.05	0.06592787
Battery discharge current	$I_{dis,b}$	0.01 kA	0.017 kA
Supercapacitor discharge current	$I_{dis,sc}$	0.090.08 kA	0.0801 kA

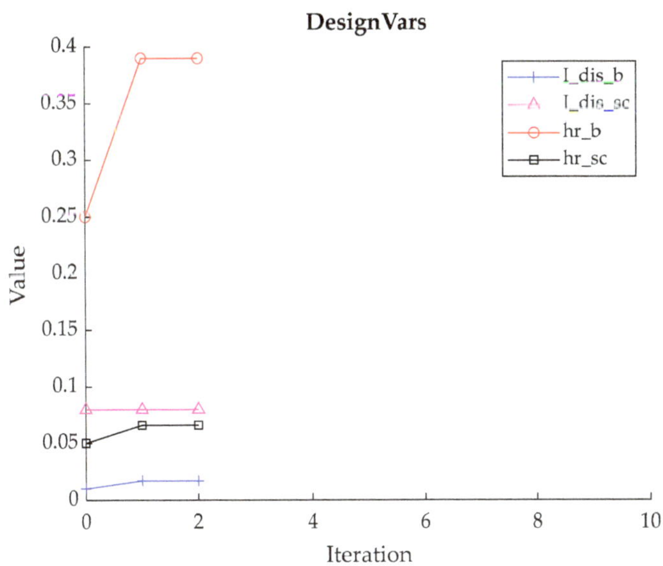

Figure 7. Iteration of parameters.

The optimal values were calculated through 2 iterations. Figure 7 depicts the iteration process.

With the optimal parameters for the proposed hybrid system, the consumption of hydrogen is 8 kg over a distance of 53.71 km, which the train travels in 2717 s. During acceleration or when driving uphill, energy management uses the power of the fuel cell, the battery, and the supercapacitor together. When cruising, the supercapacitor is first completely discharged, and then the battery is used. In order not to discharge both the battery and the supercapacitor at the same time, energy management determines the optimal hybridization ratio. Likewise, if only the supercapacitor is pulling the train, the fuel cell charges the battery. The hybridization ratio of the fuel cell is obtained by $hr_{fc} = 1 - hr_b - hr_{sc}$, which determines the power of the fuel cell and can be said to be optimal.

All that data can be graphically shown with the presented simulation model. Figure 8 shows the difference in hydrogen consumption for randomly selected and optimized parameters of the propulsion system. The optimized solution yields a decrease in fuel consumption of 14.7%. That is a saving of 1.373 kg of hydrogen compared to the initial solution.

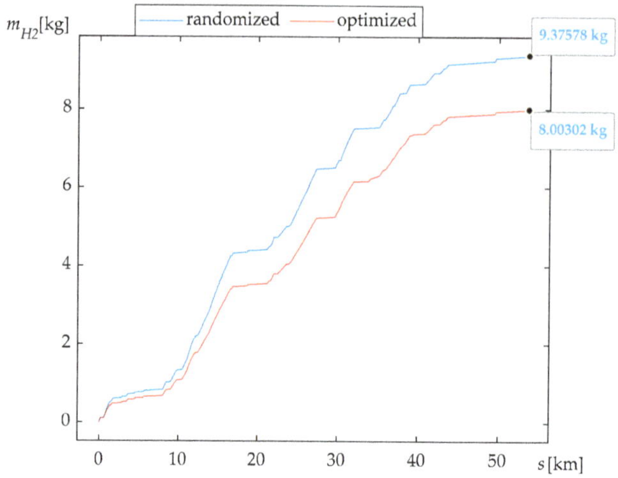

Figure 8. Consumption of the hydrogen (red is optimized, blue is randomized).

Since hydrogen consumption is a consequence of energy consumption, the consumed energy for the optimized solution is 216.99 kWh (Figure 9). Compared to the initial solution of 283.15 kWh, the optimized system results in a decrease in energy consumption of 23.37%. Due to the optimal selection of parameters and energy management that uses power sources according to given conditions, savings are possible. The profile of the railway with pronounced uphills and downhills enables regenerative braking, which is optimally used to discharge and charge the energy storage and thus save energy (Figure 10).

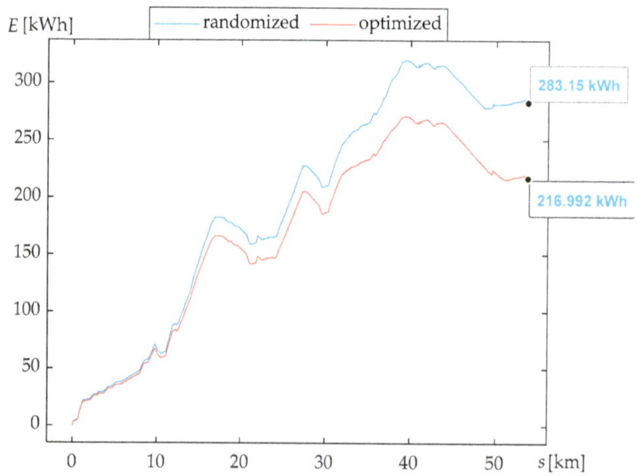

Figure 9. Demand of the energy (red is optimized, blue is randomized).

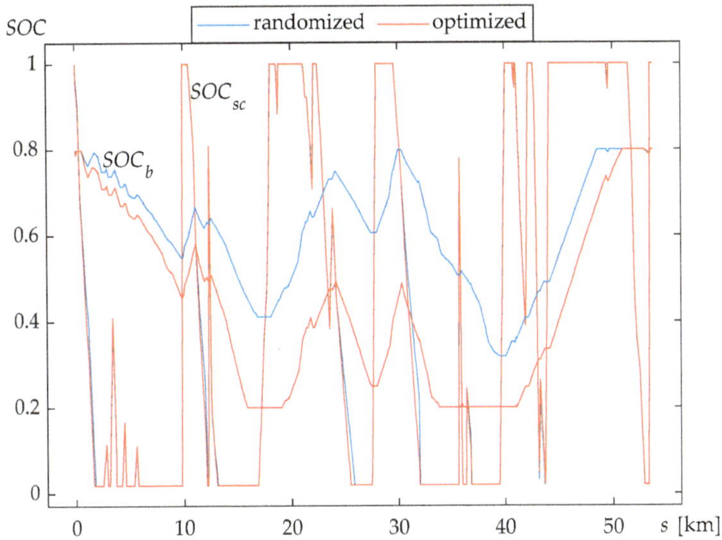

Figure 10. SOC of the battery and the supercapacitor (red is optimized, blue is randomized).

Due to the different losses in the devices, according to the hybridization ratios, the total power of all power sources will not always be the same. Consequently, with the optimal hybridization ratio, a lower required total power to propel the train was achieved. By comparing the required train power on the modeled railway for the initial and optimized drive system, this was revealed during acceleration and regenerative braking (Figure 11).

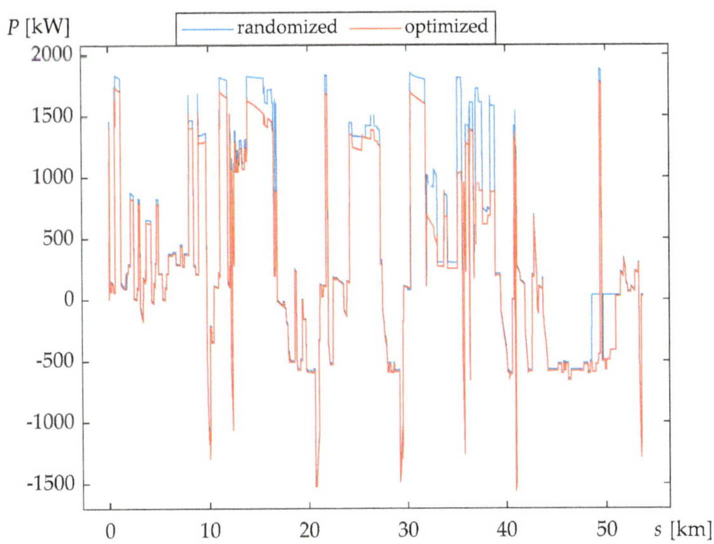

Figure 11. Demand for power (red is optimized, blue is randomized).

According to all losses, the highest power of all power sources is 1772 kW. According to the optimal hybridization ratio, the fuel cell consumes 964 kW. This power requires 7 stacks of fuel cells, which have a total mass of 2828 kg. The fuel cell with 7 stacks has 1050 kW, which means that the rest of the optimal value is used to propel auxiliary devices.

Since the hybridization ratio and discharge current of the battery package are found through optimization, the battery package has a power of 691 kW. The operating voltage of each power source is 800 V, so 243 cells will be connected in series in the battery package. According to the discharge current in the battery package, 52 cells will be connected in parallel. The mass is 960 kg.

The same types of parameters affect the supercapacitor. According to the operating voltage, 296 cells are connected in series in a supercapacitor package. By optimizing the hybridization ratio and the discharge current, the supercapacitor package has a power of 117 kW, and 2 cells are connected in parallel. According to this connection, the mass is 302 kg.

4. Conclusions

This paper presents the development of a hybrid power train system with fuel cells and presents the parameters of each component. A new energy management strategy based on railway loads and the state of charge of the energy storage is proposed. Then, the system was optimized using the SQP method, and the hydrogen consumption was calculated. By optimizing using Matlab/Simulink, it was shown that the mass of the train and the consumption of hydrogen could be reduced.

Because the railway is sharp and demanding, the energy management had to be different than what is usually set. Big uphill transits consume a lot of energy, but traveling downhill can be used for energy regeneration, which is why supercapacitor charging is used when traveling downhill, while the battery is charged via the fuel cell only when the supercapacitor conducts the traction itself. Nevertheless, the simulation showed that the train could overcome such a railway and, at the same time, find the lowest hydrogen consumption with optimal hybridization parameters and discharge currents. The result is a 14.7% decrease in hydrogen fuel consumption and 23.37% less energy consumed.

The current fuel cell systems do not leave much space for optimization. Only by optimally choosing the hybridization ratios of the battery and the supercapacitor can the power of the fuel cell be said to be optimal. The batteries and supercapacitors are delivered in smaller cells, and they give the possibility of optimization.

On the existing train prototype, the total mass of 3 diesel engines and 3 alternators is around 7300 kg. By optimizing the hybrid train with fuel cells, the mass of the power sources (including converters and inverter) was obtained at less than 4091 kg, which indicates that a fuel cell hybrid train is favorable for mass reduction. Therefore, it can be argued that a fuel cell hybrid train can have the same acceleration as a diesel train, which is important in passenger traffic.

The presented simulation model could be a useful tool in the conceptual development phase of future hybrid train propulsion systems and train modifications and shorten the development time of future environmentally friendly railway systems.

Author Contributions: Conceptualization, M.M.; methodology, R.T.; software, M.M.; validation, M.S., M.L. and R.T.; formal analysis, R.T.; investigation, M.M.; writing—original draft preparation, M.M.; writing—review and editing, M.S. and R.T.; visualization, M.L.; All authors have read and agreed to the published version of the manuscript.

Funding: This research received no external funding.

Institutional Review Board Statement: Not applicable.

Informed Consent Statement: Not applicable.

Data Availability Statement: Battery: https://lithiumwerks.com/ (accessed on 1 February 2023). Supercapacitor: https://maxwell.com/products/ultracapacitors/ (accessed on 1 February 2023). Fuel cell: https://www.ballard.com/ (accessed on 1 February 2023).

Conflicts of Interest: The authors declare no conflict of interest.

References

1. Cui, Z.; Chen, J.; Li, C.; Yin, Y.; Hao, M. The Calculation and Analysis of Carbon Emission in Traction Power Supply System of High-Speed Rail. *Tech. Gaz.* **2023**, *30*, 226–234.
2. Shang, W.; Yu, S.; Zhang, G.; Li, Q.; Chen, W. Fuel cell hybrid locomotive system based on equivalent consumption minimization strategy. In Proceedings of the Chinese Automation Con-Gress (CAC), Jinan, China, 20–22 October 2017. [CrossRef]
3. Abdelrahman, A.; Attia, Y.; Woronowicz, K.; Youssef, M. Hybrid Fuel Cell/Battery Rail Car: A Feasibility Study. *IEEE Trans. Transp. Electrif.* **2016**, *2*, 493–503. [CrossRef]
4. International Union of Railways. *UIC Code 624-Exhaust Emission Tests for Diesel Traction Engines*; International Union of Railways: Paris, France, 2017; ISBN 978-2-7461-2575-9.
5. Sumpavakup, C.; Kulworawanichpong, T. Multi-Train Movement Simulation Using MATLAB Object-Oriented Programming. *Appl. Mech. Mater.* **2015**, *763*, 153–158. [CrossRef]
6. Sarma, U.; Ganguly, S. Modelling and cost-benefit analysis of PEM fuel-cell-battery hybrid energy system for locomotive application. In Proceedings of the Technologies for Smart-City Energy Security and Power (ICSESP), Bhubaneswar, India, 28–30 March 2018.
7. Saw, L.; Somasundaram, K.; Ye, Y.; Tay, A. Electro-thermal analysis of Lithium Iron Phosphate battery for electric vehicles. *J. Power Sources* **2014**, *249*, 231–238. [CrossRef]
8. Tsukahara, K.; Kondo, K. A study on methods to design and select energy storage devices for Fuel Cell hybrid powered railway vehicles. In Proceedings of the 39th Annual Conference of the IEEE Industrial Electronics Society, Vienna, Austria, 10–13 November 2013.
9. Guo, L.; Yedavalli, K.; Zinger, D. Design and modeling of power system for a fuel cell hybrid switcher locomotive. *Energy Convers. Manag.* **2011**, *52*, 1406–1413. [CrossRef]
10. Malvezzi, M.; Pugi, L.; Conti, R.; Toni, P.; Tesi, S.; Meli, E.; Rindi, A. A Tool for Pre-diction and Optimization of Railway Traction Systems with Respect to an Expected Mis-sion Profile. *Chem. Eng. Trans.* **2013**, *33*, 7.
11. Angalaeswari, S.; Jamuna, K. Optimal Energy Management Using Sequential Quadratic Programming Algorithm for Stand Alone PV System. *Int. J. Appl. Eng. Res.* **2017**, *12*, 12250–12255.
12. Liu, X.; Xun, J.; Ning, B.; Liu, T.; Xiao, X. Moving Horizon Optimization of Train Speed Profile Based on Sequential Quadratic Programming. In Proceedings of the International Conference on Intelligent Rail Transportation (ICIRT), Singapore, 12–14 December 2018.
13. Khalik, Z.; Padilla, G.; Romijn, T.; Donkers, M. Vehicle Energy Management with Ecodriving: A Sequential Quadratic Programming Approach with Dual Decomposition. In Proceedings of the Annual American Control Conference (ACC), Milwaukee, WI, USA, 27–29 June 2018.
14. Ning, Q.; Xuan, D.; Nan, Y.; Kim, Y. Modeling and Simulation for Fuel Cell-Battery Hybrid Electric Vehicle. In Proceedings of the International Conference on Computer Modeling and Simulation, Washington, DC, USA, 20–22 February 2009.
15. Barbir, F. *PEM Fuel Cells: Theory and Practice*; Elsevier Inc.: Waltham, MA, USA, 2013.
16. *Croatian Railways-Infrastructure, Izvješće o Mreži*; Croatian Railways-Infrastructure: Zagreb, Croatia, 2022.
17. Furuta, R.; Kawasaki, J.; Kondo, K. Hybrid Traction Technologies with Energy Storage Devices for Nonelectrified Railway Lines. *IEEJ Trans. Electr. Electron. Eng.* **2010**, *5*, 291–297. [CrossRef]
18. Sehrili, E.; Cetinceviz, Y. Comparasion of Average Curent Controlled PFC SEPIC and CUK Converter Feeding Current Controlled SRM. *Tech. Gaz.* **2020**, *29*, 1789–1795. [CrossRef]
19. Lai, C.-M. Development of a Novel Bidirectional DC/DC Converter Topology with High Voltage Conversion Ratio for Electric Vehicles and DC-Microgrids. *Energies* **2016**, *9*, 410. [CrossRef]
20. Serdar, J. *Lokomotive-Opci Dio*; Sveucilisna Naklada Liber: Zagreb, Croatia, 1977.
21. Mench, M. *Fuel Cell Engines*; John Wiley & Sons, Inc.: Hoboken, NY, USA, 2008.
22. Hong, Z.; Han, Y.; Li, Q.; Chen, W. Design of Energy Management System for Fuel Cell/Supercapacitor Hybrid Locomotive. In Proceedings of the IEEE Vehicle Power and Propulsion Conference (VPPC), Hangzhou, China, 17–20 October 2016. [CrossRef]
23. Beninati, S.; Damen, L.; Mastragostino, M. Fast sol–gel synthesis of LiFePO4/C for high power lithium-ion batteries for hybrid electric vehicle application. *J. Power Sources* **2009**, *194*, 1094–1098. [CrossRef]
24. Tashima, D.; Samantara, A. *Supercapacitors for the Next Generation*; IntechOpen: London, UK, 2022.
25. Kyoungcheol, O.; Jeongmin, K.; Dongho, K.; Donghoon, C.; Hyunsoo, K. Optimal power distribution control for parallel hybrid electric vehicles. In Proceedings of the IEEE International Con-Ference on Vehicular Electronics and Safety, Xi'an, China, 14–16 October 2005.
26. Ghandriz, T.; Jacobson, B.; Murgovski, N.; Nilsson, P.; Laine, L. Real-Time Predic-tive Energy Management of Hybrid Electric Heavy Vehicles by Sequential Programming. *IEEE Trans. Veh. Technol.* **2021**, *70*, 4113–4128. [CrossRef]

27. Shivappriya, S.N.; Karthikeyan, S.; Prabu, S.; Pérez de Prado, R.; Parameshachari, B.D. A Modified ABC-SQP-Based Combined Approach for the Optimization of a Parallel Hybrid Electric Vehicle. *Energies* **2020**, *13*, 4529. [CrossRef]
28. Jeong, K.; Oh, B. Fuel economy and life-cycle cost analysis of a fuel cell hybrid ve-hicle. *J. Power Sources* **2022**, *105*, 58–65. [CrossRef]

designs

MDPI

Article

Energy-Efficient Control of a Gas Turbine Power Generation System

Marwan Al-Shami, Omar Mohamed * and Wejdan Abu Elhaija

Department of Electrical Engineering, King Abdullah I School of Graduate Studies and Scientific Research, Princess Sumaya University for Technology, Amman 11941, Jordan; marwan-996@hotmail.com (M.A.-S.); elhaija@psut.edu.jo (W.A.E.)
* Correspondence: o.mohamed@psut.edu.jo

Abstract: Gas turbines are used in the energy sectors as propulsion and power generation technologies. Despite technological advances in power generation and the emergence of numerous energy resources, gas turbine technology remains important due to its flexibility in load demand following, dynamical behavior, and the ability to work on different fuels with minor design changes. However, there would be no ambitious progress for gas turbines without reliable modeling and simulation. This paper describes a novel approach for modeling, identifying, and controlling a running gas turbine power plant. A simplified nonlinear model structure composed of s-domain transfer functions and nonlinear blocks represented by rate limiters, saturations, and look-up tables has been proposed. The model parameters have been optimized to fit real-world data. The verified model was then used to design a multiple PI/PD control to regulate the gas turbine via the inlet guide vane and fuel vales. The aim is to raise and stabilize the compressor's differential pressure or pressure ratio, as well as raise the set-point of the temperature exhausted from the combustion turbine; as a result, energy efficiency has been improved by an average of 237.16 MWh saving in energy (or 8.96% reduction in fuel consumption) for a load range of 120 MW to 240 MW.

Keywords: gas turbine; whale optimizer; modeling; identification; control

Citation: Al-Shami, M.; Mohamed, O.; Abu Elhaija, W. Energy-Efficient Control of a Gas Turbine Power Generation System. *Designs* **2023**, *7*, 85. https://doi.org/10.3390/designs7040085

Academic Editors: Wenbin Yu and Guang Zeng

Received: 5 June 2023
Revised: 27 June 2023
Accepted: 28 June 2023
Published: 3 July 2023

1. Introduction

1.1. Background and Motivation

Natural gas, hydrogen, fuel oil, and biogas are all fuels that can be used to power a gas turbine (GT), if the GT is properly configured or manufactured [1,2]. Apart from this input flexibility, the GT is classified as a flexible energy resource due to its output flexibility to follow the dynamics of electricity demand in a stable and efficient manner [3,4]. Therefore, investing more research effort into development of more efficient gas turbines can be very useful in terms of fuel, energy, and the environment. The issue of improvement in such devices is an endless research problem that requires continuous work by recent and future engineers with multidisciplinary objectives or various collaborative disciplines, such as Electrical Engineering, Artificial Intelligence, Mechanical Engineering, Chemical Engineering, and so on. The recent energy crisis caused by instabilities in the Middle East and Arabian region has made research even more difficult. Flexibility is thus an inherent feature of GTs due to its thermodynamic operational characteristic. The thermodynamics of a very basic GT is demonstrated below, where Figure 1 represents the basic components of a gas turbine and the corresponding temperature–entropy diagram is shown in Figure 2. The air compression is demonstrated by the isentropic process (1–2), in the isobaric process (2–3), the air/fuel mixture is combusted, then expanded in the isentropic gas turbine process (3–4) to create useful work for power production. In process (4–1), the output heat at constant pressure is then either rejected into the atmosphere or injected to a heat recovery steam generator (HRSG) to produce more power from the exhausted gas.

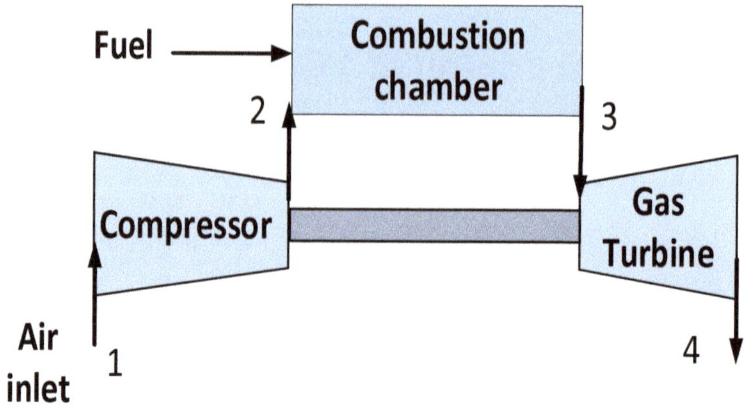

Figure 1. Main components of gas turbine unit [3].

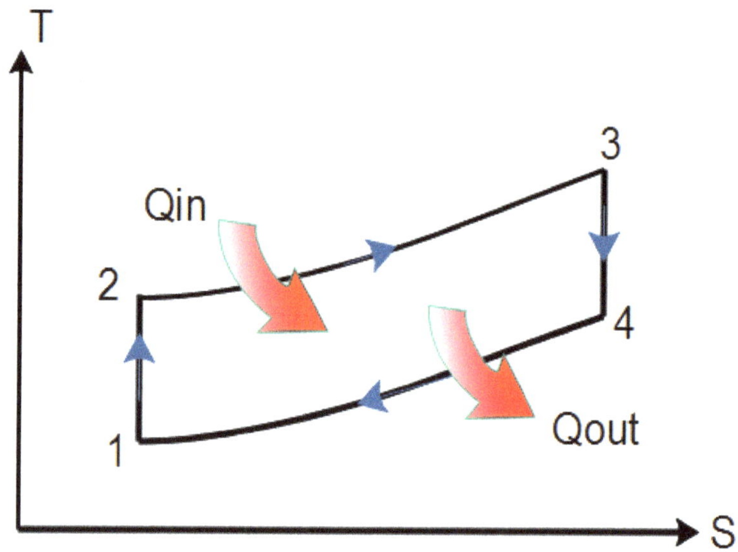

Figure 2. Temperature–Entropy (or T–S) diagram [3].

Despite technological advances in intensive research of control system technology, the directed research methodology continues to hold analogous steps towards successful control system implementation. The case under study may be far more complex for dual fuel GTs, which has been adopted in this paper, in which there are two main operating modes: the premix mode where the pilot valve and natural gas valve supply gas to the combustion chamber, and the diffusion mode where the fuel oil is permitted through the pilot valve from the fuel boosting system. In the case of low loads, the diffusion mode is better for stable combustion before, whereas the premix mode is preferable for high, peak, or near peak loads for more efficient operation. However, the situation could be further improved in the premix mode, which is the most likely situation, by proper control system design. The steps for control system design for energy-efficient improvement are depicted in Figure 3 with emphasis on our own research, which has the major steps to be explained throughout the paper context.

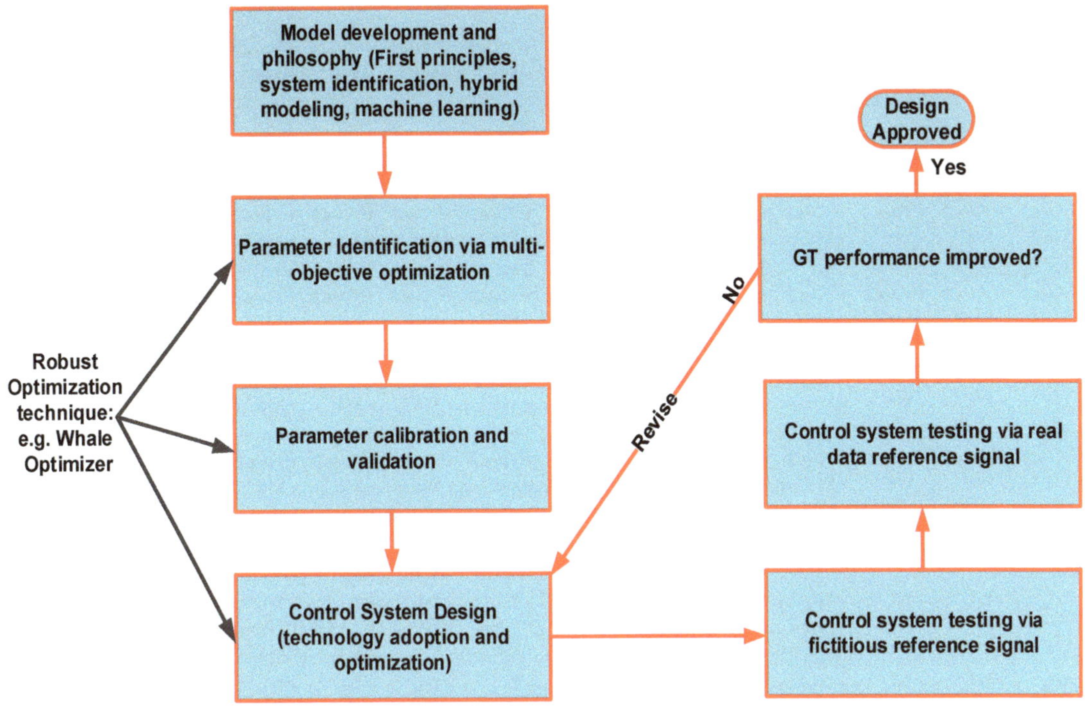

Figure 3. Directed research steps for GT control system development.

The importance of modeling GTs can be further clarified by the literature review and the explained paper contribution to theory, knowledge, and practice. This is detailed in the next subsection.

1.2. Literature Review and Paper Contributions

The control system design of Gas Turbines (GTs) has become a challenge across the world due to its intricate requirements, as its multi-timescale dynamics are naturally nonlinear and its process requirements are critical. Having multiple constraints add a huge computational load on classical control algorithms, therefore, its control requirements have raised the bar and require careful attention and intensive research.

Renewable energy sources are becoming increasingly popular; however, their intermittent nature poses a problem when integrated into the grid because it causes fluctuations in the levels of electricity generated; thus, controlling such fluctuations necessitates radical control techniques. Having the load change instantly necessitates the use of a controller that allows the power plant to respond to load changes without violating or exceeding the predefined limits or set-points, while keeping the power plant's efficiency as high as possible and meeting all of these requirements in a timely manner.

Fast load following operations, which are becoming more popular, have recently reduced plant efficiencies and shortened equipment life as load following capabilities have become more critical. This can have disastrous consequences because it increases mechanical and thermal stresses when the plant is not operating at full capacity. As a result, more intricate design requirements have been added in order to satisfy both load following capabilities and power efficiency.

The most recent control schemes are focused on increasing efficiency, lowering capital costs and running costs, reducing emissions, and enhancing operational flexibility. Being able to design and implement an optimizer that solves metaheuristic algorithms such as implementing the whale optimizer algorithm can be a potential solution.

The use of whale optimizer for of enhancing load following capabilities will ultimately minimize operating costs while explicitly fulfilling and satisfying the physical constraints associated with these power plants without causing huge burdens on the computational load. However, such a controller could require generous investment into the initial investment to fulfill these requirements.

The control of gas turbines has proven to be challenging due to their nonlinear multi-timescale dynamics, along with intrinsic constraints that need to be satisfied. Nowadays, classical control methods are adopted such as PID controllers [5], since PID controllers have the ability to offset the proportional controller which is considered a major advantage of such controllers along with their quicker response time. However, their limitations are highlighted by their high steady-state error and precincts in feedback loop stability.

The main components of a GT that need to be included in the design and analyzed are the compressor, combustion chamber, turbine, and generator. The models that have been used recently for these nonlinear systems are obtained from identification techniques only around a certain operating region. Thus, requiring improved control strategies that minimize operating costs and consider the physical constraints that exist within such a system is needed. One option which was analyzed in one of previous studies was the model predictive controller (MPC) [5–7].

The natural plant's operation begins with the compressor. The compressor is in charge of drawing filtered air from its surroundings and pressurizing it to the appropriate levels. Pressurized gas is important because having a higher-pressure gas increases the kinetic energy stored in the gas particles, thus the potential to produce more energy, but the levels to which the pressure can be leveled up must be carefully maintained. The compressed air then enters the heart of the operation, the combustion chamber, where it is mixed with natural gas, allowing it to ignite and reach high temperatures and pressures. In some cases, fuel oil must be fed into the system. It is then expanded through the turbine, causing the turbine's blades to rotate and the generator to begin producing electricity. With its nonlinear dynamics, all of these processes require careful thought and design because they are interrelated and affect one another.

The firing temperature and gas flow control the turbine's output. The inlet guide vane, which is connected to the compressor and controls the amount of gas entering the combustion chamber, is another important mechanism. The desired load is determined by the firing temperature, the inlet guide vane (IGV), and the exhaust temperature. Furthermore, the amount of fuel fed into the system and ignited has a direct impact on the load.

The gas turbine has six main controllers, each are interconnected and interdependent on one another, which are [5]:

1. Starting controller, which sets the right amount of fuel for ignition.
2. Run up controller takes over; this controller will begin during start-up and until the right speed is reached where the next controller takes over.
3. Frequency and load controller, which takes control of the turbine speed before reaching the synchronous speed, also known as full load.
4. Maximum load controller, as the name suggests, limits the maximum active power generated.
5. Temperature controllers, controls the inlet and outlet turbine temperature by controlling the variable guide vane (VGV).
6. Maximum Turbine Inlet Temperature Limiter: its main function is to limit the inlet temperature of the turbine in times of malfunction and in times of rapid load changes.

One of the most important components and working parts of the system is the variable guide vanes, which are manually handled by controlling the firing angle to acquire the desired mass flow rate; they are, in other words, responsible for regulating the amount of mass flow into the combustion chamber, and thus will ultimately control the temperature and this is crucial for the thermodynamics of the power plant. The other major components are the combustion chamber and the turbine, the fuel mass flow rate is controlled by a separate controller [7,8].

Overall, the mechanical output power of the gas turbine is the difference between the generated power of the turbine and the power consumed by the compressor. The energy is converted from chemical energy which is stored in the fuel that will be ignited, to mechanical energy, which is yielded for the turbine which will in turn rotate the shaft connected to the generator, which will in turn converts the energy to electrical energy.

It is important that a control system be designed in order to maintain constant power when the demand is so, execute the demand changes that are required, and keep a stable output voltage. A controller design was inspired from Montazer Ghaem gas power plant. They developed a controller to control the rotor speed during the start-up and changes in power demand.

Furthermore, Shete and Jape [9] have stated that during operations, gas turbines are typically operating in high-temperature and high-speed environments. The paper then has introduced Fuzzy Modified Model Reference Adaptive Controller for the speed by controlling the input fuel flowing into the combustion chamber.

Previously, PID controllers were the solution for this system because they can respond to such robust responses and can be easily tuned, but the system's efficiency becomes a concern. An alternative solution, however, has been proposed that uses a fuzzy modified model reference adaptive controller (FMMRAC) to adapt to these responses. This model is based on Rowen's model [9], a thermodynamics-based model that represents an improved and well-established GT model.

Another study simulates the behavior of a high-efficiency gas turbine with advanced cycles obtained through the use of a regenerator, an intercoder, an economizer, and steam or water injection. These components necessitate research into efficiency optimization in design, off-design, and especially transient conditions [10]. Although the addition of these components improves the overall efficiency of the system, they change the system dynamics and present some challenges to existing control schemes because these additions affect system dynamics and increase the computational load.

Gas turbines exist as single shafted or multi-shafted, where multi-shafted are assumed to be superior because they are able to adapt better, have better cost effectiveness with load variations, the mechanical inertia is much lower, and even the electromechanical time constants are at least half or a quarter of that of a single shafted gas turbine [11]. Therefore, even though multi-shafted turbines need a more generous investment and the constraints may be amplified, the benefits of implementing multi-shafted turbines enhance flexibility and working conditions.

The rotational speed is usually controlled mostly by the change in the amount of fuel supply or fuel mass flow rate to the combustion chamber which will ignite. When using gas turbine plants in isolated power system stations, frequency and voltage deviations may be neglected which will greatly affect the power quality.

Another valid solution suggested is the use of a prognostic algorithm that introduces a forecast parameter which has the ability to perfect classical control methods with minimal costs. The use of Automatic Speed Regulator (ASR) or Automatic Voltage Regulator (AVR) will limit the overshooting and transient time of voltage which provides some supremacy to PID controllers, and will maintain the rotor rotational speeds without the need for costly complicated adjustments to be implemented to control the rotational speed in the system [12], which is vital for safe performance of the GT.

The AVR and ASR results, as well as the conclusions that can be drawn from them, allows additional loads to be connected to the system in the event of load shredding abruptly.

Nowadays, industrial heavy-duty gas turbines have proven to be more reliable and more widely spread [13], especially using a hybrid system along with a pilot valve that provides fuel oil when needed, makes the gas turbines more efficient to use.

Gas turbines are composed of many interrelated systems: thermal, mechanical, and electrical to be exact. Their interaction and their dependence require deep understanding of modelling since this interaction complicates the requirements to reach an acceptable modelling accuracy.

Two main control loops are needed in the system, one is the speed governor loop for load frequency control, and the exhaust temperature control loop for energy efficiency preservation, these control the output torque and the exhaust temperature, in which these directly control the turbine behavior. The maximum allowable harvested power depends on many factors including the frequency of the system and the exhaust temperature. The temperature control limits the exhaust temperature through the cooling air system.

The complications of gas turbines basically stem from the concept of having working fluids inside them, as they are operating under high temperature and pressure, and are working under thermodynamic processes such as combustion and expansion, which have been already described as the Brayton cycle. The high temperature and pressure add to the thermal stresses that the components of the GT will be exposed to, and will affect the lifespan of the GT.

The cycle begins with the compression of air in the compressor to increase its pressure, then the compressed air is mixed with the fuel generating a high-temperature flue, which is finally expanded in the turbine in order to produce mechanical energy, which is responsible for producing the electrical power.

Other considerations that fuel control systems require in aerospace engineering are as follows [14]:

- Provides the necessary fuel for the combustion chamber;
- Controls the fuel requirements for the start-up process;
- Limits the maximum speed of the gas turbine; and
- Limits the maximum fuel flow.

Other studies that focus on power generation application of gas turbines, including this study in the present paper, usually express a combined cycle that also incorporates a gas turbine along with a heat recovery steam generator and speed, temperature, and IGV control. These combined cycles provide higher efficiency, lower unit cost, quicker construction, and have less emissions. These combined cycles differ from conventional plants in terms of their dynamic performance [15].

The existing temperature control is reflected in the open-loop system as a proportional and integral controller (PI) which limits the fuel request. It compares the exhaust temperature with the reference temperature, where the difference represents the error variable which will instigate the proper action on the controller.

Other research analyzes the transient cases where simulation and modelling of a gas power plant is important, which are start-up, variable loads, and unexpected shutdowns. As a result, the control schemes that are implemented must be chosen with great care and caution [16]. The effect of fast load following operation is basically negative to plant efficiency and reduces the equipment lifespan. As renewable sources of energy become more widespread, along with the knowledge of renewable sources of energy having an intermittent property, conventional plants must cycle their loads more often and follow the fluctuations in energy demand [17].

Natural gas combined cycles have been recognized to be more efficient, have fewer capital costs, produce less emissions, and have higher operational flexibility than coal. However, the impacts of load following capabilities disturbed by renewables may negatively affect plant efficiency, and increase the thermal and mechanical stresses on the equipment compared to operating on the base conditions, because load following focuses on decreasing the deviation between the reference and the actual without prioritizing the capabilities of the actual GT. The outage probability increases during load following, and thus increases the maintenance costs.

To promote plant efficiency, dynamic optimization must be maximized under load following. However, there are a few points to keep in mind:

1. Stresses should be kept within certain limits.
2. Temperatures should also be kept within a narrow range.
3. Maximum overall cycle efficiency should be maintained.

These constraints, which are also imposed on the multi-objective function, make it more difficult to generate a feasible solution, particularly with a high ramp rate; thus, relaxing the ramp rate is more necessary and required to solve this dilemma. It is assumed that, first, the average ramp rate is satisfied rather than the instantaneous ramp rate, and that, second, the average ramp rate can be relaxed if necessary. However, these assumptions or considerations are dependent on the system's state and operation, whether ramping up or down. When ramping up, the deviation of the average ramp rate from the desired rate is minimized to maximize efficiency. When the optimizer sees a whole ramping down, a larger relaxation, it will relax as much as possible because lowering the ramp rate improves efficiency. The desired ramp rate is the most important parameter to satisfy when using the lexicographic approach during load following operations, because all other operational objectives are naturally ordered in terms of priority. As load increases during ramp-up operations, the efficiency of the optimizer will be also improved, so simply solving a single objective optimization problem by maximizing efficiency leads to the minimization of ramp rate relaxation. The multi-objective problem becomes apparent during ramp-down operations and must be resolved.

The tradeoff that can be concluded from this reference is between the relaxation in the ramp rate and the thermal efficiency of the power plant, which means that it is difficult to satisfy both at the same time at maximum levels, but there could be some operating point where they can be both satisfied at acceptable levels. The optimal MIMO controllers may emphasize different aspects in control theory and practical characteristics, such as load following capability [18], H_2 and H_∞ for GT control as a subsystem of a combined cycle unit [19], or decentralized active disturbance rejection [20]. The latter is applicable to other power plants, fueled even by coal [21].

From a deep investigation of the literature, Whale Optimizer (WO) is still not applied and evaluated in the field of gas turbines' modeling and control. Although WO has been applied for a coal unit [22], it is worth investigating specifically on GT control system because of the high number of differences in the design, characteristics, and practical viability between the two types of power plants, which offers an opportunity for more novel achievements in this research area. The contributions of this paper are then stated as follows:

- A simplified nonlinear model of a practically operating GT has been developed and the parameters are identified by WO. The model accurately captures the turbine dynamics from 120 MW to 220 MW. The issue of petameters' calibration has been supported by the results over a wide range of settings. Moreover, the effect of relaxation of parameters on the model robustness has been investigated for the first time, which leads to high accuracy for a broader range of power changes.
- A MIMO PI/PD controller has been optimized and incorporated into the model of the existing GT as additional loops and the controller parameters have been tuned and calibrated by WO to improve the existing controller performance in terms of fuel consumption, and hence the energy efficiency. The likely operation of the adopted GT is the premix mode. Therefore, in light of this practically feasible assumption, the overall efficiency is found to be improved with significant reduction in gas consumption. This aspect has been validated through simulations of the lower natural gas consumption for the same power trends from data of existing GTs.

The rest of the paper is organized as follows: Section 1.3 presents an overview on WO, Section 2 explains the modeling approach and depicts the model simulation results, Section 3 shows the control strategy and verifies the practical feasibility, and Section 4 concludes the paper with some research findings and future recommendations.

1.3. An Overview on Whale Optimization Algorithm

There are various levels of algorithm problems, one of which is metaheuristic problems. Such problems necessitate the use of sophisticated optimization techniques. Whale optimization, which has been used in this study, is a metaheuristic algorithm that implements the hunting and feeding techniques used by whales. It can thus be implemented into power plant control schemes because it requires simple concepts, does not require gradient information, has the ability to bypass local optima, which is necessary in many scenarios, and is more important [23]. Whale optimization technique involves two main processes:

1. Exploration; which is basically a general search, where the optimizer includes all information in the search area.
2. Exploitation; which is basically an explicit search, where it investigates details in promising areas in the region of the local search.

The challenge lies in finding a balance between both processes and finding the optima in the least amount of time. "Why study whales?" one may ask. Whales are actually one of the smartest creatures scientifically as they possess twice the amount of spindle cells as humans do, and the methods they use to acquire or hunt for food are extraordinary. They use a method called bubble-net feeding either in an upward spiral or double loops. Basically, they begin by creating a bubble shape, in one of the two directions mentioned previously, around the prey and then swim up to the surface. A mathematical model has been done in one of the papers for the three main stages of their "feeding" method.

1. Encircling the prey. It is basically suggested as the first or closest value to the optima or "first guess", where then the best search agent is defined, and other search agents will update their position towards the best search agent. One of the variables is a random vector which allows the search to go beyond and search all possible regions.
2. Bubble-net attacking method, also known as exploitation. This stage consists of two approaches: either shrinking encircling mechanism or spiral updating position, each having equal probability of occurring at any interval.
3. Search for the prey (exploration), this stage basically begins the search in other promising regions.

The basic difference between exploitation and exploration is the concept of searching either locally or globally, as the exploration stage does, which is essentially determined by the value of the vector A being higher or less than 1. The basic flow chart of WO is shown in Figure 4, where X_i and X^* are the initial population and the best search agent, respectively.

According to the results of this paper, the WO tends to exploit extensively in the early stages because the first guess should be close to the optimal value, but it does lean into the early stages to switch abruptly between exploration and exploitation. Another critical parameter is the death penalty function, which considers the main objective function that must be resolved and optimized while ignoring infeasible solutions. Because of the aforementioned literature and brief background on WO, the method of modeling, identification, and control proposed in this research is expected to be superior to other techniques discussed above; the only way to determine this is to apply previous techniques of tuning the models on our developed model and observe the differences. This will involve using GA and GWO and comparing them to WO in terms of model accuracies and control system performance, which will eventually lead to valuable contributions in the field of gas turbine modeling and control.

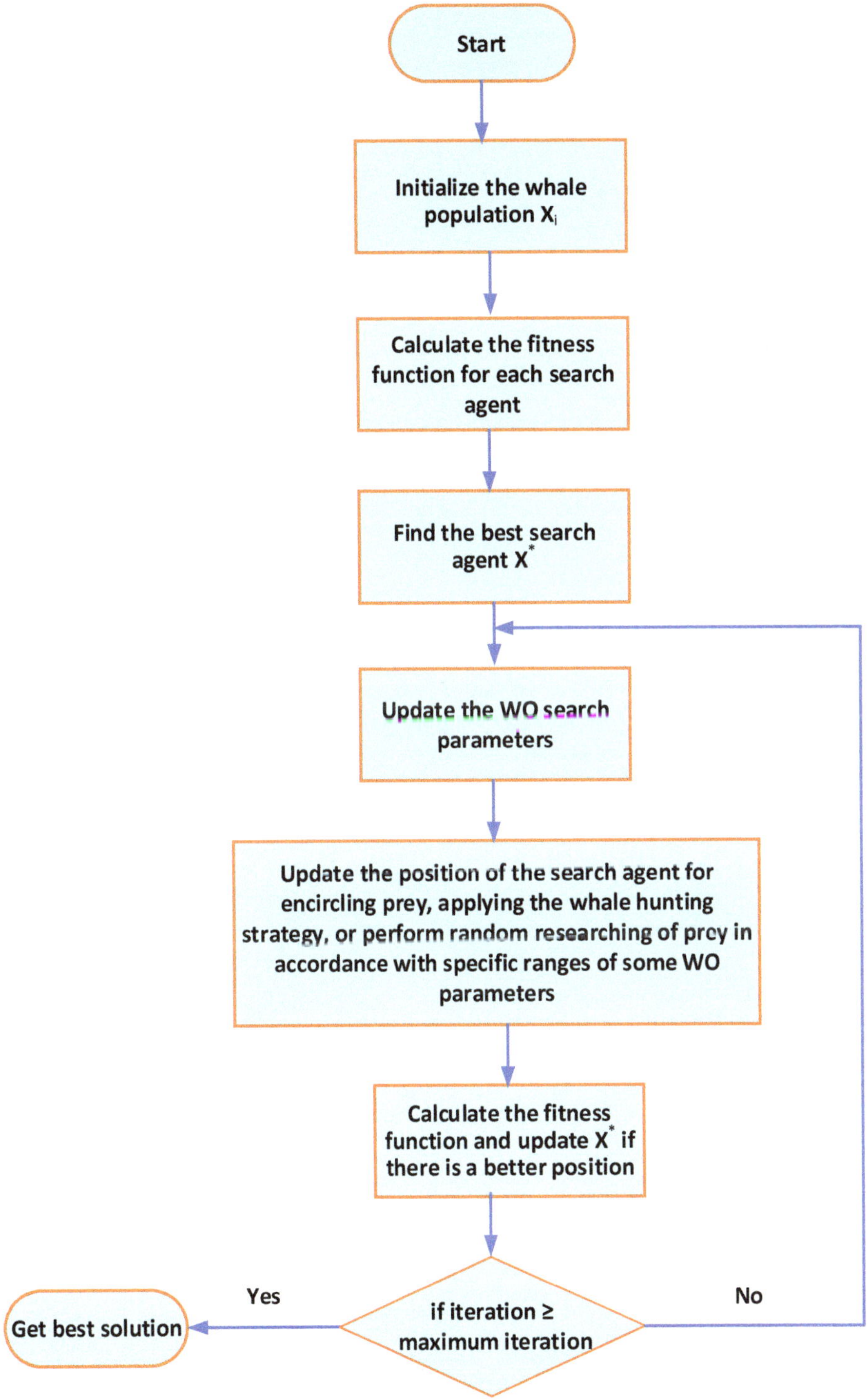

Figure 4. The flow chart of the basic working mechanism of a whale optimizer.

2. Modeling and Optimum Parameter Identification via WO

This section focuses on modeling the GT and MATLAB simulations for the approach of the GT modeling and identifying its unknown parameters. The power plant is fueled by natural gas and fuel oil which may be regulated via inlet guide vanes, and the data covers the operation from 120 MW to about 240 MW, which have been reproduced from Open Access previous publications of the corresponding author [6].

The inputs of the power plant are as follows: Natural gas valve, pilot valve, and compressor output pressure ratio. They are all represented by their normalized percentages of opening. The outputs of the power plant are active power measured in megawatts (MW) and exhaust temperature measured in degree Celsius. The final model structure is shown in Figure 5

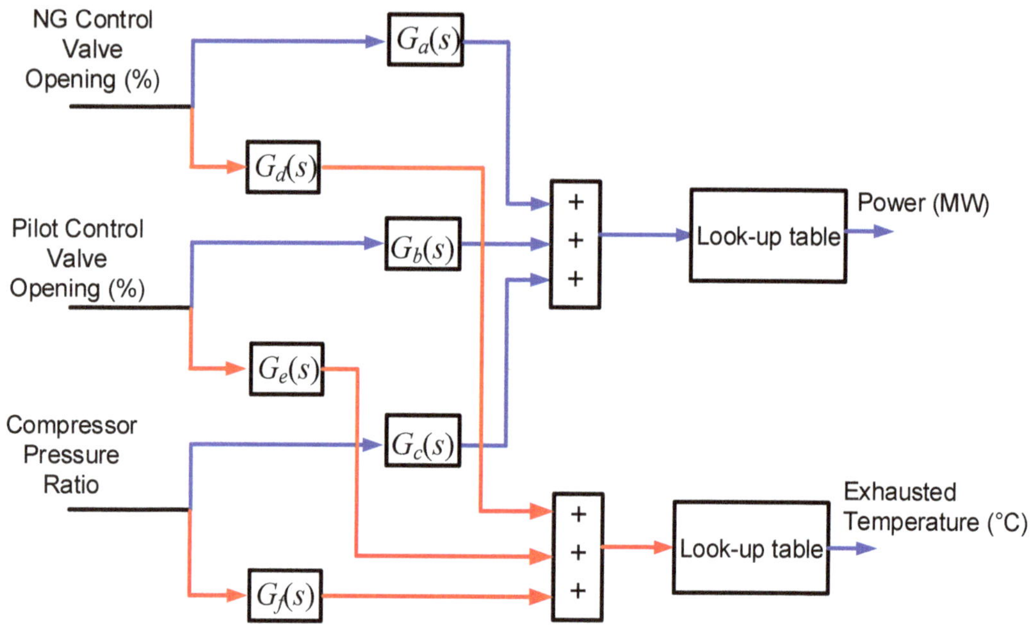

Figure 5. The Mathematical Model of the Gas Turbine, Red: the input signal flows through the system to the exhausted temperature, Blue: the input signal flows through the system to the output power.

The transfer functions of the system in Figure 5 are adopted to be as follows:

$$G_a(s) = \frac{a_1 s + a_2}{a_3 s^2 + a_4 s + a_5} \tag{1}$$

$$G_b(s) = \frac{b_1 s^2 + b_2 s + b_3}{b_4 s^2 + b_5 s + b_6} \tag{2}$$

$$G_c(s) = \frac{c_1 s^2 + c_2 s + c_3}{c_4 s^2 + c_5 s + c_6} \tag{3}$$

$$G_d(s) = \frac{d_1 s + d_2}{d_3 s^2 + d_4 s + d_5} \tag{4}$$

$$G_e(s) = \frac{e_1 s + e_2}{e_3 s^2 + e_4 s + e_5} \tag{5}$$

$$G_f(s) = \frac{f_1}{f_2 s^2 + f_3 s + f_4} \tag{6}$$

The chosen order of the numerator and denominator for each function is selected through several trails and comparison before inclusion of the WO optimizer in order to ensure realistic dynamical influences of the three inputs to the two outputs. Furthermore, the nonlinear region of operation has been emulated through look-up tables, which represents the nonlinear components in the model. Then, the parameters of every transfer function have been tuned by WO by tightening and relaxation of the bounds of the parameters.

First, data for an actual dual fuel gas turbine power plant has been taken, where it was resampled into intervals of 30 s, having a total of 2040 samples, which represents operating time of exactly 17 h. The model parameters were initially guessed by trial and error, then the implemented model was embedded into the code that represents the cost function. The cost function is the squared error between measured and simulated results. WO, with carefully chosen settings, was used to compute the optimum set of parameters. In order to ensure that the parameters are able to yield close enough results to the actual values of the model, the error has been calculated several times with changing WO setting parameters and the lower and upper bounds of every model parameter. It has been done for 10 iterations through 30 iterations of the whale optimizer. The constraints on the parameters were set as 0.05. To analyze the difference in the results when the whale optimizer constraints are relaxed and tightened, the simulation was done when the constraints were tightened to 0.005 and then relaxed to 0.1. The effect of parameters' relaxation is shown in Table 1 and the optimum set of parameters, based on the best case, are given in Table 1, and the optimum set of parameters is shown in Table 2. Simulation results for the selected set of data have been depicted in Figures 6–10. The next stage is dedicated for controller development. The simulated trajectory has the ability to follow the actual trajectory of the power plant, however, there is some clear deviation at some points, which proves the need for a modified version of the power plant. A controller should be embedded in the design to improve the input and output. The parameters were identified offline, and it took several hours to obtain the final optimal set of parameters.

Table 1. The effect of WO iterations and relaxation of the bounds of model parameters.

Iterations	Root Mean Square Error	
	Tightened to 0.005	**Relaxed to 0.1**
Ten	0.0878	0.0880
Twenty	0.0886	0.0947
Thirty	0.0891	0.0880

Table 2. Optimum set of parameters.

Function	Parameters						
$G_a(s)$	a_1–a_5	2.8957	9.9945	0.1080	1.8984	2.6918	-
$G_b(s)$	b_1–b_6	0.4933	3.0072	0.4927	0.1004	0.2988	0.2991
$G_c(s)$	c_1–c_6	0.8924	0.7027	0.9986	0.1070	0.3047	0.4919
$G_d(s)$	d_1–d_5	0.5063	0.6916	0.9079	0.5970	0.2011	-
$G_e(s)$	e_1–e_5	0.9963	1.1993	0.3048	0.9044	0.1900	-
$G_f(s)$	f_1–f_4	0.8568	0.8017	0.6997	0.0962	-	-

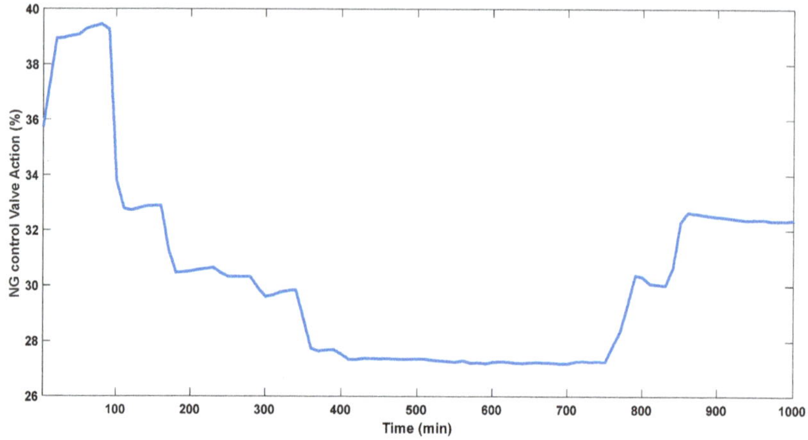

Figure 6. Input NG control valve opening.

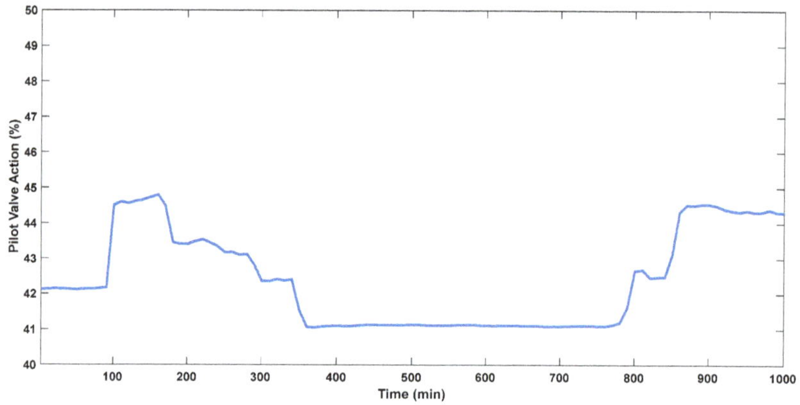

Figure 7. Input Pilot control valve opening.

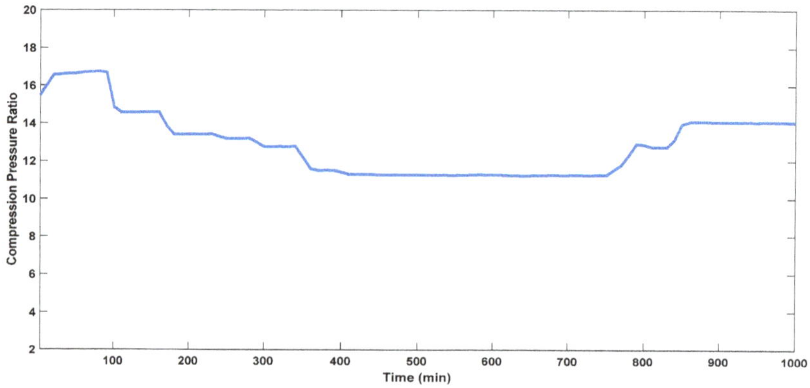

Figure 8. Compression pressure ratio.

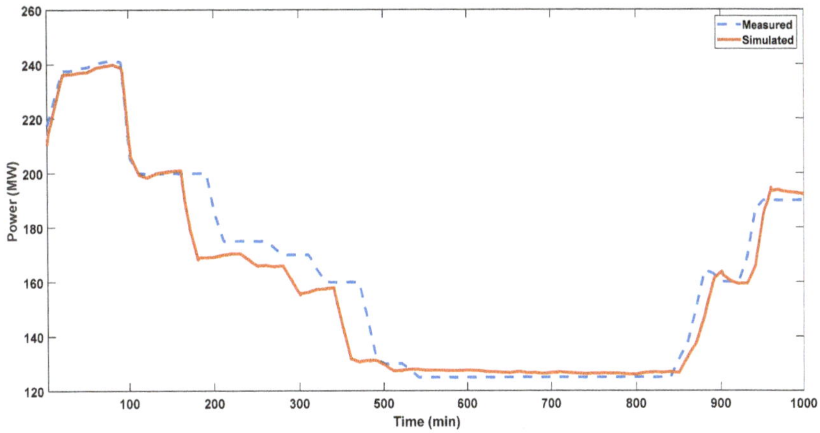

Figure 9. Measured and simulated output power responses.

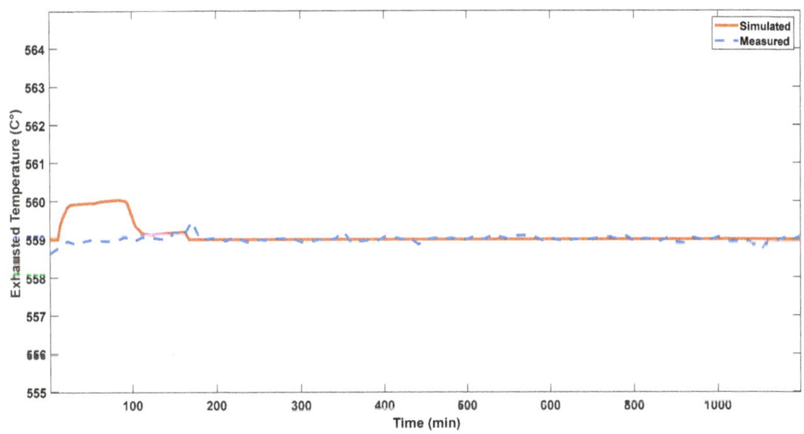

Figure 10. Measured and simulated exhausted temperature responses.

3. Control System Design and Testing via WO

The control system configuration has been assumed to correct the action over the existing control system [3]. It is well known that the identification that has been applied in Section 2 is a closed-loop identification, therefore, one must ensure the proposed controller will not interfere with existing control. From control theory point of view, nothing ensures that except authentic simulations of the proposed controller and comparison–that is rooted from experience–with existing performance. The MIMO controller has been assumed to have two PI controllers, one to regulate the NG and Pilot control valves together and the other to control the compression ratio through the IGV. The coupling control element between the two PI controllers has been chosen to be the PD controller. This structure has been widely accepted for thermal power plant control in general with different control philosophies and parameter tuning [3,6,19–21]. The proposed control system is shown in Figure 11. The rate limiters and saturations help avoid undesirable stresses that may result from extensive changes in the control signals.

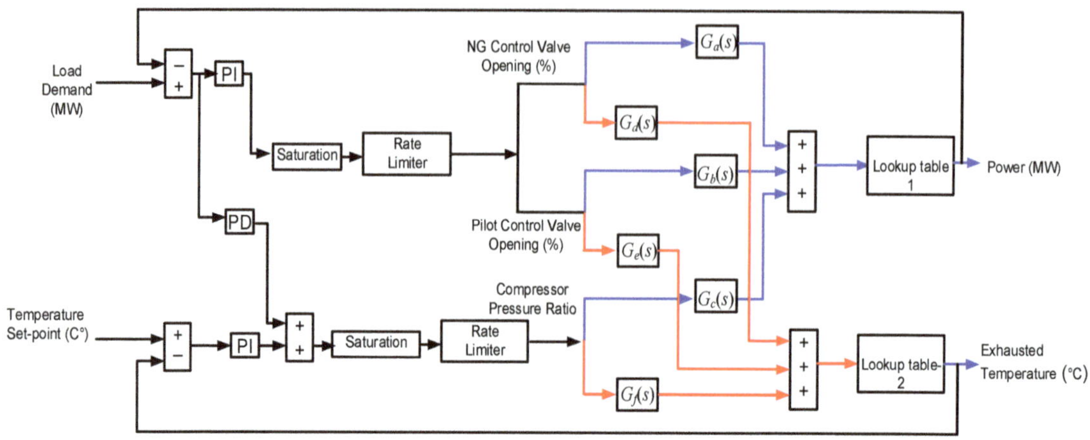

Figure 11. Simulink model with the suggested controller.

The controller has been tuned by WO to minimize the error between set-point and output from the model. It has been tested first with a fictitious pulse load signal that varies from 120 MW to 240 MW, whereas the temperature signal has been constant at 560 °C. The two signals have been implemented into the model, with the addition of loop controllers, along with rate limiters and saturation blocks. The PI/PD controllers have been optimized to follow the signals introduced to the model, with their appropriate parameters. Two proportional-integral (PI) controllers were needed along with one proportional-derivative (PD) controller. Table 3 shows the optimal set of the controller parameters.

Table 3. Controller Parameters.

	PI Controller Parameters for the Fuel Preparation System	PI Controller Parameters for the Compressor	Coupling PD Parameters
K_P	3	0.35	2.65
K_I	10.1	2	-
K_D	-	-	0.2526

The reason why the signals were designed as such: signal one demonstrates the change in active power, and this will prove whether the controller can satisfy the load following capabilities, and whether the exhaust temperature can be controlled in such a manner that would keep it at a constant value. As can be seen in Figures 12 and 13, after introducing the controller with their appropriate parameters, the power plant is capable of following the load demand signal rapidly, even though the signal increased by 120 MW as it would during sudden load application and rejection, the power plant is able to satisfy this change in a timely manner.

The power plant is also capable of stabilizing the exhaust temperature. Keep in mind that it has been able to do so with no considerable changes to the exhaust temperature and was kept at 560 degrees Celsius. This proves the success and validity of the controller in a mathematical sense; however, another practical test is required to ensure the energy-efficient aspect of the controller. One way to ensure this is to apply the data signal of the load demand as a reference set-point to the controller with a little bit higher temperature reference signal. As a consequence, the control system decisions for the fuel preparation system have shown uneven changes in the control actions. Figures 14 and 15 shows higher pilot valve action requirement while the NG valve action is lower. The improvement could be decided with full confidence through the total consumed fuel during premix mode, which is depicted in Figure 16. The average fuel reduction has been 8.96% fuel consumption, which is equivalent to about 237.16 MWh average energy savings for the entire time window of operation.

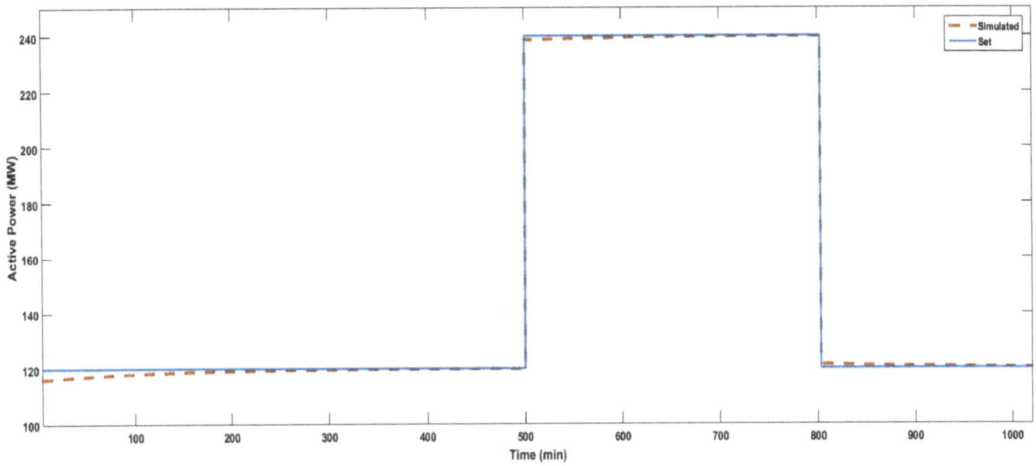

Figure 12. Active power simulated after implementing the controller (dashed orange) along with the actual active power signal that should be followed (blue).

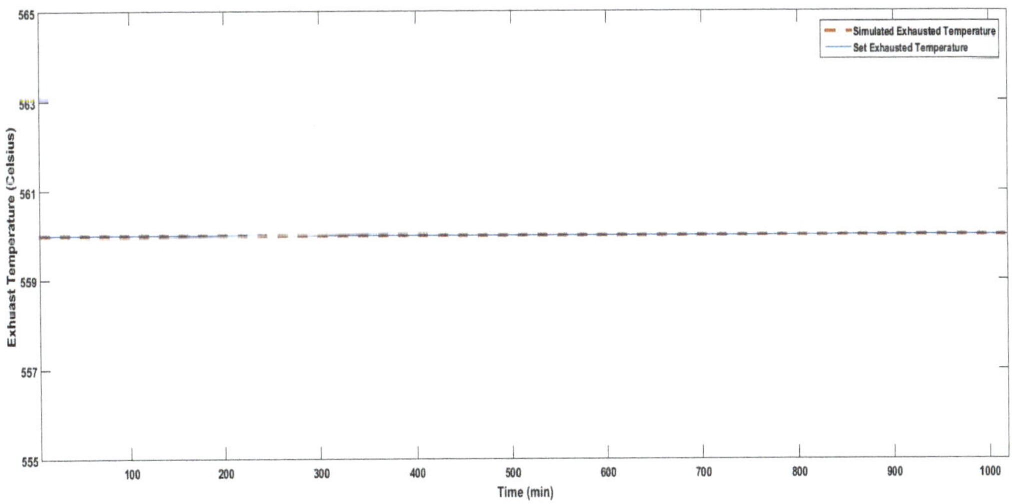

Figure 13. Exhaust temperature simulated after implementing the controller (dashed orange) along with the actual exhaust temperature signal that should be followed (blue).

Overall, there are fewer running costs and less use of raw materials which is necessary and a non-ending goal for the design of most power plants. Figure 17 shows that high and constant compression is required for compressor pressure output ratio response with the controller, where the value was constant during the simulation period which fundamentally proves that the efficiency of the plant has been improved, but higher input should be invested through the compressor. Figure 17 shows that more kinetic energy should be invested to the compressor through higher and constant compression ratio over the existing or measured case, in which the energy conservation principle can be proved. As can be seen in Figures 18 and 19, the active power was followed during the whole 17 h, almost perfectly taking into consideration that the total input has dropped from the actual value yet the yield of active power stayed the same. The exhaust temperature shows almost a 4 degrees Celsius increase over the actual value with no fluctuations, which is also significant to prove the controller has improved overall.

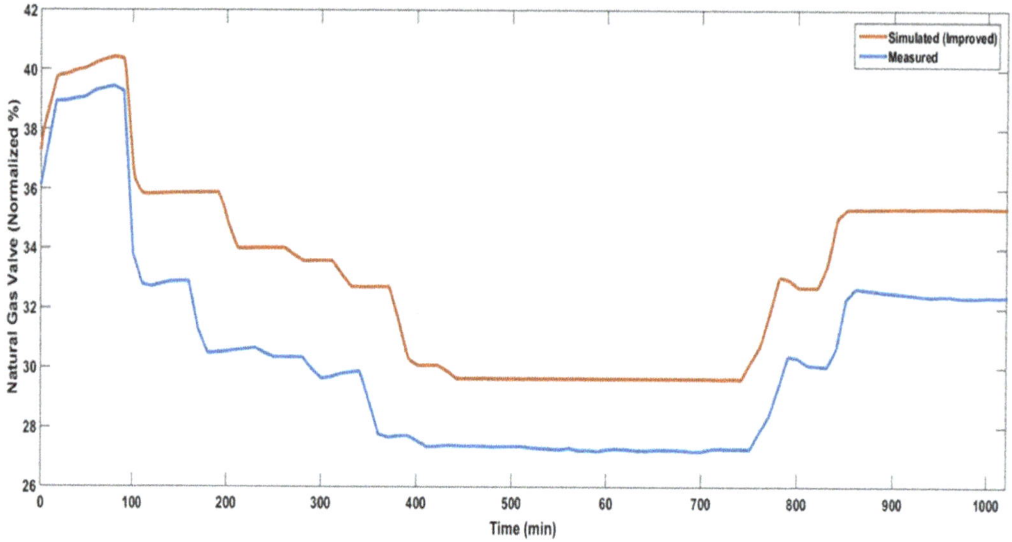

Figure 14. The Pilot valve dynamics, showing the existing and actual fuel oil (diffusion mode)/natural gas (premix mode) consumption (blue) and the simulated (improved dynamics (orange)). Notice: higher fuel consumption from the Pilot valve.

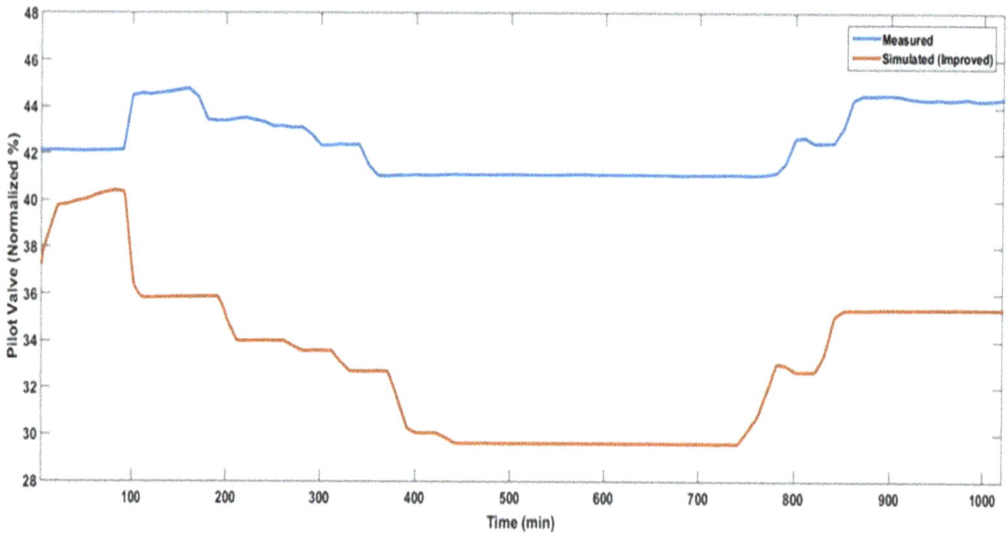

Figure 15. The NG valve dynamics, showing much lower fuel consumption with the simulated (improved with the controller) (orange) and the existing case (blue).

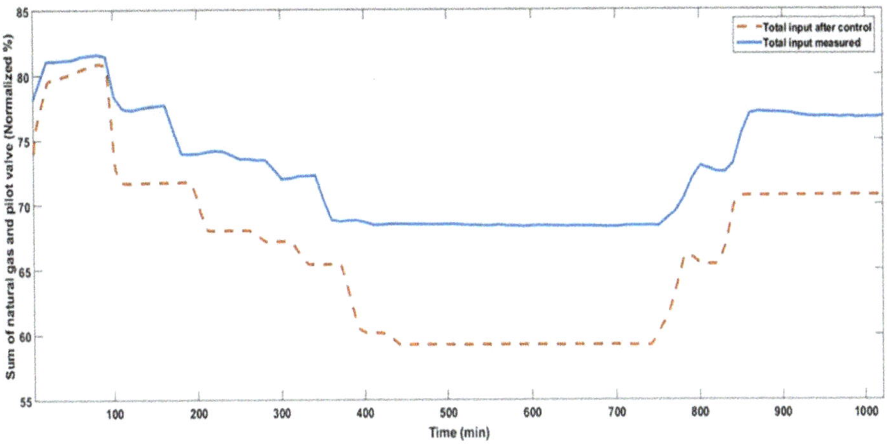

Figure 16. Total fuel input to the plant (pilot valve flow + NG flow) showing significant reduction in fuel or natural gas consumption (dashed orange) if operated totally on NG (i.e., premix mode) over the existing operation fuel consumption (blue).

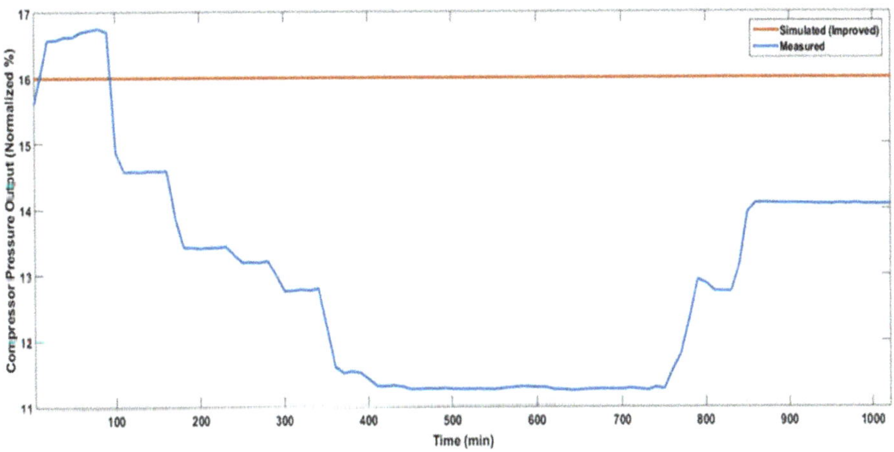

Figure 17. Compressor ratios in the improved (ornage) and existing case (blue).

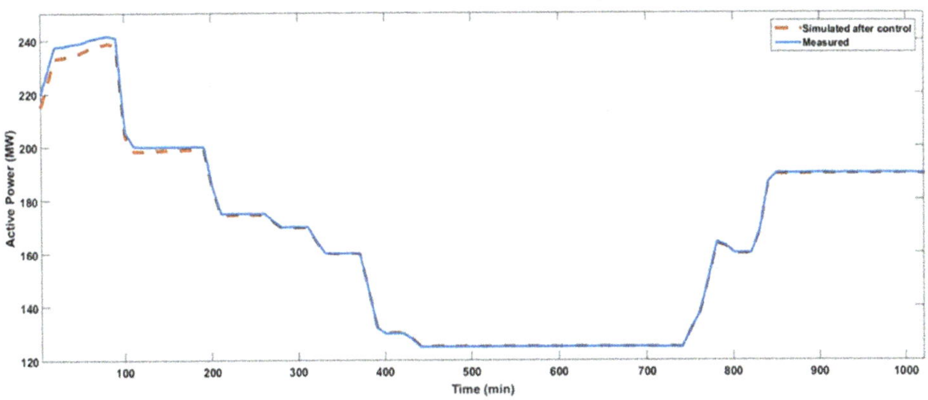

Figure 18. GT output (dashed orange) following the load demand (blue solid).

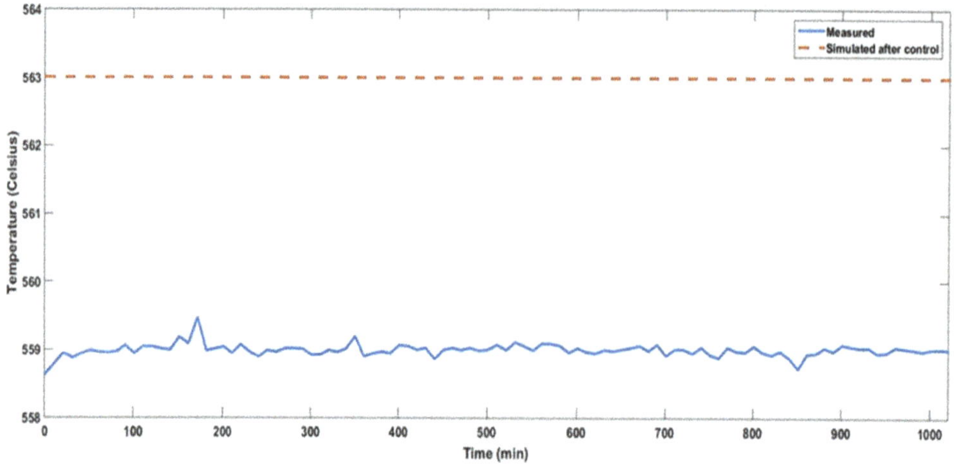

Figure 19. Exhaust temperature, the actual exhausted temperature (blue) along with the improved exhausted temperature response (dashed orange).

The next section concludes the research findings and recommends future points.

4. Conclusions

In this paper, a new simplified model for capturing the essential dynamical performance for a heavy-duty GT has been presented. The model embeds linear and nonlinear components, and the unknown parameters have been optimized using WO with sufficient relaxation of the bounds of the parameter. The model outputs have been verified from a load range of 120 MW up to 240 MW with reasonable accuracy. In addition, a MIMO PI/PD controller has been designed with optimal adjustments of the control parameters by WO. It has been shown that the controller regulates the essential outputs of the plant with more efficient operation. Thereby, it has been proven that WO is a robust optimizer for GT modeling, identification, and control.

There are some future opportunities to investigate, for example, using different metaheuristic optimizers for quantified comparison with WO in terms of accuracy and computation requirements. Different and modern GTs could also be tested with this control strategy, which are fueled by hydrogen or biogas. An economic feasibility study of the control practical implementation in real time is also a scientific merit that could be achieved.

Author Contributions: Conceptualization, O.M.; methodology, M.A.-S., O.M. and W.A.E.; software, M.A.-S.; validation, M.A.-S.; formal analysis, M.A.-S. and O.M.; investigation, O.M. and W.A.E.; resources, O.M. and W.A.E.; data curation, O.M.; writing—original draft preparation, M.A.-S. and O.M.; writing—review and editing, O.M. and W.A.E.; supervision, O.M. and W.A.E. All authors have read and agreed to the published version of the manuscript.

Funding: This research received no external funding.

Data Availability Statement: Data available on request due to preferability to be on request to ensure suitable usage by specialists.

Conflicts of Interest: The authors declare no conflict of interest.

References

1. Mom, J.A. Introduction to gas turbines. In *Modern Gas Turbine Systems*; Jansohn, P., Ed.; Woodhead Publishing Series in Energy; Woodhead Publishing: Cambridge, UK, 2013; pp. 3–20.
2. Prade, B. Gas turbine operation and combustion performance issues. In *Modern Gas Turbine Systems*; Jansohn, P., Ed.; Woodhead Publishing Series in Energy; Woodhead Publishing: Cambridge, UK, 2013; pp. 383–422, 423e.
3. Mohamed, O.; Khalil, A. Progress in Modeling and Control of Gas Turbine Power Generation Systems: A Survey. *Energies* **2020**, *13*, 2358. [CrossRef]
4. Alsarayreh, M.; Mohamed, O.; Matar, M. Modeling a Practical Dual-Fuel Gas Turbine Power Generation System Using Dynamic Neural Network and Deep Learning. *Sustainability* **2022**, *14*, 870. [CrossRef]
5. Shahin, S.; Elkhashap, A.; Abdelgelil, M.; Hamad, M.S. Nonlinear Model Predictive Control for Power Generation Gas Turbines. In Proceedings of the 2021 International Telecommunications Conference (ITC-Egypt), Alexandria, Egypt, 13–15 July 2021; pp. 1–6.
6. Mohamed, O.; Wang, J.; Khalil, A.; Limhabrash, M. Predictive control strategy of a gas turbine for improvement of combined cycle power plant dynamic performance and efficiency. *SpringerPlus* **2016**, *5*, 980. [CrossRef] [PubMed]
7. Tayyah, A.; AlShami, M.; Daboubi, O. Design of a Discrete Time Model Predictive Control for a Fossil Fueled Power Plant. B.Sc. project document, Princess Sumaya University for Technology, Amman, Jordan, 2019.
8. Kanashev, E.A.; Rozhko, E.V. Experimental Research of Gas Turbine Unit Control Systems Using Simulation Model. In Proceedings of the 2018 International Russian Automation Conference (RusAutoCon), Sochi, Russia, 9–16 September 2018; pp. 1–6.
9. Shete, S.A.; Jape, V.S. Design of a Fuzzy Modified Reference Adaptive Controller for a Gas Turbine Rotor Speed Control using T-S Fuzzy Mechanism. In Proceedings of the 2017 International Conference on Technological Advancements in Power and Energy (TAP Energy), Kollam, India, 21–23 December 2017.
10. Rowen, W.I. Simplified Mathematical Representations of Heavy-Duty Gas Turbines. *ASME J. Eng. Gas Turbines Power* **1983**, *105*, 865–869. [CrossRef]
11. Camporeale, S.M.; Fortunato, B. Dynamic Analysis and Control of Turbo-Gas Power Plant. In Proceedings of the IECEC-97 Proceedings of the Thirty-Second Intersociety Energy Conversion Engineering Conference (Cat. No.97CH6203), Honolulu, HI, USA, 27 July–1 August 1997; Volume 3, pp. 1702–1707.
12. Bulatov, Y.N.; Kryukov, A.V.; Van Huan, N. Simulation of Gas Turbine Power Plants with Voltage and Speed Prognostic Regulators. In Proceedings of the 2020 International Russian Automation Conference (RusAutoCon), Sochi, Russia, 6–12 September 2020; pp. 160–164.
13. Patrascioiu, C.; Popescu, M. Control System for Natural Gas Compression of Sinca Plant. In Proceedings of the 2020 12th International Conference on Electronics, Computers and Artificial Intelligence (ECAI), Bucharest, Romania, 25–27 June 2020.
14. Talah, D.; Bentarzi, H. Modeling and Analysis of Heavy-Duty Gas Turbine Based on Frequency Dependent Model. In Proceedings of the 2020 International Conference on Electrical Engineering (ICEE), Istanbul, Turkey, 25–27 September 2020; pp. 1–4.
15. Ma, S.; Tan, J.; Ning, Y.; Gao, Z. Modeling and Simulation of Gas Turbine Starter and Fuel Control System. In Proceedings of the 2017 36th Chinese Control Conference (CCC), Dalian, China, 26–28 July 2017; pp. 2149–2154.
16. Zhang, Q.; So, P.L. Dynamic Modelling of a Combined Cycle Plant for Power System Stability Studies. In Proceedings of the 2000 IEEE Power Engineering Society Winter Meeting. Conference Proceedings (Cat. No.00CH37077), Singapore, 23–27 January 2000; Volume 2, pp. 1538–1543.
17. Al-Dalwi, K.K.; Vural, A.M. Modeling and Simulation of Bazyan Gas Power Plant. In Proceedings of the 2017 4th International Conference on Electrical and Electronic Engineering (ICEEE), Ankara, Turkey, 8–10 April 2017; pp. 163–169.
18. Wang, Y.; Bhattacharyya, D.; Turton, R. Multiobjective Dynamic Optimization for Optimal Load-Following of Natural Gas Combined Cycle Power Plants under Stress Constraints. *Ind. Eng. Chem. Res.* **2021**, *60*, 14251–14270. [CrossRef]
19. Haji Haji, V.; Fekih, A.; Monje, C.A.; Fakhri Asfestani, R. H_2, H_∞, H_2/H_∞, and *μ-synthesis* controllers for the speed and temperature control of a real gas turbine unit in a combined cycle power plant. *Energy Sci. Eng.* **2019**, *7*, 2205–2222. [CrossRef]
20. Shi, G.; Wu, Z.; He, T.; Li, D.; Ding, Y.; Liu, S. Decentralized active disturbance rejection control design for the gas turbine. *Meas. Control* **2020**, *53*, 1589–1601. [CrossRef]
21. Wu, Z.; Li, D.; Xue, Y.; Chen, Y. Gain scheduling design based on active disturbance rejection control for thermal power plant under full operating conditions. *Energy* **2019**, *185*, 744–762. [CrossRef]
22. Qasem, M.; Mohamed, O.; Abu Elhaija, W. Parameter Identification and Sliding Pressure Control of a Supercritical Power Plant Using Whale Optimizer. *Sustainability* **2022**, *14*, 8039. [CrossRef]
23. Mirjalili, S.; Lewis, A. The Whale Optimization Algorithm. *Adv. Eng. Softw.* **2016**, *95*, 51–67. [CrossRef]

sustainability

MDPI

Review

Carbon Footprints of Active and Non-Active Transport Modes: Hierarchy and Intergenerational Narrative Analyses

Nestor Asiamah [1,2,*], Kofi Awuviry-Newton [3], Whitney Nesser [4] and Evelyn N. Alvarez [5]

1 Division of Interdisciplinary Research and Practice, School of Health and Social Care, University of Essex, Colchester CO4 3SQ, UK
2 Department of Gerontology, Africa Centre for Epidemiology, Accra North P.O. Box AN 16284, Ghana
3 College of Sport, Health and Engineering, Victoria University, Melbourne, VIC 8001, Australia; kofi.awuviry-newton@vu.edu.au
4 Department of Applied Clinical and Educational Sciences, Indiana State University, 401 North Seventh Street, Terre Haute, IN 47809, USA; whitney.nesser@indstate.edu
5 Department of Public Health, California State University, 5151 State University Drive, ST 305, Los Angeles, CA 90032, USA; evelyna@calstatela.edu
* Correspondence: n.asiamah@essex.ac.uk; Tel.: +44-7763489544

Abstract: This paper aimed to (1) develop a hierarchy for understanding the impacts of active and non-active transport modes on the environment and (2) analyse the adoption of active transportation between older and younger people. A narrative review with two parts was adopted to develop the hierarchy. In the first part, a framework was adopted to map active and non-active transport modes onto three operational boundaries of greenhouse gas emission to develop the hierarchy. In the second part, an intergenerational theoretical framework was developed to analyse the adoption of active transportation between older and younger people. The review suggests that the only active transport modes with no or negligible carbon footprint are walking, running, and swimming without a product that adds to atmospheric greenhouse gases. The evidence that younger people perform higher active transportation behaviour is inconsistent and is, therefore, inconclusive. This review suggests a need for manufacturers to prioritise the production of active vehicles (e.g., wheelchairs and scooters) that are biodegradable, recyclable, and small.

Keywords: carbon footprint; active transportation; older adults; generations; health

Citation: Asiamah, N.; Awuviry-Newton, K.; Nesser, W.; Alvarez, E.N. Carbon Footprints of Active and Non-Active Transport Modes: Hierarchy and Intergenerational Narrative Analyses. *Sustainability* **2023**, *15*, 12795. https://doi.org/10.3390/su151712795

Academic Editors: Wenbin Yu and Guang Zeng

Received: 30 July 2023
Revised: 21 August 2023
Accepted: 22 August 2023
Published: 24 August 2023

1. Introduction

Research assessing carbon dioxide equivalent emissions often called a carbon footprint [1] has gained momentum in recent years in response to an increase in global greenhouse gas emissions from individuals [1,2]. A parallel development is the acceleration of research on the health-sustainability dimension of transportation, with an emphasis on avoiding or decreasing per capita carbon footprint through active transportation, defined as walking or cycling to a place [3]. This definition undermines other forms of active transportation, resulting in active transportation being operationally defined as moving to places in ways involving physical activity but not involving the combustion of fossil fuels. This definition was informed by the above health-sustainability research agenda that emphasises the role of active transportation in health and environmental protection [1,2]. This agenda ought to progress since one-fifth of greenhouse gas emissions come from transportation involving the combustion of fossil fuels alone [1,2].

Walking, for example, may involve a negligible emission of greenhouse gases each time it is performed. In this review, any such travel behaviour is treated as an active transport mode with a zero-carbon footprint. The term "transport mode" has been used in the literature [4] to refer to different ways to travel between places (e.g., walking, bicycling, and driving). In this paper, therefore, we use this phrase to refer to various ways to travel.

The literature to date suggests that walking may be the ultimate physical activity for older adults because it requires less physical strength and energy expenditure [5,6]. As such, it can be sustained over the life course by all age groups, an idea that recalls a recent debate in gerontology about the role of older adults in environmental activism [7,8]. This debate has portrayed older adults as victims of ageism championed by younger adults who are concerned about climate change and the future [9]. Younger generations are concerned that older adults are responsible for climate change since older adults generally lead environmental policy interventions that have been unproductive [7,8]. An aspect of the literature also suggests that older adults have had more time to contribute to the emission of greenhouse gases and are less interested in pro-environmental behaviours, such as active transportation [9]. These ageist views imply older adults contribute less to environmental sustainability through active transportation.

Ageist views about older adults threaten the solidarity needed between older and younger generations to fight climate change [7–9]. Pro-environment behaviours (e.g., walking) are potentially the best ways to reduce carbon footprint and achieve sustainability goals [10,11], but their positive influence on the environment depends on how many people practice them. When ageist views are prevalent in climate crises and sustainability discussions, initiatives become divisive and undermine the significant role older adults can play in overcoming the climate crisis. This is particularly important, especially in a world where the population is rapidly ageing [12] and sustainability initiatives include many older adults.

Given the above concerns, the authors aimed to develop a heuristic for understanding the carbon footprints of active and non-active transport modes. This heuristic is needed because, though studies suggest active modes of transportation are the best ways to minimise per capita emission of greenhouse gases [10,11], there is no framework describing their respective carbon footprints. The authors further analysed the adoption of active transportation between older and younger people through a theoretical framework delineating active transportation behaviour across four generations (i.e., children, adolescents, adults, and older adults).

This review is significant for some reasons. Though studies have reported active and non-active transport modes with their potential carbon footprints, this review is the first to put these forms of transportation on a hierarchy, enabling stakeholders to better appraise the role active transportation plays in campaigns for a safer environment. The hierarchy may serve as a model for empirically investigating the relative impacts of transport modes on the environment. It is generally assumed that active modes of transportation protect the environment, but this review suggests otherwise. With this review, individuals may consider ways to use active vehicles (e.g., bicycles, scooters, and wheelchairs) without generating a carbon footprint. The hierarchy can encourage manufacturers to consider opportunities for designing vehicles to make them more active. Our intergenerational analysis may correct the assumption that the adoption of active transport modes is not necessarily higher among younger people. This contribution of the review is a way to better value the role of older adults in pro-environment campaigns, encourage stakeholders to include older adults in such campaigns, and ensure that as many older adults as possible are included in campaigns encouraging active transportation adoption, given that there may be more older people than younger ones in the world in the coming decades.

2. Search Methodology in Brief

A narrative review was adopted, and the PRISMA (Preferred Reporting Items for Systematic Reviews and Meta-Analyses) guideline was followed to search and review the literature. The search aimed to identify up-to-date documents reporting active or non-active transportation concerning greenhouse gas emissions and its personal as well as psychosocial predictors. Appendix A is the review work plan showing the inclusion and exclusion criteria, search terms, and SPIDER (Sample, Phenomenon of Interest, Design, Evaluation, Research type) tool used. Appendix B is the PRISMA flowchart reached using the inclusion and exclusion criteria to select suitable documents. The databases searched were PubMed, ProQuest, PsychInfo, CINAHL, Google Scholar, and Scopus. MeSH (Medical Subject Headings) terms were identified and developed into a search string using PubMed. These databases were searched twice as shown in Appendix A.

The steps taken in the review were (1) systematic searches; (2) screening of titles and abstracts of 5% of the studies twice to pilot the inclusion criteria; (3) screening of full texts by one of the authors; and (4) checking of data extraction for 20% of studies. Two researchers independently piloted abstract and title screening on 5% of the records downloaded. Inconsistencies in the pilot results were discussed and resolved before proceeding to the next stage. The titles and abstracts of the remaining 95% of the records were screened by the individual researchers. Subsequently, the full texts of the records selected through title and abstract screening were assessed for inclusion in the review against the inclusion criteria. The lists of references of included studies were examined to select relevant articles that had not been downloaded into the bank of records realised from our screening of titles and abstracts. Appendix B is the PRISMA flowchart resulting from the review. To ensure that quality documents were reviewed, we focused on peer-reviewed journals indexed in Web of Science, Scopus, or PubMed.

Only 26 documents were deemed appropriate for this review (see Appendix B), though other complementary documents outside the scope of the search were used. Data were extracted independently by one author and a research assistant with a piloted data extraction Excel sheet. The author and research assistant discussed in person to resolve minor disagreements in data extraction. Seven of the studies [13–19] reported active transport modes, namely, walking, running (i.e., jogging), swimming, bicycling, skating (including skateboarding and roller skating), skiing, surfing, scooter or wheelchair use (including kick scooter use), and rowing. Non-active transport modes dependent on fossil fuels include motorcycling, driving a car, ship travel, train travel, and air travel. A study [1] reported a framework that could be used to assess the carbon footprint of transport modes. Some studies [19–29] also assessed the relationship between age, pro-environment behaviour, and active transportation adoption.

3. A Framework for Assessing Carbon Footprint

This review focused on the carbon footprint of individuals and how this can be reduced or avoided through active transportation. To meet this aim, a carbon footprint is defined as the exclusive total amount of carbon dioxide emissions that are directly or indirectly caused by an activity or accumulated over the lifespan of a product [1]. This definition suggests that a carbon footprint can be generated directly or indirectly by an individual through daily behaviours. A direct example is driving a petrol- or diesel-dependent car, which directly releases greenhouse gases into the atmosphere. Indirect examples are producing non-biodegradable waste through the consumption of products (e.g., a canned drink) or felling down trees to provide services or products. Non-biodegradable waste produces greenhouse gases [30], whereas the felling of trees would increase the concentration of greenhouse gases in the atmosphere by reducing the proportion of trees absorbing these gases while releasing oxygen.

The foregoing definition makes Wicker's framework [1] for assessing carbon footprint ideal for the current review. It comprises three operational boundaries or scopes that specify whether some behaviours generate a carbon footprint. These behaviours are within three scopes. Scope 1 comprises direct emissions resulting from onsite fuel consumption, including all emissions from combustions relating to the use of vehicles. This includes behaviours causing emissions from travelling to a destination, with a typical example being driving a car. Scope 2 encompasses direct emissions from purchased electricity, heating, and cooling. This category includes heating or cooling a vehicle while travelling and wearing, for example, an electric jacket to keep warm while walking during the winter. Scope 3 concerns indirect emissions occurring during the lifespan of a product, including emissions resulting from the production and distribution of a product and management of waste.

Indirect emissions relate to the production of products requiring a supply chain dependent on the transportation of goods and individuals.

To use the above framework [1], the authors decided whether individual transport behaviours can directly or indirectly produce any greenhouse gas per unit of time. Each transport behaviour was mapped onto all three operational scopes with a "yes" (i.e., scope applicable) or "no" (i.e., scope not applicable) decision, which allowed us to determine whether the behaviour generates a carbon footprint directly or indirectly. To achieve reliable results, two researchers with expertise in transportation research performed independent mappings, which produced consistent findings. A zero-carbon footprint was achieved if a transport behaviour, hereby referred to as absolute active transportation, did not result in a greenhouse gas emission across the three scopes. Any active transport behaviour that was associated with emission for at least one scope had a carbon footprint and could be referred to as partial active transportation.

Whether an individual would use or adopt an active transport mode depends on several factors, such as the social and physical environment, as well as age [27,31]. In view of these factors, the adoption of active transportation between older and younger people is analysed through a theoretical framework explaining unique opportunities and barriers to active transportation across four generations. Children between 0 and 12 years who cannot make transport decisions for themselves are the first generation, whereas teenagers and adolescents aged 13–17 years who can make transport decisions but are dependent on parents are the second generation. Adults aged 18–49 years who can make transport decisions and may be independent of their parents are the third generation. The minimum for what is considered old age differs between countries; the United Kingdom (UK), for instance, sets the minimum old age at 65 years [32], whereas Ghana sets it at 60 years [33]. Globally, the minimum old age is 50 years [32,34]. Although the minimum age of 50 is not a good indicator of the individual's health and physiological conditions [34], it is a globally acceptable baseline. Thus, older people are operationally defined as individuals aged 50 years or higher and are the fourth generation.

4. Carbon Footprint and a Hierarchy of Active Transport Modes

The hierarchy of active transport modes is the pyramidal heuristic showing the relative impacts of transport modes on the environment. This framework was developed by mapping identified transport modes onto the operational scopes, which are recalled and operationalised as follows:

Scope 1—direct emissions resulting from onsite fuel consumption, including all emissions from combustions relating to the use of vehicles.

With this scope, any transport behaviour not involving the combustion of fossil fuel and not emitting a greenhouse gas does not generate a carbon footprint. As such, any transport behaviour that involves the combustion of fossil fuel applies to this scope and is mapped onto it with "yes" (with red colour).

Scope 2—direct emissions from purchased electricity, heating, and cooling. These emissions come from the use of air-conditioning systems that may be part of vehicles.

This scope does not require the direct combustion of fossil fuel in transportation but involves heating or cooling through air conditioning, which results in the emission of greenhouse gases [35]. Individuals with pro-environment behaviours may decide to drive an electric car, but they may use heating or cooling systems in the car (e.g., an air-conditioner) which produce greenhouse gases. Someone walking during the winter may wear a jacket with an inbuilt or mobile heating system, which may generate a carbon footprint. Therefore, any transport behaviour that uses a heating or cooling system and could emit greenhouse gases applies to this scope and is mapped onto it with "yes".

Scope 3—emissions that occur during the lifespan of a product, including those from the production and distribution of a product and management of waste from this product.

Any product whose production indirectly increases the concentration of greenhouse gases in the atmosphere is considered environmentally unfriendly. For instance, the production of products dependent on wood requires the felling of trees that absorb some greenhouse gases, such as carbon dioxide. From this perspective, the use of biodegradable products (e.g., a bicycle made of wood) indirectly generates a carbon footprint. Secondly, the use of any product that can become a part of waste in its production or consumption indirectly generates a carbon footprint. This assumption is premised on research [30] indicating that waste is a major source of greenhouse gases, such as methane. The quantity of greenhouse gases emitted partly depends on the size of a product; larger products that are not biodegradable or cannot be recycled would add more waste to the environment and may, therefore, generate a higher carbon footprint. Biodegradable waste, compared to non-biodegradable waste (e.g., plastics), has a shorter lifespan, so its carbon footprint can be expected to be short-lived. Similarly, recyclable waste would generate a smaller footprint.

Table 1 shows the results of mapping all transport modes onto the three operational scopes. Mapping was based on whether the transport behaviour involves the use of a product that could be harmful to the environment, depends on a utility or energy source that emits greenhouse gases, and whether the product is small, biodegradable, or recyclable. It was also assumed that greenhouse gas emissions across the lifespan of fuel-dependent transport modes (i.e., motorcycle, car, ship, train, and aeroplane) are more than emissions across the lifespan of active transport modes. Only walking, running, and swimming with no or negligible greenhouse gas emissions constitute absolute active transportation. "Walking (PS)" in the table may be associated with a significant emission of greenhouse gases and may, thus, has a carbon footprint. A study [1] has revealed that individuals may drive to convenient destinations before performing sporting activities or active transportation behaviours. Such individuals directly generate a carbon footprint before performing an active transportation behaviour at the chosen destination. Others might use canned energy drinks and other products during active transportation (e.g., walking) which may add up to waste, especially if not properly disposed of. The use of products, especially non-biodegradable ones, in active transportation can have a significant detrimental impact on the environment in the long term.

Figure 1 (based on Table 1) depicts the heuristic of walking as the most environment-friendly active transportation behaviour. The non-active transport modes are at the base of the framework, which signifies that transportation involving the combustion of fossil fuels has the highest carbon footprint. Walking is above running on the pyramid for two reasons. Firstly, research has suggested that walking, compared to running, is more sustainable across the lifespan because it requires less energy expenditure and is part of daily routines [5]. This being so, more people can be expected to perform walking behaviours and impact the environment positively. Secondly, whether people would sustain walking or running as a behaviour depends on their connectedness to nature [36], hereby defined as the amount of time spent observing lawns, forests, gardens, wildlife, rivers, and other natural attributes of the physical environment. People who walk may be better engaged with nature because they can more closely observe and admire nature. In running, people hurriedly observe nature, so their nature-driven motivation to keep fit through running would be low, compared with people who walk. Swimming is set below running in the framework because it is less relaxing and, if conducted in an indoor or artificial facility, provides limited nature connectedness. Worth noting is the idea that all individuals can contribute to environmental sustainability through active transportation, an idea substantiated by the following theoretical analysis of the adoption of this travel behaviour across four generations.

Table 1. The authors' mapping of key active and non-active transport modes onto the three operational scopes or boundaries.

SN	Transport Mode	Operational Boundaries			Attribute(s)	Description
		Scope 1	Scope 2	Scope 3		
				Active modes of transportation		
1	Walking (EF)	No	No	No	Eco-friendly *	Walking without using any supporting product (e.g., canned energy drink or car)
2	Walking (PS)	No	Yes	Yes	Less eco-friendly **	Walking while using a product or driving to a point before starting to walk
3	Running (EF)	No	No	No	Eco-friendly	Running without using any supporting product
4	Running (PS)	No	Yes	Yes	Less eco-friendly	Running while using a product or driving to a point before starting to run
5	Swimming (EF)	No	No	No	Eco-friendly	Swimming without using any supporting product
6	Swimming (PS)	No	Yes	Yes	Less eco-friendly	Using a product while swimming or driving **** to a point before engaging in swimming
7	Skiing/surfing (EF)	No	Yes	Yes	Eco-friendly	Skiing or surfing without any supporting product
8	Skiing/surfing (PS)	No	Yes	Yes	Less eco-friendly	Using a product while surfing or skiing or driving to a point before surfing or skiing
9	Biking (EF)	No	Yes	Yes	Eco-friendly	Using a bicycle that is made of biodegradable or recyclable materials
10	Biking (LEF and PS)	No	Yes	Yes	Less eco-friendly	Using a bicycle that is made of traditional materials ***
11	Skating, skateboarding, roller skating (EF)	No	Yes	Yes	Eco-friendly	Using equipment that is made of biodegradable or recyclable materials
12	Skating, skateboarding, roller skating (LEF and PS)	No	Yes	Yes	Less eco-friendly	Using equipment that is made of traditional materials that are less eco-friendly or can result in non-biodegradable waste
13	Scooter, kick scooter/wheelchair (EF)	No	Yes	Yes	Eco-friendly	Using equipment that is made of biodegradable or recyclable materials
14	Scooter, kick scooter/wheelchair (LEF)	No	Yes	Yes	Less eco-friendly	Using equipment that is made of traditional materials that are less eco-friendly or can result in non-biodegradable waste
15	Rowing (EF and PS)	No	Yes	Yes	Eco-friendly	Using equipment that is eco-friendly and can, therefore, result in less or biodegradable waste
16	Rowing (LEF and PS)	No	Yes	Yes	Less eco-friendly	Using equipment that is made of traditional materials that are less eco-friendly or can result in non-biodegradable waste
				Non-active modes of transportation		
17	Motorbike, car, ship, train, and aeroplane (EF)	Yes	Yes	Yes	Eco-friendly	A motorcycle made of recyclable/biodegradable materials and is 100% electric
18	Motorbike, car, ship, train, and aeroplane (NEF and PS)	Yes	Yes	Yes	Not eco-friendly	A vehicle that uses fossil fuels and is made of materials not biodegradable or recyclable

Note: Active transport modes shown (i.e., 1–16) do not involve the combustion of fossil fuels; the numbers 1–18 do not represent ranks or an order; mapping of transport modes onto the three operational boundaries was based on whether the transport behaviour involves the use of a product or vehicle, depends on a utility or energy source that emits greenhouse gases, and whether the productive involved is small, biodegradable, or recyclable; mapping was also based on the assumption that greenhouse gas emissions across the lifespan of fuel-dependent transport modes are more than emissions across the lifespan of active transport modes; "No" (i.e., colour green) means the boundary or scope does not apply to the corresponding transport type, and this suggests a zero or negligible footprint of the transport type; "Yes" (i.e., colour red) means the boundary applies to the corresponding transport mode; SN—serial number; PS—product-supported; EF—eco-friendly; LEF—less eco-friendly; NEF—not eco-friendly; * biodegradable (e.g., made of wood) or recyclable; ** not biodegradable or recyclable; *** traditional materials are raw or processed materials that are not recyclable or biodegradable; **** driving a vehicle that involves the combustion of a fossil fuel.

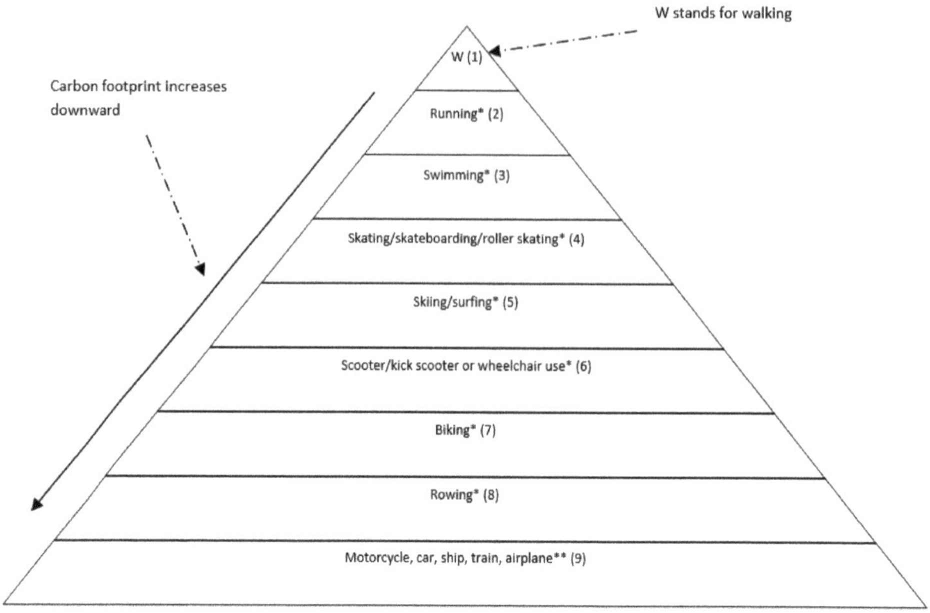

Figure 1. A hierarchy of potential environmental impact of active and non-active transport modes. Note: Active transport modes shown (i.e., 1–8) do not involve the combustion of fossil fuels; the hierarchy was developed based on whether the transport behaviour involves the use of a product or vehicle, depends on a utility or energy source that emits greenhouse gases, and whether the productive involved is small, biodegradable, or recyclable; the hierarchy also assumes that greenhouse gas emissions across the lifespan of fuel-dependent transport modes are more than emissions across the lifespan of active transport modes; size of the vehicle, equipment, or product is assumed to increase down the pyramid; ** Represent non-active or fossil fuel-dependent modes of transportation; * Active modes of transportation.

5. Theoretical Framework

The literature [27,29,31] to date suggests that active transportation behaviour is influenced by three categories of factors, namely, demographic (e.g., age, income, and gender), psychosocial (e.g., neighbourhood trust, safety, and social cohesion), and physical environmental factors (e.g., street connectivity and mixed land use). Income, for example, determines car ownership and whether one will choose driving over walking [28,37]. Neighbourhoods with highly interconnected streets are more likely to encourage walking and bicycling [29,38], and those with psychosocial factors such as safety offer a better contextual advantage for active transportation [29,39]. Yet, the extent to which these factors affect active transportation differs among age groups due to changes in living conditions experienced by the individual in ageing [39]. The Bioecological Systems Theory (BST) developed by Unrie Bronfenbrenner [40,41] implies that the onset of these changes starts in childhood.

The BST is a multi-level framework for understanding the influence of the above categories of factors on active transportation. The primary part of this system is the microsystem where young children begin life by developing relationships with parents and other close relatives. It provides a social climate where family norms and values are transferred by older ones to children in a gradual way, making it possible for younger ones to learn and apply family traditions. Children may grow to appreciate and enjoy biking to school owing to their exposure to a longstanding family tradition of biking to school. Studies have confirmed that children with active parents who travel to work through active transportation are more likely to walk or cycle to school [42,43]. Though the BST suggests that several other factors (e.g., family income) can influence the active travel of children, it implies that children begin to develop behaviours and habits through their subjection to relationships and norms in their immediate family environment. The main disadvantage at this stage is that children may not grow up with healthy behaviours (e.g., walking) if their immediate families do not value these behaviours. If family norms favour driving over walking, children would be influenced to cultivate the habit of using non-active transportation.

Beyond the microsystem, there are the mesosystem and exosystem that encompass a system of external relationships (e.g., teachers, and neighbours) intertwined with the child's immediate family [41]. These systems provide a wider social and physical environment, hereby referred to as the community, where knowledge and habits can diffuse between the family system, neighbours, and service providers, such as the school. From this viewpoint, active transportation among children may be co-influenced by the family, neighbours, and service providers. The wider social environment may support active travel through its qualities of cohesiveness, reciprocity, and safety [39], whereas teachers and social networks (e.g., friends and classmates) can encourage active travel [39,44] depending on norms within the community and their immediate families. As mentioned earlier, children and adolescents who are dependent on their parents are subject to influences from their family and community, so they would not perform active transportation behaviour if it were not a value in these social settings.

Before entering the third generation, children who grow into the second generation may enjoy enhanced autonomy and flexibility in decision-making because their parents may begin to recognise their improving maturity. In the third generation, therefore, parents may allow their adolescent children to make some decisions, including transport decisions. Thus, the second generation is in a stage where there is an onset of opportunities to exercise free will, monitored by mature members of the micro and mesosystems. If the child was well embedded in a family tradition of active transportation, for example, they might exercise free will in ways that translate into active transportation [43,44]. Yet, individuals in this generation may not have absolute autonomy, possibly because their parents are ambivalent about their life experiences.

In adulthood, individuals may have started an independent life, but they can maintain relationships across the three systems through regular communication and commuting with family members, workmates, and business partners. Autonomy and flexibility in decision-making may have reached an optimum level, enabling the individual to decide whether to pursue a lifestyle influenced by the family and other social systems in the first and second generations. Members of this generation are generally a working class who exercise control over their earnings and how to spend them through, for example, car ownership. Adults who grew up in a family or environment where active transportation was a shared hobby are likely to own a bicycle or similar equipment for active transportation. Such individuals, depending on how the three systems including the educational system influenced their pro-environment behaviour [19,27], may own cars but may only use them to travel long distances. The main disadvantage at this stage is that individuals may have new commitments (e.g., work) that may deprive them of social support and time for active transportation behaviour.

Between the first and third generations, the individual can adapt life experiences from social ties spanning the three systems. For instance, those who had left their family and community but were positively influenced to practice walking as a habit may continue to utilise past experiences to maintain walking. The opportunity to draw on the three systems (implicit in social ties and experiences from the previous community) is recognised by the Activity Theory of Ageing (ATA) [45,46]. The ATA asserts that ageing people can adapt their past experiences (e.g., cycling to school) and social networks to maintain physical activity. This process of life course adaptation occurs in a social context where skills and life experiences from social networks (e.g., parents or friends) are acquired through learning, observation, communication, and role modelling.

On the other hand, the Disengagement Theory of Ageing (DTA) argues that the ability to adapt past experiences to maintain physical activity and possibly stick to an acquired active transport behaviour dwindles in the ageing process due to a decline in the individual's resources (e.g., social ties and income) and physical functional ability in later life [46]. The DTA, thus, suggests that resources and physical abilities that may support active transportation may be insufficient in the fourth generation. As such, habits such as walking and cycling learned through the initial life stages may be discouraged by a decline in social networks (e.g., through the death of social ties) and physical functional capacity. Some researchers [39,46] and the ATA suggest, nevertheless, that individuals who maintain physical activity over the life course avoid this decline and maintain physical activity into later life. Individuals may maintain active transportation in the fourth generation if they started an active lifestyle earlier, ideally in the first generation and maintained it through the remaining stages. The ATA also insinuates that people can maintain autonomy in the fourth generation if they adapt past experiences rooted in the three systems of the BST across the lifespan.

Another factor that may influence active transportation across the four generations is a change in life goals necessitated by the ageing process [47]. For example, people in the fourth generation may decide to spend more time with closer social ties such as grandchildren and in-laws and avoid less important activities. This decision stems from older adults' future time perspective [48], which is about awareness of how short their remaining life is and a need to spend time on only activities and people who matter to them. This concept of future time perspective originates with the Socioemotional Selectivity Theory [49], which asserts that older adults may limit their social and physical environment through social disengagement by focusing on only a few valued relations and social activities associated with these relations. This behaviour may terminate the positive influence of demographic, psychosocial, and environmental factors on active transportation in later life. To explain, an older adult may give up social ties and activities in the community to spend more time with grandchildren through childcare at home.

Depending on the lifestyle of their social ties, nevertheless, older adults can maintain engagement with life through active transportation. Older adults who are psychologically and emotionally attached to their valued active grandchildren, in-laws, or surviving spouses may continue to perform physical activity (e.g., walking and cycling) necessitated by social activities. If individuals grew old in a family where active transportation was a tradition, their future time perspective would rather support the maintenance of active transportation [43], especially if this behaviour enhances their longevity, which they need to spend more time with their loved ones. An active family tradition makes it more likely for people in the fourth generation to maintain the willingness and ability to sustain active transportation. The above theoretical deductions suggest that all generations face barriers to active transportation and that even younger adults do not have a perfect chance to adopt active transportation behaviour.

6. Active Transportation Adoption: Are Older Adults Laggards?

Ageism against older adults in environmental activism is partly premised on the notion that older adults perform less pro-environment behaviour (i.e., active transportation) [19,23] and have generated a higher carbon footprint linked to non-active transportation over the life course [9]. As Table 2 suggests, however, every generation has unique barriers and opportunities for active transportation; opportunities for performing active transportation behaviour are uniquely counteracted in each generation, which is why the sustainability of active transportation over the life course is not necessarily higher in generations under 50 years. As the above theoretical framework suggests, the way opportunities and barriers in Table 2 play out in practice depends on context, characterised by the family and wider community from which learning and adaptive behaviour take place. In this section, the authors emphasise empirical evidence supporting this reasoning by reviewing studies on the relationship between age, pro-environmental behaviour, and active transportation.

Table 2. The authors' visualisation of the core characteristics of the four generations within the theoretical framework.

Group	Description	Core Attribute(s)	Possible Barriers	Possible Opportunities	Implications *
Children (generation 1)	Infants and other young children aged 0–12 years who cannot make decisions	Members live with parents or guardians and are subject to parents	(1) Little or no autonomy, and (2) dependence **** on parents that may limit active transportation	(1) Teachableness, and (2) opportunities to start learning from family, networks (e.g., teachers), and community **	Children do not make their own decisions, so their parents and immediate social environment may prevent them from choosing active transportation if they do not value this travel behaviour
Adolescents (generation 2)	Adolescents and teenagers aged 13–17 years who are living with parents or guardians	Members live with parents or guardians and are subject to parents but with improved autonomy vias-a-vis stage 1	(1) Improved but limited autonomy, and (2) insufficient independence from parents, which can prevent active transportation	(1) Youthful vigour or physical strength and (2) learning opportunities through mentoring, formal education, and positive norms (e.g., walking regularly for health)	Adolescents can draw on their physical strength to perform active transportation behaviour if their family and community provide relevant norms and model behaviours
Adults (generation 3)	Individuals aged between 18 and 49 years	Members are likely working, have optimum autonomy, and can make and act on personal decisions	(1) May leave family as well as the community and networks one grew up with, and (2) new commitments (e.g., work) necessitated by independence may prevent active transportation	(1) Independence, (2) income from employment, and (3) optimum autonomy	Adults can make personal decisions, but the pursuit of new goals (e.g., using a car) can prevent them from choosing active transportation, especially in the absence of support *** from previous networks
Older adults (generation 4)	Individuals aged 50 years or higher	Members may have retired; functional ability may decline, and autonomy may reduce due to a disability	May lose supportive social networks, income, or functional abilities due to ageing	(1) Rich life experience, (2) a future time perspective that may support active behaviours, and (3) close ties (e.g., grandchildren) to support engagement with life	Older adults may lose the physical functional ability and resources (e.g., previous social networks) needed to perform active transportation behaviour

* Implications of the barriers and opportunities for the individual's active transportation choices; ** The community represents the multilevel exosystem implicit in the BST that includes institutions (e.g., school and teachers), neighbourhood social environment, and the physical built environment; *** Support refers to positive experiences (e.g., family norms, positive influence from school and teachers, and favourable environmental conditions, such as safety and social support) that encourage or allow the individual to develop and grow with positive behaviours, such as active transportation; **** Dependence on parents is only positive if the family environment provides positive norms and values; otherwise, parents who do not value some behaviours (e.g., active transportation) may prevent their children from choosing it as a travel method.

The empirical literature to date provides mixed evidence regarding the relationship between age and pro-environmental behaviour. A cross-sectional study in Spain found pro-environmental behaviour higher among older adults [21], but two other studies in different countries found this behaviour is higher in younger adults [20]. In China, a cross-sectional study utilising data from 31 countries produced mixed findings, affirming that pro-environmental behaviour is not always higher in younger adults [23]. In a systematic review [20], 31 out of 33 studies reported mixed findings about the association between age and pro-environment behaviour, further affirming that younger adults do not perform higher pro-environment behaviour. This review revealed that older adults can perform higher pro-environment behaviour depending on personal and socio-environmental factors, which is congruent with the foregoing theoretical framework.

A study in the UK [19] found older adults, compared with younger adults, reported less active transportation time, but another study in Germany [26] reported older adults were more likely to perform active transportation behaviours (i.e., walking and cycling), compared with younger adults. In Taiwan, older adults were found to report more active transportation (i.e., cycling) [27] and two other studies [28,29] in Ghana and New Zealand have found older children are more likely to walk or cycle. Thus, the idea that younger people, compared with older people, perform higher active transportation behaviour is also not empirically supported.

Deductively, there is no consistent evidence that active transportation is higher in younger adults because of contextual differences (e.g., some may not grow up in a family or community with opportunities to perform positive behaviours), and differences in people's ability to learn and maintain active transportation over the life course. Because of this, ageing would limit or favour active transportation depending on the context. If so, older adults are not laggards when it comes to the adoption of active transportation, which means ageism against older adults in active transportation and environmental activism has no empirical basis.

7. Discussion

This review aimed to (1) develop a hierarchy for understanding the impacts of active and non-active transport modes on the environment and (2) analyse the adoption of active transportation between older and younger people.

Our analyses suggest that walking, running, and swimming are the only active transport modes with no or negligible carbon footprints. These three modes of transportation are on top of the pyramid or hierarchy, which indicates that they are the most environmentally friendly way to travel. Studies [50,51] recognise these modes of transportation as some of the best ways to travel without adversely impacting the environment, but our review adds to this recognition by proposing the hierarchy and implying that these three transportation methods would produce the least carbon footprint. The hierarchy suggests that active transport modes can generate carbon footprints depending on their size and how they are designed and used. Given this understanding, researchers are encouraged to rather promote active transportation with no or negligible emissions of greenhouse gases and avoid implying that all forms of active transportation behaviour are protective of the environment. Researchers have generally framed active transportation as a pro-environment behaviour [37,50,51], but our analyses reveal a need for them to acknowledge the limitations of active transport modes with a carbon footprint.

The general perception that older adults adopt active transportation less and would, therefore, contribute less to a safer environment is based on mixed and inconclusive evidence. As such, there is no basis for the ageist views reported in the literature [7,9] against older adults in environmental activism. If so, more generalisable empirical evidence is needed on the relationship between active transportation adoption and age, and any future studies assessing this relationship ought to consider the relative carbon footprints of the different modes of active transportation, without which it would be impossible to accurately determine each generation's transportation-related carbon footprint and its impact on the environment. As our review suggests, the adoption of active transportation is not impossible among older adults, though this segment of the population may have the least physical functional ability for performing active transportation behaviour [39,46]. Stakeholders are, therefore, encouraged to include older adults in transportation-related initiatives against climate change. Older adults may contribute unique and complementary experiences for advancing these initiatives.

7.1. Implications for Practice

According to the theoretical framework, communities comprising the family system, service providers (e.g., schools and hospitals), built environment, social networks, and psychosocial factors (e.g., safety and social cohesion) are the embodiment of micro-, meso-, and exosystem factors that enable people to learn and maintain positive behaviours over the life course. Some of the specific factors within these systems that support active transportation across the lifespan are family income, norms favouring healthy behaviours, walkability, and services, such as formal education [27,29,31]. Stakeholders should create inclusive communities and provide services enabling and empowering individuals to overcome barriers to active transportation across their lifespan. Public health education can also be implemented to influence families for developing pro-environment traditions through which their ageing members learn to maintain active transportation behaviour as a habit.

As the ATA suggests, older adults can maintain active transportation in later life if the above measures provide contexts where they can adapt their life experiences across the lifespan. Both the ATA and DTA agree that people become inactive in the ageing process because of a decline in resources and functional ability, but this decline can be avoided through lifelong adaptation of positive experiences acquired through learning in one's family and community. Learning in this vein can be encouraged through public health education or promotion in which individuals are conscientised to practice healthy behaviours, such as walking and cycling [52]. There is, thus, a need for scaling up public health promotion and education programmes intended to create awareness about the role of absolute active transportation in the fight against climate change.

As the hierarchy suggests, people's carbon footprint depends on whether vehicles or products used in their active transportation are biodegradable, recyclable, or large. If possible, manufacturers should prioritise products that are biodegradable, recyclable, and small. They may also consider avoiding the dependence of these products on utilities or energy sources that emit greenhouse gases. If products worn or used during active transportation do not emit any greenhouse gas or eventually become part of environmental waste, most or all active transport modes would generate a zero-carbon footprint. This recalls a need for individuals to properly dispose of waste during active transportation or use environment-friendly products during active transportation.

7.2. Limitations and Future Research

This review included 26 documents from two searches, but an analysis of the relationship between these documents was beyond the scope of our narrative review. Future systematic reviews discussing this relationship are needed. To meet our review goal, we focused on documents exclusively reporting a transportation type and a carbon footprint relating to it. Hence, the number of articles included in this review may be smaller than the number of studies based on active transportation in general. The active transportation types on top of the pyramid (e.g., walking) are not necessarily carbon-free as humans may wear clothes that produce greenhouse gases when walking or running. By referring to them as methods with a "zero-carbon footprint", we meant that they produce the least greenhouse gases and are the most sustainable alternatives for the earth. A more objective way to develop the hierarchy is to rank the transport modes based on their estimated carbon footprints. The literature, however, does not provide standard carbon footprints for transport modes, a shortcoming that future research should remedy. Similarly, this review does not estimate the carbon footprints of the various transport modes due to the non-availability of relevant data. Future researchers are encouraged to provide these estimates, preferably using objectively generated data. Future studies may also assess the validity of the hierarchy by comparing the carbon footprints of transportation types within the hierarchy. The authors' evaluation of product size may not be consistent across contexts and manufacturer niche markets. For instance, a bicycle for adults may be larger than a scooter for young children. Decision-makers should consider these inconsistencies in assessing and using the pyramid.

8. Conclusions

Active transportation can add to atmospheric greenhouse gases and is, therefore, not always environmentally friendly. Active transport modes that may have a zero-carbon footprint are walking, running, and swimming without a product. There are mixed and inconclusive findings regarding the potential effect of age on active transportation and pro-environment behaviour; hence, ageist stereotypes against older adults in environmental activism and active transportation are unwarranted and would weaken the impetus needed to overcome climate change. Younger and older people can avoid a carbon footprint if stakeholders can design the built environment and roll out policies that maximise the diffusion of knowledge and positive active transportation experiences in the family and community. Policies need to be rolled out to encourage families and communities to adopt active transportation behaviour as a culture. Public health education programmes aimed at encouraging individuals to practice active transportation behaviour over the life course are imperative and should be infused into public health policy. Manufacturers are encouraged to prioritise the production of active vehicles (e.g., scooters) that are biodegradable, recyclable, and small.

Author Contributions: Conceptualisation, N.A.; methodology, N.A.; software, N.A. and K.A.-N. validation, K.A.-N., W.N. and E.N.A.; formal analysis, N.A.; investigation, N.A., W.N. and E.N.A.; data curation, N.A. and W.N.; writing—original draft, N.A.; writing—review and editing, N.A., K.A.-N., W.N. and E.N.A.; visualisation, N.A. and K.A.-N.; supervision, N.A.; project administration, N.A. All authors have read and agreed to the published version of the manuscript.

Funding: This research received no external funding.

Institutional Review Board Statement: Not applicable.

Informed Consent Statement: As this study was a review, no informed consent was needed.

Data Availability Statement: Data were not used for this study.

Conflicts of Interest: The authors declare no conflict of interest.

Abbreviations

ATA	Activity Theory of Ageing
BST	Bioecological Systems Theory
DTA	Disengagement Theory of Ageing
MeSH	Medical Subject Headings
PRISMA	Preferred Reporting Items for Systematic Reviews and Meta-Analyses
SPIDER	Sample, Phenomenon of Interest, Design, Evaluation, Research type
UK	United Kingdom

Appendix A

Table A1. The review workplan.

Review title	Active and Non-Active Transportation and Associated Carbon Footprints
Start and end date	1–15 April 2023, (2 searches were performed; the first one was performed on the 1st of April and the second one on the 15th of April).
Research question	What are the potential carbon footprints of active and non-active transport modes?
Condition being studied	Transport modes (i.e., air, land, and sea) and their associated carbon footprints
Search Strategy	
Eligibility criteria (based on SPIDER)	
Sample	All individuals and age groups (to make an intergenerational analysis possible)
Phenomenon of interest	Transport modes accompanying information about their carbon footprints or carbon-dioxide-related emissions
Design	Mixed (qualitative and quantitative)
Evaluation	The relative amount of greenhouse gases produced by each transportation type
Research type	Reviews, primary studies, studies using secondary data, and narratives
Language	English

Table A1. *Cont.*

Date restrictions	No date restriction
Exclusion criteria	Documents published in other languages apart from English, not peer-reviewed, not reporting a transportation type and its carbon footprint, and not published by journals indexed by SCOPUS, Web of Science, or PubMed
Inclusion criteria	Published in English
	Reported transportation type linked to its carbon footprint or greenhouse gas emission
	Peer-reviewed
	Published by journals indexed by Scopus, Web of Science, or PubMed
Geographical scope	Documents from anywhere in the world
Databases	
Essential	PubMed, CINAHL, PsychInfo, ProQuest
As relevant to the subject:	Google Scholar, SCOPUS
Search terms	Transportation, "active transportation", "carbon footprint", "greenhouse gas emissions", association, health, age
Search results	Search 1 = 205; Search 2 = 2

Note: SPIDER—Sample, Phenomenon of Interest, Design, Evaluation, Research type.

Appendix B

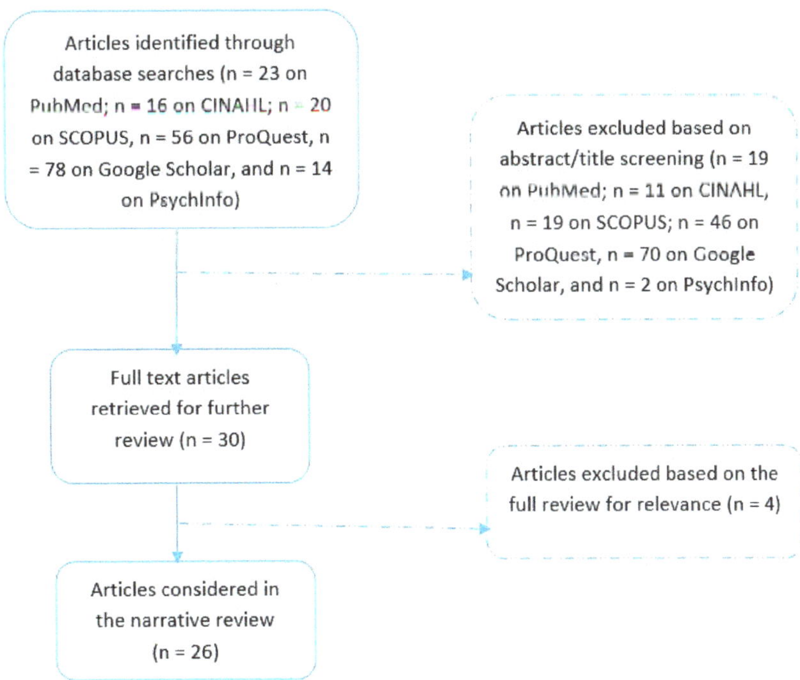

Figure A1. The PRISMA flowchart.

References

1. Wicker, P. The carbon footprint of active sport participants. *Sport Manag. Rev.* **2019**, *22*, 513–526. [CrossRef]
2. Frank, L.D.; Greenwald, M.J.; Winkelman, S.; Chapman, J.; Kavage, S. Carbonless footprints: Promoting health and climate stabilization through active transportation. *Prev. Med.* **2010**, *50*, S99–S105. [CrossRef]

3. Broberg, A.; Salminen, S.; Kyttä, M. Physical environmental characteristics promoting independent and active transport to children's meaningful places. *Appl. Geogr.* **2013**, *38*, 43–52. [CrossRef]
4. Ermagun, A.; Samimi, A. Promoting active transport modes in school trips. *Transp. Policy* **2015**, *37*, 203–211. [CrossRef]
5. Bempong, A.E.; Asiamah, N. Neighbourhood walkability as a moderator of the associations between older Ghanaians' social activity, and the frequency of walking for transportation: A cross-sectional study with sensitivity analyses. *Arch. Gerontol. Geriatr.* **2022**, *100*, 104660. [CrossRef]
6. Goel, R.; Oyebode, O.; Foley, L.; Tatah, L.; Millett, C.; Woodcock, J. Gender differences in active travel in major cities across the world. *Transportation* **2022**, *50*, 733–749. [CrossRef]
7. Ayalon, L.; Roy, S. The Role of Ageism in Climate Change Worries and Willingness to Act. *J. Appl. Gerontol.* **2022**, *42*, 1305–1312. [CrossRef] [PubMed]
8. Ayalon, L.; Ulitsa, N.; AboJabel, H.; Engdau, S. Older Persons' Perceptions concerning Climate Activism and Pro-Environmental Behaviors: Results from a Qualitative Study of Diverse Population Groups of Older Israelis. *Sustainability* **2022**, *14*, 6366. [CrossRef]
9. Ayalon, L.; Roy, S.; Aloni, O.; Keating, N. A Scoping Review of Research on Older People and Intergenerational Relations in the Context of Climate Change. *Gerontologist* **2022**, *63*, 945–958. [CrossRef] [PubMed]
10. Glazener, A.; Sanchez, K.; Ramani, T.; Zietsman, J.; Nieuwenhuijsen, M.J.; Mindell, J.S.; Fox, M.; Khreis, H. Fourteen pathways between urban transportation and health: A conceptual model and literature review. *J. Transp. Health* **2021**, *21*, 101070. [CrossRef]
11. Scheepers, C.E.; Wendel-Vos, G.C.W.; den Broeder, J.M.; van Kempen, E.E.M.M.; van Wesemael, P.J.V.; Schuit, A.J. Shifting from car to active transport: A systematic review of the effectiveness of interventions. *Transp. Res. Part A Policy Pract.* **2014**, *70*, 264–280. [CrossRef]
12. Dixon, A. The United Nations Decade of Healthy Ageing requires concerted global action. *Nat. Aging* **2021**, *1*, 2. [CrossRef] [PubMed]
13. Toner, A.; Lewis, J.S.; Stanhope, J.; Maric, F. Prescribing active transport as a planetary health intervention–benefits, challenges and recommendations. *Phys. Ther. Rev.* **2021**, *26*, 159–167. [CrossRef]
14. Hoj, T.H.; Bramwell, J.J.; Lister, C.; Grant, E.; Crookston, B.T.; Hall, C.; West, J.H. Increasing active transportation through e-bike use: Pilot study comparing the health benefits, attitudes, and beliefs surrounding e-bikes and conventional bikes. *JMIR Public Health Surveill* **2018**, *4*, e10461. [CrossRef]
15. Young, D.R.; Cradock, A.L.; Eyler, A.A.; Fenton, M.; Pedroso, M.; Sallis, J.F.; Whitsel, L.P. Creating Built Environments That Expand Active Transportation and Active Living across the United States: A Policy Statement from the American Heart Association. *Circulation* **2020**, *142*, E167–E183. [CrossRef]
16. Crawford, S.; Garrard, J. A combined impact-process evaluation of a program promoting active transport to school: Understanding the factors that shaped program effectiveness. *J. Environ. Public Health* **2013**, *2013*, 816961. [CrossRef]
17. Silveira-Santos, T.; Manuel Vassallo, J.; Torres, E. Using machine learning models to predict the willingness to carry lightweight goods by bike and kick-scooter. *Transp. Res. Interdiscip. Perspect.* **2022**, *13*, 100568. [CrossRef] [PubMed]
18. Shephard, R.J. Is active commuting the answer to population health? *Sports Med.* **2008**, *38*, 751–758. [CrossRef]
19. Adams, J. Prevalence and socio-demographic correlates of "active transport" in the UK: Analysis of the UK time use survey 2005. *Prev. Med.* **2010**, *50*, 199–203. [CrossRef]
20. Diamantopoulos, A.; Schlegelmilch, B.B.; Sinkovics, R.R.; Bohlen, G.M. Can socio-demographics still play a role in profiling green consumers? A review of the evidence and an empirical investigation. *J. Bus. Res.* **2003**, *56*, 465–480. [CrossRef]
21. Casaló, L.V.; Escario, J.J. Heterogeneity in the association between environmental attitudes and pro-environmental behavior: A multilevel regression approach. *J. Clean. Prod.* **2018**, *175*, 155–163. [CrossRef]
22. Chen, X.; Peterson, M.N.; Hull, V.; Lu, C.; Lee, G.D.; Hong, D.; Liu, J. Effects of attitudinal and sociodemographic factors on pro-environmental behaviour in urban China. *Environ. Conserv.* **2011**, *38*, 45–52. [CrossRef]
23. Wang, Y.; Hao, F.; Liu, Y. Pro-environmental behavior in an aging world: Evidence from 31 countries. *Int. J. Environ. Res. Public Health* **2021**, *18*, 1748. [CrossRef]
24. Sánchez, M.; López-Mosquera, N.; Lera-López, F. Improving Pro-environmental Behaviours in Spain. The Role of Attitudes and Socio-demographic and Political Factors. *J. Environ. Policy Plan.* **2016**, *18*, 47–66. [CrossRef]
25. Patel, J.; Modi, A.; Paul, J. Pro-environmental behavior and socio-demographic factors in an emerging market. *Asian J. Bus. Ethics* **2017**, *6*, 189–214. [CrossRef]
26. Finger, J.D.; Varnaccia, G.; Gabrys, L.; Hoebel, J.; Kroll, L.E.; Krug, S.; Manz, K.; Baumeister, S.E.; Mensink, G.B.M.; Lange, C.; et al. Area-level and individual correlates of active transportation among adults in Germany: A population-based multilevel study. *Sci. Rep.* **2019**, *9*, 16361. [CrossRef] [PubMed]
27. Liao, Y.; Wang, I.T.; Hsu, H.H.; Chang, S.H. Perceived environmental and personal factors associated with walking and cycling for transportation in Taiwanese adults. *Int. J. Environ. Res. Public Health* **2015**, *12*, 2105–2119. [CrossRef]
28. Agyeman, S.; Cheng, L.; Alimo, P.K. Determinants and dynamics of active school travel in Ghanaian children. *J. Transp. Health* **2022**, *24*, 101304. [CrossRef]
29. Ikeda, E.; Stewart, T.; Garrett, N.; Egli, V.; Mandic, S.; Hosking, J.; Witten, K.; Hawley, G.; Tautolo, E.S.; Rodda, J.; et al. Built environment associates of active school travel in New Zealand children and youth: A systematic meta-analysis using individual participant data. *J. Transp. Health* **2018**, *9*, 117–131. [CrossRef]

30. Mohareb, E.A.; Maclean, H.L.; Kennedy, C.A.; Mohareb, E.A.; Maclean, H.L.; Kennedy, C.A.; Mohareb, E.A.; Maclean, H.L.; Kennedy, C.A. Greenhouse Gas Emissions from Waste Management—Assessment of Quantification Methods Greenhouse Gas Emissions from Waste Management—Assessment of Quantification Methods. *J. Air Waste Manag. Assoc.* **2011**, *2247*, 480–493. [CrossRef]
31. Oliver, M.; Badland, H.; Mavoa, S.; Witten, K.; Kearns, R.; Ellaway, A.; Hinckson, E.; Mackay, L.; Schluter, P.J. Environmental and socio-demographic associates of children's active transport to school: A cross-sectional investigation from the URBAN Study. *Int. J. Behav. Nutr. Phys. Act.* **2014**, *11*, 70. [CrossRef] [PubMed]
32. Allin, S.; Masseria, C.; Mossialos, E.; Allin, S.; Masseria, C.; Mossialos, E. Equity in health care use among older people in the UK: An analysis of panel data. *Appl. Econ.* **2011**, *43*, 2229–2239. [CrossRef]
33. Kpessa-Whyte, M. Aging and Demographic Transition in Ghana: State of the Elderly and Emerging Issues. *Gerontologist* **2018**, *58*, 403–408. [CrossRef] [PubMed]
34. Ouchi, Y.; Rakugi, H.; Arai, H.; Akishita, M.; Ito, H.; Toba, K.; Kai, I. Rede fi ning the elderly as aged 75 years and older: Proposal from the Joint Committee of Japan Gerontological Society and the Japan Geriatrics Society. *Geriatr. Gerontol. Int.* **2017**, *17*, 1045–1047. [CrossRef]
35. Dong, Y.; Coleman, M.; Miller, S.A. Greenhouse Gas Emissions from Air Conditioning and Refrigeration Service Expansion in Developing Countries. *Annu. Rev. Environ. Resour.* **2021**, *46*, 59–83. [CrossRef]
36. Rickard, S.C.; White, M.P. Barefoot walking, nature connectedness and psychological restoration: The importance of stimulating the sense of touch for feeling closer to the natural world. *Landsc. Res.* **2021**, *46*, 975–991. [CrossRef]
37. Lee, R.E.; Kim, Y.; Cubbin, C. Residence in unsafe neighborhoods is associated with active transportation among poor women: Geographic Research on Wellbeing (GROW) Study. *J. Transp. Health* **2018**, *9*, 64–72. [CrossRef]
38. Carlson, J.A.; Saelens, B.E.; Kerr, J.; Schipperijn, J.; Conway, T.L.; Frank, L.D.; Chapman, J.E.; Glanz, K.; Cain, K.L.; Sallis, J.F. Association between neighborhood walkability and GPS-measured walking, bicycling and vehicle time in adolescents. *Health Place* **2015**, *32*, 1–7. [CrossRef]
39. Asiamah, N.; Bateman, A.; Hjorth, P.; Khan, H.T.A.; Danquah, E. Socially active neighborhoods: Construct operationalization for aging in place, health promotion and psychometric testing. *Health Promot. Int.* **2023**, *38*, daac191. [CrossRef]
40. Yang, C. Moral education in mainland China today: A bio-ecological systems analysis. *J. Moral Educ.* **2021**, *50*, 529–543. [CrossRef]
41. Oke, A.; Butler, J.E.; O'Neill, C. Identifying Barriers and Solutions to Increase Parent-Practitioner Communication in Early Childhood Care and Educational Services: The Development of an Online Communication Application. *Early Child. Educ. J.* **2021**, *49*, 283–293. [CrossRef]
42. Davison, K.K.; Werder, J.L.; Lawson, C.T. Children's Active Commuting to School: Current Knowledge and Future Directions. *Prev. Chronic Dis.* **2008**, *5*, A100. [PubMed]
43. Henne, H.M.; Tandon, P.S.; Frank, L.D.; Saelens, B.E. Parental factors in children's active transport to school. *Public Health* **2014**, *128*, 643–646. [CrossRef]
44. Rothman, L.; Macpherson, A.K.; Ross, T.; Buliung, R.N. The decline in active school transportation (AST): A systematic review of the factors related to AST and changes in school transport over time in North America. *Prev. Med.* **2018**, *111*, 314–322. [CrossRef] [PubMed]
45. Havighurst, R.J. Successful aging. *Gerontologist* **1961**, *1*, 8–13. [CrossRef]
46. Asiamah, N. Social engagement and physical activity: Commentary on why the activity and disengagement theories of ageing may both be valid. *Cogent Med.* **2017**, *4*, 1289664. [CrossRef]
47. Asiamah, N.; Conduah, A.K.; Eduafo, R. Social network moderators of the association between Ghanaian older adults' neighbourhood walkability and social activity. *Health Promot. Int.* **2021**, *36*, 1357–1367. [CrossRef]
48. Bardach, S.H.; Rhodus, E.K.; Parsons, K.; Gibson, A.K. Older Adults' Adaptations to the Call for Social Distancing and Use of Technology: Insights from Socioemotional Selectivity Theory and Lived Experiences. *J. Appl. Gerontol.* **2021**, *40*, 814–817. [CrossRef]
49. Löckenhoff, C.E.; Carstensen, L.L. Socioemotionol selectivity theory, aging, and health: The increasingly delicate balance between regulating emotions and making tough choices. *J. Pers.* **2004**, *72*, 1395–1424. [CrossRef]
50. Rosa, C.D.; Collado, S. Enhancing Nature Conservation and Health: Changing the Focus to Active Pro-environmental Behaviours. *Psychol. Stud.* **2020**, *65*, 9–15. [CrossRef]
51. Alipour, D.; Dia, H. A Systematic Review of the Role of Land Use, Transport, and Energy-Environment Integration in Shaping Sustainable Cities. *Sustainability* **2023**, *15*, 6447. [CrossRef]
52. Cochrane, T.; Davey, R.C. Increasing uptake of physical activity: A social ecological approach. *Sage J.* **2008**, *128*, 31–40. [CrossRef] [PubMed]

energies

MDPI

Review

A Review of Current Advances in Ammonia Combustion from the Fundamentals to Applications in Internal Combustion Engines

Fei Ma [1], Lingyan Guo [1], Zhijie Li [1], Xiaoxiao Zeng [1], Zhencao Zheng [2], Wei Li [1,*], Feiyang Zhao [2,*] and Wenbin Yu [2]

1 State Key Laboratory of Engine and Powertrain System, Weichai Power Co., Ltd., Weifang 261061, China
2 School of Energy and Power Engineering, Shandong University, Jinan 250100, China
* Correspondence: liwei09@weichai.com (W.L.); fyzhao@sdu.edu.cn (F.Z.)

Abstract: The energy transition from hydrocarbon-based energy sources to renewable and carbon-free energy sources such as wind, solar and hydrogen is facing increasing demands. The decarbonization of global transportation could come true via applying carbon-free fuel such as ammonia, especially for internal combustion engines (ICEs). Although ammonia has advantages of high hydrogen content, high octane number and safety in storage, it is uninflammable with low laminar burning velocity, thus limiting its direct usage in ICEs. The purpose of this review paper is to provide previous studies and current research on the current technical advances emerging in assisted combustion of ammonia. The limitation of ammonia utilization in ICEs, such as large minimum ignition energy, lower flame speed and more NO_x emission with unburned NH_3, could be solved by oxygen-enriched combustion, ammonia–hydrogen mixed combustion and plasma-assisted combustion (PAC). In dual-fuel or oxygen-enriched NH_3 combustion, accelerated flame propagation speeds are driven by abundant radicals such as H and OH; however, NOx emission should be paid special attention. Furthermore, dissociating NH_3 in situ hydrogen by non-noble metal catalysts or plasma has the potential to replace dual-fuel systems. PAC is able to change classical ignition and extinction S-curves to monotonic stretching, which makes low-temperature ignition possible while leading moderate NO_x emissions. In this review, the underlying fundamental mechanism under these technologies are introduced in detail, providing new insight into overcoming the bottleneck of applying ammonia in ICEs. Finally, the feasibility of ammonia processing as an ICE power source for transport and usage highlights it as an appealing choice for the link between carbon-free energy and power demand.

Keywords: ammonia; internal combustion engines; combustion; emissions

Citation: Ma, F.; Guo, L.; Li, Z.; Zeng, X.; Zheng, Z.; Li, W.; Zhao, F.; Yu, W. A Review of Current Advances in Ammonia Combustion from the Fundamentals to Applications in Internal Combustion Engines. *Energies* **2023**, *16*, 6304. https://doi.org/10.3390/en16176304

Academic Editor: Sergey M. Frolov

Received: 26 July 2023
Revised: 27 August 2023
Accepted: 28 August 2023
Published: 30 August 2023

1. Introduction

Global transition from traditional fossil to renewable resources has been a concern for years to mitigate greenhouse gases. To a large extent, renewable fuels are regarded as promising energy carries particularly adapted for long-distance and high-powered mobility or long-term energy storage. Hydrogen, a carbonless fuel, has been recognized as the most promising fuel and clean energy carrier for automotive, marine and power generation [1]. A global hydrogen-based economy will be a sunrise for the energy issue after solving the bottleneck regarding the transportation and storage of hydrogen with reliable safety and reasonable cost. Therefore, closer attention has been paid to studying H_2-carrier fuel in the form of different chemical substances.

Ammonia (NH_3) is presently receiving a surge of attention as an energy carrier of high gravimetric hydrogen density (17.8% hydrogen content by mass), and is considered a carbon-free fuel that can be directly used in both combustion and fuel cell systems. Ammonia is a colorless gas with a very pungent odor at room temperature and can be dissolved in water. Ammonia has trigonal pyramidal molecule geometry with three hydrogen atoms, as well as an unshared pair of electrons attached to a nitrogen atom, and molecule polar covalent bond formation takes place between nitrogen and hydrogen [2]. Due to the strong hydrogen bonds between ammonia molecules, it is easily liquefied. Liquefaction of ammonia is able to happen at 10 bar at room temperature or −33.4 °C at atmospheric pressure, while hydrogen can only be liquefied under −253 °C. Ammonia is able to be stored in liquid form under suitable pressure conditions, ensuring a comparable energy density with other fuels, and it has competitive lower heating value (LHV) [3]. In terms of energy density in storage conditions, liquid ammonia is more than 40% denser than liquid hydrogen or more than twice that of compressed gas hydrogen.

Ammonia is the second-largest chemical products (after sulfuric acid) in the world (over 200 million tons per annum globally with more than USD $60 billion market value) [4,5]. Approximately 80% of global ammonia is utilized in agriculture as fertilizer, with around 5% for explosives, and the balance for industrial cooler refrigerant and chemical commodities. Currently, little ammonia is utilized for energy carriage, but there is definitely great potential for ammonia to be consumed as a renewable fuel in gas turbines, internal combustion engines or fuel cells without a carbon footprint in the years to come. The industrial production of ammonia (NH_3) from N_2 and H_2 is mainly dominated by the Haber–Bosch (H-B) process in the presence of metal catalyst at high temperature (~700 K) and 10–25 MPa, responsible for 1.2% of the global anthropogenic CO_2 emissions [6]. The energy consumption of green H-B is within the range of 27.4–31.8 GJ $t_{NH3}{}^{-1}$, and improvement in overall energy efficiency up to 65% [7]. Furthermore, ammonia could be manufactured from renewable energy sources such as biomass, wind or solar. Achieving a CO_2-free, energy-efficient, low-capital Haber–Bosch synthesis loop is under investigation by electrically driven [8] or electrochemical [9] power sources. In addition, because the absorption and desorption reactions of ammonia is fully reversible, ammonia is able to be released from a solid complex such as $Mg(NH_3)_6Cl_2$ upon heating and compacted into a dense shape, which makes storage simple and safe [10]. Some of the metal ammine complexes show promising results for storing hydrogen, for example, $Mg(NH_3)_6Cl_2$ can store over 9% hydrogen by weight in its solid form. Therefore, ammonia is considered a superb fuel due to its carbon-free structure, safe storage and transportation and low production cost, but with high hydrogen gravimetric density.

Ammonia has regained a great deal of interest from governments and institutes, since it not only enables the vital delivery of nitrogen needed for crop growth but also serves as a chemical that is capable of producing cooling, heating, power, and propulsion with minimum storage cost [11]. When ammonia is labeled as one of the carbon-free alternative fuels, the interest in deploying ammonia is in fast growth. Herbinet et al. [12] summarized an interesting comparison between the advantages and disadvantages of ammonia when used as power solutions. Gas turbines are the system of choice for large-scale ammonia utilization, while solid oxide fuel cells perform better at small scales (below 10 MW). Ammonia can also be directly used in both spark ignition engines or compression ignition engines [12,13]. Predicted by life cycle analysis, it will reduce greenhouse gas emissions by three times through ammonia-fueled vehicles as an alternative to conventional gasoline [14]. The UK Department for Transport proposed taking action on clean maritime growth by placing "a group of hydrogen or ammonia powered domestic vessels in operation" [15]. Indeed, ammonia has a higher octane number than gasoline and natural gas (Table 1), thus allowing a higher compression ratio applied in engine operation. Even though the energy density of compressed ammonia is comparably less than that of gasoline and diesel, it is still much higher than that of compressed natural gas or liquid hydrogen. Research on ammonia for internal combustion engines (ICEs) is in its early stages. Ammonia has extreme low combustion intensity with difficult flammability, reflected by its narrow flammability limits and low laminar burning velocity (Table 1).

Table 1. The main characteristics of some fuels [16,17].

Properties	Units	Ammonia	Hydrogen	Hydrogen	Natural Gas	Gasoline	Diesel	Methanol	Ethanol
Molecular formula		NH_3	H_2	H_2	CH_4	C_5H_{12}-$C_{12}H_{26}$	C_4H_{100}-$C_{12}H_{26}$	CH_3OH	C_2H_5OH
Density	@NTP (kg/cm^3)	0.73		0.083	0.66	700–780	830–855	791	785
Boiling point	(°C)	−33		−253	−161.5	33–190	180–370	64.7	78
Evaporation latent heat	(kJ/kg)	1370		446	511	305	230–250	1160	840
Low heating value	(MJ/kg)	18.8	120.0	120.0	50.0	44.0	42.6	19.6	26.8
Laminar flame velocity	@NTP (cm/s)	7		100–1000	36.53	50	86.5	50	47
Carbon atomic mass fraction	(wt%)	0	0	0	75	86.5	86.3	37.5	52.1
Hydrogen atomic mass fraction	(wt%)	17.6	100	100	25	13.5	13.1	12.5	13
Storage method		Compressed Liquid	Compressed Liquid	Compressed Gas	Compressed Liquid	Liquid	Liquid		
Storage temperature	(K)	298	20	298	298	298	298		
Energy density under storage	(MJ/L)	11.5	8	4.8	9.7	32.0	35.2		21.3
Autoignition temperature	(K)	924	844/824	844/824	723/885	573	503/499–506	737	696
Minimum ignition energy	(mJ)	8		0.02	0.28	0.2–0.3		0.14	0.28
Octane rating	(RON)	110/≥130	>130/≥130	>130/≥130	107/125	90–98/92–98		109	109

NTP: Normal temperature and pressure condition: P = 1 bar, T = 293.15 K.

2. Method of Ammonia Production

Ammonia production can be classified as brown (or gray) ammonia, blue ammonia and green ammonia based on the carbon emissions from the manufacturing process [18]. Current brown ammonia is massively produced by reforming hydrocarbons such as methane using proven Haber–Bosch industrial technology [19], which requires a high-temperature, high-pressure environment as well as addition of N_2, as shown in Figure 1. Blue ammonia is able to be generated using hydrogen from natural gas in conjunction with CCUS (carbon capture, utilization, and storage), while the most desirable green ammonia can be made using hydrogen from water electrolysis without CO_2 emissions. Overall, the majority of ammonia (around 86% to 96%) is manufactured through the traditional Haber–Bosch process, although it is energy-intensive with low conversion efficiency (25% at 30 MPa) and high carbon emissions [20]. In addition, the remaining commercial ammonia comes from alternative technologies such as electrochemical or thermochemical processes etc. [19,20] powered by low pressure and no catalysts with carbon-free emissions [21]. Moreover, the electrochemical process can be performed over a wide range of temperatures, depending on the electrolyte applied during the process [19].

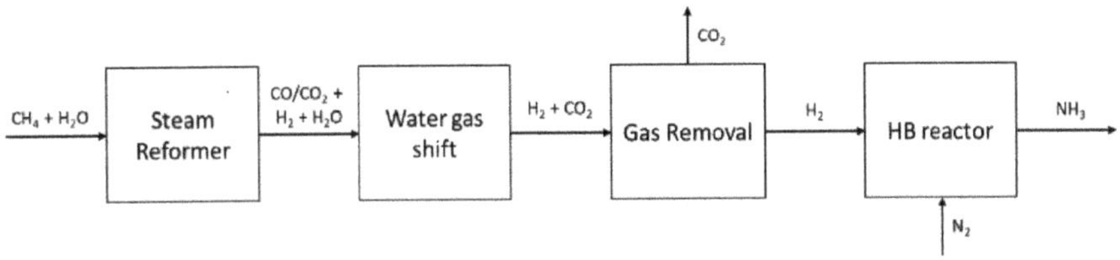

Figure 1. Simplified NH_3 production progress with CH_4 as the main raw material [19].

Plasma-assisted ammonia synthesis is alternative option to produce NH_3 from H_2 and N_2 via the formation of NH radicals. Many studies have concentrated on the application of plasma catalysts at ambient temperature. Li et al. [22] found that the reaction rate of ammonia synthesis was facilitated by the interaction between catalyst and plasma, which was double the pure catalyst load and 30 times that of plasma only. Andersen et al. [23] and Sun et al. [24] used the open-source code ZDPlasKin to numerically figure out the underlying kinetic processes of plasma-assisted ammonia synthesis. They both found surface interactions such as Eley–Rideal (E-R) and Langmuir–Hinshelwood (L-H) reactions contributed to the production of ammonia significantly. The dominant reaction mechanisms of plasma-catalyzed ammonia synthesis are illustrated in Figure 2 and were summarized by Engelmann et al. [25]. Due to electron collisions, a large number of N_2 and H_2 molecules were ionized, excited, adsorbed and dissociated, followed by the adsorption of free radicals (N, H, NH and NH_2) on the surface of the catalyst. After that, these particles in the adsorbed state either underwent E-R reactions with groups in the gas phase or LH reactions on the catalyst surface due to atomic diffusion. Eventually, the resulting NH_3 was released from the catalyst surface. Unlike the traditional H-B process, breakage of the NN triple bond can happen at room temperature and pressure via plasma-assisted ammonia synthesis. However, the efficiency of plasma-assisted ammonia synthesis alone is not high because the strength of the NN bond (NN bond = 946 kJ/mol) is much greater than that of the NH bond (NH bond = 389 kJ/mol) at room temperature. Therefore, the choice of proper catalysts to improve the selectivity of the reaction products is also a hot topic of research.

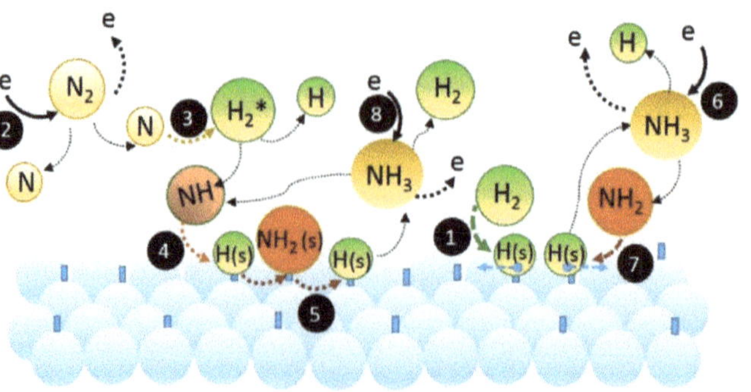

Figure 2. The main reaction process of plasma ammonia synthesis [25].

An efficient catalyst should exhibit not only strong adsorption of reactants but also low dissociation energy of products, which is contradictory to the catalyst itself. However, it is possible to modulate the bonding strength of N_2 and catalyst by elaborate design of alloy catalysts based on the surface bonding energy between metal atoms and N atoms. Xie et al. [26] prepared a high-entropy alloy (HEA) containing five common metallic elements that became saturated at 600 °C (with 100% conversion), and it was much more effective in catalysis than the noble metal Ru (with 73% conversion).

3. Limitations of Ammonia Fuel in ICEs

3.1. Ignition Energy

The minimum ignition energy (MIE) is typically defined as the minimum energy required to ignite the combustible gas [27], and a lower MIE indicates that a stable fire nucleus can be formed more easily in the early stages of combustion. Although ammonia has a high hydrogen content, pure ammonia combustion is inherently difficult to ignite. An ignition energy of 2.8 J is still required in spite of mixing with hydrogen fuel under a modified ignition system, which is two orders of magnitude higher than hydrocarbon fuel and four orders of magnitude higher than hydrogen [28]. Xin et al. [29] performed experimental studies on combustion and emission properties of a hydrogen/ammonia-fueled engine at part-load operating conditions. As illustrated in Figure 3a,b, the addition of ammonia changed the combustion characteristics by prolonging ignition delay times (duration of CA0-10) and flame development periods (duration of CA10-90) due to its higher ignition energy. Meanwhile, the combustion phase was still controllable by modulation of ignition timing for improved indicated thermal efficiency (ITE) and acceptable NOx emission.

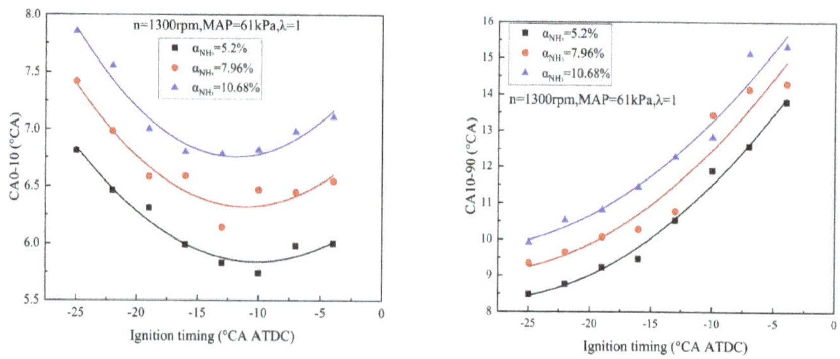

(a) (b)

Figure 3. Combustion durations versus ignition timing at different ammonia levels [29]. (**a**) duration of CA0-10 for varied addition of ammonia (**b**) duration of CA10-90 for varied addition of ammonia.

3.2. Flame Speed

The primary limitation on the practical usage of ammonia as an engine fuel is the relatively slow flame speeds. The laminar burning velocity (LBV or Su) is an important parameter to characterize the premixed combustion process. Investigations on accelerating NH_3 efficient combustion with fuel additives have attracted the interest of many researchers. Numerical studies on the performance of premixed combustion of NH_3/H_2/air mixtures was conducted by Wang et al. [28] using Cantera open-source code. It was shown that the properties of NH_3/H_2/air mixture combustion could be comparable to that of hydrocarbon fuels under engine-relevant conditions; therefore, a high compression ratio is tolerable because of the excellent knock-resistance ability of the NH_3/H_2 mixture. In addition, the increase in Su with H_2 addition contributes to facilitated diffusion, intensified reactivity, and increased flame temperature. At a compression ratio of 10, Su is observably improved with more H_2 added. However, for the stoichiometric combustion of NH_3/H_2, there still needs to be a 39% hydrogen doping ratio to reach a comparable Su level as CH_4 combustion.

Lhuillier et al. [30] found the same phenomenon when studying ammonia blends under engine-relevant turbulent conditions: enrichment of hydrogen in ammonia leads to an earlier, more intense heat release. To figure out the effects of additives such as H_2, CO and CH_4 on the Su of ammonia blend combustion, Han et al. [31] proposed normalized enhancement parameters for quantitative analysis of the enhancement level. It was concluded that the effects of H_2 enhancement were exponential, while non-monotonic with CO and almost linear with CH_4 at mixing ratios greater than 0.2. Very low Su is constrained with pure CO/air combustion since almost no H or OH radical is accumulated during combustion, as shown in Figure 4, but when blended with moderate CO in NH_3, a rapid increase in Su is achieved by decomposed H and OH radicals. The trends of Su almost coincided with the curves of the maximum H molar fraction, which implied the effect of H radicals on Su was dominant. Similar findings with concentration of H and OH radicals were also reported in the flame combustion of ammonia/methane/air [32].

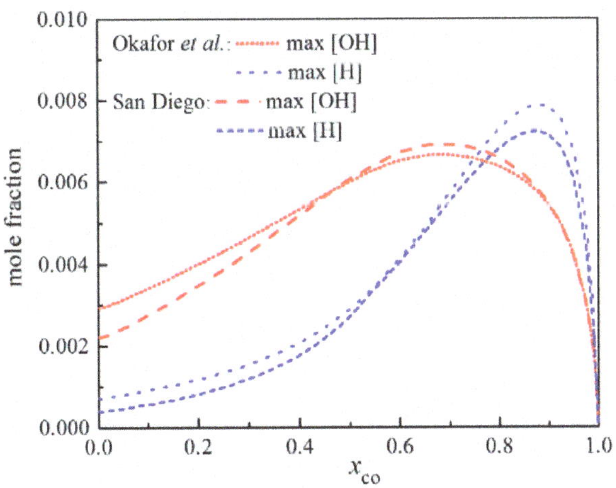

Figure 4. Maximum molar fraction of H and OH radicals of stoichiometric NH_3/CO/air flame [31].

3.3. NOx and Unburned NH_3 Emissions

The use of ammonia fuels is effective in reducing hydrocarbon emissions, but relatively it also increases the content of NOx and unburned NH_3 in exhaust. Representing significant progress in microscopic combustion kinetic reactions, chemical kinetic mechanisms are widely used to understand how quickly or slowly chemical reactions occur in nature. Numerous robust and concise chemical kinetic mechanisms have been proposed for ammonia oxidation under a wide range of conditions after verification of combustion characteristics and NOx emission. Since ammonia chemistry is less complex than that of hydrocarbons, the most recent ammonia oxidation mechanisms in different studies are generally fairly compatible. However, the differences are largely limited to the choice of rate constants or branching ratios for specific elementary reaction groups in varied application circumstances [33]. It is concluded that NO generation in burning lean ammonia depends heavily on OH radicals and HNO relative reactions, but under rich-ammonia combustion, rich H radicals promote formation of NH_x radicals [34]. However, as concluded by Li et al. [35], development of an accurate mechanism to model ammonia-based flame is urgently necessary. Apart from NOx emission, the carbon capture from low-carbon fuel-assisted ammonia flame is still worth studying.

Wang et al. [28] studied the performance of premixed hydrogen–ammonia combustion by simulations. The integrated rate of production (ROP) of NO and the species molar fraction with varied hydrogen molar fraction α are displayed in Figure 5. NO was considered the main source of NOx and two competitive mechanisms of NO production were analyzed: on the one hand, as more H content involved, the flame temperature and concentrations of reactive O/H radical were both increased, leading to a growth in NO generation as well as NH_3 conversion; on the other hand, the decrease in the reactant NH_3 suppressed the NO-related reaction rates in turn. Unlike fuel NO, which was formed with NH_3, the hot NO was more likely to be formed with the increasing of flame temperature caused by O/H radicals. Moreover, NO_2 was rapidly generated by conversion from NO in the flame region, then reconverted back to NO in the post-flame region via $NO_2 + O \rightarrow NO + O_2$, which finally reduced the total amount of NO_2 [36,37]. Jin et al. [38] investigated the effect of different ammonia-to-energy ratios (AER) on the combustion and emission characteristic of an ammonia–diesel dual-fuel engine. Compared to pure diesel engines, the NO emission was decreased instead with increasing AER because of the denitrification of amine, forming more stable N_2 via reduction reaction. However, the incomplete combustion of NH_3 was increased significantly due to the low combustion temperature.

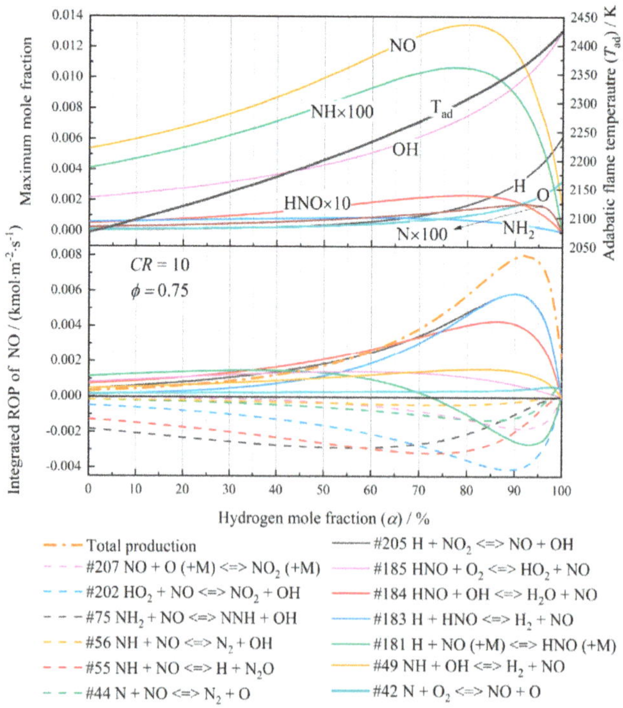

Figure 5. Temperature distribution and integrated production rate such as NO varies with α, φ = 0.75, CR = 10 [28].

As mentioned above, obstacles to the further development of ammonia-fueled engines are large minimum ignition energy, lower flame speed, and more NOx emission with unburned NH_3. However, some new technologies expanded on the next chapter are expected to overcome the difficulties.

4. Current Technologies of Ammonia-Fueled Engines

4.1. Oxygen-Enriched Combustion

It has been confirmed that oxygen-enriched combustion has the ability to lower fuel ignition points, speed up reactions, widen combustion limits and raise flame temperature. Therefore, some scholars studied the combustion characteristics of ammonia under oxygen-enriched conditions. For internal combustion engines, the advantages of oxygen-enriched technology solve the problems of further oxidation of CO and unburned hydrocarbons, but bring higher NOx emissions [39–44]. In order to overcome the shortcomings caused by oxygen enrichment, Liang et al. [45] applied an emulsification technique to diesel, which made the fuel unstable and then disperse throughout the combustion chamber after micro-explosion during the compression stroke in the engine. As a result, the combination of water diesel emulsion and oxygen-enriched combustion reduced the combustion temperature in the cylinder, thus leading to less NOx emission, while the output power of the engine was lower than the normal level. Karimi et al. [46] investigated the effects of oxygen enrichment on the combustion and emission characteristics of a hydrogen–diesel dual fuel (HDDF) engine under low load. Compared with traditional HDDF engines, the oxygen-enriched conditions improved the characteristics of low combustion temperature and laminar flame speed under low load, thus reducing ignition delay. However, more NOx was emitted than a diesel-only engine, but this could be reduced through the EGR technique. Since ammonia engines are not widespread, most of the research on ammonia has been on basic combustion characteristics. An experimental and kinetic modeling investigation on the laminar flame propagation of ammonia was conducted by Mei et al. [47] under oxygen-enrichment conditions. It was found the laminar burning velocity increased with the increasing oxygen content in both experimental and numerical studies, which was driven by the enriched concentrations of key radicals H, OH and NH_2. However, the oxygen enrichment also caused more NOx emissions in turn. Wang [3] investigated the basic characteristics of ammonia fuel in a constant-volume bomb, and the modeling-predicted laminar burning velocities by Cantera code under oxygen enrichment are given in Figure 6. The laminar flame velocity in the ammonia-burning system increased first and then declined as the equivalence ratio changed from 0.6 to 1.5, predicted by varied mechanisms. The greater the oxygen enrichment, the faster the laminar flow flame speed. Meanwhile, there was an interesting phenomenon: the equivalent ratio corresponding to the maximum laminar flame velocity moved towards the lean combustion side with increasing oxygen enrichment Ω. Similar findings were discovered for NOx emissions. As a result, when considering oxygen enrichment to assist ammonia combustion, NOx emissions should be given special attention.

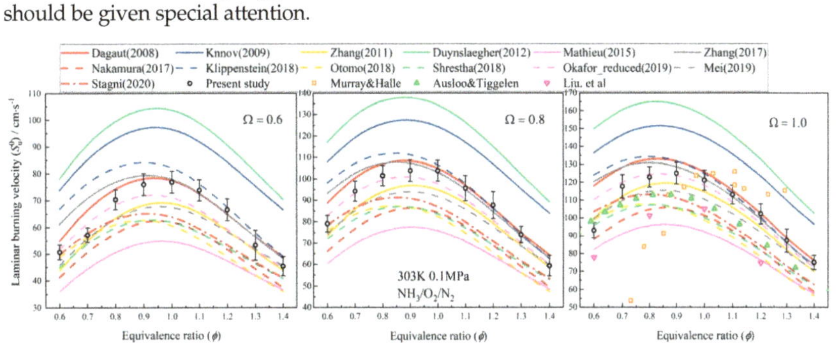

Figure 6. Comparison of predicted laminar flame velocity of ammonia with low oxygen-rich concentration [3].

4.2. Ammonia–Hydrogen Mixed Combustion

To improve the combustion characteristics of ammonia, it is effective to blend hydrogen with ammonia fuel because hydrogen laminar combustion is faster while having a wider combustion limit compared with ammonia. Differently from dual-fuel combustion or oxygen-enriched combustion, addition of hydrogen derived from partially cracked ammonia typically improves the combustion properties under the conditions of high temperature, high pressure and suitable catalyst. A basic investigation on ammonia decomposition was conducted by Ganley et al. [48], who evaluated the activity of metal catalysts: Ru > Ni > Rh > Co > Ir > Fe > Pt > Cr > Pd > Cu >> Te, Se, Pb. Comptti et al. [49] developed a hydrogen generation system (HGS) for ammonia–hydrogen-fueled internal combustion engines using a commercial ruthenium-based catalyst named ACTA 10010. The HGS heated by exhausted combustion gases performed well in hydrogen production, engine brake thermal efficiency and net heat release, but poorly for NOx emissions. Similar results can be found in the work by Ryu et al. [50], and it is emphasized that only a low ammonia flow rate would improve combustion performance.

Apart from dissociating NH_3 in situ from hydrogen, another method of preparing an NH_3/H_2 mixture is to inject hydrogen and ammonia into the intake manifold in the gaseous phase separately [51]. Zhang et al. [52] conducted a numerical investigation on the effects of hydrogen-rich reformate addition on the combustion and emission characteristics of an ammonia engine. It was found that the in-cylinder pressure and heat release rate increased almost linearly with the increasing reformate blending ratio (R_{re}) in stoichiometric cases. This is because more H and OH radicals were generated by the reaction $H_2 + O \rightarrow H + OH$ and the production would promote the consumption of NH_3. The increase in the combustion temperature resulting from greater R_{re} also reduced the emission of unburned NH_3 and N_2O, but too high a reforming ratio would appear undesirable, and thus they recommended a ratio of 7.5–10% near stoichiometry. Li et al. [53] investigated the effects of an ammonia–hydrogen mixture on combustion stability in a single-cylinder, four-value optical SI engine. The results showed that compared with a pure ammonia engine, the misfire phenomenon could be avoided by 5% hydrogen addition, 7.5% was the best thermal efficiency, and more than 20% would lead to unstable combustion. This is because a small amount of hydrogen would increase the combustion temperature in the cylinder and accelerate the ammonia oxidation, while further increasing the hydrogen content would cause heat loss, resulting in a decrease in the indicated thermal efficiency.

To understand the effect of cracking ratio on ammonia combustion, Mei et al. [54] performed both experimental and kinetic numerical investigations on laminar flame propagation of partially cracked NH_3/air mixtures. It was reported the laminar flame speed of the mixture was improved as the cracking ratio increased. The combustion was comparable to methane combustion with a cracking ratio of 40% at atmospheric pressure. Wang [3] investigated the variations in laminar flame speeds with different ammonia oxidation mechanisms at low hydrogen-doping ratios under ambient temperature and pressure conditions. It was found that the laminar flame speed increased significantly with increasing hydrogen-doping ratio α in both simulation and available experiments, since more O/H radicals produced by hydrogen accelerated the oxidation reaction of ammonia. Nevertheless, more NO was formed because NH radicals converted directly to NO without HNO oxidation and the conversion of NO to N_2O would be reduced by higher flame temperature.

Lhuiller et al. [55] investigated the behaviors of an ammonia-fueled engine at different H_2 concentrations, equivalence ratios and boosted pressures. It was found that an engine with high hydrogen concentrations performed well under lean combustion conditions, since the addition of H_2 promoted ignition and combustion stability. Wang et al. [56] studied the combustion characteristics of a NH_3/H_2 mixture with high hydrogen doping ratio (30%) in a medium-speed marine engine. As the equivalence ratio increased, it failed to ignite unless the initial intake temperature and pressure increased, but this limitation could be overcome when moderate proportion of hydrogen doping was added, for example, the peak flame temperature exceeded 2150 K at a 40% hydrogen doping ratio. However, if more hydrogen is required to achieve the assisted-ignition effect [57], it brings new challenges to ensure safe engine operation, since a large amount of additional hydrogen needs to be stored for fuel supply. In general, dual-fuel compression ignition engine operation with ammonia is dependent on the cetane number of pilot fuel and its injection strategies. The concerns of high unburned ammonia and NOx emissions because of the fuel-bound nitrogen are expected to be mitigated by an aftertreatment system. Thus, ammonia can be positively seen as a feasible solution as an alternative fuel for ICEs, without significant engine refit.

5. Plasma-Assisted Combustion Technology

5.1. Principle of Plasma-Assisted Combustion

Plasma is the fourth kind of matter distinguished from solid, liquid and gas, consisting of a large number of charged particles and neutral particles, which are electrically neutral. It can be classified into thermal equilibrium plasma and nonequilibrium plasma. Thermal or equilibrium plasmas are characterized by high energy density and equality between the temperature of heavy particles and electrons, while conversely a nonequilibrium plasma characteristic is a lower-pressure plasma with low ion and neutral temperatures, as shown in Figure 7. Nonequilibrium plasma plays a similar role to a catalyst to activate ambient gas molecules at room temperature and pressure through colliding and dissociating chemical bonding with the help of high-energy electrons [58,59]. Plasma generation methods commonly used are dielectric barrier discharge (DBD), atmospheric pressure plasma jet (APPJ) and sliding arc discharge plasma (GAD), while pulsed power is the most commonly used excitation power source for atmospheric low-temperature plasma [60].

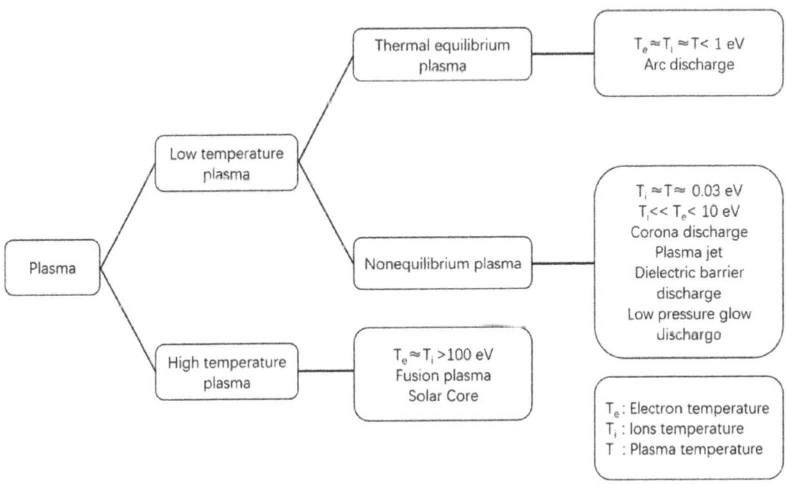

Figure 7. Plasma classification.

The enhancement of plasma-assisted combustion is achieved mainly through three pathways, as illustrated in Figure 8: thermodynamic effects, kinetic effects and transport [61]. Thermodynamically, the increase in overall gas and electron temperatures is ascribed to the energy conversion of the plasma discharge and the exotherm of chemical reactions, thus further accelerating the reactions. Kinetically, a large number of high-energy electrons and active particles involved in plasma not only facilitate chemical reactions due to the reduced activation energy but also introduce new branched reactions in the original system. For transport, plasma discharge will form ionic wind, which can accelerate the mixing process of reaction gases.

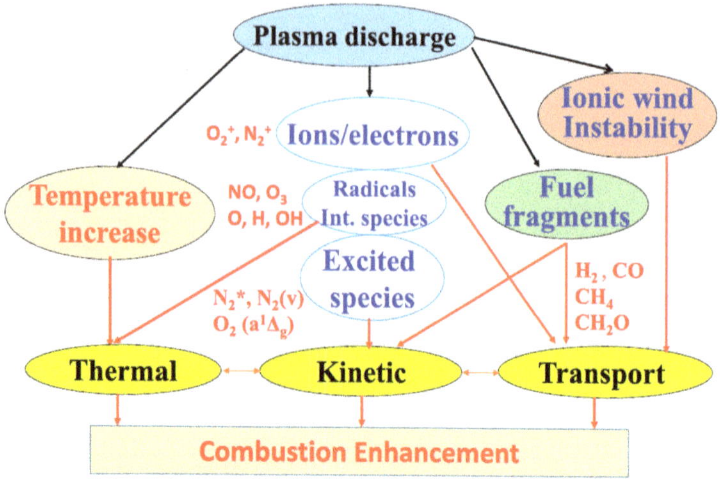

Figure 8. Principle of plasma-assisted combustion [61].

As mentioned, plasma-assisted combustion involves complicated plasma dynamics and combustion kinetics, so it is difficult to decouple the plasma effect and thermal effect from the perspective of conventional experiments due to cross-disciplinary complexity. Although it has been verified that plasma can play a significant role in enhancing combustion, modulating emissions and fuel reforming, the underlying mechanisms are a worthy ongoing topic in different application scenarios.

5.2. PAC Ignition Enhancement

It was found plasma-assisted combustion (PAC) can overcome the flame extinction limit and that PAC ignition enables lower-temperature combustion in contrast to high-temperature ignition, while it must follow the ignition S-curves in conventional ignition process, as shown in Figure 9 [61]. For internal combustion engines, plasma generated by different types of discharge such as microwave [62], radio frequency [63], laser ignition [64] and nanosecond pulses (NRP) [65,66], has been used to assist ignition and combustion. Among these, NRP has a significant advantage applied in low-energy ignition. To clarify the cumulative effect of repetitive NRP pulse number on the PAC, Barleon et al. [67] investigated the ignition of a premixed methane–air mixture by NPR in pin-pin configuration. The conclusion was that there was less total energy required as the pulse frequency increased. Compared with traditional spark ignition, Cathey et al. [65] found that it caused shorter ignition delay, higher peak pressure and greater net heat release in a single-cylinder gasoline engine through NRP. In fact, the expected minimum ignition energy is controlled by the minimum flame radius [61], which has a strong relationship to the Lewis number of the mixture, the fuel reactivity (activation energy) and the flame thickness. Therefore, the reduced ignition delay and lowered ignition energy by PCA were to a great extent due to the increase in the diversity of the reaction system (reduced Lewis number) and the decrease in activation energy.

The reduced electric field E/N is the most significant parameter to control the distribution of energy deposited to the different excitation modes and then generate active particles. Research to figure out the best reduced electric field to minimize the ignition time of CH_4/Air/He mixture was conducted by Mao et al. [68] using a combination of nanosecond pulses and DC discharges. In their study, a DC electric field of 5 Td minimized the time used for ignition, as shown in Figure 10, and it was revealed that the low-DC electric field promoted the excitation of CH_4(v) and O_2(v) effectively, as well as O_2(a1Δg), which played a positive role in ignition enhancement through pathway flux analysis.

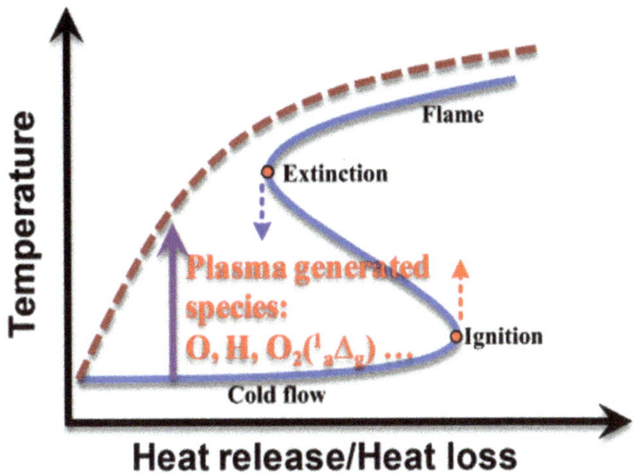

Figure 9. Schematic diagram of the plasma-assisted transition from the classical ignition and extinction S-curve (blue solid line) to the monotonic stretching S-curve (red dashed line) [61].

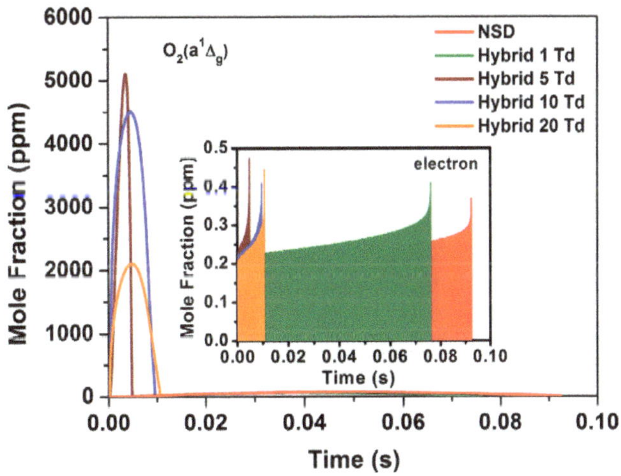

Figure 10. Time evolution of O_2(a1Δg) for NSD and mixed discharges with different DC field strengths at 900 K [68].

Furthermore, the time scale of plasma kinetics is within the order of nanoseconds, while the combustion kinetics fall in ranges from micro- to milliseconds [61], as illustrated in Figure 11. Therefore, the PAC process is the coupling of long-life plasma species with active radicals from fuel pyrolysis via energy transfer and kinetic interaction. From this point of view, the bridge linking fast plasma kinetics and combustion kinetics is critical to fully explore the PAC principles. This motivates the development of an efficient numerical model for multi-timescale PAC. Recently, the zero-dimensional plasma kinetic solver ZDplaskin incorporated with combustion kinetic solver CHEMKIN has been the most widely used numerical tool for PAC. ZDplaskin integrated with Boltzmann equation solver (BOLSIG+) is used to predict time evolution of neural radical and active species. Rate coefficients are formulated based on the incident energy by plasma discharges and collisional cross sections data of electron-associated reaction in LXCat format. Many studies have employed the ZDplaskin-CHEMKIN solver to study flame stability, contemplating the effects of plasma discharges on the flame propagation characteristics and ignition delay times [68,69]. However, there remain challenges in developing plasma kinetic mechanisms accommodating low-temperature and high-pressure conditions. At present, 2-D and 3-D numerical tools are not available for PAC modeling.

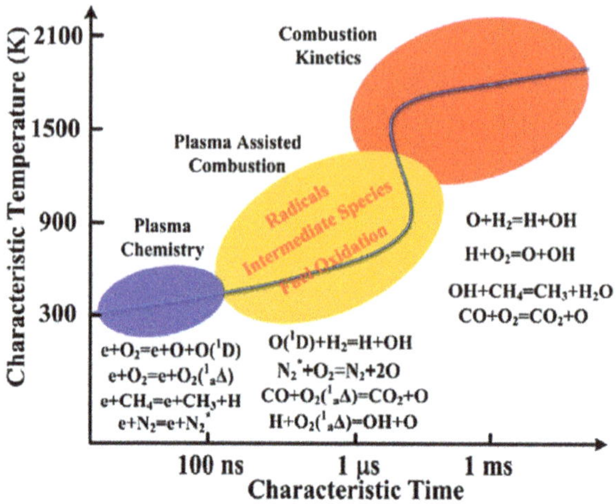

Figure 11. Schematic of timescales for PAC kinetics [61].

The dilution of different types of inert gases also affects electron energy distribution function in plasma, thus influencing the mixture combustion reactivity. A numerical investigation on the effects of methane pyrolysis with different diluents (N_2/Ar/He) was conducted by Mao et al. [70] using the zero-dimensional solver ZDPlasKin coupled with CHEMKIN. The results indicated that the quenching of electronic excited states and relaxation of the vibrational state contributed greatly to the pyrolysis of CH_4, for which the addition of N_2 as diluent performed best. As a result, it is concluded the excited states of equilibrium gases play a dominant role in physicochemical mechanisms and fuel oxidation efficiency through collisional quenching reactions. Similar findings were reported by Snoeckx et al. [71].

The commercial applications of plasma on ICEs are still immature, but many investigations are being carried out. Hwang et al. [72] developed a new microwave-assisted plasma ignition system and applied it to a direct injection gasoline engine. The new ignition system has a shorter ignition delay and a more advanced combustion stage than conventional spark ignition systems. It also broadened the thin combustion limit since it can form a larger fire nucleus by providing abundant reactive radicals over a larger area. A laser-induced plasma (LIP) ignition system has been evaluated to assist the ignition of a diesel–gasoline mixture in a CI engine by José et al. [73]. It was reported that the LIP ignition system can effectively solve the problem of automatic ignition of mixed fuels under different operations, and in addition, the variable height ignition technology improved combustion efficiency and reduced hydrocarbon emissions. L. et al. [74] investigated the ignition behavior of a methane–air mixture in an optically equipped setup consisting of a double chamber under repeated pulse discharge (NRPD). It showed that NRPD has a higher ignition success rate and average flame front propagation compared to standard inductive ignition. It can achieve lean combustion by adjusting the energy and number of pulses, but it did not work after inflammation.

6. Plasma-Assisted Ammonia Ignition

As mentioned above, addition of hydrogen derived from partially cracked ammonia typically improves the combustion properties. Direct decomposition of ammonia to hydrogen is a typical endothermic reaction with an enthalpy change of 91.2 kJ/mol. The decomposition of ammonia is able to be achieved at high temperature in the presence of a classic selective catalyst. Nevertheless, plasma catalysis is a promising approach to decompose ammonia at low temperature (<450 °C), but with a relatively high conversion rate. The synergistic effect of plasma and catalyst refers to the truth that it affects the catalytic process more than the sum of plasma and catalyst alone [75]. Although plasma catalytic ammonia synthesis is well understood, plasma catalytic ammonia decomposition still needs further study, and there is no commercial application available so far. Research on plasma catalytic ammonia decomposition has been centered on experiments, most of which have focused on designing high-efficient catalyst to improve hydrogen conversion. Wang et al. [76] measured the effect of DBD plasma synergism with Fe-based catalysts on ammonia decomposition efficiency through optical diagnostics and isotope scanning techniques. It was revealed the excited-state NH_3* and NH* played an important role in N atoms desorption, thus influencing conversion efficiency as the reaction temperature reduced by 100–140 °C. In their subsequent study [77], it was found the plasma promoted the chemisorption of NH_3*, leading to increased conversion by nearly 40% at 550 °C. The underlying physicochemical mechanisms of plasma catalytic kinetics should be further explored in depth to facilitate its practical usage in on-site hydrogen generation.

A number of simulation studies were carried out by using the kinetic model under wide conditions of temperatures ranging from 600 to 1500 K, and the characteristics of plasma-assisted ammonia combustion were predicted with varying pulsation frequencies and pulse numbers. Faingold and Lefkowitz [78] employed ZDplaskin-CHEMKIN with an assembled kinetic model for the oxidation of ammonia/oxygen/helium to perform pulse repetition frequencies (PRFs) on ignition delay times (IDTs). It was found IDTs could be reduced by 40–60%, as shown in Figure 12, under a moderate number of pulses. Higher PRFs promote an expanding radical pool, whereas lower PRFs favor the radical recombination between pulses.

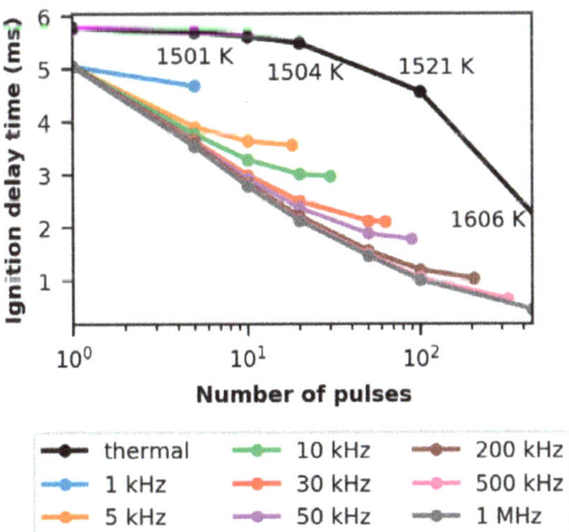

Figure 12. IDTs for different PRFs and number of pulses [78].

A series of experiments on plasma-assisted ammonia combustion were carried out because of the great advantages of plasma-assisted combustion as mentioned before. Lin et al. [79] studied the performance of plasma-assisted combustion via a new gliding arc plasma (GAP) generator combined with a cyclonic burner. The operation map of ammonia combustion flames with the GAP on or off is displayed in Figure 13, in which the ammonia combustion limitation was widened under both air and ammonia GAP. However, if the equivalence ratio exceeded 2.2 under a high air flow rate, the ammonia combustion flame would become unstable. Furthermore, the physicochemical mechanisms of discharged air and ammonia worked differently. The particles (OH*, H* and O*) played a dominant role in the dehydrogenation reaction of ammonia through kinetic mechanisms for the former (air medium) while the latter (ammonia medium) improved the combustion limit by the hydrogen produced from ammonia directly. A similar investigation on ammonia combustion was conducted by Choe et al. [80] using a nanosecond high-pressure pulse generator. It was concluded that plasma was able to extend the lean blowoff limits of ammonia flames and reduce NOx emissions simultaneously, but further research needs to be focused on the interaction between plasma dynamics and combustion kinetics.

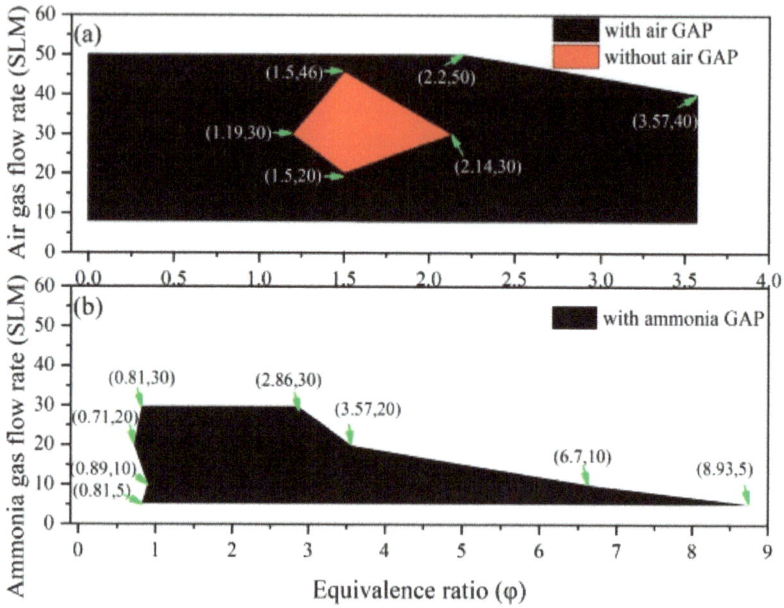

Figure 13. An operation map of ammonia combustion flames under the following conditions: (**a**) air GAP working on/off and (**b**) ammonia GAP working on.

Plasma-assisted ammonia combustion has great capacity for shortening ignition delay timing and extending combustible ranges while reducing NOx emissions. It will definitely promote the use of ammonia in engines. This has attracted lots of attention from scholars dedicated to figuring out the underlying physicochemical mechanisms via theoretical analysis and numerical simulation. Taneja et al. [81] found that PAC achieved the fastest ignition in a lean fuel mixture because of the accumulation of OH radicals through the reactivity-inhibiting reactions between plasma pulses, while an inversely proportional impact on ignition delay times was exhibited on plasma pulse frequency and energy density deposited. Moreover, the reforming of ammonia to nitrogen resulted in lower production of NOx with plasma. Another similar investigation also discovered PAC changed the conventional ignition and extinction characteristics, and the S-shaped curves were replaced by the monotonic and stretched ignition curves, which made low-temperature ignition possible [78]. Shahsavar et al. [82] figured out the most effective range of a reduced electric field on ammonia equivalent ratio combustion: 250–400 Td. This greatly promoted ignition time by increasing the reduced electric field in the early stage. With a more reduced electric field, a large fraction of the energy in the plasma system would be utilized in the ionization reactions of the diluent, thus neutralizing the effective excitations of fuel and oxidizer species, as shown in Figure 14.

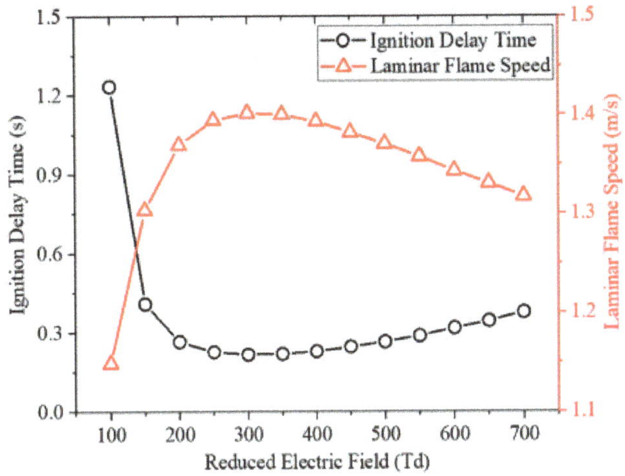

Figure 14. Effects of reduced electric field on ignition delay times and laminar flame velocities for plasma-assisted $NH_3/O_2/N_2$ combustion [82].

Finally, when projecting the adoption of plasma as an encouraging approach to boost ammonia decomposition and ignition, this review summarizes the latest innovations in the field of plasma catalysis and PAC, including progress in both numerical models and experimental studies. Great interest in improving ignition delay timing, increasing flame speed, and extending flammability limits while reducing NOx emissions will promote fast application in practical ICEs.

7. Conclusions

The outlook of independence from conventional fossil fuels is decarbonization in the automotive, marine and power generation sectors. Towards this goal, running with carbon-free fuel such as ammonia in ICEs is drawing more attention in research activities. Currently, ammonia can be ignited with diesel or any other high-reactivity fuel in dual-fuel mode, especially in marine engines and heavy-duty engines. Moreover, the addition of hydrogen is able to be derived from partially cracked ammonia with suitable catalyst in situ. However, the optimal hydrogen ratio in ammonia–hydrogen mixed combustion should not exceed 10%, considering thermal efficiency and combustion stability. A promising technology of plasma-assisted combustion to overcome the bottleneck of ammonia limits has attracted positive interest. The fundamental mechanisms of possible technical advances emerging in assisted combustion of ammonia are reviewed in this study.

(1) The laminar burning velocity of ammonia combustion increased with the increasing oxygen content, driven by the enriched concentrations of key radicals H, OH and NH_2. The maximum laminar flame velocity corresponds to equivalent ratio combustion, but the maximum NOx emissions moved towards lean combustion with increased oxygen enrichment.

(2) Investigations of ammonia-fueled engines have been widely carried out with varied additions of hydrogen, but modulations of inlet temperature and pressure are still necessary, especially for lean combustion. If more hydrogen is needed, it must be a new challenge to ensure operation safety with hydrogen supply.

(3) Plasma-assisted combustion enables lower-temperature combustion and has ability to overcome the flame extinction limit while reducing NOx emissions, due to the increase in diversity via active particles in the reaction system and the decrease in activation energy. The underlying physicochemical mechanisms of plasma-assisted ammonia combustion are rarely reported. The accumulation of OH radicals through reactivity-inhibiting reactions in lean fuel combustion could be stimulated through thermodynamic and kinetic effects via proper plasma generator configuration.

Funding: This research was funded by Open Funds of State Key Laboratory of Engine Reliability grant number [skler-202109], and Shandong Provincial Natural Science Foundation grant number [2022HWYQ-061, ZR2021ME212].

Data Availability Statement: Not applicable.

Conflicts of Interest: The authors declare no conflict of interest.

References

1. Dawood, F.; Anda, M.; Shafiullah, G.M. Hydrogen production for energy: An overview. *Int. J. Hydrogen Energy* **2020**, *45*, 3847–3869. [CrossRef]
2. Gorky, F.; Lucero, J.M.; Crawford, J.M.; Blake, B.A.; Guthrie, S.R.; Carreon, M.A.; Carreon, M.L. Insights on cold plasma ammonia synthesis and decomposition using alkaline earth metal-based perovskites. *Catal. Sci. Technol.* **2021**, *11*, 5109–5118. [CrossRef]
3. Wang, D.; Ji, C.; Wang, Z.; Wang, S.; Zhang, T.; Yang, J. Measurement of oxy-ammonia laminar burning velocity at normal and elevated temperatures. *Fuel* **2020**, *279*, 118425. [CrossRef]
4. Bartels, J.R. A Feasibility Study of Implementing an Ammonia Economy. Master's Thesis, Iowa State University, Ames, IA, USA, 2008.
5. Giddey, S.; Badwal, S.P.S.; Munnings, C.; Dolan, M. Ammonia as a renewable energy transportation media. *ACS Sustain. Chem. Eng.* **2017**, *5*, 10231–10239. [CrossRef]
6. Smith, C.; Hill, A.K.; Torrente-Murciano, L. Current and future role of Haber–Bosch ammonia in a carbon-free energy landscape. *Energy Environ. Sci.* **2020**, *13*, 331–344. [CrossRef]
7. Hasan, M.H.; Mahlia, T.M.I.; Mofijur, M.; Fattah, I.M.R.; Handayani, F.; Ong, H.C.; Silitonga, A.S. A comprehensive review on the recent development of ammonia as a renewable energy carrier. *Energies* **2021**, *14*, 3732. [CrossRef]
8. Schiffer, Z.J.; Manthiram, K. Electrification and decarbonization of the chemical industry. *Joule* **2017**, *1*, 10–14. [CrossRef]
9. Kyriakou, V.; Garagounis, I.; Vourros, A.; Vasileiou, E.; Stoukides, M. An electrochemical Haber-Bosch process. *Joule* **2020**, *4*, 142–158. [CrossRef]
10. Elmøe, T.D.; Sørensen, R.Z.; Quaade, U.; Christensen, C.H.; Nørskov, J.K.; Johannessen, T. A high-density ammonia storage/delivery system based on $Mg(NH_3)_6Cl_2$ for SCR–DeNOx in vehicles. *Chem. Eng. Sci.* **2006**, *61*, 2618–2625. [CrossRef]
11. Rouwenhorst, K.H.R.; Elishav, O.; Mosevitzky Lis, B.; Grader, G.S.; Mounaïm-Rousselle, C.; Roldan, A.; Valera-Medina, A. Future Trends. In *Techno-Economic Challenges of Green Ammonia as an Energy Vector*; Elsevier: Amsterdam, The Netherlands, 2021; pp. 303–319.
12. Herbinet, O.; Bartocci, P.; Dana, A.G. On the use of ammonia as a fuel—A perspective. *Fuel Commun.* **2022**, *11*, 100064. [CrossRef]
13. Dimitriou, P.; Javaid, R. A review of ammonia as a compression ignition engine fuel. *Int. J. Hydrogen Energy* **2020**, *45*, 7098–7118. [CrossRef]
14. Bicer, Y.; Dincer, I. Life cycle assessment of ammonia utilization in city transportation and power generation. *J. Clean. Prod.* **2018**, *170*, 1594–1601. [CrossRef]
15. Brown, T. UK Department of Transport Recommends Launch of Ammonia/Hydrogen Powered Vessels within 5–15 Years [EB/OL]. Ammonia Energy. 2019. Available online: https://www.ammoniaenergy.org/articles/uk-department-of-transport-recommends-launch-of-ammonia-hydrogen-powered-vessels-within-5-15-years/ (accessed on 26 June 2023).
16. Valera-Medina, A.; Amer-Hatem, F.; Azad, A.K.; Dedoussi, I.C.; de Joannon, M.; Fernandes, R.X.; Glarborg, P.; Hashemi, H.; He, X.; Mashruk, S.; et al. Review on ammonia as a potential fuel: From synthesis to economics. *Energy Fuels* **2021**, *35*, 6964–7029. [CrossRef]
17. Dimitriou, P.; Tsujimura, T. A review of hydrogen as a compression ignition engine fuel. *Int. J. Hydrogen Energy* **2017**, *42*, 24470–24486. [CrossRef]
18. Cardoso, J.S.; Silva, V.; Rocha, R.C.; Hall, M.J.; Costa, M.; Eusébio, D. Ammonia as an energy vector: Current and future prospects for low-carbon fuel applications in internal combustion engines. *J. Clean. Prod.* **2021**, *296*, 126562. [CrossRef]
19. Juangsa, F.B.; Irhamna, A.R.; Aziz, M. Production of ammonia as potential hydrogen carrier: Review on thermochemical and electrochemical processes. *Int. J. Hydrogen Energy* **2021**, *46*, 14455–14477. [CrossRef]
20. Berwal, P.; Kumar, S.; Khandelwal, B. A comprehensive review on synthesis, chemical kinetics, and practical application of ammonia as future fuel for combustion. *J. Energy Inst.* **2021**, *99*, 273–298. [CrossRef]
21. Gálvez, M.E.; Frei, A.; Halmann, M.; Steinfeld, A. Ammonia Production via a Two-Step Al_2O_3/AlN Thermochemical Cycle. *Ind. Eng. Chem. Res.* **2007**, *46*, 2047–2053. [CrossRef]
22. Li, K.; Chen, S.; Wang, H.; Wang, F. Plasma-assisted ammonia synthesis over Ni/LaOF: Dual active centers consisting of oxygen vacancies and Ni. *Appl. Catal. A Gen.* **2023**, *650*, 118983. [CrossRef]
23. Andersen, J.A.; Holm, M.C.; van't Veer, K.; Christensen, J.M.; Østberg, M.; Bogaerts, A.; Jensen, A.D. Plasma-catalytic ammonia synthesis in a dielectric barrier discharge reactor: A combined experimental study and kinetic modeling. *Chem. Eng. J.* **2023**, *457*, 141294. [CrossRef]

24. Sun, J.; Chen, Q.; Zhao, X.; Lin, H.; Qin, W. Kinetic investigation of plasma catalytic synthesis of ammonia: Insights into the role of excited states and plasma-enhanced surface chemistry. *Plasma Sources Sci. Technol.* **2022**, *31*, 094009. [CrossRef]
25. Engelmann, Y.; Kevin Yury Gorbanev Neyts, E.C.; Schneider, W.F.; Bogaerts, A. Plasma Catalysis for Ammonia Synthesis: A Microkinetic Modeling Study on the Contributions of Eley–Rideal Reactions. *ACS Publ.* **2021**, *9*, 13151–13163. [CrossRef]
26. Xie, P.; Yao, Y.; Huang, Z.; Liu, Z.; Zhang, J.; Li, T.; Wang, G.; Shahbazian-Yassar, R.; Hu, L.; Wang, C. Highly efficient decomposition of ammonia using high-entropy alloy catalysts. *Nat. Commun.* **2019**, *10*, 4011. [CrossRef] [PubMed]
27. Cui, G.; Zeng, W.; Li, Z.; Fu, Y.; Li, H.; Chen, J. Experimental study of minimum ignition energy of methane/air mixtures at elevated temperatures and pressures. *Fuel* **2016**, *175*, 257–263. [CrossRef]
28. Wang, D.; Ji, C.; Wang, S.; Yang, J.; Wang, Z. Numerical study of the premixed ammonia-hydrogen combustion under engine-relevant conditions. *Int. J. Hydrogen Energy* **2021**, *46*, 2667–2683. [CrossRef]
29. Xin, G.; Ji, C.; Wang, S.; Meng, H.; Chang, K.; Yang, J. Effect of different volume fractions of ammonia on the combustion and emission characteristics of the hydrogen-fueled engine. *Int. J. Hydrogen Energy* **2022**, *47*, 16297–16308. [CrossRef]
30. Lhuillier, C.; Brequigny, P.; Contino, F.; Mounaïm-Rousselle, C. Experimental investigation on ammonia combustion behavior in a spark-ignition engine by means of laminar and turbulent expanding flames. *Proc. Combust. Inst.* **2021**, *38*, 5859–5868. [CrossRef]
31. Han, X.; Wang, Z.; Costa, M.; Sun, Z.; He, Y.; Cen, K. Experimental and kinetic modeling study of laminar burning velocities of NH_3/air, NH_3/H_2/air, NH_3/CO/air and NH_3/CH_4/air premixed flames. *Combust. Flame* **2019**, *206*, 214–226. [CrossRef]
32. Jin, B.; Deng, Y.F.; Li, G.; Li, H. Experimental and numerical study of the laminar burning velocity of NH_3/H_2/air premixed flames at elevated pressure and temperature. *Int. J. Hydrogen Energy* **2022**, *47*, 36046–36057. [CrossRef]
33. Cai, T.; Zhao, D.; Gutmark, E. Overview of fundamental kinetic mechanisms and emission mitigation in ammonia combustion. *Chem. Eng. J.* **2023**, *458*, 141391. [CrossRef]
34. Chai, W.S.; Bao, Y.; Jin, P.; Tang, G.; Zhou, L. A review on ammonia, ammonia-hydrogen and ammonia-methane fuels. *Renew. Sustain. Energy Rev.* **2021**, *147*, 111254. [CrossRef]
35. Li, J.; Lai, S.; Chen, D.; Wu, R.; Kobayashi, N.; Deng, L.; Huang, H. A review on combustion characteristics of ammonia as a carbon-free fuel. *Front. Energy Res.* **2021**, *9*, 760356. [CrossRef]
36. Westlye, F.R.; Ivarsson, A.; Schramm, J. Experimental investigation of nitrogen based emissions from an ammonia fueled SI-engine. *Fuel* **2013**, *111*, 239–247. [CrossRef]
37. Mounaïm-Rousselle, C.; Bréquigny, P.; Dumand, C.; Houillé, S. Operating Limits for Ammonia Fuel Spark-Ignition Engine. *Energies* **2021**, *14*, 4141. [CrossRef]
38. Jin, S.; Wu, B.; Zi, Z.; Yang, P.; Shi, T.; Zhang, J. Effects of fuel injection strategy and ammonia energy ratio on combustion and emissions of ammonia-diesel dual-fuel engine. *Fuel* **2023**, *341*, 127668. [CrossRef]
39. Song, J.; Zello, V.; Boehman, A.L.; Waller, F.J. Comparison of the Impact of Intake Oxygen Enrichment and Fuel Oxygenation on Diesel Combustion and Emissions. *Energy Fuels* **2004**, *18*, 1282–1290. [CrossRef]
40. Maxwell, T.; Setty, V.; Jones, J.; Narayan, R. *The Effect of Oxygen Enriched Air on the Performance and Emissions of an Internal Combustion Engines*; SAE Technical Paper 932804; SAE International: Warrendale, PA, USA, 1993. [CrossRef]
41. Subramanian, K.; Ramesh, A. *Experimental Investigation on the Use of Water Diesel Emulsion with Oxygen Enriched Air in a DI Diesel Engine*; SAE Technical Paper 2001-01-0205; SAE International: Warrendale, PA, USA, 2001. [CrossRef]
42. Salzano, E.; Basco, A.; Cammarota, F.; Di Sarli, V.; Di Benedetto, A. Explosions of Syngas/CO_2 Mixtures in Oxygen-Enriched Air. *Ind. Eng. Chem. Res.* **2011**, *51*, 7671–7678. [CrossRef]
43. Baskar, P.; Senthilkumar, A. Effects of oxygen enriched combustion on pollution and performance characteristics of a diesel engine. *Eng. Sci. Technol. Int. J.* **2016**, *19*, 438–443. [CrossRef]
44. Li, S.-Q.; Guan, Q.; Zhang, W.-H. Test Study on Combustion Process of Oxygen-enriched Gasoline Engine. *J. Highw. Transp. Res. Dev.* **2008**, *3*, 25.
45. Liang, Y.; Shu, G.; Wei, H.; Zhang, W. Effect of oxygen enriched combustion and water–diesel emulsion on the performance and emissions of turbocharged diesel engine. *Energy Convers. Manag.* **2013**, *73*, 69–77. [CrossRef]
46. Karimi, M.; Wang, X.; Hamilton, J.; Negnevitsky, M. Numerical investigation on hydrogen-diesel dual-fuel engine improvements by oxygen enrichment. *Int. J. Hydrogen Energy* **2022**, *47*, 25418–25432. [CrossRef]
47. Mei, B.; Zhang, X.; Ma, S.; Cui, M.; Guo, H.; Cao, Z.; Li, Y. Experimental and kinetic modeling investigation on the laminar flame propagation of ammonia under oxygen enrichment and elevated pressure conditions. *Combust. Flame* **2019**, *210*, 236–246. [CrossRef]
48. Ganley, J.C.; Thomas, F.S.; Seebauer, E.G.; Masel, R.I. A Priori Catalytic Activity Correlations: The Difficult Case of Hydrogen Production from Ammonia. *Catal. Lett.* **2004**, *96*, 117–122. [CrossRef]
49. Comotti, M.; Frigo, S. Hydrogen generation system for ammonia–hydrogen fuelled internal combustion engines. *Int. J. Hydrogen Energy* **2015**, *40*, 10673–10686. [CrossRef]
50. Ryu, K.; Zacharakis-Jutz, G.E.; Kong, S.C. Performance enhancement of ammonia-fueled engine by using dissociation catalyst for hydrogen generation. *Int. J. Hydrogen Energy* **2014**, *39*, 2390–2398. [CrossRef]
51. Frigo, S.; Gentili, R.; Doveri, N. *Ammonia Plus Hydrogen as Fuel in a S.I. Engine: Experimental Results*; SAE Technical Paper 2012-32-0019; SAE International: Warrendale, PA, USA, 2012. [CrossRef]

52. Zhang, H.; Li, G.; Long, Y.; Zhang, Z.; Wei, W.; Zhou, M.; Belal, B.Y. Numerical study on combustion and emission characteristics of a spark-ignition ammonia engine added with hydrogen-rich gas from exhaust-fuel reforming. *Fuel* **2023**, *332*, 125939. [CrossRef]
53. Li, J.; Zhang, R.; Pan, J.; Wei, H.; Shu, G.; Chen, L. Ammonia and hydrogen blending effects on combustion stabilities in optical SI engines. *Energy Convers. Manag.* **2023**, *280*, 116827. [CrossRef]
54. Mei, B.; Zhang, J.; Shi, X.; Xi, Z.; Li, Y. Enhancement of ammonia combustion with partial fuel cracking strategy: Laminar flame propagation and kinetic modeling investigation of $NH_3/H_2/N_2$/air mixtures up to 10 atm. *Combust. Flame* **2021**, *231*, 111472. [CrossRef]
55. Lhuillier, C.; Brequigny, P.; Contino, F.; Mounaïm-Rousselle, C. Experimental study on ammonia/hydrogen/air combustion in spark ignition engine conditions. *Fuel* **2020**, *269*, 117448. [CrossRef]
56. Wang, Y.; Zhou, X.; Liu, L. Theoretical investigation of the combustion performance of ammonia/hydrogen mixtures on a marine diesel engine. *Int. J. Hydrogen Energy* **2021**, *46*, 14805–14812. [CrossRef]
57. Pochet, M.; Dias, V.; Moreau, B.; Foucher, F.; Jeanmart, H.; Contino, F. Experimental and numerical study, under LTC conditions, of ammonia ignition delay with and without hydrogen addition. *Proc. Combust. Inst.* **2019**, *37*, 621–629. [CrossRef]
58. Lin, Q.F.; Jiang, Y.M.; Liu, C.Z.; Chen, L.W.; Zhang, W.J.; Ding, J.; Li, J.G. Instantaneous hydrogen production from ammonia by non-thermal arc plasma combining with catalyst. *Energy Rep.* **2021**, *7*, 4064–4070. [CrossRef]
59. Gu, J.G.; Zhao, P.; Zhang, Y.; Wang, H.Y.; Jiang, W. Discharge Enhancement Phenomenon and Streamer Control in Dielectric Barrier Discharge with Many Pores. *Catalysts* **2020**, *10*, 68. [CrossRef]
60. Mei, D.H.; Fang, C.; Shao, T. Recent Progress on Characteristics and Applications of Atmospheric Pressure Low Temperature Plasmas. *Chin. J. Electr. Eng.* **2020**, *40*, 1339–1358+1425.
61. Ju, Y.; Sun, W. Plasma assisted combustion: Dynamics and chemistry. *Prog. Energy Combust. Sci.* **2015**, *48*, 21–83. [CrossRef]
62. Wolk, B.; DeFilippo, A.; Chen, J.Y.; Dibble, R.; Nishiyama, A.; Ikeda, Y. Enhancement of flame development by microwave-assisted spark ignition in constant volume combustion chamber. *Combust. Flame* **2013**, *160*, 1225–1234. [CrossRef]
63. Mariani, A.; Foucher, F. Radio frequency spark plug: An ignition system for modern internal combustion engines. *Appl. Energy* **2014**, *122*, 151–161. [CrossRef]
64. Morsy, M.H. Review and recent developments of laser ignition for internal combustion engines applications. *Renew. Sustain. Energy Rev.* **2012**, *16*, 4849–4875. [CrossRef]
65. Cathey, C.; Tang, T.; Shiraishi, T.; Urushihara, T.; Kuthi, A.; Gundersen, M.A. Nanosecond Plasma Ignition for Improved Performance of an Internal Combustion Engine. *IEEE Trans. Plasma Sci.* **2007**, *35*, 1664–1668. [CrossRef]
66. Wang, F.; Liu, J.B.; Sinibaldi, J.; Brophy, C.; Kuthi, A.; Jiang, C.; Ronney, P.; Gundersen, M. Transient plasma ignition of quiescent and flowing air/fuel mixtures. *IEEE Trans. Plasma Sci.* **2004**, *33*, 844–849. [CrossRef]
67. Barleon, N.; Cheng, L.; Cuenot, B.; Vermorel, O.; Bourdon, A. Investigation of the impact of NRP discharge frequency on the ignition of a lean methane-air mixture using fully coupled plasma-combustion numerical simulations. *Proc. Combust. Inst.* **2022**, *39*, 5521–5530. [CrossRef]
68. Mao, X.; Rousso, A.; Chen, Q.; Ju, Y. Numerical modeling of ignition enhancement of CH_4/O_2/He mixtures using a hybrid repetitive nanosecond and DC discharge. *Proc. Combust. Inst.* **2019**, *37*, 5545–5552. [CrossRef]
69. Wang, Y.; Guo, P.; Chen, H.; Chen, Z. Numerical modeling of ignition enhancement using repetitive nanosecond discharge in a hydrogen/air mixture I: Calculations assuming homogeneous ignition. *J. Phys. D Appl. Phys.* **2020**, *54*, 065501. [CrossRef]
70. Mao, X.; Chen, Q.; Guo, C. Methane pyrolysis with N_2/Ar/He diluents in a repetitively-pulsed nanosecond discharge: Kinetics development for plasma assisted combustion and fuel reforming. *Energy Convers. Manag.* **2019**, *200*, 112018. [CrossRef]
71. Snoeckx, R.; Cha, M.S. Inevitable chemical effect of balance gas in low temperature plasma assisted combustion. *Combust. Flame* **2021**, *225*, 1–4. [CrossRef]
72. Hwang, J.; Kim, W.; Agarwal, A.K.; Choe, W.; Cha, J.; Woo, S. Application of a novel microwave-assisted plasma ignition system in a direct injection gasoline engine. *Appl. Energy* **2017**, *205*, 562–576. [CrossRef]
73. José Martín Pastor Garcia-Oliver, J.M.; García, A.G.; Micó, C. Combustion improvement and pollutants reduction with diesel-gasoline blends by means of a highly tunable laser plasma induced ignition system. *J. Clean. Prod.* **2020**, *271*, 122499.
74. Merotto, L.; Balmelli, M.; Vera-Tudela, W.; Soltic, P. Comparison of ignition and early flame propagation in methane/air mixtures using nanosecond repetitively pulsed discharge and inductive ignition in a pre-chamber setup under engine relevant conditions. *Combust. Flame* **2022**, *237*, 111851. [CrossRef]
75. Zhang, S.; Oehrlein, G.S. From thermal catalysis to plasma catalysis: A review of surface processes and their characterizations. *J. Phys. D Appl. Phys.* **2021**, *54*, 213001. [CrossRef]
76. Wang, L.; Zhao, Y.; Liu, C.; Gong, W.; Guo, H. Plasma driven ammonia decomposition on a Fe-catalyst: Eliminating surface nitrogen poisoning. *Chem. Commun.* **2013**, *49*, 3787–3789. [CrossRef]
77. Wang, L.; Yi, Y.; Guo, Y.; Zhao, Y.; Zhang, J.; Guo, H. Synergy of DBD plasma and Fe-based catalyst in NH_3 decomposition: Plasma enhancing adsorption step. *Plasma Process. Polym.* **2017**, *14*, 1600111. [CrossRef]
78. Faingold, G.; Lefkowitz, J.K. A numerical investigation of NH_3/O_2/He ignition limits in a non-thermal plasma. *Proc. Combust. Inst.* **2021**, *38*, 6661–6669. [CrossRef]

79. Lin, Q.; Jiang, Y.; Liu, C.; Chen, L.; Zhang, W.; Ding, J.; Li, J. Controllable NO emission and high flame performance of ammonia combustion assisted by non-equilibrium plasma. *Fuel* **2022**, *319*, 123818. [CrossRef]
80. Choe, J.; Sun, W.; Ombrello, T.; Carter, C. Plasma assisted ammonia combustion: Simultaneous NOx reduction and flame enhancement. *Combust. Flame* **2021**, *228*, 430–432. [CrossRef]
81. Taneja, T.S.; Johnson, P.N.; Yang, S. Nanosecond pulsed plasma assisted combustion of ammonia-air mixtures: Effects on ignition delays and NOx emission. *Combust. Flame* **2022**, *245*, 112327. [CrossRef]
82. Shahsavari, M.; Konnov, A.A.; Valera-Medina, A.; Jangi, M. On nanosecond plasma-assisted ammonia combustion: Effects of pulse and mixture properties. *Combust. Flame* **2022**, *245*, 112368. [CrossRef]

sustainability

MDPI

Article

Formulation of a Jet Fuel Surrogate and Its Kinetic Chemical Mechanism by Emulating Physical and Chemical Properties of Real Jet Fuel

Guangze Li [1,2,3], Boxuan Cui [1], Chenglin Zhang [3], Liuyong Chang [1,2,3,*] and Longfei Chen [1,*]

[1] School of Energy and Power Engineering, Beihang University, Beijing 100191, China; liguangze@buaa.edu.cn (G.L.)
[2] Beihang Hangzhou Innovation Institute Yuhang, Hangzhou 310023, China
[3] Zhongfa Aviation Institute of Beihang University, Hangzhou 311115, China
* Correspondence: changliuyong@buaa.edu.cn (L.C.); chenlongfei@buaa.edu.cn (L.C.)

Abstract: The application of jet fuel in gas turbines and diesel engines adheres to the Army's single-fuel forward policy, streamlining supply chains. To ensure precise engine combustion numerical studies, surrogate fuels and mechanisms should faithfully replicate real fuel properties and combustion traits. In this work, a new four-component jet fuel surrogate containing 39.05% n-dodecane/21.79% isocetane/11.49% decalin/27.67% toluene by mole fraction is formulated based on a property optimizer. The new-formulated fuel surrogate can satisfactorily emulate the chemical and physical properties of real jet fuel, including cetane number (CN), threshold sooting index (TSI), molecular weight (MW), lower heating value (LHV), the ratio of hydrogen and carbon (H/C), liquid density, viscosity, and surface tension. Furthermore, a reduced and robust kinetic chemical mechanism (containing 124 species and 590 reactions) that could be directly employed in practical engine combustion simulations has also been developed for the proposed surrogate jet fuel. The mechanism is validated through comprehensive experimental data, including ignition delay time (IDT) determined in shock tubes and rapid compression machines (RCMs), species mole fractions measured in premixed flames and jet stirred reactors (JSRs), and laminar flame speeds. Generally, the property deviations of the jet fuel surrogate are less than 2% except for MW (10.73%), viscosity (5.88%), and surface tension (8.71%). The comparison results between the predictions and measurements are in good agreement, indicating that the current kinetic mechanism is capable of reflecting the oxidation process of real jet fuel. The current mechanism can accurately capture variations in the ignition delay time in the negative temperature coefficient (NTC) region as well. In the future, the proposed surrogate jet fuel could be applied in practical engine computational fluid dynamic (CFD) simulations.

Keywords: fuel property; ignition delay time; species profile; laminar flame speed; reduced chemical mechanism

Citation: Li, G.; Cui, B.; Zhang, C.; Chang, L.; Chen, L. Formulation of a Jet Fuel Surrogate and Its Kinetic Chemical Mechanism by Emulating Physical and Chemical Properties of Real Jet Fuel. *Sustainability* **2023**, *15*, 13792. https://doi.org/10.3390/su151813792

Academic Editor: Talal Yusaf

Received: 7 July 2023
Revised: 4 September 2023
Accepted: 12 September 2023
Published: 15 September 2023

1. Introduction

Jet fuel has been widely adopted as a general fuel for aircraft and ground vehicles in battlefields under the auspices of the U.S. army single-fuel policy [1], which aims to significantly reduce the costs of fuel supply. Furthermore, jet fuel is expected to reduce NOx and particle matter emissions in diesel engines, and thus it would be beneficial to investigate the effects of using jet fuel on the combustion and emission characteristics of diesel engines. However, due to the limitation of current computational capacity, the real jet fuel, which comprises hundreds of compounds, was unable to be directly employed in a practical engine simulation [2,3]. Therefore, surrogate jet fuels, which normally consist of several well-characterized compounds, were proposed [4–6]. The research on surrogate jet fuels is expected to achieve sustainability in the aviation fuel industry.

Previous research studies have been conducted on the development of jet fuel surrogate and their chemical mechanisms. In order to reproduce real fuels as well as possible, surrogates are required to closely emulate the thermo-physical properties and combustion characteristics of target jet fuels. Violi et al. [7] developed a JP-8 surrogate that contained six hydrocarbons and could emulate the distillation curve and thermo-physical properties of the target fuel. A detailed chemical mechanism of the surrogate was also presented in their work and validated by experimental data. Vasu et al. [8] evaluated the aforementioned chemical mechanism by comparing the measured and simulated IDT in a shock tube, and they reported that the mechanism could provide accurate IDT predictions above a temperature of 1000 K while predicting IDT poorly at low temperatures. Dagaut et al. [9] compared the kinetic modelling results of four jet fuel surrogates (one- and three-component mixtures of n-decane, n-propylbenzene, and n-propylcyclohexane) with JSR experimental data. The comparison results showed that the three-component surrogate was the most reliable substitute for the target fuel among all the test fuel surrogates. Gokulakrishnan et al. [10] developed a detailed kinetic model for the four-component kerosene fuel surrogate (n-decane/n-propylcyclohexane/n-propylbenzene/decene), which exhibited good performance in predicting the measured species concentrations. However, the IDT validations were not comprehensively conducted by them. Eddings et al. [11] proposed two six-component jet fuel surrogates named Hex-11 and Hex-12 to emulate a Jet-8 pool fire. The comparison results indicated that the surrogates could accurately capture the burning rate, emissive power, flame height, and puffing frequency of the steady-state pool fire. Dooley et al. [12] formulated a three-component jet fuel surrogate, 'MURI1' (n-decane/iso-octane/toluene), based on chemical group theory. The CN and H/C of the target fuel were regarded as the target properties of MURI1. Several devices, including a flow reactor and shock tube, were employed to validate the combustion properties of the proposed surrogate. They further developed an improved jet fuel surrogate named MURI2 comprising n-dodecane, iso-octane, n-propylbenzen, and 1,3,5-trimethylbenzene [13]. Apart from CN and H/C, TSI and MW were also selected as the emulating metrics. According to the ignition characteristic study conducted by Malewicki et al. [14], MURI2 exhibited a satisfactory performance on predicting the species profiles of small molecules including oxygen, carbon monoxide, carbon dioxide, and C1-C3 hydrocarbons. Kim et al. [15] proposed two four-component aviation fuel surrogates named UM1 and UM2 based on a model-based optimizer. The physical and chemical properties of the target fuel were reproduced by the two surrogates to some extent.

Despite the extensive research efforts directed towards the creation of various jet fuel surrogates and their accompanying kinetic mechanisms, notable limitations remain apparent in this scientific domain: (1) A recurring challenge is the inability of these surrogates to comprehensively replicate the intricate amalgamation of the physical and chemical properties characteristic of the target fuel. Achieving a harmonious equilibrium between these multifaceted attributes remains an elusive goal. (2) Furthermore, the chemical kinetic mechanisms developed for these surrogates have often exhibited tendencies towards impracticality. They tend to be excessively voluminous or overly intricate, thus impeding their direct application within the confines of practical engine combustion simulations. (3) In addition, the validation of these proposed surrogate mechanisms has not been exhaustive, especially when it comes to crucial parameters such as ignition delay time (IDT), species mole fractions, and flame speeds. The inadequacy of rigorous validation procedures raises pertinent concerns about the reliability and robustness of these surrogates in real-world applications. Addressing these limitations represents an imperative for the advancement of this field.

With the aim of addressing the mentioned issues, a four-component fuel surrogate incorporating the physical and chemical properties of real jet fuel was developed using a property optimizer. Eight properties, including CN, TSI, MW, LHV, H/C, liquid density, viscosity, and surface tension, were selected as the target physical and chemical properties. A reduced kinetic mechanism, comprising 124 species and 590 reactions, for the surrogate was created which can prove advantageous in practical engine combustion simulations. Massive amounts of experimental data, including IDT, species concentration, and laminar flame speed, were collected to validate the mechanism from two aspects: the surrogate mixture and its components.

2. Formulation of Jet Fuel Surrogate

2.1. Target Fuel and Its Properties

The fuel combustion process inside an engine is very complex and always involves intricate physical and chemical processes. As shown in Figure 1, the fuel jet is injected into the cylinder via an injector, breaks down to small liquid droplets under the action of air shear force, and then evaporates and mixes with air to form a combustible fuel–air mixture. The processes are heavily affected by fuel physical properties such as viscosity, liquid density, and surface tension [16,17]. After that, the spontaneous ignition and combustion processes of the mixed fuel/air occur, where the chemical properties of the fuel have a significant influence on the events. Herein, CN dominates fuel ignitability, and MW plays a critical role in the diffusive transport process between liquid and the vaporized fuel. H/C influences the local ratio of fuel/air and adiabatic flame temperature, while LHV represents the energy generated from the fuel oxidation. Considering the above processes, eight properties, including CN, TSI, MW, LHV, H/C, liquid density, viscosity, and surface tension, were chosen as target properties for the surrogate in order to comprehensively reproduce the chemical and physical characteristics of real fuel in engine.

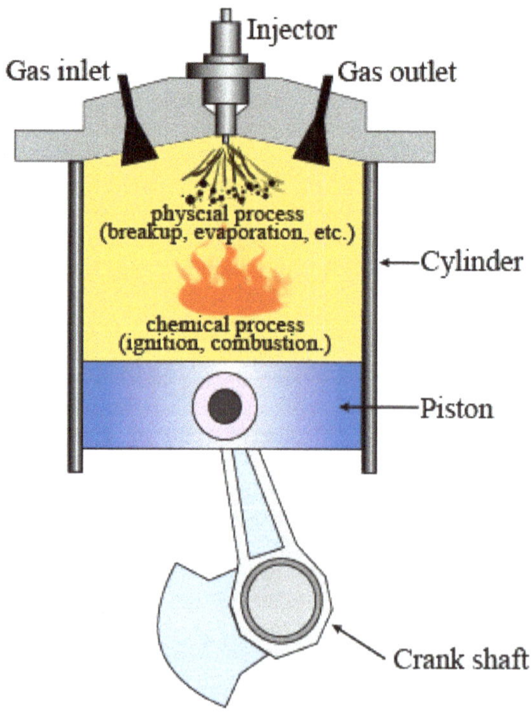

Figure 1. The physical and chemical processes in compression ignition engine.

In this work, Jet-A POSF-4658 (supplied by Edwards, AFRL-WP, Wright, OH, USA) was selected as the target real fuel for two reasons: First, it is a representative jet fuel comprising different jet-A batches, and its components and various properties have been extensively studied. Second, the experimental data on the IDT, laminar flame speed, and species concentration profiles of the fuel are available in the literature, so comparisons can be more readily conducted. However, since the data for LHV and surface tension of POSF-4658 are not available, the LHV of JP-8 fuel and surface tension of Jet-A were used instead as references in this study, similar to the previous studies [15,18]. The relevant properties of POSF-4658 were listed in Table 1.

Table 1. The properties of the target jet fuel [18].

Target Properties	Jet-A (POSF-4658)
CN *	47.1
MW	142 kg/kmol
H/C	1.957
LHV	43.23 MJ/kg
TSI	21.4
Liquid density	Temperature-dependent (See Figure 2)
Viscosity	Temperature-dependent (See Figure 2)
Surface tension	Temperature-dependent (See Figure 2)

* Derived cetane number (DCN).

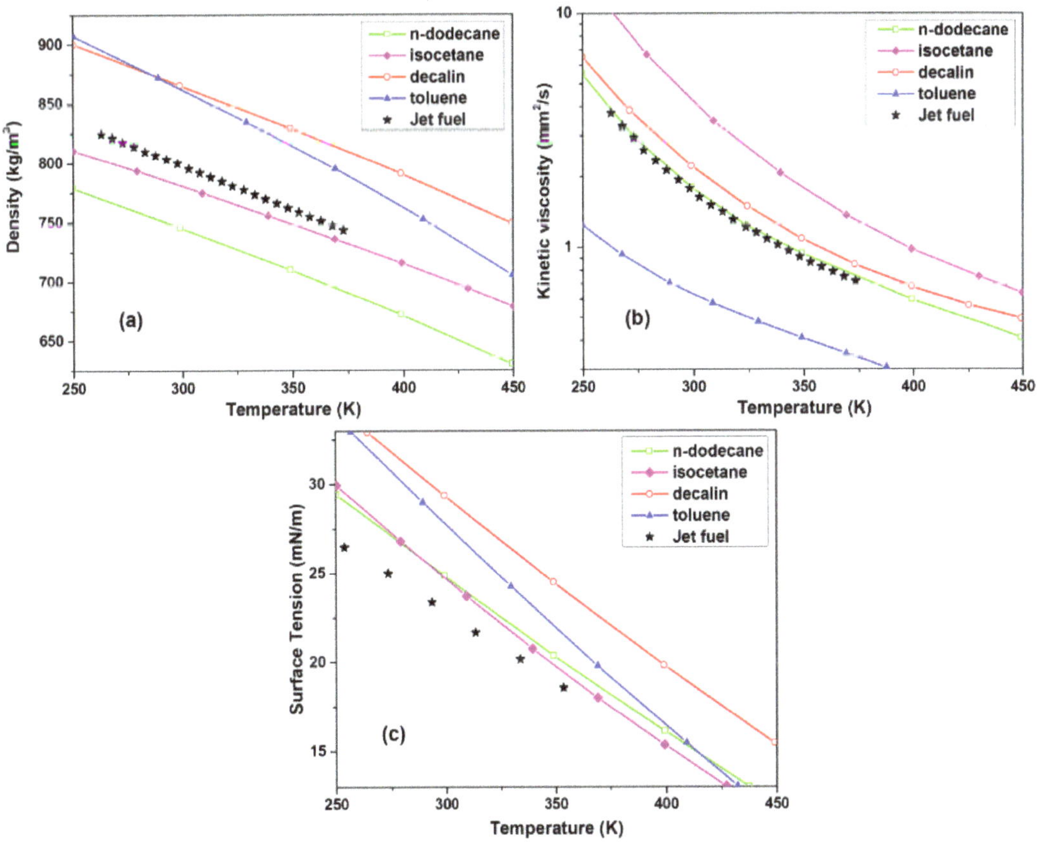

Figure 2. The temperature-dependent properties of the components: (**a**) density, (**b**) kinematic viscosity, and (**c**) surface tension.

2.2. Surrogate Fuel Components

Although real jet fuels contain a large number of compounds, according to Ref. [19], conventional jet fuels normally comprise 60% chain paraffins, 20% cycloalkanes, and 20% aromatics. Hence, it is reasonable to adopt the well-characterized compounds recognized in real jet fuels as the main surrogate components. Moreover, the components of jet fuel mainly belonged to C9–C16 hydrocarbons, which indicates that HC class and molecule size are also important criteria for surrogate components. The components whose chemical kinetic mechanisms have been well established are more preferred in our study. Based on these considerations, four components, including n-dodecane, isocetane, decalin, and toluene, were finally selected. The properties (CN, TSI, MW, LHV, and H/C) and formulae of the four components are listed in Table 2. The temperature-dependent properties, density, viscosity, and surface tension, are plotted in Figure 2. It can be easily seen that a single component is unable to represent real jet fuel because of the significant property differences between the single component and real fuel.

Table 2. The properties of the surrogate components.

Hydrocarbon Class Name	N-Alkane n-Dodecane	Iso-Alkane Isocetane	Cycloalkane Decalin	Aromatic Toluene
Formula	$C_{12}H_{26}$	$C_{16}H_{34}$	$C_{10}H_{18}$	C_7H_8
CN [20]	82.5	15	46.5	7.4
MW (g/mol)	170.33	226.44	138.25	92.14
LHV [21] (MJ/kg)	44.11	44.85	42.58	40.53
TSI [22]	7.0	22	22	40
H/C	2.17	2.13	1.8	1.1

2.3. Formation of Jet Fuel Surrogate

In this work, we improved the optimization algorithm from our previous study [18] to obtain the optimum component proportions of a jet fuel surrogate. The calculations for the target properties are showed in Table 3.

As reported in [20], the CN of the mixture can be calculated using the volume fraction average of the component CNs. The TSI of a mixture is obtained by calculating the mole fraction average of the pure component's TSI [23], while the LHV is estimated using the mass fraction average of the LHV. H/C and MW were directly determined by the component formula. The mixture density is also computed by the volume fraction average of component density. The viscosity is obtained using the Grunberg–Nissan equation [24,25], whereas surface tension is derived from the parachor correlation [26]. Finally, the equations proposed by Kim et al. [15] were employed as the merit functions:

$$MeritFunc = \sum_{i=1}^{Num_{target}} V_i \tag{1}$$

$$V_i = \frac{\sum_{j=1}^{Num_{data,i}} \left(\frac{Q_{i,j,cal} - Q_{i,j,exp}}{Q_{i,j,exp}} \right)^2}{Num_{data,\ i}} \tag{2}$$

where i and j represent the numbers of properties and experimental data points, respectively. $Num_{data,i}$ refers to the total number of measurements of the ith property, and Num_{target} represents the total number of target properties. $Q_{i,j,cal}$ and $Q_{i,j,exp}$ are the computed and measured properties, respectively.

Table 3. The estimation methods [18] of the properties of fuel surrogate.

Properties	Estimation Approaches
MW	Average of mole fraction: $\sum MW_{mix} = \sum X_i MW_i$ X_i is mole fraction of component i,
H/C	$\frac{H}{C_{mix}} = \sum X_i \frac{H_i}{\sum X_i} C_i$ C_i is the number of carbon atoms of component i
TSI	Average of mole fraction: $TSI_{mix} = \sum X_i TSI_i$
LHV	Average of mass fraction: $LHV_{mix} = \sum Y_i LHV_i$
CN	Average of volume fraction: $\sum CN_{mix} = \sum V_i CN_i$ V_i is volume fraction of component i
Density	Average of volume fraction: $\rho(T)_{mix} = \sum V(T)_i \rho(T)_i$
MW	Average of mole fraction: $\sum MW_{mix} = \sum X_i MW_i$
H/C	$\frac{H}{C_{mix}} = \sum X_i \frac{H_i}{\sum X_i} C_i$ H_i is the number of hydrogen atoms of component i;
Viscosity	Grunberg–Nissan equation [24]: $In(\mu(T)_{mix}) = \sum X_i In(\mu(T)_i) + 0.5 \sum\sum X_i X_j G_{ij}$ G_{ij} is the binary interaction parameter
Surface tension	Parachor correlation: $\sigma(T)_{mix} = \left(P_{L,mix}\rho(T)_{L,mix,molar}\right)^4$ $P_{L,mix} = 0.5 \sum\sum X_i X_j \left(P_i + P_j\right)$ $\sigma(T)_{mix}$ is liquid surface tension, P is parachor, $\rho(T)_{L,mix,molar}$ is liquid mixture molar density

A flowchart of the whole optimization process is given in Figure 3. It should be noted that the component will be removed when its mole fraction is less than 0.3% during the optimization process, the same as in our previous study [18]. Based on the above calculations, the optimum jet fuel surrogate was achieved, and it comprised 39.05% n-dodecane, 21.79% isocetane, 11.49% decalin, and 27.67% toluene by mole fraction and was abbreviated as JFS.

Table 4 shows the comparison results among the JFS surrogate and other jet fuel surrogates, including UM1, UM2, MURI2, S5 [27], and HEX12 [11]. In general, the property deviations of the surrogate JFS are less than 2% except for MW (10.73%), viscosity (5.88%), and surface tension (8.71%). The high deviations of MW and surface tension can be regarded as acceptable compared to other surrogates. As shown in Table 4, the deviation of the MW of S5 is 12.11%, and the averaged deviations of surface tension of UM2, Hex12, and S5 are larger than 15%, which are all larger than the corresponding property deviations of the surrogate JFS. It is also observed that the CN, MW, H/C and, LHV of the surrogate UM1 show good agreement with the target fuel; however, the deviations of TSI (−21.52%) and viscosity (−21.2%) are too large. The surface tension of UM2 deviates from the target property by 15.8%, even though CN, LHV, and density are well captured by the surrogate UM2. The LHV, MW, CN, and H/C are well emulated by MURI2; however, the deviation of viscosity (34.608%) is too large. As for the properties of HEX12 and S5, the majority of them greatly deviate from the target properties. Overall, the surrogate JFS performs better than other surrogates at reproducing the chemical and physical properties of real jet fuel; hence, it is more likely to be adopted for practical engine combustion simulation.

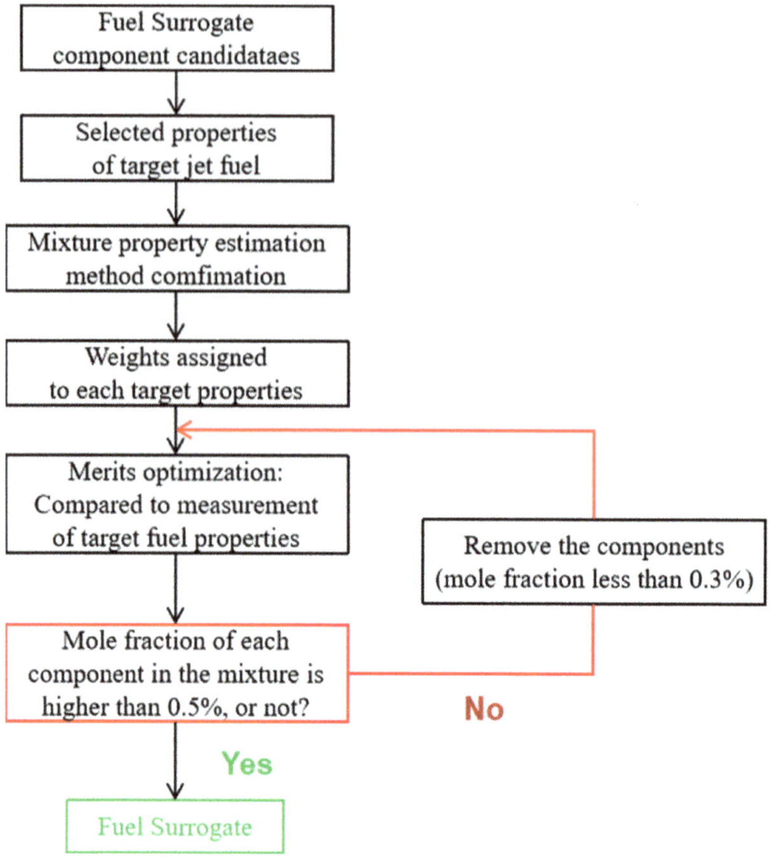

Figure 3. The flow chart of optimization process for the jet fuel surrogate.

Table 4. The comparison results among JFS and other jet fuel surrogates.

Jet Fuel	Surrogate	JFS	UM1	UM2	MURI2	S5	HEX12
CN 47.1	Val	46.93	46.8	46.7	48.5	32.1	60.5
	Dev (%)	−0.35	−0.64	−0.85	2.97	−31.8	28.45
H/C 1.957	Val	1.94	1.967	1.881	1.950	1.807	1.856
	Dev (%)	−0.86	0.51	−3.88	−0.36	−7.66	−5.16
MW 142	Val	157.23	143.5	148.6	138.7	159.2	152.2
	Dev (%)	10.73	1.06	4.65	2.32	12.11	7.18
LHV	Val	43.61	43.62	43.36	43.55	43.02	44.6
	Dev (%)	0.87	0.90	0.30	0.74	−0.49	3.17
TSI 21.4	Val	21.12	16.79	22.14	20.4	34.61	25.0
	Dev (%)	−1.29	−21.52	3.45	−4.67	61.72	16.84
Density	Average dev (%)	1.84	−3.4	0.6	−5.518	2.392	1.423
Viscosity	Average dev (%)	5.88	−21.2	−3.6	−34.608	18.167	5.077
Surface tension	Average dev (%)	8.71	9.1	15.8	3.131	19.1	18.774

3. Kinetic Modelling

3.1. Methodology

In this study, to facilitate the simulation of a practical engine combustion, a so-called decoupling methodology [5,18] was adopted for developing the JFS chemical mechanisms. The chemical kinetic mechanism was established through a systematic formation process, starting from C16 reactions and progressing towards $H_2/O_2/C1$ reactions. The subsequent steps involved incorporating a NOx sub-mechanism and polycyclic aromatic hydrocarbon (PAH). The reduction and optimization procedures employed in this process are briefly outlined as follows:

(1) The initial stage of the reduction and optimization process involves conducting a reaction pathway analysis to identify the key reactions. Subsequently, unimportant species and reactions are eliminated from the initial kinetic model. Simultaneously, the rate of production (ROP) and sensitivity analyses are performed to evaluate the remaining species and reactions. This allows for a clear understanding of the impact of each reaction on the oxidation process.

(2) Subsequently, the reaction rate constants were optimized to improve the agreement between the simulated and experimental data of fuel ignition delay time (IDT).

(3) Afterward, the concentrations of species and laminar flame speeds predicted by the reduced mechanism were compared to the corresponding measurements. This allowed for references to fine-tune the reaction rate constants further.

(4) Steps 1 to 3 were iteratively repeated until the desired size and accuracy of the mechanism were attained.

3.2. Toluene Sub-Mechanism

The current chemical kinetic model mainly comprises four sub-mechanisms, including the toluene sub-mechanism, decalin sub-mechanism, n-dodecane sub-mechanism, and isocetane sub-mechanism (Figure 4).

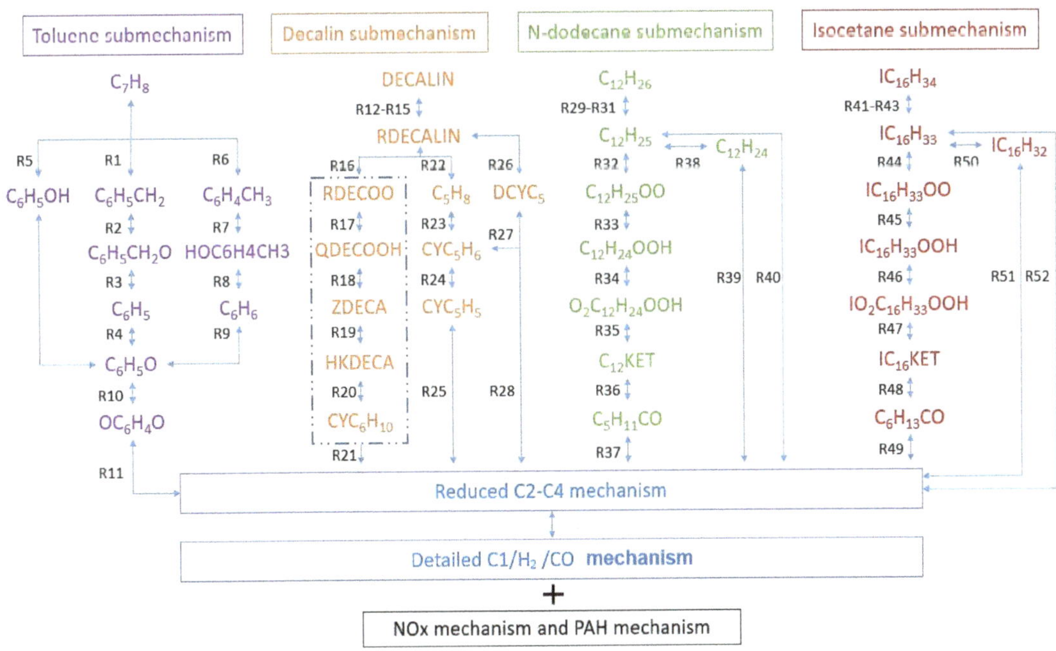

Figure 4. The major reaction pathways for the JFS kinetic mechanism.

The toluene sub-mechanism was tuned from the reduced TRF-PAH mechanism [28], and it was employed as the initial mechanism based on two reasons: (1) The size of the TRF-PAH mechanism is compact, which is advantageous for three-dimensional engine combustion simulation; and (2) The mechanism has good performance in predicting species concentrations and flame speeds.

Figure 4 illustrates the principal reactions of the reduced toluene mechanism. Toluene is predominantly consumed through three pathways: the H-atom abstraction reactions leading to the production of $C_6H_4CH_3$ (R6), $C_6H_5CH_2$ (R1), and the reaction R5. Among these, the formation of $C_6H_5CH_2$ is the primary pathway for toluene consumption, while C6H5OH ranks as the second most abundant product. Through some intermediate species, such as $C_6H_5CH_2O$, C_6H_5, and C_6H_5O, the toluene molecule ultimately converts into the small molecules. Some species, such as $C_6H_5CH_2OO$, C_6H_5CO, and $OC_6H_4CH_3$, which are absent in Figure 4, were also incorporated into the toluene sub-mechanism.

The NOx and PAH sub-mechanisms of the original TRF-PAH mechanism were retained in the current mechanism. The NOx sub-mechanism includes four species and 12 reactions, and it contains thermal reactions and N_2O-intermediate reactions. The PAH mechanism is able to reproduce PAH formation up to a four-ring PAH. Herein, H-atom abstraction reactions, methyl substitution reactions, and hydrogen abstraction acetylene addition (HACA) make great contributions to PAH formation, and hence the reactions of MAHs (e.g., A_1C_2H and $A_1C_2H_3$) were considered. As reported in Ref. [28], the predictions obtained from the PAH mechanism demonstrated outstanding agreement with the measured data collected from eight different flames, which indicated that the present PAH mechanism is satisfactory to some extent.

3.3. Decalin Sub-Mechanism

The decalin mechanism was reduced from the detailed decalin oxidation mechanism proposed by Dagaut et al. [29]. As shown in Figure 4, the current decalin mechanism contains both low- and high-temperature mechanisms, and so the mechanism exhibits the capability to accurately describe decalin oxidation across a broad temperature range. The reactions framed by the dashed box (Figure 4) belong to the low-temperature mechanism.

Decalin radicals can be diverse considering the symmetrical structure of a decalin molecule. Three $C_{10}H_{17}$ radicals lumped as 'RDECALIN' were produced via decalin consumption reactions with small molecules including O_2, H, and HO_2, as shown in Figure 4 (R12–R15). In accordance with Yu et al. [30], the low-temperature and high-temperature mechanisms exhibit significant differences in their core reactions. In the low-temperature regime, the primary consumption pathway for RDECALIN involves the O_2 addition reaction (R16), while other consumption reactions are negligible. The formation of alkylperoxy radicals (RDECOO) via R16 is followed by their isomerization to hydroperoxy alkyl (QDECOOH). QDECOOH, in turn, undergoes a conversion to ZDECA (lumped $C_{10}H_{17}O_4$ isomers) through the reaction R18. ZDECA is then consumed, leading to the formation of $C_{10}H_{16}O_3$ isomers (KHDECA), and the decalin chain-branching sequence is completed by the reaction R20, which represents the KHDECA decomposition at low temperatures. Conversely, in the high-temperature regime, the decomposition of RDECALIN primarily occurs through reactions that tend to generate C_5H_8 species and cyclopentyl-cyclopentene (DCYC5). Then C_5H_8 decomposes to cyclopentadiene (CYC_5H_6) via the H-abstraction reaction (R23). Eventually, the big hydrocarbon molecules are converted into small C2–C4 molecules.

3.4. N-Dodecane Sub-Mechanism

The n-dodecane mechanism was reduced via our previous work [31] and was briefly introduced here. As shown in Figure 4, n-dodecane is consumed by three reactions (R29–R31) to produce dodecyl radicals ($C_{12}H_{25}$) at low temperatures. $C_{12}H_{25}$ then transforms to peroxy dodecyl radicals ($C_{12}H_{25}OO$) via the pathway reaction R32, which is significant for the formation of the negative temperature coefficient (NTC) region. Hydroperoxy dodecyl radicals ($C_{12}H_{24}OOH$) are produced by a isomerization reaction (R33). $C_{12}H_{24}OOH$ transforms into $O_2C_{12}H_{24}OOH$ by reacting with oxygen (R34). $C_{12}KET$ radicals are produced from the reaction R35, and they further decompose into a few species, including $C_5H_{11}CO$, with the reaction R36. The small species will be generated through the decomposition reaction R37. However, at high temperatures, dodecyl radicals first convert to $C_{12}H_{24}$ by a reaction with oxygen (R38) and then decompose into C2–C4 molecules via the reaction R39 or directly produce the small hydrocarbon molecules by R40.

3.5. *Isocetane Sub-Mechanism*

The isocetane sub-mechanism was also taken from our previous work [32]. A short depiction of the isocetane sub-mechanism is given here. The consumption of isocetane is completed by the reactions R41–R43 at low temperatures (Figure 4). Ketohydroperoxide is produced via the reactions R44–R47, while the decomposition of ketohydroperoxide is completed by the reactions R48 and R49. At high temperatures, all reactions are represented by the three reactions R50, R51, and R52.

4. Results and Discussion

The experimental data, including IDTs, species concentrations, and laminar flame speeds determined by various devices, were used for validating the current kinetic mechanism. The mechanism was first validated for each component and then for the surrogate fuel mixture. The validation of the individual components will give a more comprehensive assessment on the accuracy and reliability of the JFS mechanism. The simulations were performed on CHEMKIN-PRO Version-19.0 software [33].

4.1. *Verifications of Toluene*

4.1.1. IDT

It is necessary to validate the current mechanism on IDT because the fuel autoignition behaviors characterized by IDT are important for engine combustion and emission performances [34]. Figure 5 exhibits a comparison of the measured and simulated IDTs for toluene in a shock tube under varying conditions of temperature, pressure, and equivalence ratios. The measurements were determined by Shen et al. [35], and the predicted IDT with the current mechanism and the based mechanism [28] were exhibited. Both the reduced mechanism and the based mechanism demonstrated strong agreement with the measurements. The current mechanism showed better performance under an equivalence ratio of 1.0. Compared to the based mechanism, the IDTs predicted by the current mechanism are closer to the measurements at low pressure. In addition, the predicted IDTs of toluene are compared with the determined IDTs in an RCM [36]. The corresponding results are presented in Figure 6. The toluene IDTs were slightly underestimated, but the discrepancy is acceptable.

4.1.2. Species Concentration

As emphasized by Pitz et al. [37], it is equally crucial to validate the profiles of major species concentrations alongside the validation of fuel IDT. Hence, the current mechanism was validated for the major species profiles during the oxidation process. In Figure 7, the concentration profiles of critical species in the premixed toluene flame are displayed. These species include reactants (toluene and oxygen), an inactive gas (argon), and products (such as carbon monoxide and hydrogen). Li et al. [38] experimentally investigated premixed toluene flames at three equivalence ratios (Φ = 0.75, 1.0, and 1.5) under a pressure of 10 atm. A satisfactory agreement was observed between the predicted and observed species profiles for all flames, with a notable consistency observed in the case of the stoichiometric toluene flame. The argon species concentration initially decreased and then stabilized further downstream of the burner. The model accurately predicted this trend, demonstrating its capability to replicate the mole expansion effects observed in the toluene flame, as reported in reference [38]. Additionally, the model successfully captured the overall profiles of the final products, specifically water and carbon dioxide, and the maximum deviations between the measurements and predictions for final products are less than 10%. Carbon monoxide displayed a profile resembling that of hydrogen, gradually surpassing carbon dioxide as the predominant carbon product. Despite the discrepancies in the predicted species concentrations of toluene and oxygen compared to the corresponding measurements at equivalence ratios of 0.75 and 1.5, the model successfully represented the depletion positions of these species with accuracy.

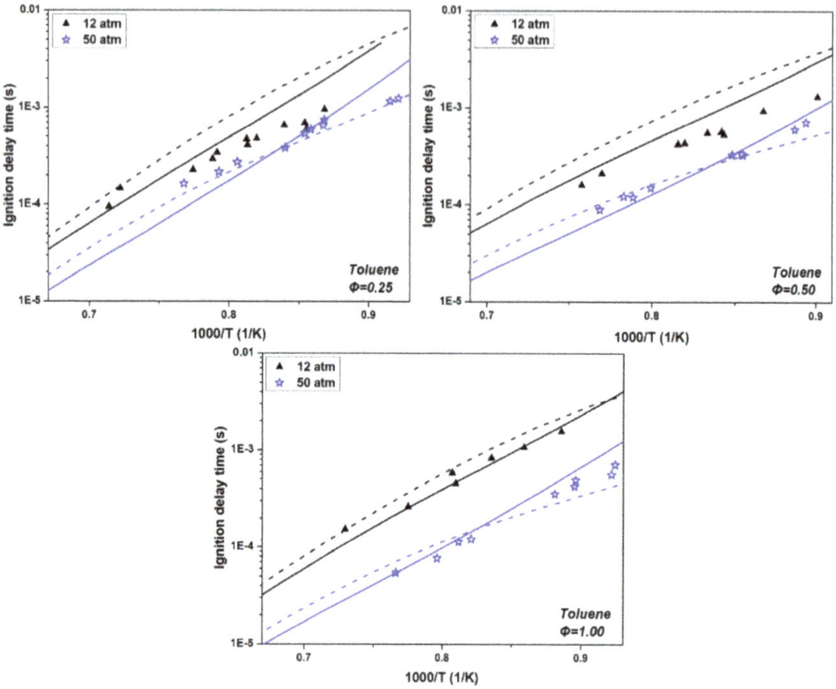

Figure 5. The measured (symbols [35]) and predicted (solid lines: current mechanism, dash lines: the based mechanism [28]) toluene IDTs determined in shock tube.

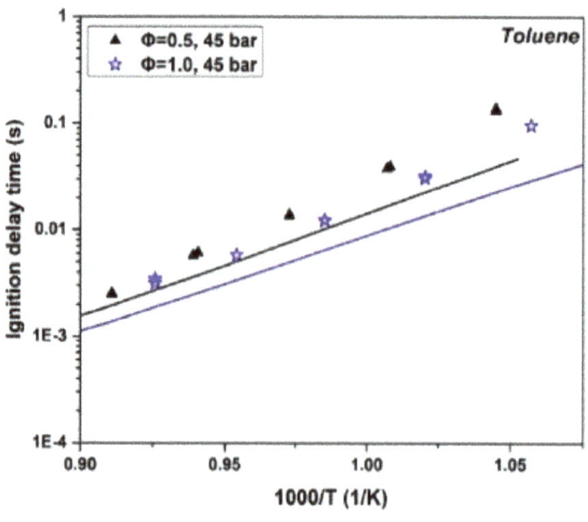

Figure 6. The measured (symbols) and simulated (lines) toluene IDTs in RCM.

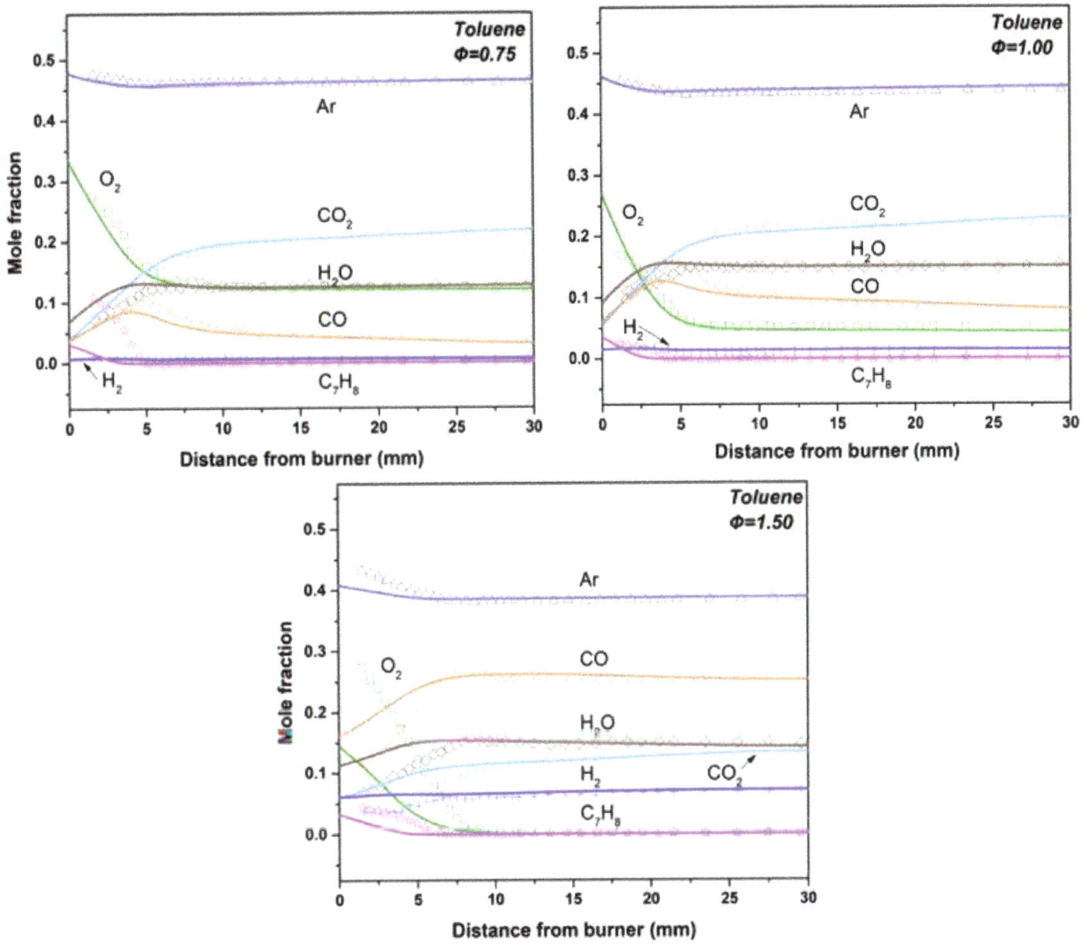

Figure 7. The major species profiles of premixed toluene flames under three equivalence ratios (0.75, 1.0, and 1.5) and a pressure of 10 atm. The determined results [38] are represented by symbols, while the simulated values from the current mechanism are depicted by lines.

Figure 8 shows the toluene oxidation process in a JSR [39] at a residence time of 0.6 s and a pressure of 10 atm. The concentration evolution of both the major species (CO, CO_2, and C_7H_8) and the minor species (C_2H_2 and C_4H_4) are presented in Figure 8. The current mechanism rendered good performance in predicting the concentration profiles of C_7H_8, CO, and CO_2. C_2H_2 is esteemed as a significant precursor of PAH and soot formation [40]. Moreover, the proposed model exhibits its capability to predict the toluene concentration evolution. Furthermore, the mechanism successfully predicted the species profiles of C_4H_4 under different conditions, except for the scenario with an equivalence ratio of 1.5.

4.1.3. Laminar Flame Speed

The laminar flame speed (LFS) is a valuable parameter that provides insights into the reactivity, heat release rate, and propagation characteristics of actual fuels. Figure 9 shows the comparison between the simulated and measured LFSs of counter-flow toluene flames. The measurements were conducted under two conditions: p = 1 atm, Tu = 298 K [41], and p = 3 atm, Tu = 450 K [42], where Tu denotes the temperature of the unburned mixture. It can be observed that the mechanism performed well in predicting the flame behavior under atmospheric conditions. However, the discrepancy of the predictions and measurements were larger under fuel-rich conditions compared to oxygen-rich conditions. Therefore, we can conclude that the current mechanism is more suitable for modeling lean-burn engine combustion.

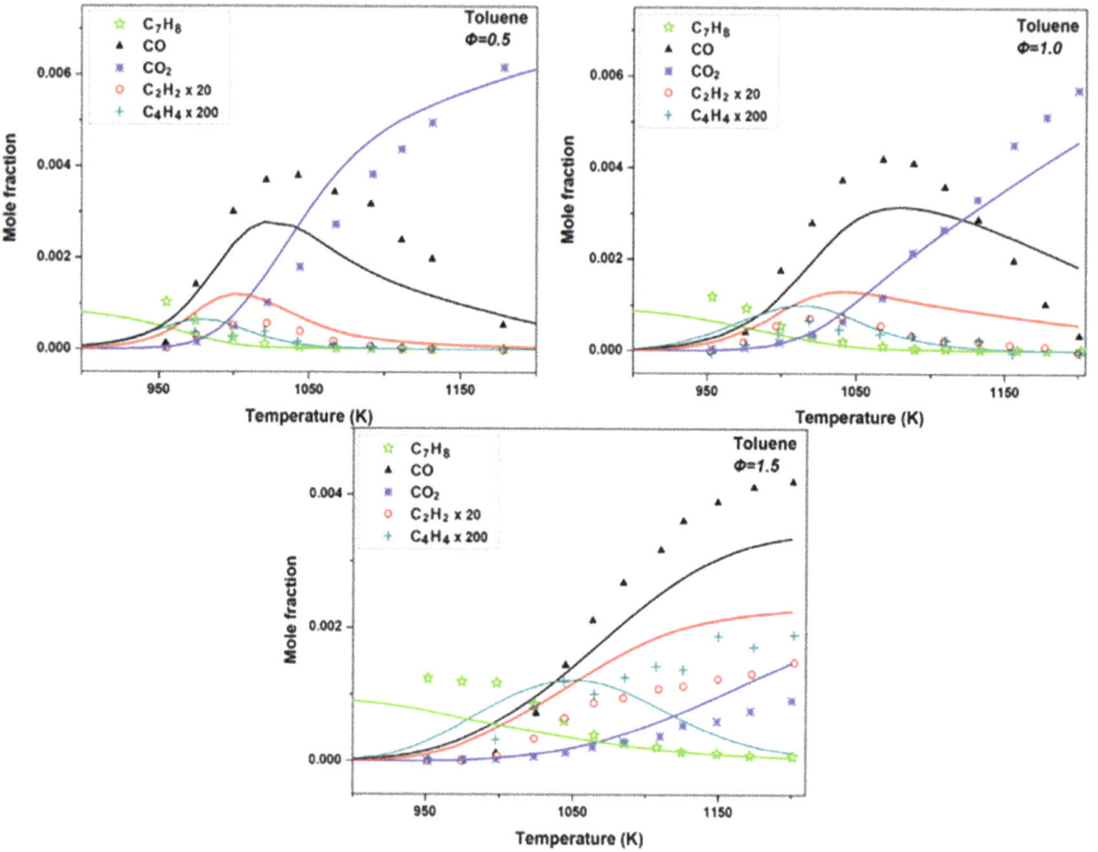

Figure 8. The concentrations of species resulting from the oxidation of toluene in the JSR are presented for various equivalence ratios at 0.6 s residence time and a pressure of 10 atm. The measured values [39] are denoted by symbols, while the simulation results obtained by the current mechanism are represented by lines.

4.2. Verifications of Decalin

4.2.1. IDT

Figure 10 displays the comparisons of the simulated and measured ignition delay times for decalin in a shock tube. The experimental data were obtained from previous studies [43,44], while the simulation results were obtained with both the present mechanism and the detailed mechanism [29]. Except for the case of $\Phi = 0.5$ and $p = 12$ atm, the current mechanism consistently outperformed the detailed mechanism across all other conditions, as demonstrated in Figure 10. However, considering the results achieved by the detailed mechanism, the level of discrepancy observed under the mentioned condition is still within an acceptable range. These results confirm that the current mechanism effectively captures the dependence of IDT on pressure and equivalence ratio, particularly at two specific conditions: (1) $p = 20$ atm, $\Phi = 0.5$; (2) $p = 40$ atm, $\Phi = 1.0$.

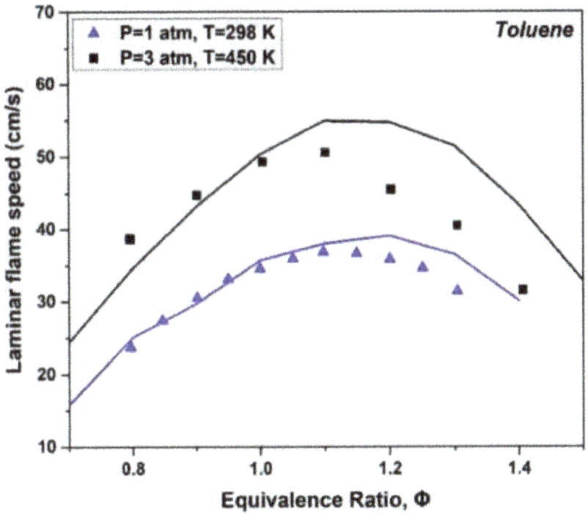

Figure 9. The simulated and measured LFSs in counter-flow toluene flames.

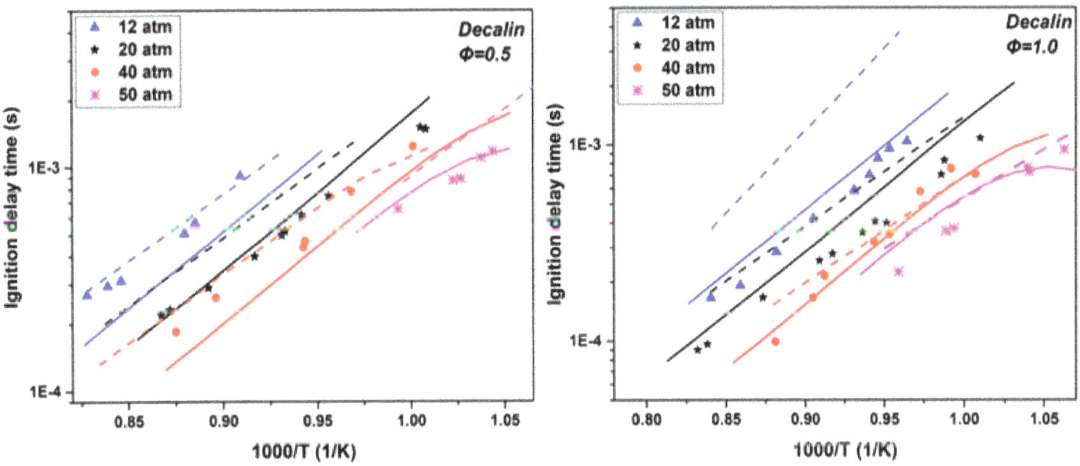

Figure 10. The measured (symbols) and predicted IDTs (solid lines: current mechanism; dash lines: the detailed mechanism) for decalin in a shock tube.

Figure 11 depicts the ignition delay time (IDT) of decalin over a temperature range of 770 K to 1250 K. Non-thermal combustion (NTC) behaviors were observed within this temperature range. At low-pressure conditions (p = 20 atm), the NTC region started at around 910 K, while at high temperatures, the NTC region initiated at a higher temperature of approximately 940 K. The shift in the NTC region suggests that pressure significantly influences the NTC behavior, which aligns with similar findings reported in Ref. [45].

In Figure 12, the decalin IDT is presented as determined in a rapid compression machine over a broad temperature range. The measurements were conducted at various equivalence ratios and a 15 atm compression pressure [30]. The results clearly demonstrate that IDTs decrease with increasing equivalence ratios. The variations in IDTs among different conditions were relatively small within the temperature range of 631 to 750 K, indicating a weak influence of equivalence ratio on IDTs at low temperatures. As the temperature increased, the differences in the ignition delay times (IDTs) became more pronounced. This can be attributed to the intensified chain-branch reactions occurring between the decalin molecules and hydroperoxyl radicals in both the negative-temperature-coefficient and high-temperature zones. The current mechanism successfully captures these observed behaviors.

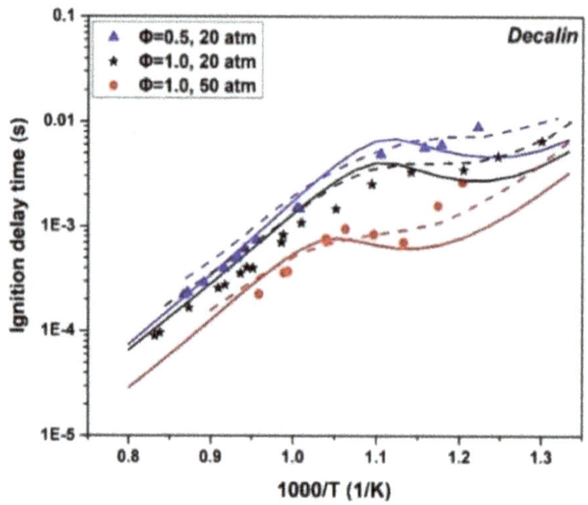

Figure 11. The decalin IDTs' NTC behaviors in a shock tube. The measured (symbols [43,44]) and predicted IDTs (solid lines: current mechanism; dash lines: the detailed mechanism [29]) for decalin.

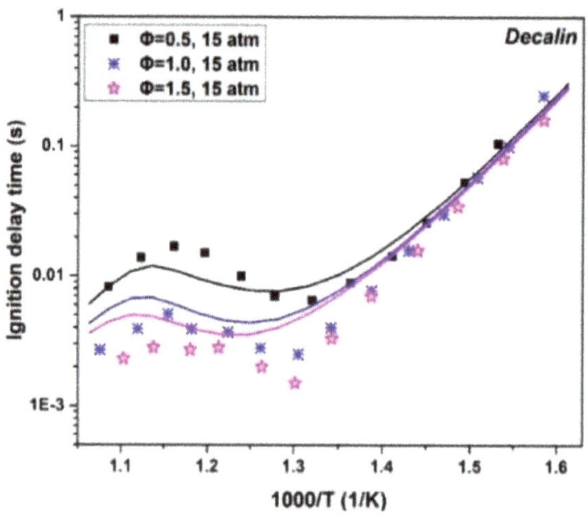

Figure 12. The measured (symbols) and predicted (lines) decalin IDTs in a RCM at three equivalence ratios under 15 atm compression pressure.

4.2.2. Species Concentration

Figure 13 exhibits the species profiles of the premixed decalin flames [46]. The predictions and measurements showed excellent agreement across various equivalence ratios. According to the ROP analysis, CO is predominantly generated through reactions involving HCO and HCCO, while its consumption occurs through its reaction with hydroxyl radicals, leading to the formation of CO_2. The formation of H_2O is notably influenced by H-atom abstraction reactions as well as the reaction between OH and H_2. H_2 is produced through a reaction between decalin and formaldehyde (CH_2O). The positions of depletion for decalin species varied with different equivalence ratios, and in the case of the stoichiometric decalin flame, the depletion position was observed to be closer to the burner surface. Similar observations were reported in the referenced study [38].

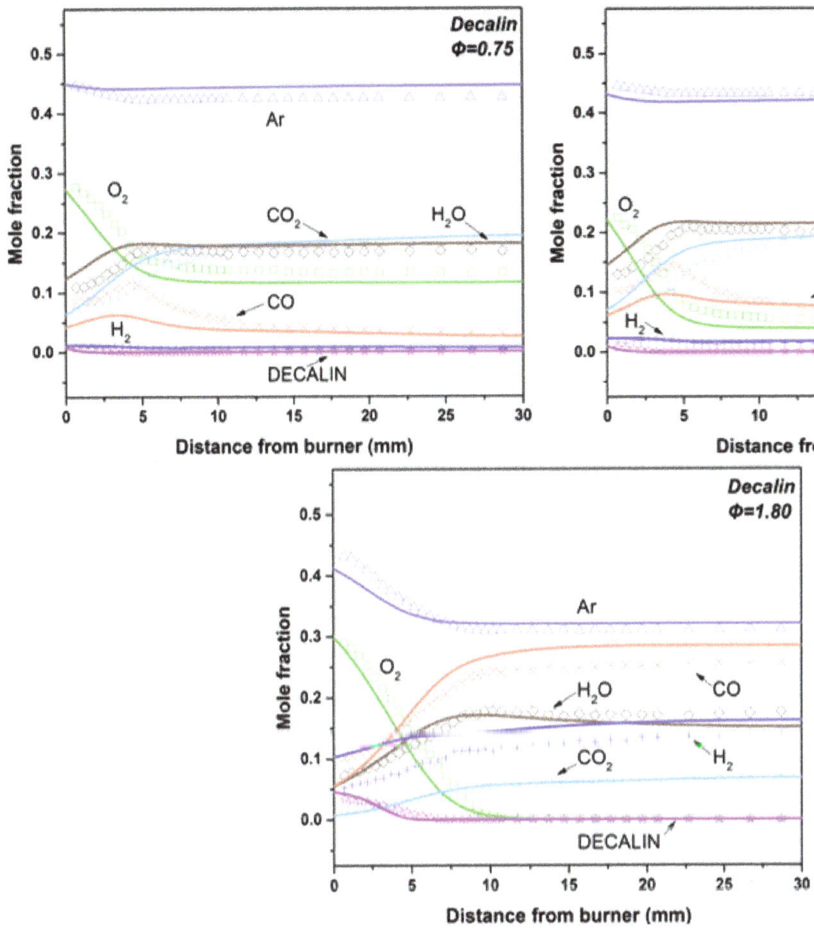

Figure 13. The premixed decalin flames species profiles at various equivalence ratios. The measured values [46] are denoted by symbols, while the simulated values from the current mechanism are represented by lines.

The species concentration profiles of decalin were also examined in the JSR [29]. C_2H_2 and CH_4 were selected for validating the accuracy of the current mechanism since they are the significant intermediate hydrocarbons produced during decalin oxidation. The current mechanism accurately predicted the concentration variations of C_2H_2 and CH_4, although it slightly overestimated the decalin concentration. It was observed that the concentration of CO initially increased and then reached a steady level above 900 K. The concentration of CO_2 exhibited a continuous increase with rising temperature. Overall, the current mechanism demonstrated good performance in reproducing these trends. Furthermore, the concentration profiles of CH_2O, a major aldehyde produced during the oxidation process of decalin, were also accurately captured by the current mechanism (Figure 14).

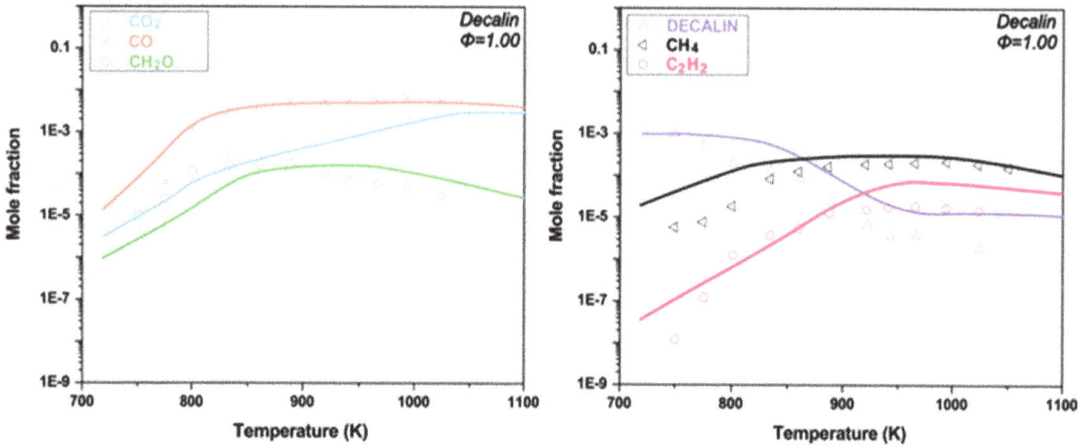

Figure 14. The measured (symbols [29]) and predicted (lines) concentration profiles of species resulting from the oxidation of decalin in the JSR at an equivalence ratio of 1.0 and a compression pressure of 10 atm.

4.2.3. Laminar Flame Speed

Figure 15 illustrates the determined and predicted LFSs for the counter-flow decalin flame at atmospheric pressure [47]. The initial mixture temperature for the flame is 443 K. It is evident that the current mechanism accurately predicts the laminar flame speed of decalin, although the deviation between measured and predicted LFSs becomes larger under high equivalence ratios.

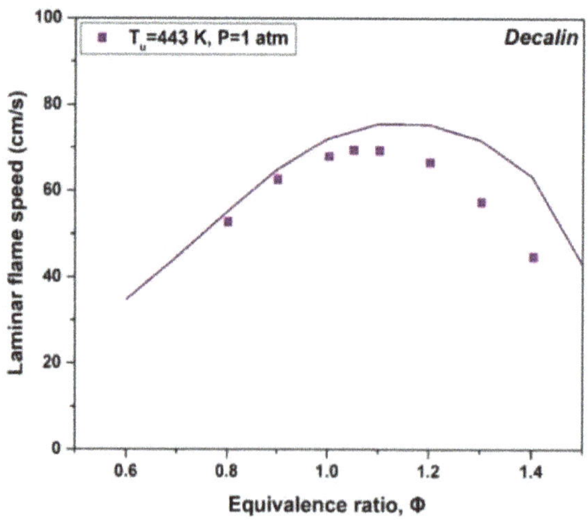

Figure 15. The laminar flame speeds of the counter-flow decalin flame. The line stands for the simulations, and dots represent the experimental data [47].

4.3. Verifications of n-Dodecane

4.3.1. IDT

Figure 16 shows the predicted and measured n-dodecane IDTs at two equivalence ratios (0.5 and 1.0) under three pressures (14 atm, 20 atm, and 40 atm). The shock tube measurements were taken from the Refs. [48–50]. The results confirmed that the current mechanism was capable of capturing the variation trends of the measured IDTs, except in the case of p = 40 atm. The NTC behaviors observed in the temperature range from 750 K to 900 K were well reflected by the mechanism. In addition, the experimental phenomenon that IDT decreases with an increase in pressure was also captured by the mechanism.

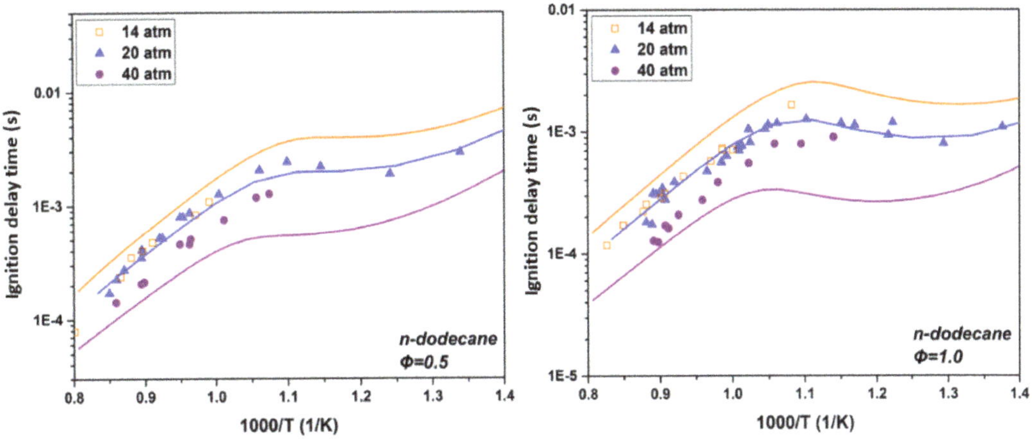

Figure 16. The measured (symbols) and predicted (lines) IDTs of n-dodecane in shock tube.

4.3.2. Species Concentration

Malewicki et al. [14] conducted JSR experiments to investigate the n-dodecane oxidation process. The experiments were carried out at a residence time of 1 s, a pressure of 10 atm, and three equivalence ratios. The comparison between the computed and measured results is shown in Figure 17. The prediction for the variation trend of O_2 and n-dodecane ($C_{12}H_{26}$) with temperature was consistent with the measurements under all conditions. H_2O was slightly underestimated, but the deviation is still at an acceptable level. There were some deviations between the simulated and determined mole fractions of C_2H_2 and CO_2; nevertheless, the overall simulated profiles of C_2H_2 and CO_2 were satisfactory.

4.3.3. Laminar Flame Speed

The measured data of the n-dodecane laminar flame speed were adopted here from the experiment conducted by Kumar et al. [51]. The measurement was performed at atmospheric pressure and unburned mixture temperatures of 400 K and 470 K. It can be easily seen from Figure 18 that the predictions showed fairly good agreement with the experimental data under all conditions.

4.4. Verifications of Isocetane

4.4.1. IDT

The IDTs of isocetane were measured in a shock tube under two pressures (10 atm and 40 atm) and three equivalence ratios (0.5, 1.0, and 1.5) by Oehlschaeger et al. [52]. Figure 19 presents the predictions and measurements of isocetane IDT, and the maximum deviations between the measured and predicted IDT appeared under the condition of p = 10 atm and Φ = 1.0, confirming the derived mechanism can accurately predict IDT. The decreasing trend of IDT with increasing pressure was perfectly captured by the mechanism. Due to the absence of experimental data, the NTC behaviors were not obviously observed at low pressure and low equivalence ratios. Therefore, it is necessary to do further IDT validations for isocetane in the future.

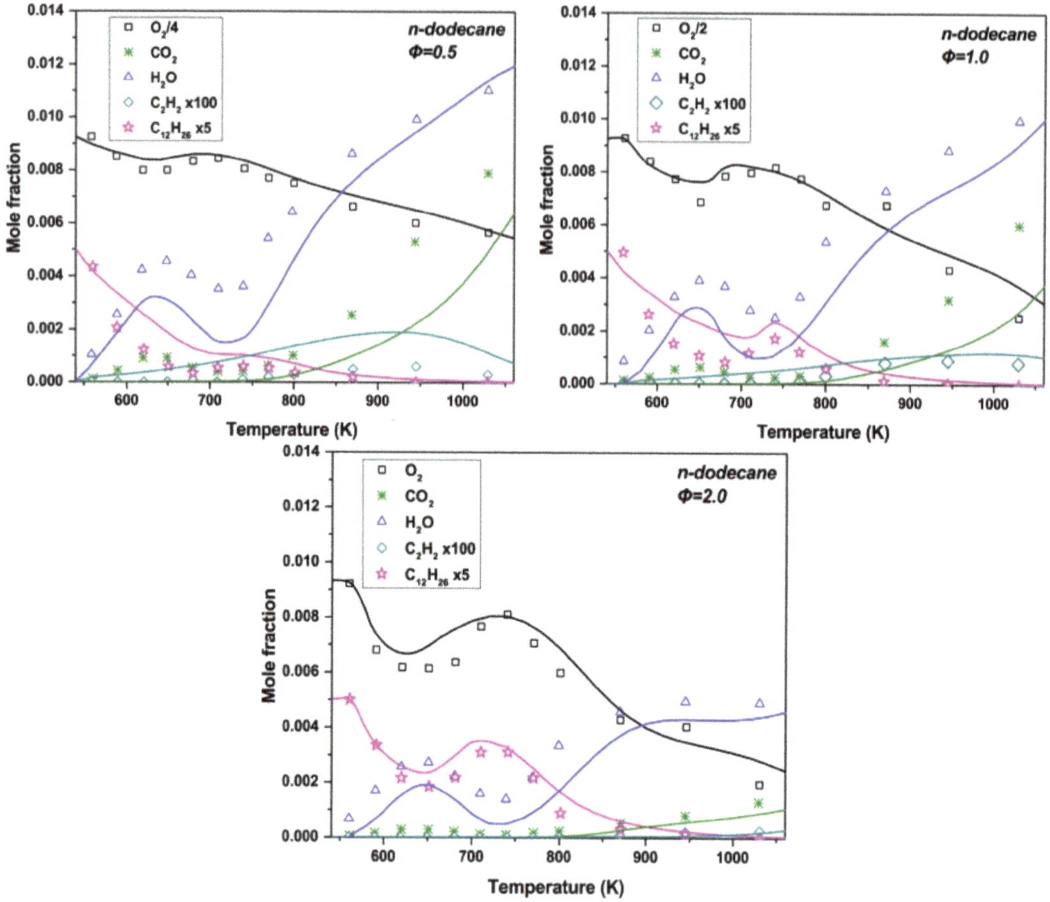

Figure 17. The measured (symbols [14]) and predicted (lines) mole fractions of the species during n-dodecane oxidation in JSR experiments at different equivalence ratios.

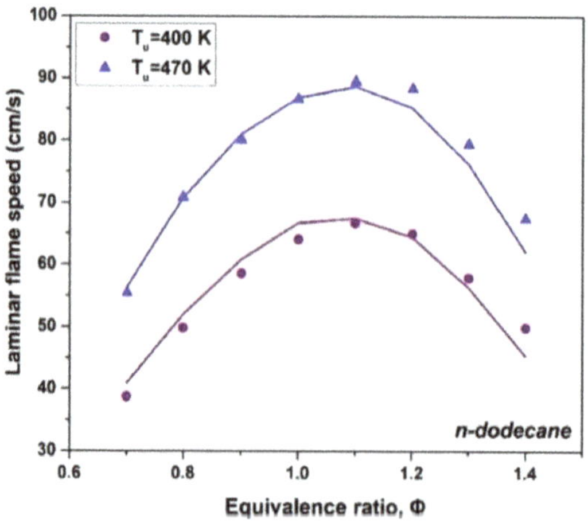

Figure 18. The laminar flame speeds of the n-dodecane flame. The line stands for the simulations, and dots represent the experimental data [51].

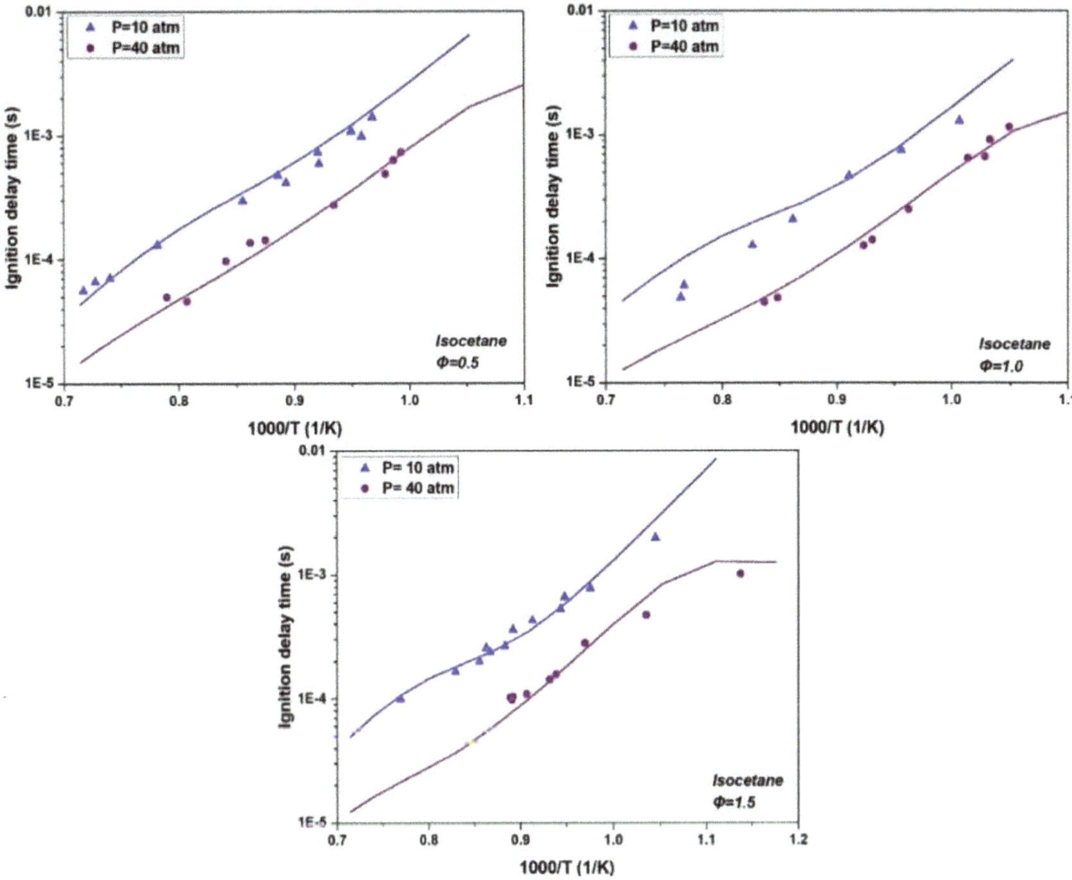

Figure 19. The measured (symbols [52]) and predicted IDTs (lines) of isocetane in the shock tube.

4.4.2. Species Concentration

Figure 20 shows the concentrations profiles of the major species produced during isocetane oxidation in a JSR [53]. The measurements were performed at a residence time of 1 s, under a pressure of 10 atm, and under three equivalence ratios (0.5, 1.0, and 2.0). It can be seen that the mechanism nicely predicts the consumption trend of isocetane under all conditions. The species mole fractions of H_2O and CO were also well predicted. The concentration of isocetane was overestimated in the high temperature range when $\Phi = 0.5$. For intermediate hydrocarbon products, the species profiles of CH_4 were not excellently reproduced by the mechanism, especially in the case of $\Phi = 1.0$.

4.5. Verifications of Jet Fuel

In this section, comparisons of ignition delay times, species mole fractions, and laminar flame speeds were further carried out for the newly developed JFS surrogate and the target jet fuel POSF-4658.

4.5.1. IDT

The shock tube experiments of jet fuel (POSF-4658) were conducted by S. Vasu et al. [8] and Wang et al. [54], as shown in Figure 21.

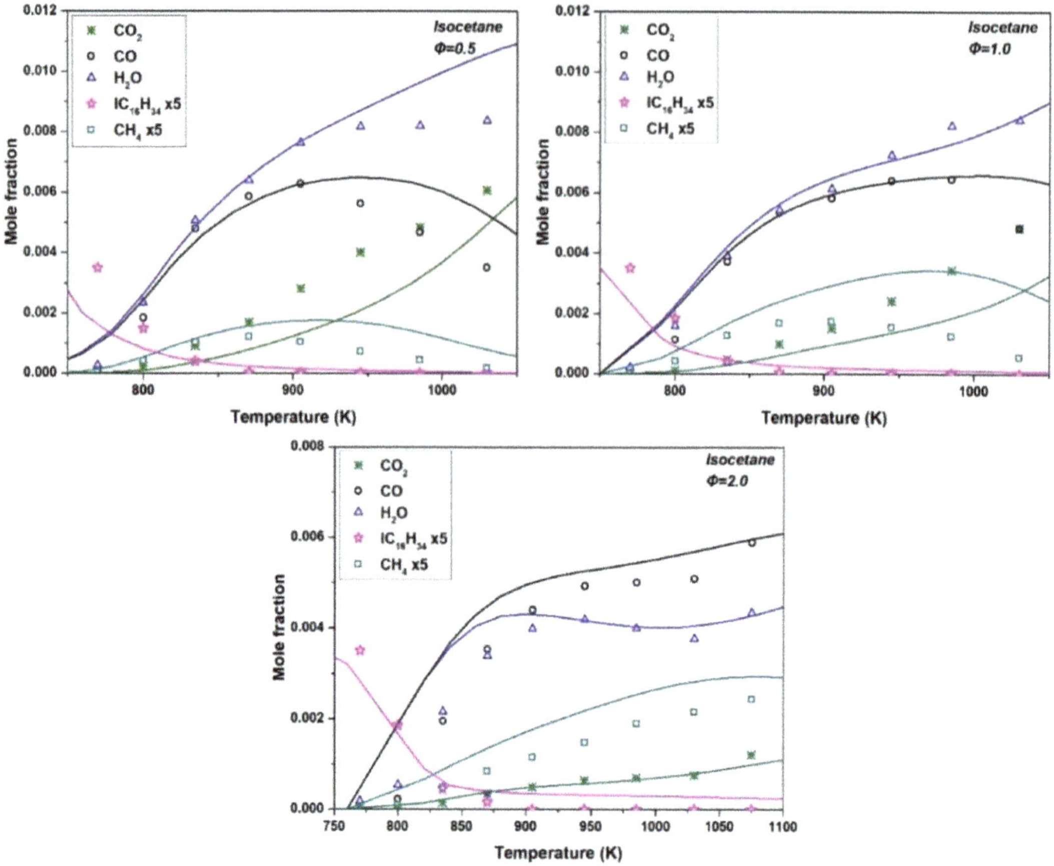

Figure 20. The measured (symbols [53])and predicted (lines) major species profiles of isocetane in JSR experiments at different equivalence ratios.

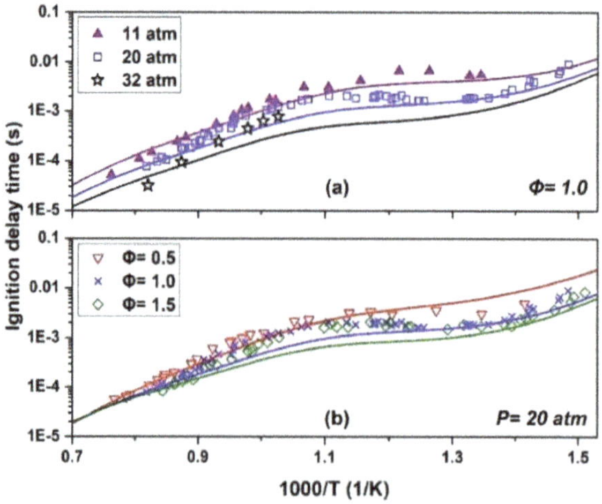

Figure 21. The comparisons between the determined and simulated IDTs of jet fuel under different (**a**) pressures and (**b**) equivalence ratios in a shock tube.

The comparisons between the determined and simulated IDTs of jet fuel at three pressures are exhibited in Figure 21a, while Figure 21b shows the results at three equivalence ratios: Φ = 0.5, 1.0, and 1.5. The overall trends of the jet fuel IDTs were well reflected by the current mechanism, especially at low pressure and low equivalence ratios, which can be attributed to the reason that the H/C, CN, and LHV of the proposed JFS surrogate are similar to those of the target fuel. Although the predicted NTC region underwent a slight shift to the left under the conditions of Φ = 1.5 and 20 atm pressure, the NTC behavior was satisfactorily captured via the developed mechanism. Furthermore, the maximum deviation between the predicted and measured IDTs is less than 0.001 s, and the small deviations between the predicted and measured IDTs can be attributed to two reasons: (1) Some reactions and species are removed from the reduced mechanism, which inevitably makes the mechanism unable to reproduce the measurements as well as the detailed mechanism. (2) There are still some differences between the properties of the JFS surrogate and the target fuel, which in turn influence the performance of the associated mechanism.

Figure 22 shows the comparison results between the measured IDTs and the IDTs predicted from the current JFS mechanism as well as the UM1 and UM2 mechanisms at Φ = 1.0 and p = 20 atm. Both the surrogates UM1 and UM2 contain four components, and the components of UM2 and the current JFS surrogate are identical. The detailed mechanisms of UM1 and UM2 were developed by Kim et al. [15]. In general, all of the mechanisms give good predictions of the measured IDTs of jet fuels. The predictions still suffer some deviations in IDT against the experimental data within the intermediate temperature range (from 855 to 1030 K), even with the detailed mechanisms, and the maximum deviation between the predicted and measured IDTs is about 0.0015 s. It is worth mentioning that the reduced JFS mechanism has a comparable emulation capability to the UM1 and UM2 mechanisms, although the UM1 and UM2 mechanisms are much more detailed compared to the current mechanism.

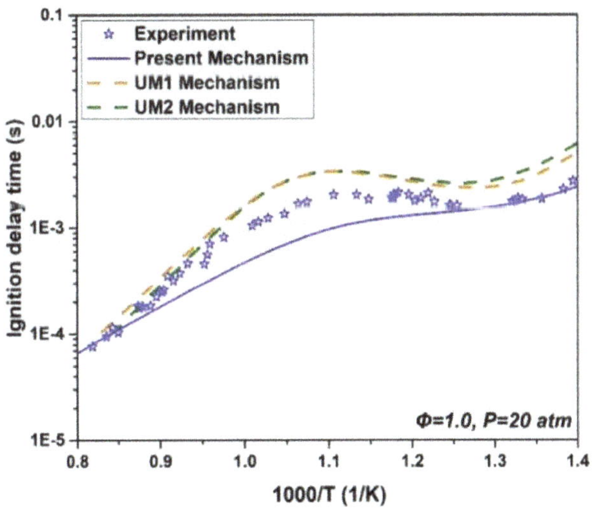

Figure 22. The comparison between the determined IDTs of jet fuel [12] and the simulated IDTs by the current mechanism, as well as the UM1 and UM2 mechanisms [15].

4.5.2. Species Concentration

Figure 23 shows the species profiles of the JFS surrogate in comparison with the experimental data determined in a shock tube at the temperature range of 890 K to 1680 K by Malewicki et al. [14].

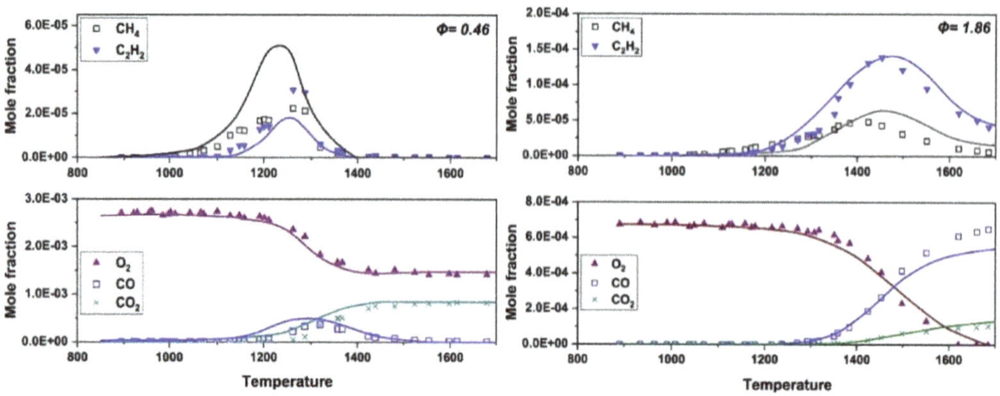

Figure 23. The measured and predicted species profiles of jet fuel. Symbols are the experimental data [14]; lines are the predictions made by the current mechanism.

The concentration profiles of CH_4 and C_2H_2 were well captured by the current mechanism, except for the slight overprediction for CH_4 at the lean condition (Φ = 0.46), which implies that the mechanism is reliable in predicting the soot and PAH formulation in engine combustion. The decaying trend of O_2 and the increasing trend of CO_2 were both greatly reflected by the mechanism. As shown in Figure 23, the current mechanism also showed good performance in predicting the CO concentration except for a slight underestimation under the rich- and high-temperature condition (Figure 22b). Overall, the current mechanism exhibited a satisfactory prediction of the major species concentrations of the JFS surrogate, although the maximum deviation between the measurements and the predictions of CH_4 was larger than the measured value. In fact, similar deviations were also observed even using a detailed mechanism, as reported by Malewicki et al. [14].

4.5.3. Laminar Flame Speed

The comparison results between the simulated and measured laminar flame speeds is shown in Figure 24. The laminar flame speeds of jet fuel were measured by Dooley et al. [13] at atmospheric pressure and the unburned mixture temperatures of 400 K and 470 K. The predicted values showed a surprising consistency with the measurements, and the maximum discrepancy between them was less than 9.4%. According to Li et al. [55], H_2, CO, and small hydrocarbons have decisive impacts on flame propagation; hence, great consistency can be attained through the adoption of a detailed $C1/H_2/CO$ sub-mechanism in the current JFS mechanism.

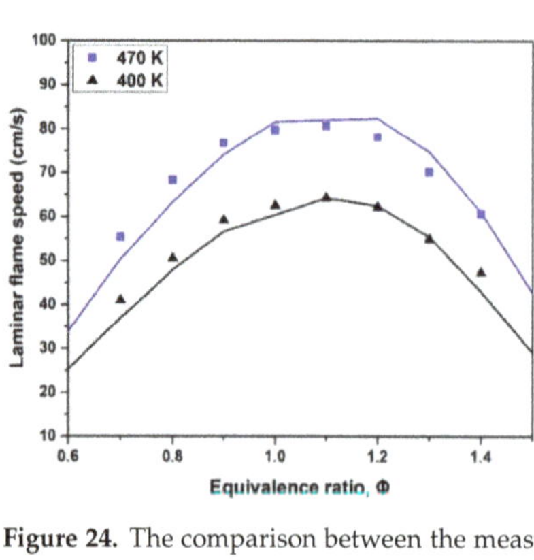

Figure 24. The comparison between the measured and predicted laminar flame speeds. Symbols represent the experimental data [13], lines are the values predicted by the current mechanism.

5. Conclusions

In this study, a new four-component jet fuel surrogate (n-dodecane/iso-octane/isocetane/decalin/toluene) was formulated with a property optimizer. The chemical and physical properties of the target jet fuel were considered during the formulation of the fuel surrogate. Compared to the previous-existing jet surrogates, the current jet fuel surrogate showed better performance in emulating the real jet fuel properties. The deviations in CN, TSI, LHV, and density are −0.35, −1.29, 0.87, and 1.84, which are extremely in a low level. The property deviations of MW (10.73%), viscosity (5.88%), and surface tension (8.71%) are slightly larger, but these remain within acceptable ranges.

A compact reduced kinetic mechanism consisting of 124 species and 590 reactions was proposed for the newly formulated surrogate. This mechanism is highly suitable for practical engine combustion simulations due to its reduced size. The mechanism is composed of four sub-mechanisms covering toluene, decalin, n-dodecane, and isocetane. It was systematically developed, starting from the C5–C16 reactions and transitioning to the C2–C4 reactions, followed by the $H_2/O_2/C1$ reactions, PAH reactions, and NOx reactions. To validate the proposed mechanism, experimental data including ignition delay time (IDT), species concentrations, and laminar flame speeds were utilized, encompassing both the surrogate mixture and its individual components. The results demonstrated a favorable concordance between the proposed mechanism and the experimental data. The current mechanism could accurately capture the variations of the ignition delay time in the negative temperature coefficient (NTC) region. Consequently, the proposed mechanism emerges as a viable choice for effective utilization in practical engine simulations.

Author Contributions: Conceptualization, G.L.; methodology, G.L.; software, B.C.; validation, B.C.; data curation, C.Z.; writing—review and editing, L.C. (Longfei Chen); supervision, L.C. (Liuyong Chang). All authors have read and agreed to the published version of the manuscript.

Funding: This work was supported by the National Natural Science Foundations of China (Grant No. 52306181 and Grant No. 52306184) and National Key Research and Development Project of China (Grant No. 2022YFB2602002).

Institutional Review Board Statement: Not applicable.

Informed Consent Statement: Not applicable.

Data Availability Statement: The data used to support the findings of this study are available from the corresponding author upon request.

Conflicts of Interest: The authors declare no conflict of interest.

References

1. Wolfowitz, P. *DoD Management Policy for Energy Commodities and Related Services*; Department of Defense: Washington, DC, USA, 2004; p. 4140.25.
2. Sarathy, S.; Farooq, A.; Kalghatgi, G.T. Recent progress in gasoline surrogate fuels. *Prog. Energy Combust. Sci.* **2018**, *65*, 67–108. [CrossRef]
3. Zhen, X.; Yang, W.; Daming, L. An overview of the chemical reaction mechanisms for gasoline surrogate fuels. *Appl. Therm. Eng.* **2017**, *124*, 1257–1268. [CrossRef]
4. Kim, D.; Violi, A. Hydrocarbons for the next generation of jet fuel surrogates. *Fuel* **2018**, *228*, 438–444. [CrossRef]
5. Yu, W.; Zhao, F.; Yang, W.; Tay, K.; Xu, H. Development of an optimization methodology for formulating both jet fuel and diesel fuel surrogates and their associated skeletal oxidation mechanisms. *Fuel* **2018**, *231*, 361–372. [CrossRef]
6. Dooley, S.; Won, S.H.; Dryer, F.L. Chapter 10—Surrogate fuels. In *Computer Aided Chemical Engineering*; Faravelli, T., Manenti, F., Ranzi, E., Eds.; Elsevier: Amsterdam, The Netherlands, 2019; pp. 513–602.
7. Violi, A.; Yan, S.; Eddings, E.G.; Sarofim, A.F.; Granata, S.; Faravelli, T.; Ranzi, E. Experimental formulation and kinetic model for JP-8 surrogate mixtures. *Combust. Sci. Technol.* **2002**, *174*, 399–417. [CrossRef]
8. Vasu, S.S.; Davidson, D.F.; Hanson, R.K. Jet fuel ignition delay times: Shock tube experiments over wide conditions and surrogate model predictions. *Combust. Flame* **2008**, *152*, 125–143. [CrossRef]
9. Dagaut, P.; El Bakali, A.; Ristori, A. The combustion of kerosene: Experimental results and kinetic modelling using 1- to 3-component surrogate model fuels. *Fuel* **2006**, *85*, 944–956. [CrossRef]

10. Gokulakrishnan, P.; Gaines, G.; Currano, J.; Klassen, M.S.; Roby, R.J. Experimental and Kinetic Modeling of Kerosene-Type Fuels at Gas Turbine Operating Conditions. *J. Eng. Gas Turbines Power* **2006**, *129*, 655–663. [CrossRef]
11. Eddings, E.G.; Yan, S.; Ciro, W.; Sarofim, A.F. Formulation of a surrogate for the simulation of jet fuel pool fires. *Combust. Sci. Technol.* **2005**, *177*, 715–739. [CrossRef]
12. Dooley, S.; Won, S.H.; Chaos, M.; Heyne, J.; Ju, Y.; Dryer, F.L.; Kumar, K.; Sung, C.-J.; Wang, H.; Oehlschlaeger, M.A.; et al. A jet fuel surrogate formulated by real fuel properties. *Combust. Flame* **2010**, *157*, 2333–2339. [CrossRef]
13. Dooley, S.; Won, S.H.; Heyne, J.; Farouk, T.I.; Ju, Y.; Dryer, F.L.; Kumar, K.; Hui, X.; Sung, C.-J.; Wang, H.; et al. The experimental evaluation of a methodology for surrogate fuel formulation to emulate gas phase combustion kinetic phenomena. *Combust. Flame* **2012**, *159*, 1444–1466. [CrossRef]
14. Malewicki, T.; Gudiyella, S.; Brezinsky, K. Experimental and modeling study on the oxidation of Jet A and the n-dodecane/iso-octane/n-propylbenzene/1, 3, 5-trimethylbenzene surrogate fuel. *Combust. Flame* **2013**, *160*, 17–30. [CrossRef]
15. Kim, D.; Martz, J.; Violi, A. A surrogate for emulating the physical and chemical properties of conventional jet fuel. *Combust. Flame* **2014**, *161*, 1489–1498. [CrossRef]
16. Verhelst, S.; Turner, J.W.; Sileghem, L.; Vancoillie, J. Methanol as a fuel for internal combustion engines. *Prog. Energy Combust. Sci.* **2019**, *70*, 43–88. [CrossRef]
17. Bae, C.; Kim, J. Alternative fuels for internal combustion engines. *Proc. Combust. Inst.* **2017**, *36*, 3389–3413. [CrossRef]
18. Yu, W.; Yang, W.; Tay, K.; Zhao, F. An optimization method for formulating model-based jet fuel surrogate by emulating physical, gas phase chemical properties and threshold sooting index (TSI) of real jet fuel under engine relevant conditions. *Combust. Flame* **2018**, *193*, 192–217. [CrossRef]
19. McDaniel, A.; Dickerson, T.; Luning-Prak, D.; Hamilton, L.; Cowart, J. *A Technical Evaluation of New Renewable Jet and Diesel Fuels Operated in Neat Form in Multiple Diesel Engines*; SAE Technical Paper; SAE International: Warrendale, PA, USA, 2016. [CrossRef]
20. Murphy, M.J.; Taylor, J.D.; McCormick, R.L. *Compendium of Experimental Cetane Number Data*; National Renewable Energy Laboratory: Golden, CO, USA, 2004.
21. *DIPPR Project 801—Full Version*; Design Institute for Physical Property Research/AIChE: New York, NY, USA, 2012.
22. Mensch, A.; Santoro, R.; Litzinger, T.; Lee, S.-Y. Sooting Characteristics of Surrogates for Jet Fuels. *Combust. Flame* **2010**, *157*, 1097–1105. [CrossRef]
23. Gill, R.J.; Olson, D.B. Estimation of Soot Thresholds for Fuel Mixtures. *Combust. Sci. Technol.* **1984**, *40*, 307–315. [CrossRef]
24. Grunberg, L.; Nissan, A.H. Mixture Law for Viscosity. *Nature* **1949**, *164*, 799–800. [CrossRef]
25. Poling, B.E.; Thomson, G.H.; Friend, D.G.; Rowley, R.L.; Wilding, W.V. *Perry's Chemical Engineers' Handbook*; McGraw-Hill Publishing: New York, NY, USA, 2008.
26. Hugill, J.; Van Welsenes, A.J. Surface tension: A simple correlation for natural gas+ conensate systems. *Fluid Phase Equilibria* **1986**, *29*, 383–390. [CrossRef]
27. Agosta, A.; Cernansky, N.; Miller, D.; Faravelli, T.; Ranzi, E. Reference components of jet fuels: Kinetic modeling and experimental results. *Exp. Therm. Fluid Sci.* **2004**, *28*, 701–708. [CrossRef]
28. Wang, H.; Yao, M.; Yue, Z.; Jia, M.; Reitz, R.D. A reduced toluene reference fuel chemical kinetic mechanism for combustion and polycyclic-aromatic hydrocarbon predictions. *Combust. Flame* **2015**, *162*, 2390–2404. [CrossRef]
29. Dagaut, P.; Ristori, A.; Frassoldati, A.; Faravelli, T.; Dayma, G.; Ranzi, E. Experimental and semi-detailed kinetic modeling study of decalin oxidation and pyrolysis over a wide range of conditions. *Proc. Combust. Inst.* **2013**, *34*, 289–296. [CrossRef]
30. Yu, L.; Wu, Z.; Qiu, Y.; Qian, Y.; Mao, Y.; Lu, X. Ignition delay times of decalin over low-to-intermediate temperature ranges: Rapid compression machine measurement and modeling study. *Combust. Flame* **2018**, *196*, 160–173. [CrossRef]
31. Chang, Y.; Jia, M.; Liu, Y.; Li, Y.; Xie, M.; Yin, H. Application of a Decoupling Methodology for Development of Skeletal Oxidation Mechanisms for Heavy n-Alkanes from n-Octane to n-Hexadecane. *Energy Fuels* **2013**, *27*, 3467–3479. [CrossRef]
32. Fan, W.; Jia, M.; Chang, Y.; Xie, M. Understanding the relationship between cetane number and the ignition delay in shock tubes for different fuels based on a skeletal primary reference fuel (n-hexadecane/iso-cetane) mechanism. *Energy Fuels* **2015**, *29*, 3413–3427. [CrossRef]
33. CHEMKIN-PRO RJI, San Diego, CA. 15112, Reaction Design. 2011. Available online: https://www.ansys.com/products/fluids/ansys-chemkin-pro (accessed on 6 July 2023).
34. Wang, L.; Wu, Z.; Ahmed, A.; Badra, J.A.; Sarathy, S.M.; Roberts, W.L.; Fang, T. Autoignition of direct injection spray of light naphtha, primary reference fuels, gasoline and gasoline surrogate. *Energy* **2019**, *170*, 375–390. [CrossRef]
35. Shen, H.-P.S.; Vanderover, J.; Oehlschlaeger, M.A. A shock tube study of the autoignition of toluene/air mixtures at high pressures. *Proc. Combust. Inst.* **2009**, *32*, 165–172. [CrossRef]
36. Kukkadapu, G.; Kang, D.; Wagnon, S.W.; Zhang, K.; Mehl, M.; Monge-Palacios, M.; Wang, H.; Goldsborough, S.S.; Westbrook, C.K.; Pitz, W.J. Kinetic modeling study of surrogate components for gasoline, jet and diesel fuels: C7-C11 methylated aromatics. *Proceeding Combust. Inst.* **2019**, *37*, 521–529. [CrossRef]
37. Pitz, W.J.; Mueller, C.J. Recent progress in the development of diesel surrogate fuels. *Prog. Energy Combust. Sci.* **2019**, *37*, 330–350. [CrossRef]
38. Li, Y.; Cai, J.; Zhang, L.; Yuan, T.; Zhang, K.; Qi, F. Investigation on chemical structures of premixed toluene flames at low pressure. *Proceeding Combust. Inst.* **2011**, *33*, 593–600. [CrossRef]

39. Yuan, W.; Li, Y.; Dagaut, P.; Yang, J.; Qi, F. Investigation on the pyrolysis and oxidation of toluene over a wide range conditions. I. Flow reactor pyrolysis and jet stirred reactor oxidation. *Combust. Flame* **2015**, *162*, 3–21. [CrossRef]
40. Zhao, F.; Yang, W.; Zhou, D.; Yu, W.; Li, J.; Tay, K.L. Numerical modelling of soot formation and oxidation using phenomenological soot modelling approach in a dual-fueled compression ignition engine. *Fuel* **2017**, *188*, 382–389. [CrossRef]
41. Davis, S.; Wang, H.; Breinsky, K.; Law, C.K. Laminar flame speeds and oxidation kinetics of benene-air and toluene-air flames. Conference Laminar flame speeds and oxidation kinetics of benene-air and toluene-air flames. *Symp. (Int.) Combust.* **1996**, *26*, 1025–1033. [CrossRef]
42. Johnston, R.; Farrell, J. Laminar burning velocities and Markstein lengths of aromatics at elevated temperature and pressure. *Proc. Combust. Inst.* **2005**, *30*, 217–224. [CrossRef]
43. Zhu, Y.; Davidson, D.; Hanson, R. Pyrolysis and oxidation of decalin at elevated pressures: A shock-tube study. *Combust. Flame* **2014**, *161*, 371–383. [CrossRef]
44. Oehlschlaeger, M.A.; Shen, H.-P.S.; Frassoldati, A.; Pierucci, S.; Ranzi, E. Experimental and Kinetic Modeling Study of the Pyrolysis and Oxidation of Decalin. *Energy Fuels* **2009**, *23*, 1464–1472. [CrossRef]
45. Curran, H.; Gaffuri, P.; Pitz, W.; Westbrook, C. A Comprehensive Modeling Study of n-Heptane Oxidation. *Combust. Flame* **1998**, *114*, 149–177. [CrossRef]
46. Zeng, M.; Li, Y.; Yuan, W.; Li, T.; Wang, Y.; Zhou, Z.; Zhang, L.; Qi, F. Experimental and kinetic modeling study of laminar premixed decalin flames. *Combust. Flame* **2017**, *36*, 1193–1202. [CrossRef]
47. Li, B.; Zhang, H.; Egolfopoulos, F.N. Laminar flame propagation of atmospheric iso-cetane/air and decalin/air mixtures. *Combust. Flame* **2014**, *161*, 154–161. [CrossRef]
48. Vasu, S.; Davidson, D.; Hong, Z.; Vasudevan, V.; Hanson, R. n-Dodecane oxidation at high-pressures: Measurements of ignition delay times and OH concentration time-histories. *Proc. Combust. Inst.* **2009**, 32, 173–180. [CrossRef]
49. Shen, H.-P.S.; Steinberg, J.; Vanderover, J.; Oehlschlaeger, M.A. A Shock Tube Study of the Ignition of n-Heptane, n-Decane, n-Dodecane, and n-Tetradecane at Elevated Pressures. *Energy Fuels* **2009**, *23*, 2482–2489. [CrossRef]
50. Haylett, D.R.; Davidson, D.F.; Hanson, R.K. Ignition delay times of low-vapor-pressure fuels measured using an aerosol shock tube. *Combust. Flame* **2012**, *159*, 552–561. [CrossRef]
51. Kumar, K.; Sung, C.J. Laminar flame speeds and extinction limits of preheated n-decane/O2/N2 and n-dodecane/O2/N2 mix-tures. *Combust. Flame* **2007**, *151*, 209–224. [CrossRef]
52. Oehlschlaeger, M.A.; Steinberg, J.; Westbrook, C.K.; Pitz, W.J. The autoignition of iso-cetane at high to moderate temperatures and elevated pressures: Shock tube experiments and kinetic modeling. *Combust. Flame* **2009**, *156*, 2165–2172. [CrossRef]
53. Dagaut, P.; Hadj-Ali, K. Chemical Kinetic Study of the Oxidation of Isocetane (2,2,4,4,6,8,8-Heptamethylnonane) in a Jet-stirred Reactor: Experimental and Modeling. *Energy Fuels* **2009**, *23*, 2389–2395. [CrossRef]
54. Wang, H.; Oehlschlaeger, M.A. Autoignition studies of conventional and Fischer–Tropsch jet fuels. *Fuel* **2012**, *98*, 249–258. [CrossRef]
55. Li, J.; Zhao, Z.; Kazakov, A.; Chaos, M.; Dryer, F.L.; Scire, J.J., Jr. A comprehensive kinetic mechanism for CO, CH_2O, and CH3OH combustion. *Int. J. Chem. Kinet.* **2007**, *39*, 109–136. [CrossRef]

World Electric Vehicle Journal

MDPI

Article

Reasons Why Only Kuwaiti Citizens Drive Electric Vehicles despite Being Only a Quarter of the Population

Sumayya Banna [1], Andri Ottesen [2,*] and Basil Alzougool [1]

[1] LSE Middle East Centre (MEC), Arab Open University (AOU), Ardiya P.O. Box 830, Kuwait; sumayya@aou.edu.kw (S.B.); balzougool@aou.edu.kw (B.A.)

[2] LSE Middle East Centre (MEC)—Sustainability Research and Consultancy (CSRC), Australian University (AU), West Mishref P.O. Box 1411, Kuwait

* Correspondence: a.ottesen@au.edu.kw

Abstract: The adoption of fully battery-powered electric vehicles (EVs) in Kuwait is currently at less than one percent of the population, and hence Kuwait has the lowest index rank among countries around the globe. Effectively, only Kuwaiti nationals have the opportunity to own EVs, as there are no fast direct-current charging stations in Kuwait, and Kuwaiti landlords (expats cannot own real estate) do not allow the installation of EV-charging home wall boxes at their rented properties. Given these facts, this paper only focuses on the preferences of Kuwaiti nationals, even though they only constitute 25 percent of the population of Kuwait. To accomplish the present objectives, a quantitative descriptive method (closed-ended questions) was used to collect data from a sample of 227 Kuwaiti nationals who were representative of the owners of half a million internal combustion engine (ICE) cars and that were categorized as early majority consumers. The findings of the present study indicate that over 50 percent of the respondents would prefer to buy an EV in the following three years when certain criteria were satisfied, including government-controlled pricing policies and recharging point availability, high-speed roads, and free EV-dedicated parking spaces. Furthermore, over 40 percent of respondents stated that they would contemplate purchasing an electric vehicle if the price of gasoline or diesel increased by 19 to 50 percent. The findings also indicated that more than 40 percent of respondents believed that EVs are fire- and crash-safe, and roughly 50 percent of the respondents would be willing to pay between 6 and 20% more for an EV because they believe that EVs are ecofriendlier vehicles and are significantly faster than conventional petrol vehicles. Additionally, respondents rewarded those vehicles with an excellent mark because of their ecological, economical, and technological attributes and benefits.

Keywords: electric vehicles; green transport; carbon footprint; Kuwaiti national customer preference

Citation: Banna, S.; Ottesen, A.; Alzougool, B. Reasons Why Only Kuwaiti Citizens Drive Electric Vehicles despite Being Only a Quarter of the Population. *World Electr. Veh. J.* **2023**, *14*, 287. https://doi.org/10.3390/wevj14100287

Academic Editors: Wenbin Yu and Guang Zeng

Received: 27 August 2023
Revised: 21 September 2023
Accepted: 28 September 2023
Published: 10 October 2023

1. Introduction and Significance

The Pathway for the Reduction of Greenhouse Gas (GHG) Emissions in Kuwait

In 2021, Kuwait emitted 25 tons of CO_2 per capita, which was the third highest in the world after its neighbors Qatar and Bahrain (36 and 27 tons of CO_2 per capita, respectively). This made Kuwait's carbon emissions more than four times higher than the average in 28 European Union countries (6.1 tons of CO_2 per capita) [1–3]. The State of Kuwait has pledged to the United Nations to reduce GHG emissions by adopting a "low carbon equivalent emission economy". In its first NDC, submitted in April 2018, Kuwait outlined its action plan to reduce greenhouse gas emissions via launching its Shayaya Renewable Energy Park, a 3.2 GWe renewable solar and wind energy compound, and the creation of a mass transit (metro) system that has yet to be realized 5 years later. There are some issues of consistency and clarity in the Kuwaiti government's pledge to reduce greenhouse gas emissions and its action plan, as "there is no discussion of what this pledge entails or how it should be implemented" [4,5]. The largest emitting sector is electricity generation

(58 percent), which is primarily used for air cooling through the use of air conditioning and for the desalination of saline water in the world's hottest and driest climate conditions. The oil and gas sector is responsible for 11% of GHG emissions, whereas the transport segment of the economy is accountable for 18% of GHG emissions [6,7]. This paper aims to provide a viable path for GHG reduction by greening the transportation sector, and specifically, by transitioning from internal combustion engine (ICE) automobiles to electric vehicles (EVs) [8].

Currently, the total ownership of EVs is only about 400 vehicles (about 300 registered in Kuwait and about 100 that are estimated to drive there using a foreign license plate) [9]. This is extremely low, especially compared to Norway, the nation with the highest rate of EV adoption and with a similar population to Kuwait; Norway has over 650,000 of fully electric passenger cars and another about 200,000 Plug-In-Hybrids on the road, that is, more than one out of every four passenger cars [10]. Hence, Kuwaiti EV adoptions are 1500 times lower than the best-in-class. At the heart of this low adoption is the lack of DC fast-charging stations; Norway has about 22,000 public charging stations compared to 42 slow alternating current (AC) and no-fast-charging direct current (DC) stations in Kuwait [11,12]. What even further limits electromobility adoption, according to our interviews with all the EV car dealers and 12 current owners of EVs in Kuwait, is that Kuwaiti landlords will not allow installations of home wall boxes for their tenants, who are mostly expats who are prohibited by law to own real estate. A total lack of fast-charging DC stations and an effective ban on the installation of home-charging equipment effectively results in only Kuwaiti nationals who own their own house being able to own and operate EVs in Kuwait.

2. Background

2.1. Lowering GHG Emissions in the Transportation Sector

Only 0.2% of transportation in Kuwait is via public transport, primarily due to a lack of bus infrastructure, cultural conventions, and extreme climate conditions. There have been no campaigns or other government initiatives to promote or encourage the general public's use of public transportation [9]. Private automobiles are predominantly utilized by the majority of Kuwaiti citizens, and hence appear to be the most prevalent mode of transportation. Only about 600,000 of the 2.3 million vehicles registered in Kuwait belong to expats, who account for 70% of residents versus 30% being citizens [10]. In addition, approximately 90 percent of the 295 EVs registered in Kuwait between 2019 and March 2023 were owned by Kuwaitis [11]. One of the reasons why expats in Kuwait do not purchase electric vehicles is the absence of fast-charging stations and the refusal of landlords to enable tenants to install EV-charging wall boxes in rented properties. Kuwaiti citizens tend to live in their own houses as a result of various government programs, whereas expats are prohibited from owning real estate in Kuwait; as a result, most expatriates are forced to rent apartments and must obtain approval from their landlords for the installation of charging equipment, which is typically denied [12,13].

2.2. The Reason Electric Vehicles Are Optimal for Zero-Emission Transportation

In terms of vehicle efficiency, electric vehicles are the best adapted to reduce the carbon footprint and they have the lowest operating expenses. Mr. Abu Dagga, the director of Powerid Germany, believes that e-mobility is the key to sustainability and the high ground against climate change. The primary reason is that e-mobility is superior to ICE transportation, regardless of whether the fuel source is petroleum, diesel, or natural gas, particularly when energy losses are considered. Front-end well-to-tank investigations reveal that EVs lose approximately 6% of their energy compared to 45% in petroleum-powered vehicles [14,15]. Additionally, rear-end tank-wheels studies reveal an additional 17% energy loss for EVs compared to a 35% energy loss for diesel-powered vehicles, and a 39% energy loss for gasoline-powered vehicles, where the majority of that energy is lost as excess heat that is not converted into energy for the wheels. EVs are up to five times more effective than gasoline-powered internal combustion engine (ICE) vehicles, proving their superiority in reducing greenhouse gas emissions (See Figure 1). The efficiency of EVs over ICEs is today three times more, but will be four times more efficient in 20 years, as EV batteries will become lighter and more efficient in terms of energy storage and transfer with new technologies, such as the solid-state batteries and EV batteries having a second life as energy storage for wind and solar farms [16]. Although EV batteries are generally disposed of with the car or replaced after their charging capacity goes below the warranty terms, which are usually from 70 to 75%, they present a cheap option in making up battery packs in a container system, as they are repurposed to store solar energy overnight when the sun is not shining. A good example of that is the 25 MWh grid-scale energy-storage system in California, where approximately 1300 disposed Honda and Nissan EV batteries are strung together in the B2U SEPV Sierra hybrid solar-storage facility, which uses slower charging time than the high-performance EVs [17,18]. The cost of installing such a system is USD 200 per kilowatt hour, as the batteries do not need refurbishing [19]. Lithium recycling factories are also coming fast online, such as the Li Cycle Rochester facility where the company claims to be able to recapture up to 95% of the battery resources for recycling using a closed-system water-based solution where the minerals are captured in a "black mass" that is then reseparated into battery-grade lithium, cobalt, and nickel, as well as other materials to be reused [20]. However, according to Powerid [16], hydrogen is not as suitable for smaller vehicles as it is for heavy-duty transport or construction vehicles, large sea vessels, or even aircrafts, due to its lack of energy efficiency, storage, and transportation complications, and a flame speed that is 10 times that of methane gas and thus more explosive [21,22]. This view is in stark contrast to that of Toyota and other Japanese automakers, who favor hydrogen as the primary source of zero-emission technology due to its rapid charging and long range, as they appear to be losing the EV world market [23].

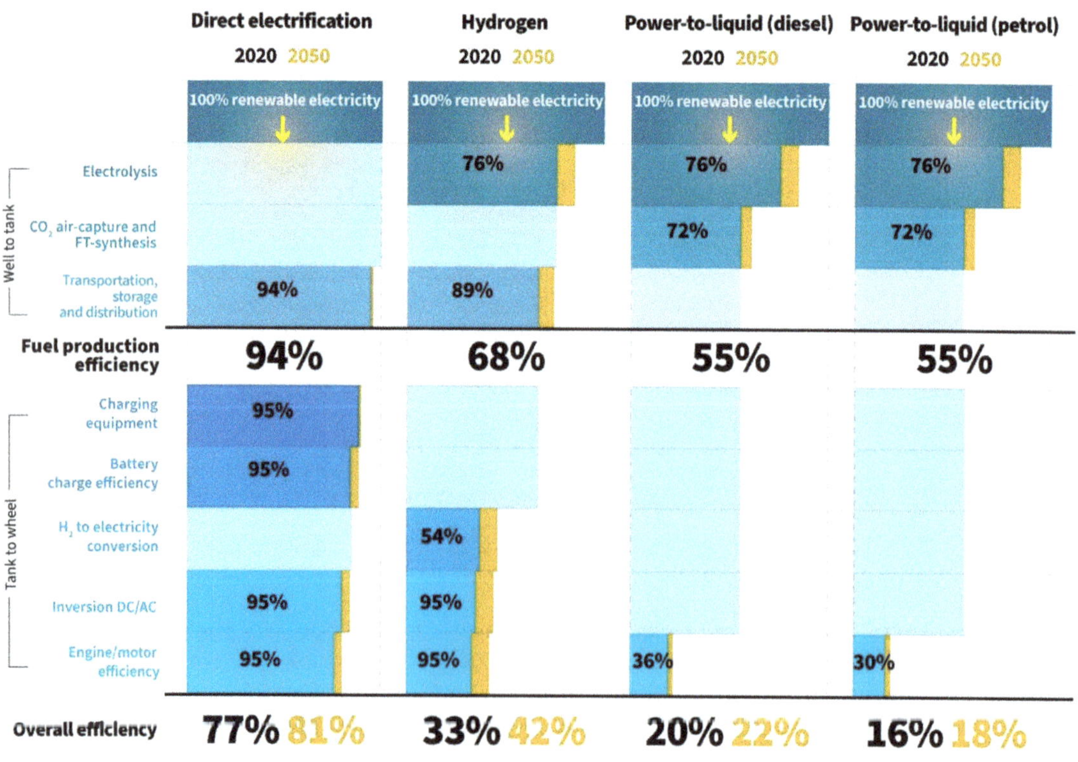

Figure 1. Vehicle fuel energy efficiency; source: Powerid. Printed with permission [16].

2.3. The Oil Savings for Each Electric Vehicle

With each transition from internal combustion engines (ICEs) to EVs, oil that would have been used for combustion is saved. This equates to approximately 10 barrels of oil equivalent (BOI) per year for a midsize car, one barrel for a motorcycle, 244 barrels for an A-class vehicle, and 274 barrels for a bus [24]. By the end of 2022, over 20 million electric vehicles will have been sold [25]. The transition from ICEs to EVs is anticipated to conserve 2.5 m petrol barrels per day by 2025, which is roughly equivalent to the everyday petrol production of Kuwait, the tenth-largest oil producer [26,27]. BloombergNEF predicts that the demand for oil will plateau in 2026, and will start to decline in 2027 as a result of the conversion from ICEs to EVs for all vehicle types. The Tesla Model-Y just took over the Toyota Corolla as the world's sales-record-breaking automobile for the first quarter for year 2023, with 267,200 vehicles sold compared to 256,400 for the Toyota Corolla [28].

2.4. Low Maintenance of Electric Vehicles

EVs have lower maintenance costs than internal combustion engine (ICE) vehicles due to the factors listed below. EVs have significantly fewer mechanical components or moving parts than ICE vehicles, with approximately 20 moving parts for EVs compared to approximately 2000 parts for ICE vehicles [29]. Fewer components translate to fewer items that can fail or wear out over time, resulting in lower maintenance costs. The US Energy Department (Office of Energy Efficiency and Renewable Energy) determined in 2021 that maintenance costs for EVs were only 6.1 cents per mile (4 cents per kilometer) compared to ICE costs of 10.1 cents per mile [30]. The following are the EERE study's explanations for this disparity: (A) Unlike ICE vehicles, EVs do not require regular lubricant replacements; (B) Electric vehicles utilize regenerative braking, which pauses the vehicle by converting the electric motor into a generator to recharge the battery. This reduces the frequency of replacing brake pads and rotors by minimizing system wear and tear. EVs lack exhaust systems, which are susceptible to corrosion and other damage in ICE vehicles. ICE vehicles require regular battery maintenance or replacement, whereas EV batteries are designed to last a very long time, often the life of the vehicle; (C) ICE vehicles require more frequent flushing of the refrigerant and other maintenance on the refrigeration system. EVs typically have less stringent cooling requirements, which further reduces their maintenance needs; (D) EVs have no gasoline pumps, fuel injectors, or fuel containers to maintain and replace; (E) Due to their simplicity and lack of moving parts, electric motors are exceptionally durable and can outlast the rest of the vehicle with minimal or no maintenance; (F) EVs have a gearless powertrain and, as a result, no (automatic) transmission (AT), which in ICE vehicles often requires costly maintenance. Altogether, these factors can save a substantial amount of money over the tenure of an EV versus an ICE vehicle.

2.5. The Durability of the EV Battery Dependence

Evidently, while routine maintenance on EVs is typically much less expensive, replacing the EV battery after the warranty expires after 8–10 years can be quite costly (Kia/Hyundai and Mercedes offer 10 years and lifetime warranties for the Hyundai Ioniq). The battery replacement cost (USD 5000–15,000) may determine the vehicle's lifespan, which is approximately 13 years for ICE vehicles [14]. EV resale value depreciation versus ICE resale value depreciation may represent the greatest operational cost if the battery does not last. Fortunately, the majority of studies indicate that EV batteries will outlive the vehicle, with over 70% of their charging capacity remaining after 13 years of use. One can anticipate a loss of 2.3% of range per charge for every year driven, which equates to approximately 13 years or 240,000 km driven before the charging capacity falls below 70% of its initial capacity [31]. According to industry sources, if the original battery is large enough, a 50% charge may suffice for the majority of usage, extending the EV's lifespan to 20 years. The Nissan Leaf has been on the market for thirteen years, and a significant number of original vehicles are still on the road. Tesla also reports empirical data of average EV utilization of 300,000 km in the United States and 244,000 km in Europe [32]. Tony Seba, a Stanford technology forecaster, states in a recent report by RethinkX from March 2023 that the longevity and autonomy of EVs will be such that sales of new automobiles will plummet by 75% over the next 20 years, giving some hope that EV depreciation will not outweigh ICEs and that operational costs (OPEX) will be lower [33]. He also predicts that the EVs will take over the ICEs much faster than predicted based on his S-curve cost-lowering predictions for emerging technology, and that Chinese-made EVs will take over as the mainstream consumer choice in the very near future because of value for money and technology and quality improvements while lowering the cost per MW of charging capacity in production of the EV battery.

On the other hand, it is a known fact that EV batteries degrade more quickly, and the majority of studies on the durability of EV batteries were conducted in Europe or North America. The Kuwait Institute of Scientific Research (KISR), the national laboratory of Kuwait, has conducted its own research with promising results [34,35]. However, due to the lack of years that EVs have existed in Kuwait, real-world empirical data are still deficient and do not provide conclusive evidence regarding how long EV batteries will last in extreme heat conditions. The Kingdom of Saudi Arabia (KSA) recently bought a controlling stake in US Lucid Motors with the intention of producing a quarter of the production in KSA or a total of half a million vehicles by 2030 and the first car by 2025. Similarly, the Investment Promotion Agency Qatar is exploring if the country can be an EV manufacturing hub for the Middle Eastern (Arab) region in cooperation with German Volkswagen and Chinese Gaussin and Yutong, with whom the Qatari government has already partnered [36]. It is clear that the issue of EV driving with atmospheric heat reaching above 50 °C has to be dealt with both in terms of battery cooling and durability of the battery. In normal driving conditions for EVs, which are between 10 °C and 40 °C, the battery is at its nominal use capacity; at higher temperatures, the use capacity decreases, and between 60 °C and 70 °C the battery could sustain permanent damage [37].

2.6. EV Policies of the Kuwait Government as Opposed to Qatar

The Government of Kuwait will have to set goals of its transitions from conventional ICE vehicles to EVs, which in turn can offer a tremendous opportunity to reduce GHG emissions. The government of the State of Kuwait has pledged to the UN to reduce its carbon footprint as a vital share of its own national vision of 2035, which in particular contains a sustainability objective that promotes urgent actions to be implemented to fight off climate change effects [38,39]. However, conversion to EVs is not mentioned in these documents.

Qatar's national vision, much like the Kuwaiti Vision 2035, serves as a guide for the country's social, economic, environmental, and human development. However, in Qatar's vision, commitment has already been made to achieving a 25% reduction in greenhouse gas emissions by 2030. The Qataris plan to achieve this goal by converting all public transportation to electric vehicles, as well as giving significant governmental supports to various initiatives to transition to green mobility. Qatar's 'Electric Vehicle Strategy 2021' anticipates that EV sales will account for 10% of total vehicle sales in Qatar by 2030, contributing to the CO_2 reduction goals of the Paris Agreement [40].

2.7. Fast-Charging Stations in Kuwait and Qatar

Apart from the need for clear governmental goals on EV transformation, the automotive dealers in Kuwait state that the biggest hurdle for rapid mass adoption of EVs is the lack of fast-charging stations. Currently, there are no fast direct current (DC) to DC charging stations installed in Kuwait, and only 38 slow alternating current (AC) to DC charging stations spread around urban areas of Kuwait. The EV-charging facility provider ChargedKw recommends as a start to install one fast-charging station at every of the 150 gas stations in Kuwait, but due to a backlog of orders with the suppliers, the bidding process completion of this installment is likely to take 2.5 years from the initial tender, and the likely cost will be around KD 2.3 million, which is equivalent to nearly USD 8 million [41,42].

Qatar, which has half the land mass and population of Kuwait and lies 800 km south of Kuwait, is considerably father ahead in the installation of fast charging stations. As of the end of 2022, there were 100 such stations installed, and 150 installations are scheduled for 2023. Mohamed Khalid al-Sharshani, the head of the technical section at Tarsheed, which is the National Program for Conservation and Energy Efficiency, states that 600–1000 such stations will be installed by 2025 [43,44]. The rise of the fast-charging station has spurred the electrification of public transport: "We have crossed the target of converting 25 percent of the public transportation into eco-friendly mode. The public and school transportation will be completely eco-friendly by 2030", states the Minister of Transport Jassim Saif Ahmed Al Sulaiti [45,46]. Apart from aiming for 100% electric public transport by 2030, Qatar authorities still have a modest goal of transforming only 10% of all new sales in all sectors of transportation to electric, as many will be reluctant to change from their ICE vehicle given the low price of gasoline of QR 2 per liter, or about USD C 50 [47].

3. Literature Review

In general, the existing literature suggests that successful transformations and transitions are required to increase the adoption of sustainable e-mobility and surmount barriers to its adoption. While practitioners, academics, and researchers have studied the EV phenomenon on an international scale and with country-specific evidence, limited evidence has been found, however, from an emerging market of Kuwait. Apparently, there is insufficient research that has been conducted among drivers of ICE vehicles that aims to know their future stated preferences for EVs. This study had three goals: (i) to examine Kuwaiti ICE vehicle drivers' indicated perceptions of EV attitudes; (ii) to identify their preferable EV features; (iii) to determine their prospective EV purchase requirements and conditions. This research is part of a larger initiative that examines the EV landscape in Kuwait and utilizes a mixed-methods approach [48].

In summary, the recent EV literature reviews emphasized the following stated attributes for EVs:

(1) Economic: upfront purchasing prices, maintenance and repairs cost, protection and insurance cost, charging costs [49,50], and guarantees costs [51,52];
(2) Technology characteristics: battery, range, top speed, and acceleration [53–55];
(3) Infrastructure characteristics, including sluggish and rapid-charge networks, commercial and charging public infrastructures, and home-recharging infrastructures [56];
(4) Financial, nonfinancial, and social attributes: free parking spots, price reduction, government subsidy policies, health policies and safety policies [57], tax-discount policies, and penalty policies for petrol-fueled vehicles [58];
(5) Brand attributes: design, brand reputation, and credibility [59].

Interestingly, few studies were found that aimed to investigate these aforementioned attributes. In the following respects, the present study complements the current literature as follows: first, the current study applies five categories of attributes, including economic, technical, infrastructure, brand, and financial and nonfinancial policy attributes, collectively and exhaustively on a single set of options for EV preferences. Importantly, the attributes have not been investigated collectively in a single study; rather, they have been analyzed in various existing studies. Second, indeed, the present investigation focuses on conventional automobile drivers that have never been investigated. Lastly, the current research closes a void in the literature on EV trends in terms of demographics and stated preferences through a large sample survey collected in Kuwait. As the six gulf countries share similar societies, cultures, economies, government policies, climates, and geography, the present study contributes to the growing body of knowledge that can be applied to consumers in other GCC countries—Oman, the Saudi Arabian Peninsula, along with Bahrain, Qatar, and the UAE—regarding markets and stated preferences for EVs. Most studies focused on markets in developed nations such as Europe, the United States, and Australia. For instance, Guerra and Daziano analyzed a large sample of respondents' EV purchase motivations in the United States. Consumers were generally willing to pay more for longer-ranging vehicles with shortened recharging periods, reduced operating costs due to scheduled and unscheduled maintenance, and shorter parking and charge facility search times, according to their findings [60].

By contrast, ICE vehicles continue to be perceived as more comfortable in terms of design and appearance than EVs, particularly among senior consumers. Moreover, women were discovered to be more environmentally conscious than males. Ziemba discovered that, in Poland, consumer EV acceptability is most strongly influenced by technological standards [61]. For instance, Vilchez conducted a large-scale survey among vehicle owners in a number of European nations, including Italy, France, Germany, Spain, Poland, and the UK, and found that half of the participants did not prefer EVs due to the higher initial capital cost (CAPEX)—a higher cost than conventional automobiles [62]. Fortunately, a recent study conducted by Temple found that the production of new batteries (lithium-metal batteries) could make EVs as affordable and convenient as gasoline-powered vehicles, indicating an anticipated increase in the adoption rates among car drivers worldwide as a result of technological development [63].

Kowalska-Pyzalska examined consumer preferences for alternative-fuel vehicles or electric vehicles (EVs) in Central and Eastern European nations and found that safety attributes are one of the most sought-after characteristics of EVs, followed by price, range, and brand [64]. Their findings require automakers and policymakers to actively promote electric vehicles in order to increase their adoption rates. Chao Ma analyzed the preferences of Chinese consumers for electric vehicles and found that purchase price, vehicle category, and fast-charging batteries were the most influential factors. Government policy is crucial in promoting the adoption of electric vehicles [65].

4. Methods

This study employed a quantitative descriptive approach consisting of mainly closed-ended questions to accomplish its research objectives by designing a large-scale questionnaire survey to provide a comprehensive, all-inclusive perspective on EVs in Kuwait. The first section of the extensive survey centered on respondents' demographic information, such as their gender, age, level of education, income ranges, occupation, ethnicity, and vehicle ownership. In order to measure the respondents' approach to willingness-to-buy or pay, the second section of the extensive survey consisted of exhaustive questionnaires about EV features, factors, and attributes that have been intensively discussed by the majority of the existing and updated strands of the literature. In addition, the large-scale survey was distributed to and evaluated by 50 full-time instructors and part-time educators, who provided their views and recommendations. The assessment phase comprised inquiries regarding objective, content, and layout of the questionnaire. We received 30 completed evaluations in total as part of the validation process. The overall response rate was near 60%. We subsequently incorporated the suggested modifications and revised the questionnaire's format and overall design proposed mainly by AOU instructors. To make the questionnaire easier to complete, we omitted several sections and treated it as a unit. In addition, we added the response option "I don't know" in order to collect honest responses and permit the free expression of opinions. The evaluation revealed that approximately 96% of the questionnaire's content is unambiguous, informative, and simple to understand. Similarly, 93 percent of questionnaire objectives are clearly specified, the content is relevant, and the order is logical. Of the respondents, 90 percent indicated that the content is correct and complete, and the layout is workable, and 86 percent agreed that the structure is straightforward. We developed two variants of the questionnaire; one version was written in the Arabic language and another in the English language. We used Google Forms to formulate the survey because it is one of the commonly used, validated tools for data collection. The questionnaire was then disseminated to diverse groups in Kuwait, including the general public, faculty members, students, and tutors. Participants had to be Kuwaiti citizens and 18 years of age and also hold a driver's license and a car in their own or family name in order to be eligible for the survey. We proposed this condition as many Kuwaitis use private drivers to get around and do not possess a driving license. We believe that purchasing behavior for a car that is driven by a driver is different than purchasing behavior for a car that is driven by the owner. We started the data-collection phase from February to May of 2022. A technique of random sampling was utilized during the data-collection phase [66]. The researchers distributed the survey links by randomly selecting the target audience from diverse groups of individual nationals in Kuwait inclusive of mature students, instructors, professionals, and others. Moreover, the study sample included a few Kuwait individuals with temporary or permanent car-ownership contracts such as rental or a company car. Those were contractual employees who were provided with ICE automobiles from their employing companies with petrol paid.

Initially, we collected a large sample size (604) in order to obtain comprehensive demographic data representative of the population in Kuwait and to draw actionable conclusions about the population. Subsequently, we eliminated around 132 surveys because our aim was not to include those who never drive a car or had no driving license, and we ended up with a total of 472 questionnaires that were analyzed, and these data were used for our current analysis. However, after observing how Kuwaiti nationals had different EV purchasing behavior from the expat population, due to strict rules and restrictions expats have to adhere to in order to obtain a driver's license within the State of Kuwait, as well as expats not being allowed to own real estate and landlords not permitting installation of home charging for EVs, and as a result almost all EV buyers to date have been Kuwaiti nationals, after careful selection, we isolated the Kuwait nationals in our sample and obtained a reduced sample of 227 Kuwaiti nationals who owned and drove conventional vehicles.

At the beginning of the data-collection phase, the purpose and aims of the study were introduced to the participants and thoroughly explained. Additionally, instructions were written on the front page to prevent confusion. Then, we invited participants to complete the survey online. Finally, descriptive statistics—frequencies and percentages—were computed using SPSS 19, applying t-test and AVOVA tables for in-depth analysis and standard error analysis. Note that this particular endeavor has certain limitations. Because the sample was gathered from a single location, the Arab Open University in Kuwait, we are precluded from generalizing to the context of all Kuwaiti drivers. In fact, additional research with a larger sample size is required on this specific topic. For example, in-depth investigations are necessary to uncover concealed insights regarding the acceptance and preferences associated with EV use among EV owners. One of the requirements for participation in the survey was automobile ownership. In 2021, approximately 3 million vehicles were registered in the country [14]. Excluding company-owned and expat-registered vehicles, we estimate that approximately 1.6 million registered vehicles were owned by Kuwaiti citizens. Here, we examined prospective buyers and, hence, we estimated that forty percent of the population, or 6,400,000, were included in the group of prospective Kuwaiti drivers. The sample consisted of 227 Kuwaiti drivers and vehicle proprietors, or approximately 0.04 percent of the total representative population, which is modest and, if not for our resource limitations, would have ideally included at least 1000 participants. Consequently, our margin of error is approximately ±7 percent, which translates to a 95% confidence level [67].

5. Findings

This investigation included the participation of two-hundred twenty-seven Kuwaiti ICE owners who possessed driving licenses. Table 1 provides a demographic characteristics summary. The questionnaire was completed equally by men (50.2%) and women (49.8%). More than three-quarters of the sample were between 26 years old and 39 years old (76.2%). Over one-third were unmarried (37.4%), while 25.6% were married with at least three children. Over fifty percent held bachelor's degrees. More than a fourth of respondents owned five or more automobiles (40.1%), and more than a quarter of them owned two automobiles (28.2%). More than half of the participants (53.3%), versus slightly above a quarter (26%), were working in the public and private sectors, respectively. Over one-third of the sample (39.6%) worked in the public and government ministries sector, and a quarter (25.6%) were intermediate managers. A total of 32.6% of the participants had a monthly income of between KD 1000 and 1499.

Table 1. Demographic characteristics summary.

Variable	Categories	N = 227	%
Gender	Male	114	50.2%
	Female	113	49.8%
Age Range	26–39 years	173	76.2%
	40–49 years	42	18.5%
	50–60 years	12	5.3%
Marital Status	Single	85	37.4%
	Married without kids	24	10.6%
	Married with 1 kid	26	11.5%
	Married with 2 kids	34	15.0%
	Married with 3 kids or more	58	25.6%
Number of Cars in household	One car	12	5.3%
	Two cars	64	28.2%
	Three cars	27	11.9%
	Four cars	33	14.5%
	Five cars & more	91	40.1%
Educational Level	Less than high school	4	1.8%
	High School diploma	46	20.3%
	Trade/Commerce degree	39	17.2%
	Bachelor's degree	114	50.2%
	Master's degree	19	8.4%
	PhD	5	2.2%
Employment	Public sector	121	53.3%
	Private sector	59	26.0%
	Unemployed	21	9.3%
	Self-employed	14	6.2%
	Family-owned business	12	5.3%
Field of employments	Government & Ministries	90	39.6%
	Other private services	32	14.1%
	Educations-government & private	25	11.0%
	Oil and Gas sector	23	10.1%
	Large Kuwaiti corporation	20	8.8%
	My family business	19	8.4%
	Military or police	11	4.8%
	Health Care–government & private	7	3.1%

Table 1. *Cont.*

Variable	Categories	N = 227	%
Which of the following best describes your role in industry?	Middle Management	58	25.6%
	Administrative Staff	46	20.3%
	Upper Management	29	12.8%
	Lower Management	22	9.7%
	Support Staff	15	6.6%
	Temporary Employee	12	5.3%
	Trained Professional expert	12	5.3%
	Self-employed/Business Partner	11	4.8%
	Student-Not working	7	3.1%
	Researcher	6	2.6%
	Skilled Laborer	5	2.2%
	A Consultant	4	1.8%
Monthly Income	Less than 500 KD	22	9.7%
	500–999 KD	44	19.4%
	1000–1499 KD	74	32.6%
	1500–1999 KD	50	22.0%
	2000 and above	37	16.3%

Table 2 displays the participants' perspectives on EVs. More than forty percent of respondents (41.8%) stated that they would be willing to pay an extra 6–20%, while one-third (33.5%) stated that they would pay no more than 5% more. More than one-third of participants (37.8%) would pay 6–20% extra for EVs, which they viewed as significantly faster than a corresponding ICE vehicle (0–100 km per hour in 4 s), while approximately one-fifth (19.4%) declared that they would pay nothing extra. More than 42% of participants (42.7% to be exact) indicated that they would seriously contemplate purchasing EVs if gasoline/fuel prices would go up by 50–199%, while about 18.5% were indifferent to gasoline price changes. Over 50% (54.7%) indicated that they would contemplate purchasing an EV if and only if the government controlled the prices of EVs and offered a price reduction, making them 10–30% less expensive than gasoline vehicles, whereas roughly one-fifth (18.9%) suggested that they would change their minds if the prices were comparable to gasoline vehicles. A third of participants (29.1%) would consider purchasing EVs under the condition of the availability of rapid-charging points every 10–25 km, and a quarter (24.2%) would consider purchasing EVs if there were fast-charging stations every 26–50 km.

More than half of respondents (57.7%) responded that they would reconsider purchasing an electric vehicle under the condition of the availability of a dedicated express lane located on major highways. More than 50% (53.3%) indicated that they would reconsider purchasing an EV under the condition of the availability of free public parking spaces for EVs in an amount comparable to handicapped spaces. Also, 56% said they would purchase an EV within the next three years. Of these, 14.1% suggested that they would definitely purchase an EV, while 41.9% said that they would likely do so. A total of 46.2 percent cared about the security of EVs regarding fire and accidents; 43.2% were concerned about being able to recharge their EVs around their residential areas; however, 30.8 percent stated that they would not be able to charge their EVs in their residential areas.

Table 2. Attitudes and preferences.

Question	Categories	N = 227	%
Preferred price increases for EV	0%	42	18.5%
	1–5%	34	15.0%
	6–10%	55	24.2%
	11–20%	40	17.6%
	21–29%	22	9.7%
	30% and more	34	15.0%
Preferred speed range (0–100 kmh in 4 s)?	0%	44	19.4%
	1–5%	40	17.6%
	6–10%	43	18.9%
	11–20%	43	18.9%
	21–29%	29	12.8%
	30% and more	28	12.3%
Preferred price increase brackets for gasoline cars that would push you to think of buying an EV.	500% and above	8	3.5%
	400–499%	8	3.5%
	300–399%	12	5.3%
	200–299%	22	9.7%
	100–199%	53	23.3%
	50–99%	44	19.4%
	Less than 50%	38	16.7%
	Indifferent	42	18.5%
Prefer that the government regulates and controls the purchasing costs of EVs to be __________.	30% cheaper costs than ICE cars	78	34.4%
	10% cheaper costs than ICE cars	46	20.3%
	The same costs as ICE cars	43	18.9%
	10% higher costs than ICE cars	26	11.5%
	Indifferent	34	15.0%
Preferred fast charging stations every ____________.	Less than 10 km	35	15.4%
	10–25 km	66	29.1%
	26–50 km	55	24.2%
	51–75 km	28	12.3%
	76 km and more	11	4.8%
	Indifferent	32	14.1%
Prefer a fast lane dedicated to EVs on major highways (such as on highways *30* and *40*)	No	45	19.8%
	Yes	131	57.7%
	Indifferent	51	22.5%
Preferred public and free parking spaces almost at the same capacity as handicapped spaces.	No	57	25.1%
	Yes	121	53.3%
	Indifferent	49	21.6%
Are you interested on purchasing an EV in the future?	Of course, not	34	15.0%
	Unlikely	66	29.1%
	Very likely	95	41.9%
	Certainly	32	14.1%

Table 2. *Cont.*

Question	Categories	N = 227	%
Level of safety concerns regarding EVs in Kuwait, in terms of fire and car crashes	Hazardous	18	7.9%
	Unsafe	22	9.7%
	Neutral	82	36.1%
	Safe	48	21.1%
	Very Safe	57	25.1%
The possibility of charging your EVs in residential areas in Kuwait	Very impossible	38	16.7%
	Impossible	32	14.1%
	Neutral	59	26.0%
	Possible	31	13.7%
	Very possible	67	29.5%

The following scale shows:

- High agreement: with a calculated mean (M) $\geq$ 2.33;
- Medium agreement: with a calculated mean of 1.67 $\leq$ M < 2.33;
- Low agreement: with a calculated mean < 1.67.

The highest agreement among contributors (Table 3) was regarding three EV-related issues: the environmental friendliness and lower CO_2 emissions of EVs, which leads to improved quality of breathable air (M = 2.43); significantly low fuel price (M = 2.35); and a silent engine (M = 2.33). Features with moderate agreement included improved fire and collision safety (M = 2.31), quicker and better A/C conditioning (M = 2.30), significantly faster acceleration (0 100 km/h) (M = 2.20), and notably reduced maintenance costs (M = 2.19).

Table 3. The most preferred features.

Types of Attributes	The Most Preferred Features	Mean	Std
Environmental	EVs are ecofriendly with low CO_2 emissions, leading to better air quality	2.43	0.780
Financial	Lower price than petrol	2.35	0.797
Technological	Quiet engine	2.33	0.810
Technological	Raised safety concerns—crash and fire	2.31	0.788
Technological	Faster and more powerful air conditioning	2.30	0.781
Technological	Fastest acceleration (0–100 kmh)	2.20	0.804
Financial	Lower maintenance and repair costs	2.19	0.856

Regarding the perceptions of EV purchasing, as indicated in Table 4, the sample highly agreed on the following five conditions: the preferred range of fully charged EVs should be at least 400 km (M = 2.40), the preferred battery guarantee should remain 10 years at least or after 150,000 km (M = 2.39), fast-charging stations must be located within a 5 km radius (M = 2.38), and preference of the coolness and uniqueness of EV designs (M = 2.38); the remaining preferred requirements had moderate agreement, such as spotting a change in air quality once people started to drive EVs (M = 2.32); preferred price ranges should be similar or equivalent to a gasoline car (M = 2.31); purchasing an EV if gasoline prices would increase threefold (M = 2.28); distinct lanes dedicated to EVs on motorways (M = 2.26); that a majority of respondents' friends or family would purchase an EV (M = 2.02).

Table 4. Perceptions regarding EV purchase (N = 227).

Attributes	The Preferred Criteria for Purchasing an EV	Mean	SD
Technological	The driving range of a fully charged EV should be min. 400 km	2.40	0.759
Financial	Battery guarantee should remain 10 years at least or after 150,000 km	2.39	0.770
Infrastructure	Fast-charging stations should be located within a 5 km radius	2.38	0.751
Brand	Coolness and uniqueness of design	2.34	0.767
Financial	Resale value should be on par with or higher than an ICE car	2.33	0.770
Social attributes—pro-environmental	Improvement in air quality	2.32	0.779
Financial	Price should be the same or lower than a gasoline car	2.31	0.832
Financial	Gasoline prices would increase three-fold	2.28	0.809
Infrastructure	Distinct EV lanes located on motorways	2.26	0.790
Social acceptance	Family and friends would buy an EV	2.02	0.801

6. Discussion

The proposed study investigated the attitudes, preferences, and needs of fuel-vehicle drivers regarding EVs. Intriguingly, the study discovered that more than half suggested that they preferred to purchase an electric vehicle over the period of the next three years. Respondents provided the following four conditions: (i) the policymakers and government should regulate and mitigate the cost of electric vehicles by 10 to 30 percent to make them more affordable than gasoline vehicles; (ii) public fast recharging points and facilities should be accessible and spread out at least within a 10 to 50 km radius everywhere in Kuwait; (iii) specific EV fast lanes should be available on the most congested freeways.

Consequently, these findings suggest that potential EVs drivers viewed Kuwaiti regulators as playing a crucial role in promoting EVs' reputation and recognition by planning appropriate EV transportation infrastructure and establishing policies for regulating prices. Intriguingly, our results largely mirror the results of prior research conducted for various nations. In addition, nearly 40 percent of respondents suggested that they contemplated purchasing EVs if gasoline prices increased by 50 to 199 percent in the future, recognizing that EVs are still safer during fires and accidents.

Furthermore, more than a third of respondents suggested that they were willing to pay 6 to 20 percent more for EVs than for petroleum vehicles, believing that EVs are more environmentally friendly and speedier than gasoline vehicles. This intriguing finding suggests that people in Kuwait prefer EVs to petroleum vehicles because EVs offer greater environmental, economic, and technological benefits. Similar results have been found in various countries. However, about a third of residents (mainly expats) in Kuwait's residential locations are unable to charge electric vehicles. This implies that infrastructure for charging EVs must be nearby and readily available in all Kuwaiti inhabited zones. In addition, drivers favored three distinct types of EV attributes, including environmental attributes (i.e., ecological; low CO_2 emissions, which improves air quality), financial or economic attributes (i.e., much cheaper fuel price than gasoline), and technological attributes (i.e., soundless engine), which concurs with this intriguing finding.

By contrast, the current Kuwaiti ICE drivers' sample would apparently be willing to acquire EVs within the near future because they believe that EVs' technological attributes (i.e., driving range might be at least 400 km), financial or economic attributes (i.e., battery guarantee lasts 10 years or more or 150,000 km; resale EV value should be similar to or higher than fueled vehicles), and infrastructure attributes (rapid-recharging points within 5 km) could justify it. This finding suggested that the current Kuwaiti ICE drivers' sample may acquire EVs in the near future if four criteria were met: the batteries' affordability and resale value, the infrastructure's proximity of fast-charging stations, the technological features' range, and the brands with cool and attractive designs. These findings are consistent with previous cases from various nations.

While the current study provides a broader perspective on the EV phenomenon in Kuwait, it does have some limitations. Because it lacks emphasis on empirical testing to provide clearer conclusions about the population, the findings are primarily descriptive and furthermore based on stated (rather than revealed) preferences. Therefore, future research should employ hypothesis testing to reach a definitive general inference in Kuwait. Second, additional research must employ in-depth investigations with existing EV possessors to investigate any obstacles, identify any future opportunities, and identify favorable EV characteristics and preferred designs. One suggestion is to use a focus group to obtain a more complete image of the desired features and services, as well as to determine the most effective incentives for purchasing EVs in Kuwait and the MENA region. In addition, comparative studies from various GCC regions may be of interest because the regions operate under a similar legal and economic umbrella. Lastly, additional research should investigate the management perspectives of car dealerships in Kuwait in order to examine the obstacles that are delaying the adoption and sustained mobility models for electric vehicles (EVs), clarify ambiguity surrounding EV adoptions in Kuwait, and provide a better explanation and rationale for the reluctance to replace conventional vehicles.

7. Conclusions and Implications

Other than eliminating all government subsidies for water, electricity, and petroleum, the mass transition from ICEs to EVs is the most effective strategy for reducing Kuwait's carbon footprint if EVs are charged with electricity from renewable energy sources. However, due to a total lack of rapid-charging EV stations (direct current to direction current, DC2DC) and Kuwaiti landlords not permitting the installation of EV charging-wall receptacles, ICE to EV transition is only an option for Kuwaiti citizens who own their homes. Due to a backlog of orders and formal tendering processes in Kuwait, it will take up to four years to construct an effective fast-charging station network. This study has therefore focused predominantly on Kuwaiti nationals' stated preferences and attitudes toward electric vehicles. The current study presented substantial evidence and diversified stakeholder perspectives for the emerging market of Kuwait. The main goal here is to investigate the sustainability of EVs. The outcome showed that potential Kuwaiti customers anticipate purchasing an EV within the near future—three years—but only under specified criteria, such as the availability and readiness of appropriate infrastructure such as recharging facilities, rapid roads, and free public parking. In addition, they were willing to purchase EVs and strongly preferred EVs for their ecological, economical, and technological attributes, but only under four conditions: battery cost and resale value, accessible fast-recharging facilities (infrastructure), technological features in terms of range per full charge, and brand value and appealing design.

In light of the preceding findings, the existing investigation has both theoretical and practical implications. Theoretically, the present research supports the scant literature on sustainable mobility in developing nations, particularly MENA-GCC and Kuwait. The findings could facilitate broader comparisons and more accurate assessments for developing nations, including those in the MENA-GCC region. In practice, the results of this study indicate that Kuwaitis prefer EVs over petroleum vehicles because EVs offer environmental advantages, economic benefits, and technological advantages. Consequently, vehicle dealership marketing campaigns should emphasize the utility of electric vehicles (EVs) when promoting EVs to their target markets, vehicle commuters and vehicle proprietors. In addition, the results indicate that Kuwaitis may be prepared to purchase EVs if and only if infrastructure related to the availability of rapid recharging stations and facilities, fast traffic lanes, and free parking spaces is in place. Hence, policymakers and government regulators are encouraged to initiate the construction of infrastructure to facilitate the rapid acceptance of electric vehicles in the country.

EVs offer substantial environmental benefits that contribute to promoting the use of renewable energy, reducing both urban air pollution and greenhouse gas (GHG) emissions, and thereby reducing human health hazards associated with GHG exposure. Therefore, we propose that the government execute knowledge policies to promote knowledge plan to better teach the people of Kuwait about preservation of the environment and sustainability issues. In addition, we propose for the policymakers in Kuwait to provide sponsorship and funding programs along with other financial assistance to EV purchasers to prevent the high price of EVs. Moreover, the government should construct accessible and appropriate infrastructure, including a wide network of recharging facilities. Thus, consumers could refuel their EVs from renewable sources of energy and prevent their EVs from running out of electricity. Additionally, highways should be enhanced and better developed for EV drivers than previously.

The study results indicate that Kuwaitis might purchase EVs in the future if infrastructure regarding rapid recharging stations were readily available and easily accessible. Therefore, policymakers must construct and provide these stations to promote the adoption of electric vehicles. Kuwaitis would apparently purchase EVs for economic reasons, particularly regarding battery life and resale value. Therefore, manufacturers of electric vehicles should develop heat-tolerant longer-lasting batteries. As Kuwaitis would apparently purchase EVs due to technological attributes relating to range per complete charge and brand attributes, manufacturers should attempt to design EVs with these characteristics. As EVs are very efficient in generating energy from the source to the wheels and efficient as a tool to reduce greenhouse gas emissions as well as other forms of air pollution, the automobile industry should gradually shift toward their implementation.

Author Contributions: A.O. is the project's Principal Investor (PI) and is responsible for its conception, methodology, and editing, as well as the acquisition and administration of funds. S.B. and B.A. were accountable for the literature review, conceptualization, synthesis, methodology, validation, formal analysis, data curation, writing preparation of the original document, and editing. All authors have read and agreed to the published version of the manuscript.

Funding: This paper is a part of wider study called "Breaking the ICE reign: mixed method study of attitudes towards buying and using EVs in Kuwait". The study was funded by the Kuwait Foundation for the Advancement of Sciences and administrated by the London School of Economics and Political Science—Middle East Center (Grant number KFAS-MEC LSE 2021 001), and received an LSE Research Ethics Committee approval.

Institutional Review Board Statement: This study has been approved by the London School of Economics and Political Science Ethics Committee (00558000004KJE9AAO, dated 24 November 2021). Consent Form Statement: As directed by the LSE Ethics Committee, an informed statement about the utilization and purpose of the study was included in the questionnaire.

Data Availability Statement: Not applicable.

Conflicts of Interest: The authors declare no conflict of interest.

References

1. Ottesen, A.; Toglaw, S.; AlQuaoud, F.; Simovic, V. How to Sell Zero Emission Vehicles when the Petrol is almost for Free: Case of Kuwait. *J. Manag. Sci.* **2022**, *9*, 1–20. [CrossRef]
2. Ottesen, A.; Banna, S.; Alzougool, B.; Damrah, S. *A Greener Kuwait—How Electric Vehicles Can Lower CO_2 Emissions*; LSE Middle East Center Kuwait Programme Paper Series 18; LSE Middle East Centre: London, UK, 2023; pp. 1–21. Available online: http://chrome-extension://efaidnbmnnnibpcajpcglclefindmkaj/https://eprints.lse.ac.uk/120091/1/A_greener_Kuwait_paper_series_18.pdf (accessed on 10 August 2013).
3. Ottesen, A.; Thom, D.; Bhagat, R.; Mourdaa, R. Learning from the Future of Kuwait: Scenarios as a Learning Tool to Build Consensus for Actions Needed to Realize Vision 2035. *Sustainability* **2023**, *15*, 7054. [CrossRef]
4. Djoundourian, S.S. Response of the Arab world to climate change challenges and the Paris agreement. *Int. Environ. Agreem. Politics Law Econ.* **2021**, *21*, 469–491. [CrossRef]
5. Sharp, D.S.; Alshamari, A.; Hameed, K. *The Quiet Emergency: Experiences and Understandings of Climate Change in Kuwait*; LSE Middle East Center Kuwait Programme Paper Series 13; LSE Middle East Centre: London, UK, 2021; Available online: http://eprints.lse.ac.uk/112491/3/The_Quiet_Emergency.pdf (accessed on 6 June 2023).
6. New Kuwait. Kuwait National Development Plant 2020–2025. 2019. Available online: https://media.gov.kw/assets/img/Ommah22_Awareness/PDF/NewKuwait/Revised%20KNDP%20-%20EN.pdf (accessed on 26 August 2023).
7. Energy Strategy: Technical Paper Summary. In *4th Kuwait Master Plan: 2040 Toward A Smart State*; Municipality of Kuwait: Kuwait City, Kuwait, 2021; p. 65.
8. Ottesen, A.; Banna, S. Why so few EVs are in Kuwait and how to amend it. *Int. J. Eng. Technol.* **2021**, *10*, 181–189. [CrossRef]

9. Navfal, M. (KIA Kuwait, Kuwait City, Kuwait). Personal communication, 2023.
10. InsideEV Norway. Plug-In Car Sales Exceeded 90% Share in June 2023. Available online: https://insideevs.com/news/675163/norway-plugin-car-sales-june2023/#:~:text=New%20passenger%20plug%2Din%20car,more%20substantial%20growth%20over%202022 (accessed on 11 September 2023).
11. Norway Electric Vehicle Association. EV Norway. 2023. Available online: https://elbil.no/english/about-norwegian-ev-association/ (accessed on 11 September 2023).
12. ChargedKW. Charging Station Location. 2023. Available online: https://www.chargedkw.com/where-to-charge (accessed on 11 September 2023).
13. Ottesen, A.; Banna, S.; Alzougool, B. Attitudes of Drivers towards Electric Vehicles in Kuwait. *Sustainability* **2022**, *14*, 12163. [CrossRef]
14. Central Statistical Bureau. Kuwait Number of Register Vehicles in Use. 2021. Available online: https://www.ceicdata.com/en/kuwait/number-of-registered-vehicles/no-of-registered-vehicles-in-use#:~:text=The%20data%20reached%20an%20all,Global%20Database\T1\textquoterights%20Kuwait%20%E2%80%93%20Table%20KW (accessed on 6 June 2023).
15. Ottesen, A.; Banna, S.; Alzougool, B. How to Cross the Chasm for the Electric Vehicle World's Laggards—A Case Study in Kuwait. *World Electr. Veh. J.* **2023**, *14*, 45. [CrossRef]
16. Abu Dagga, N. E-Mobility—Electric Vehicles Technology and Innovation. In Proceedings of the Presentation at Australian University, Technology Leading to Sustainability—The Future of Mobility, West Misref, Kuwait, 8 May 2023.
17. Anagnostopoulou, E.; Bothos, E.; Magoutas, B.; Schrammel, J.; Mentzas, G. Persuasive technologies for sustainable mobility: State of the art and emerging trends. *Sustainability* **2018**, *10*, 2128. [CrossRef]
18. Apostolaki-Iosifidou, E.; Codani, P.; Kempton, W. Measurement of power loss during electric vehicle charging and discharging. *Energy* **2017**, *127*, 730–742. [CrossRef]
19. Geng, J.; Gao, S.; Sun, S.; Liu, Z.; Zhao, F.; Hao, H. Potential of electric vehicle batteries second use in energy storage systems: The case of China. *Energy* **2022**, *253*, 124159. [CrossRef]
20. Electrek. This Solar + Storage System is Made Up of 1300 Second Life EV Batteries. 2023. Available online: https://electrek.co/2023/02/07/this-solar-storage-system-is-made-up-of-1300-second-life-ev-batteries/ (accessed on 26 August 2023).
21. McKinsey & Company. ESG Report- Creating a More Sustainable, Inclusive and Growing Future for All. 2022. Available online: https://www.mckinsey.com/about-us/social-responsibility/esg-report-overview (accessed on 26 August 2023).
22. Cole, W.; Frazier, A.W.; Augustine, C. *Cost Projections for Utility-Scale Battery Storage: 2021 Update*; National Renewable Energy Laboratory: Golden, CO, USA, 2021. Available online: https://www.nrel.gov/docs/fy21osti/79236.pdf (accessed on 26 August 2023).
23. Li-Cycle Company. A Unique and Dependable Approach to Solving the Global Battery Cycling Problem. 2023. Available online: https://li-cycle.com/technology/ (accessed on 26 August 2023).
24. Koestner, J. 6 Thing to Remember about Hydrogen vs Natural Gas. Power Engineers. Available online: https://www.powereng.com/library/6-things-to-remember-about-hydrogen-vs-natural-gas (accessed on 7 June 2023).
25. Conti, M.; Kotter, R.; Putrus, G. Energy Efficiency in Electric and Plug-in Hybrid Electric Vehicles and Its Impact on Total Cost of Ownership. In *Electric Vehicle Business Models*; Beeton, D., Meyer, G., Eds.; Lecture Notes in Mobility; Springer: Cham, Switzerland, 2015. [CrossRef]
26. Economist. How the Japan Is Losing the Global Electric-Vehicle Race. 2023. Available online: https://www.economist.com/asia/2023/04/16/how-japan-is-losing-the-global-electric-vehicle-race (accessed on 8 June 2023).
27. American Public Transportation. Public Transportation Fact Book. 2022. Available online: https://www.apta.com/research-technical-resources/transit-statistics/public-transportation-fact-book/ (accessed on 7 June 2023).
28. IEA. Global EV Outlook 2023. 2023. Available online: https://www.iea.org/reports/global-ev-outlook-2023 (accessed on 8 June 2023).
29. BloombergNEF. Electric Vehicle Outlook 2022. 2022. Available online: https://bnef.turtl.co/story/evo-2022/page/7/1 (accessed on 6 June 2023).
30. Munoz, J.F. Tesla Model Y World Best-Selling Car for Q1 2023. 2023. Available online: https://www.motor1.com/news/669135/tesla-model-y-worlds-best-selling-car-q1-2023/ (accessed on 27 August 2023).
31. Poornesh, K.; Nivya, K.P.; Sireesha, K. A Comparative study on Electric Vehicle and Internal Combustion Engine Vehicles. In Proceedings of the 2020 International Conference on Smart Electronics and Communication (ICOSEC), Trichy, India, 10–12 September 2020; pp. 1179–1183. [CrossRef]
32. Office of Energy Efficiency and Renewable Energy. Battery Electric Vehicles Have Lower Scheduled Maintenance Cost than Other Light-Duty Vehicles. 2021. Available online: https://www.energy.gov/eere/vehicles/articles/fotw-1190-june-14-2021-battery-electric-vehicles-have-lower-scheduled (accessed on 27 August 2023).
33. Forbes. By The Numbers: Comparing Electric Car Warranties. 2022. Available online: https://www.forbes.com/sites/jimgorzelany/2022/10/31/by-the-numbers-comparing-electric-car-warranties/?sh=4c832c553fd7 (accessed on 14 February 2023).
34. J.D. Power. How Long Do Electric Battery Last? 2022. Available online: https://www.jdpower.com/cars/shopping-guides/how-long-do-electric-car-batteries-last#:~:text=Generally%2C%20electric%20vehicle%20batteries%20last,not%20pair%20well%20with%20EVs (accessed on 17 June 2023).
35. EVBox. How Long Do Electric Car Battery Last? 2022. Available online: https://blog.evbox.com/uk-en/ev-battery-longevity#:~:text=According%20to%20current%20industry%20expectations,is%20nearly%20imperceptible%20to%20drivers (accessed on 17 June 2023).

36. Monta. See How Temperature Affects the Range of Your EV. 2023. Available online: https://monta.com/uk/blog/how-temperature-affects-ev-range/#:~:text=On%20average%2C%2021.5%20degree%20celsius,used%20to%20move%20the%20car (accessed on 11 September 2023).
37. RethinkX. Rethinking Transportation—Cost and Speed of Adoption. 2023. Available online: https://www.rethinkx.com/fullsummary (accessed on 17 June 2023).
38. Hamwi, H.; Alasseri, R.; Aldei, S.; Al-Kandari, M. A Pilot Study of Electrical Vehicle Performance, Efficiency, and Limitation in Kuwait's Harsh Weather and Environment. *Energies* **2022**, *15*, 7466. [CrossRef]
39. Hamwi, H.; Rushby, T.; Mahdy, M.; Bahaj, A.S. Effects of High Ambient Temperature on Electric Vehicle Efficiency and Range: Case Study of Kuwait. *Energies* **2022**, *15*, 3178. [CrossRef]
40. United States Official Website of the International Trade Administration. Qatar Electric Vehicles Challenges and Opportunities. 2022. Available online: https://www.trade.gov/market-intelligence/qatar-electric-vehicles-challenges-and-opportunities#:~:text=Qatar\T1\textquoterights%20EV%20strategy%20aims%20to,in%20motion%20in%20September%202021 (accessed on 24 April 2023).
41. Central Statistical Bureau. Kuwait Voluntary National Review. 2019 Report on the Implementation of 2030 High Level Political Forum on Sustainable Development. Available online: https://www.arabdevelopmentportal.com/publication/kuwait-voluntary-national-review (accessed on 20 August 2022).
42. Gulf Times. Kahramaa to Set Up 150 Electric Charging Stations in 2023—EV Charging Station Network to Be Expanded to about 1000 by 2025. January 2023. Available online: https://www.gulf-times.com/article/653048/qatar/kahramaa-to-set-up-150-electric-charging-stations-in-2023 (accessed on 25 April 2023).
43. Ottesen, A.; Banna, S.; Alzougool, B. Women Will Drive the Demand for EVs in the Middle East over the Next 10 Years. *Energies* **2023**, *16*, 3756. [CrossRef]
44. The Peninsula. Qatar Converts 25% Public Transit Fleet to Eco-Friendly Mode. 19 September 2022. Available online: https://thepeninsulaqatar.com/article/19/09/2022/qatar-converts-25-public-transit-fleet-to-eco-friendly-mode (accessed on 25 April 2023).
45. The Peninsula. QatarEnergy Announces Petrol and Diesel Prices for March 2023. 28 February 2023. Available online: https://thepeninsulaqatar.com/article/28/02/2023/qatarenergy-announces-petrol-and-diesel-prices-for-march-2023#:~:text=Premium%20petrol%20will%20cost%20QR,and%20diesel%20will%20cost%20QR2 (accessed on 25 April 2023).
46. Alrajhi, J.M.; Alardhi, M.; Alhaifi, K.K.; Alkhulaifi, K.; Khalfan, A.; Alhaifi, N.A.; Alazemi, J. Prediction of Electric Vehicle Charging Stations Distribution in Kuwait. *Int. J. Traffic Transp. Eng.* **2023**, *12*, 5–9. [CrossRef]
47. Kuwait Voluntary National Review. Report on the Implementation of the 2030 Agenda to the UN High-Level Political Forum on Sustainable Development, Retrieved 12.24.2021 [Online]. 2019. Available online: https://sustainabledevelopment.un.org/content/documents/23384Kuwait_VNR_FINAL.PDF (accessed on 25 April 2023).
48. Ottesen, A.; Banna, S.; Alzougool, B.; Simovic, V. Driving factors for women's switch to electric vehicles in conservative Kuwait. *J. Women's Entrep. Educ.* **2022**, *3*, 1–21. [CrossRef]
49. Contestabile, M.; Turrentine, T. Introduction: Understanding the Development of the Market for Electric Vehicles. In *Who's Driving Electric Cars: Understanding Consumer Adoption and Use of Plug-in Electric Cars*; Springer: Cham, Switzerland, 2020; pp. 1–8.
50. Zhang, Q.; Li, H.; Zhu, L.; Campana, P.E.; Lu, H.; Wallin, F.; Sun, Q. Factors influencing the economics of public charging infrastructures for EV—A review. *Renew. Sustain. Energy Rev.* **2018**, *94*, 500–509. [CrossRef]
51. Zhang, X.; Bai, X.; Shang, J. Is subsidized electric vehicles adoption sustainable: Consumers' perceptions and motivation toward incentive policies, environmental benefits, and risks. *J. Clean. Prod.* **2018**, *192*, 71–79. [CrossRef]
52. Colak, M.; Kaya, I. providing the spark: Impact of financial incentives on battery electric vehicle adoption. *J. Environ. Econ. Manag.* **2020**, *98*, 102255.
53. Li, W.; Long, R.; Chen, H.; Geng, J. A review of factors influencing consumer intentions to adopt battery electric vehicles. *Renew. Sustain. Energy Rev.* **2017**, *78*, 318–328. [CrossRef]
54. Lia, F.; Molina ETimmermans, H.; Wee, B.V. Consumer preferences for business models in electric vehicle adoption. *Transp. Policy* **2019**, *73*, 12–24. [CrossRef]
55. Liao, F.; Molin, E.; Wee, B.V. Consumer preferences for electric vehicles: A literature review. *Transp. Rev.* **2017**, *37*, 252–275. [CrossRef]
56. Lin, B.; Wu, W. Why people want to buy electric vehicle: An empirical study in first-tier cities of China. *Energy Policy* **2018**, *112*, 233–241. [CrossRef]
57. Archsmith, J.; Muehlegger, E.; Rapson, D. Future Paths of Electric Vehicle Adoption in the United States: Predictable Determinants, Obstacles and Opportunities. *Environ. Energy Policy Econ.* **2022**, *3*, 71–110. [CrossRef]
58. Kongklaew, C.; Phoungthong, K.; Prabpayak, C.; Chowdhury, M.S.; Khan, I.; Yuangyai, N.; Yuangyai, C., Techato, K. Barriers to electric vehicle adoption in Thailand. *Sustainability* **2021**, *13*, 12839. [CrossRef]
59. Haider, S.W.; Zhuang, G.; Ali, S. Identifying and bridging the attitude behavior gap in sustainable transportation adoption. *J. Ambient. Intell. Humaniz. Comput.* **2019**, *10*, 3723–3738. [CrossRef]
60. Kim, S.; Choi, J.; Yi, Y.; Kim, H. Analysis of Influencing Factors in Purchasing Electric Vehicles Using a Structural Equation Model: Focused on Suwon City. *Sustainability* **2022**, *14*, 4744. [CrossRef]
61. Ziemba, P. Multi-criteria approach to stochastic and fuzzy uncertainty in the selection of electric vehicles with high social acceptance. *Expert Syst. Appl.* **2021**, *173*, 114686. [CrossRef]

62. Vilchez, J.; Harrison, G.; Kelleher, L.; Smyth, A.; Thiel, C. *Quantifying the Factors Influencing People's Car Type Choices in Europe*; Publications Office of the European Union: Luxembourg, 2017; p. 18. Available online: ec.europa.eu/jrc/en/publication/eur-scientific-and-technical-research-reports/quantifyingfactors-influencing-people-s-car-type-choices-europe-results-stated-preference (accessed on 8 June 2023).
63. Temple, J. Lithium-Metal Batteries for Electric Vehicles. MIT Technology Review. 2021. Available online: www.technologyreview.com/2021/02/24/1018102/lithium-metal-batteries-electric-vehicle-car/ (accessed on 8 June 2023).
64. Kowalska-Pyzalska, A.; Michalski, R.; Kott, M.; Skowrońska-Szmer, A. Consumer preferences towards alternative fuel vehicles. Results from the conjoint analysis. *Renew. Sustain. Energy Rev.* **2022**, *155*, 111776. [CrossRef]
65. Chao Ma, S.; Fan, Y.; Guo, J.F.; Xu, J.H.; Zhu, J. Analysing online behaviour to determine Chinese consumers' preferences for electric vehicles. *J. Clean. Prod.* **2019**, *229*, 244–255.
66. Bell, E.; Bryman, A.; Harley, B. *Business Research Methods;* Oxford University Press: Oxfor, UK, 2018.
67. Hunter, Pamila, Margin of Error and Confidence Levels Made Simple. 2010. Available online: https://www.isixsigma.com/sampling-data/margin-error-and-confidence-levels-made-simple/ (accessed on 7 July 2023).

World Electric Vehicle Journal

MDPI

Article

Waste from Electric Vehicle: A Bibliometric Analysis from 1995 to 2023

Arief Nurdini [1,2], Rahmat Nurcahyo [1,*] and Anton Satria Prabuwono [3]

[1] Industrial Engineering Department, Faculty of Engineering, Universitas Indonesia, Depok 16424, Indonesia; arief.nurdini11@ui.ac.id

[2] Industrial Engineering Department, Faculty of Industrial Technology, Gunadarma University, Depok 16424, Indonesia

[3] Faculty of Computing and Information Technology in Rabigh, King Abdulaziz University, Rabigh 21911, Saudi Arabia; aprabuwono@kau.edu.sa

* Correspondence: rahmat@eng.ui.ac.id

Abstract: The introduction of electric vehicles (EVs) represents a promising solution for addressing urban air pollution, particularly CO_2 emissions in the transportation sector. Numerous countries are actively promoting EV adoption and the electrification of transportation systems, leading to a surge in research on EV-related topics. This study employs bibliometrics as a valuable tool to investigate the research landscape in electric vehicle waste management. Drawing from a dataset of 593 documents retrieved from SCOPUS from 1995 to 20 September 2023, this research employs descriptive analysis and bibliometric mapping techniques. Notably, China stands out as the leading contributor to publications, with Tsinghua University being a prominent research institution in this field. An examination of keyword trends reveals dynamic shifts in research focus. In 2023, the most frequently occurring topic is "closed loop". "Recycling" is the dominant keyword, appearing 681 times. Additionally, TreeMaps and VOSviewer results indicate that the most commonly used keywords are "electronic waste" and "recycling". Projections suggest that "recycling materials" will gain prominence in mid-2023, further highlighting the evolving nature of this research field. Researchers in recycling materials disciplines can leverage these insights to explore new research avenues and contribute to sustainable waste management practices in the context of electric vehicles.

Keywords: electric vehicle; electronic waste; e-waste; bibliometric; R tools; VOSviewer

Citation: Nurdini, A.; Nurcahyo, R.; Prabuwono, A.S. Waste from Electric Vehicle: A Bibliometric Analysis from 1995 to 2023. *World Electr. Veh. J.* **2023**, *14*, 300. https://doi.org/10.3390/wevj14110300

Academic Editors: Wenbin Yu and Guang Zeng

Received: 25 September 2023
Revised: 19 October 2023
Accepted: 24 October 2023
Published: 27 October 2023

1. Introduction

Over the last few decades, there has been a steady increase in the amount of carbon dioxide (CO_2) released into the atmosphere, and people's use of fossil fuels is directly related to global warming [1]. Based on the Environmental Protection Agency (EPA) report, CO_2 is the main gas content in global greenhouse gas emissions and is produced by fossil fuels in industrial activities, which account for up to 65% of the total global greenhouse gases [2]. In addition, the transportation sector contributed 22% of total CO_2 emissions in 2020 [3].

Introducing electric vehicles has been seen as an excellent opportunity to reduce urban air pollution, especially CO_2, from the transportation sector. Promoting electric vehicles to quickly replace internal combustion engine (ICE) vehicles is an essential strategy for countries worldwide [4]. Therefore, many countries are actively promoting the electrification of the transportation sector and encouraging the use of electric vehicles [5]. In addition to the desire to reduce CO_2 emissions, psychological factors also play a role in influencing people to use electric vehicles [6]. Based on data from the International Energy Agency (IEA) contained in the Global E.V. Outlook 2022, more than 16.5 million electric vehicles were sold in 2021, an increase of 6.6 million units from the previous year. However, the largest electric car market in the world is still dominated by China, followed by Europe and America.

Sales of electric vehicles will rise along with the number of waste parts from these vehicles that will be recycled, including car bodywork, tires, plastics, traction batteries, electric motors, power electronics, electronic components, etc. [7,8]. Furthermore, electric car electrical component waste is also dangerous but highly valuable [9].

Review papers and bibliometric research analysis are two ways to summarize research publications [10]. Bibliometric analysis has gained immense popularity in business research in recent years. Its popularity can be attributed to the development, accessibility, and availability of bibliometric tools like Gephi, Leximancer, and VOSviewer, as well as to the scientific databases Scopus and Web of Science. The cross-disciplinary pollination of the bibliometric methodology from information science to business research. More importantly, the popularity of bibliometric analysis in business research is not a fad but rather a reflection of its utility for (1) handling large volumes of scientific data and (2) producing high research impact [11]. By rigorously interpreting vast amounts of unstructured data, bibliometric analysis aids in deciphering and mapping the cumulative scientific knowledge and evolutionary nuances of established areas. Therefore, bibliometric studies promote innovation and create a strong foundation for meaningfully advancing research. Researchers can use bibliometric analysis to their advantage by having a single point of access to information, looking into knowledge gaps, coming up with creative research ideas, and positioning their intended contributions to the field [10]. The method used is based on keywords in the Scopus database, which will be further analyzed using R-Tools and VOSviewer. This study aimed to explore the following key questions:

1. Which publications, institutions, countries, and authors have made the most significant contributions to research on electric vehicles and the waste they produce?
2. How are the authors, organizations, and nations related to one another and work together in academic research?
3. What are the current prominent research topics in this field, how are they evolving, and what will capture attention in the future?

2. Materials and Methods

2.1. Data Collection

SCOPUS, a database of content from the esteemed journal, was used in this study. Since it contains more articles than other databases, such as Web of Science [12], the Scopus database was chosen for this study's bibliometric analysis. This study did not assess Google Scholar because there were no reliable results [13]. Five hundred ninety-three documents related to electric vehicles and electronic waste were published between 1995 and September 2023. The papers will be used for further data analysis.

2.2. Bibliometric: VOSviewer and R Tools

A study method known as "bibliometrics" uses the library and information science to provide data and analysis in various formats, including statistical and quantitative methods [14]. An essential research area is bibliometrics since it offers detailed historical data that can be used to predict future research trends [15]. In addition, a range of crucial indicators, including h-index, impact factors, citations, and current status, can be used by universities, instructors, researchers, and professors to evaluate the quality of a study. There are many bibliometric tools, like VOSviewer and R Tools. VOSviewer is a free program for creating and displaying bibliometric networks, which can be based on citations, bibliographic coupling, co-citations, or author relationships. These networks can contain journals, researchers, or individual publications. The entire bibliographic dataset was acquired in .csv format from the Scopus database. Initially, we installed and loaded the Bibliometrix R package using R Studio. We then initiated the Biblioshiny application by typing "Biblioshiny()" in the R console. Biblioshiny serves as a web-based tool that offers non-programmers access to the Bibliometrix package within R [16]. Numerous tools provided by Bibliometrix enable scholars to undertake in-depth bibliometric analyses [17]. In this paper, the tools of bibliometric VOSviewer 1.6.18 developed by Nees Jan van Eck and Ludo Waltman at Leiden University and R-Tool version 4.2.2 are combined to get a better result.

2.3. *Literature Search*

A comprehensive scientific literature search on the SCOPUS database. The search was further modified to include other terms, such as "electric vehicle" OR "electric vehicles" OR "EV" AND "electronic waste" OR "e-waste". The established period covered data from 2015 to 2023, and to improve the efficiency of the search process, the search was limited to specific terms. On 20 September 2023, the last web page visit occurred. The methodology flowchart is depicted in Figure 1.

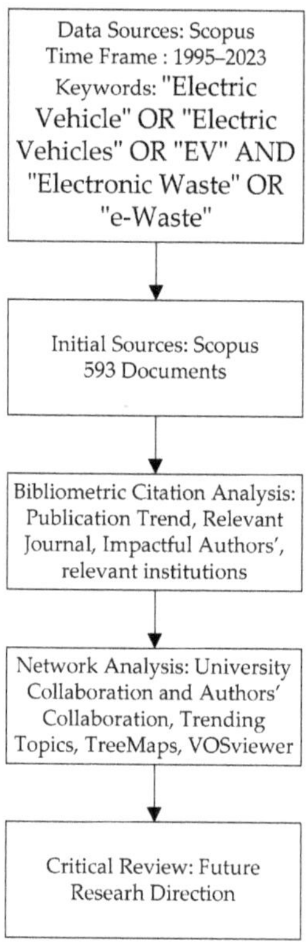

Figure 1. Methodological Flowchart for Bibliometric.

3. Results

This section may be divided into subheadings. It should provide a concise and precise description of the experimental results, their interpretation, and the experimental conclusions that can be drawn.

3.1. *Data Analysis*

The documents are from 1995 to 20 September 2023, and have 234 sources. A brief description can be seen in Table 1. A total of 593 documents were discovered, of which 363 are articles. During the time of the study, a total of 1944 authors contributed to the topic. The average age of the documents is 2.89, with the average number of citations per doc being 32.86. And then, 20,626 references are found in 593 documents. KeyWords Plus reached 4191, compared to the author's 1351 keywords. Only 17 documents with a single author's name were found. We found one book, 14 book chapters, 116 conference papers, three conference reviews, 1 note, and 95 reviews.

Table 1. Main Information about the Data.

Description	Results
MAIN INFORMATION ABOUT DATA	
Timespan	1995:2023
Sources (Journals, Books, etc.)	234
Documents	593
Annual Growth Rate %	17.71
Document Average Age	2.89
Average citations per doc	32.86
References	20,626
DOCUMENT CONTENTS	
Keywords Plus (ID)	4191
Author's Keywords (DE)	1351
AUTHORS	
Authors	1944
Authors of single-authored docs	15
AUTHORS COLLABORATION	
Single-authored docs	17
Co-Authors per Doc	4.67
International co-authorships %	23.95
DOCUMENT TYPES	
Article	363
Book	1
book chapter	14
conference paper	116
conference review	3
Note	1
Review	95

3.2. Annual Publications

The fundamental details about the publications are extracted using the Bibliometrix sub-tool Biblioshiny. We could see a significant annual growth rate of 17.71 percent in the number of publications. Only one document was published in 1995. The article describes the energy and environmental impacts of electric vehicle battery production and recycling. Based on Scopus, there are 11 citations in this paper. The topic of electronic waste increased from 2010 to 2015, then decreased in 2016. Then, there was a significant increase from 2017, with 27 articles, until 2021 and 2022, with 117 and 118 articles. For more details, see Figure 2.

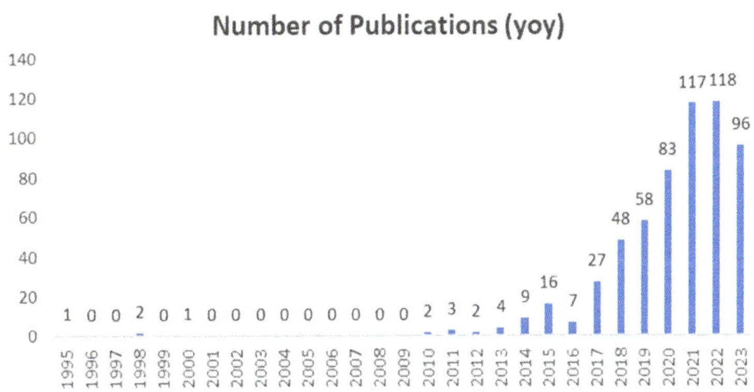

Figure 2. Number of Publications.

3.3. Annual Citations

The number of articles' citations increased as more research was conducted. Based on Figure 3, there are 11 citations for publication in 1995. Moreover, there are no citations from 1996 and 1997, 1999, and 2001 to 2009 because researchers carried out no publications. The average per doc, the citation from 1995 to 20 September 2023, is 32.86. The citations increased from 2017 with 871 citations, 2018 with 3030 citations, 2019 with 3546 citations, and 2020 with 4166 citations. We can observe that, from 1995 through early 2023, there has been a wide variety of citations per year, increasing in some and declining in others.

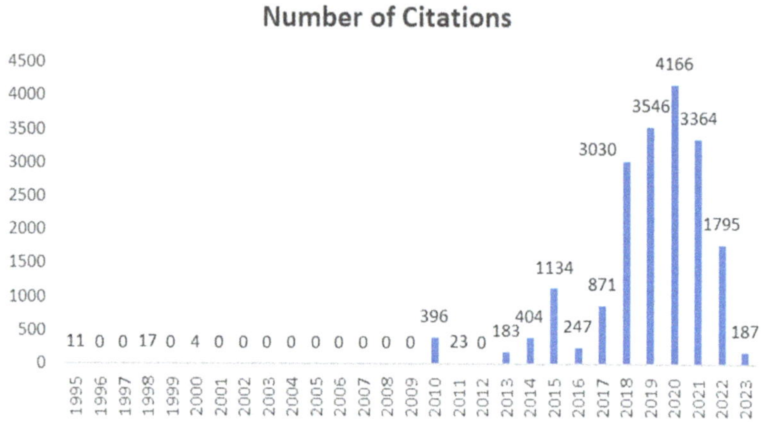

Figure 3. Number of Citations.

3.4. Highly Cited Papers

We found the most cited paper in "Lithium-ion Battery Supply Chain Considerations: Analysis of Potential Bottlenecks in Critical Metals" article by [18]. The paper was published in the Journal Joule with Cell Press Publisher. The top 30 highly cited papers can be found in Table 2.

Table 2. Top 30 Highly Cited papers.

No.	Document Title	Journal	Publisher	Year	Citation	Authors
1	Lithium-Ion Battery Supply Chain Considerations: Analysis of Potential Bottlenecks in Critical Metals	Joule	Cell Press	2017	782	[18]
2	Sustainable Recycling Technology for Li-Ion Batteries and Beyond: Challenges and Future Prospects	Chemical Reviews	American Chemical Society	2020	718	[19]
3	State-of-the-Art and Energy Management System of Lithium-Ion Batteries in Electric Vehicle Applications: Issues and Recommendations	IEEE Access	Institute of Electrical and Electronics Engineers Inc.	2018	493	[20]
4	Recycling End-of-Life Electric Vehicle Lithium-Ion Batteries	Joule	Cell Press	2019	431	[21]
5	A Mini-Review on Metal Recycling from Spent Lithium-Ion Batteries	Engineering	Elsevier Ltd.	2018	408	[22]
6	Life cycle assessment of lithium-ion batteries for plug-in hybrid electric vehicles-Critical issues	Journal of Cleaner Production	Elsevier Ltd.	2010	393	[23]

Table 2. *Cont.*

No.	Document Title	Journal	Publisher	Year	Citation	Authors
7	Recycling of spent lithium-ion batteries in view of lithium recovery: A critical review	Journal of Cleaner Production	Elsevier Ltd.	2019	392	[24]
8	Current li-ion battery technologies in electric vehicles and opportunities for advancements	Energies	MDPI AG	2019	388	[25]
9	Novel approach to recover cobalt and lithium from spent lithium-ion battery using oxalic acid	Journal of Hazardous Materials	Elsevier	2015	368	[26]
10	Examining different recycling processes for lithium-ion batteries	Nature Sustainability	Nature Publishing Group	2019	360	[27]
11	Lithium-ion batteries—Current state of the art and anticipated developments	Journal of Power Sources	Elsevier B.V.	2020	322	[28]
12	A review on the growing concern and potential management strategies of waste lithium-ion batteries	Resources, Conservation and Recycling	Elsevier B.V.	2018	287	[29]
13	A future perspective on lithium-ion battery waste flows from electric vehicles	Resources, Conservation and Recycling		2014	285	[30]
14	Pyrometallurgical options for recycling spent lithium-ion batteries: A comprehensive review	Journal of Power Sources	Elsevier B.V.	2021	254	[31]
15	Solving spent lithium-ion battery problems in China: Opportunities and challenges	Renewable and Sustainable Energy Reviews	Elsevier Ltd.	2015	247	[32]
16	Future material demand for automotive lithium-based batteries	Communications Materials	Springer Nature	2020	246	[33]
17	A critical review of lithium-ion battery recycling processes from a circular economy perspective	Batteries	MDPI	2019	227	[34]
18	Circular economy strategies for electric vehicle batteries reduce reliance on raw materials	Nature Sustainability	Nature Research	2021	180	[35]
19	Biotechnological strategies for the recovery of valuable and critical raw materials from waste electrical and electronic equipment (WEEE)—A review	Journal of Hazardous Materials	Elsevier B.V.	2019	174	[36]
20	Recent progress on the recycling technology of Li-ion batteries	Journal of Energy Chemistry	Elsevier B.V.	2020	166	[37]
21	Metallurgical and mechanical methods for recycling of lithium-ion battery pack for electric vehicles	Resources, Conservation and Recycling	Elsevier B.V.	2018	162	[38]
22	Energy and environmental assessment of a traction lithium-ion battery pack for plug-in hybrid electric vehicles	Journal of Cleaner Production	Elsevier Ltd.	2019	158	[39]
23	The Recycling of Spent Lithium-Ion Batteries: a Review of Current Processes and Technologies	Electrochemical Energy Reviews		2018	156	[40]
24	An overview of recycling and treatment of spent LiFePO4 batteries in China	Resources, Conservation and Recycling	Elsevier B.V.	2017	147	[41]
25	Progress and status of hydrometallurgical and direct recycling of Li-Ion batteries and beyond	Materials	MDPI	2020	146	[42]

Table 2. *Cont.*

No.	Document Title	Journal	Publisher	Year	Citation	Authors
26	Comparative environmental life cycle assessment of conventional vehicles with different fuel options, plug-in hybrid and electric vehicles for a sustainable transportation system in Brazil	Journal of Cleaner Production	Elsevier Ltd.	2018	146	[43]
27	Recycling and environmental issues of lithium-ion batteries: Advances, challenges and opportunities	Energy Storage Materials	Elsevier	2021	143	[44]
28	The importance of design in lithium-ion battery recycling-a critical review	Green Chemistry	Royal Society of Chemistry	2020	124	[45]
29	Eco-Efficiency Analysis of a Lithium-Ion Battery Waste Hierarchy Inspired by Circular Economy	Journal of Industrial Ecology	Blackwell Publishing	2017	137	[46]
30	Graphite Recycling from Spent Lithium-Ion Batteries	ChemSusChem	Wiley-VCH Verlag	2016	136	[47]

3.5. Most Relevant Sources

Resources are identified to determine the performance of the sources used. Based on Table 3, Resources, Conversation, and Recycling from the Netherlands are at the top, with 45 documents and an impact factor of 13.716. The Journal of Cleaner Production is a second, with a total of 30 documents and an impact factor of 11.072, and the third is the Journal of Energy Storage, with 23 documents and an impact factor of 8.907. The three prominent journals are from the Netherlands. For more details regarding the most relevant resources, see Figure 4.

Table 3. Most Relevant Sources with Impact Factor.

No.	Name of Journal	Number of Publications	Impact Factor	Rate (%)
1	Resources Conservation and Recycling	45	13.716	7.63%
2	Journal of Cleaner Production	30	11.072	5.08%
3	Journal of Energy Storage	23	8.907	3.90%
4	Energies	17	3.2	2.88%
5	Procedia CIRP	16	2.17	2.71%
6	Waste Management	14	8.816	2.37%
7	SAE Technical Papers	13	0.638	2.20%
8	ACS Sustainable Chemistry and Engineering	12	9.224	2.03%
9	Others	420		71.19%

3.6. Corresponding Author's Countries

Corresponding authors are divided into two categories: multi-country publication (MCP), which includes authors from several nations and such publications show cross-national or international collaboration; and single-country publication (SCP), where all writers are from the same country and the publication is an example of intra-national cooperation. Based on Figure 5, China has the most corresponding author documents, with 39 MCP documents and 147 SCP documents. The second country, the United States, has five MCP documents and 49 SCP documents. The third country is Germany, with eight MCP documents and 57 SCP documents.

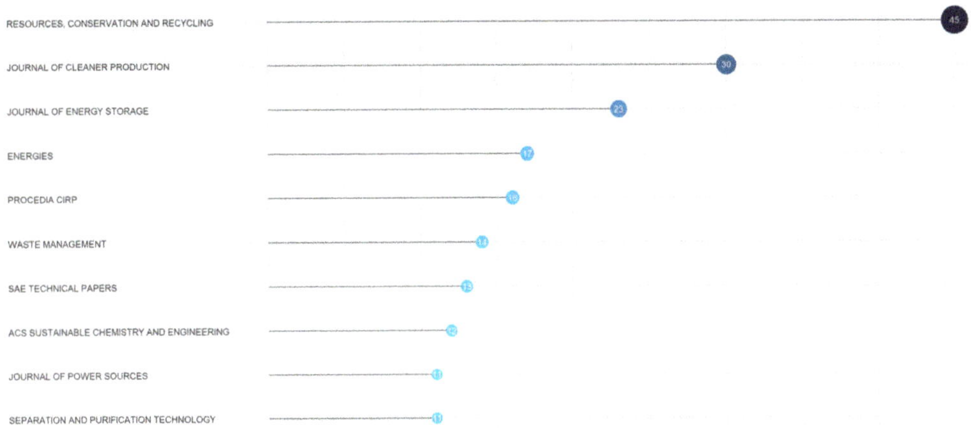

Figure 4. Most Relevant Sources.

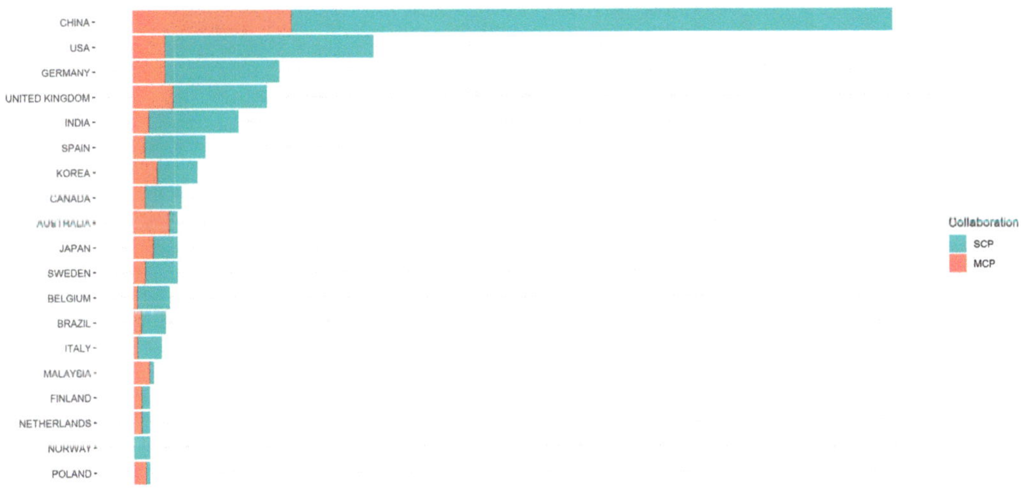

Figure 5. Corresponding Author's Countries.

3.7. Most Relevant Affiliations

The most relevant affiliations were identified to determine which university has produced the most publications on electric vehicles and electronic waste. Tsinghua University and Central South University in China are first and second, with 79 and 51 articles, respectively. The third place is Chalmers University of Technology, with 37 articles. For more details regarding the most relevant affiliations, refer to Figure 6.

3.8. Subject Categories

Subject areas are examined to determine which topics are most popular with researchers. For example, according to Figure 7, we can see that the top three research areas are engineering 22%, environmental science 19%, and energy 18%.

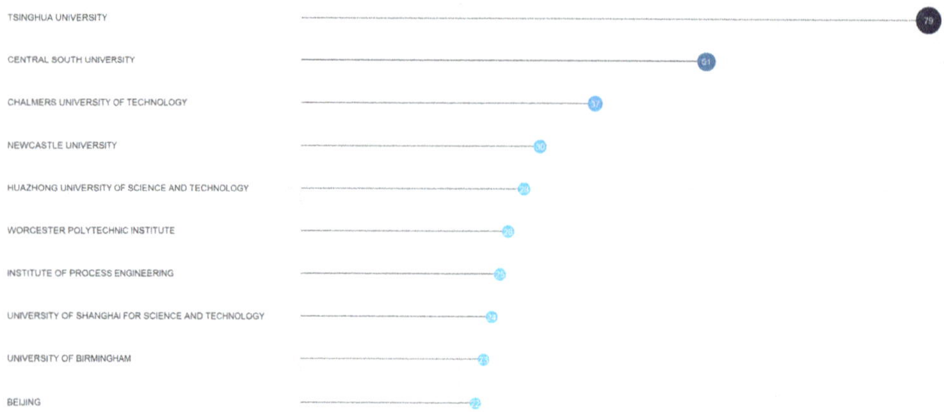

Figure 6. Most Relevant Affiliations.

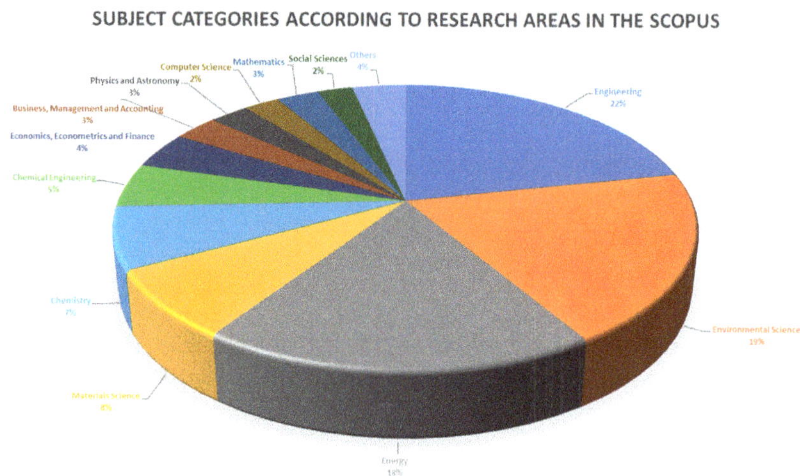

Figure 7. Subject Categories according to the Research Area.

3.9. Object Categories

The object categories explain the electronic components of electric vehicles that have been researched. In this paper, these categories are divided into four parts: batteries, charging stations, power electronics, and permanent magnets. The discussion of batteries is the most researched, as seen in [18,19,23,26,48–63]. In global electronic component waste research, there is still a dominant focus on battery waste, while research concerning electronic power component waste remains limited. For more details, please refer to Table 4.

Table 4. Object Categories.

No	Battery	Charging	Power Electronics	Permanent Magnet	Authors
1	✔				[18,19,23,26,48–63]
2	✔		✔		[63]
3			✔		[64]
4				✔	[65]
5		✔			[66,67]

3.10. Most Relevant Authors' and Production over Time

The data were analyzed to determine the most relevant authors. Based on Figure 8, Wang Y, with 18 documents, is the leading author, followed by Li J and Li Y, with 17 and 13 documents. The data were examined to determine the authors' output over time.

Figure 8. Most Relevant Authors'.

Wang Y published one article in 2016, 2017, and 2018—two in 2019 and 2021. There were three articles in 2020 and four in 2022 and 2023 (Figure 9). The dot color in Figure 9 shows the number of publications. The paper from Wang Y with the most citations is titled "Recent Progress on the Recycling Technology of Li-ion Batteries" in 2021, with 169 citations. The total citations from 18 documents are 856 citations.

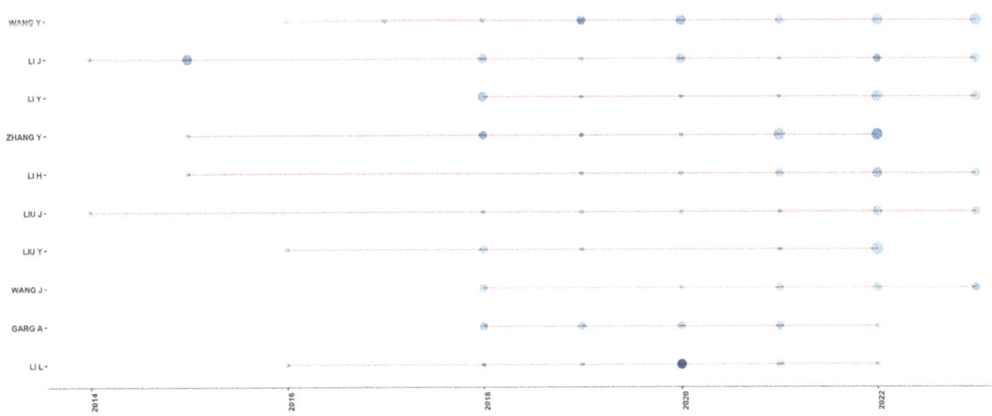

Figure 9. Author's Production over Time.

3.11. Most Frequent Words

The frequent words evolve as research advances. According to Bibliometrix, the most frequently occurring words in 1995 were "lead acid", "metal hydride electrode", and "cadmium electrode". In 2023, the most frequently used word in the literature was "recycling", totaling 678 occurrences, followed by "electronic waste", which had 603 occurrences, and "lithium-ion batteries", with 381 occurrences. The darker the color of the dot, the more citations obtained. The total number indicates that research on recycling and e-waste is evolving. For more details regarding the most relevant affiliations, see Figure 10.

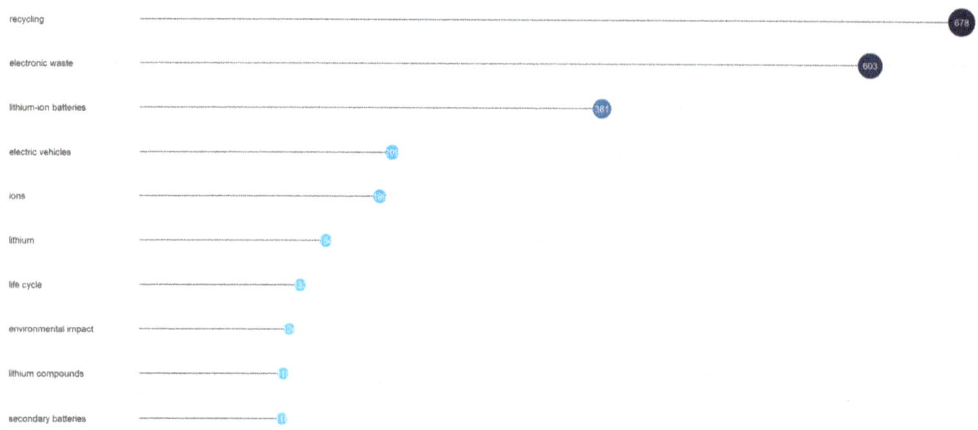

Figure 10. Most Frequent Words.

3.12. *Trending Topics*

Trending topics are analyzed based on the authors' keyword frequency. Frequency refers to the count of keyword usage in papers for a particular year. Based on Table 5, in 2010, the keyword "Vehicle Technology" became trending with a total frequency of 4. In 2014, the keywords "Integrated Approach", "Regenerative Braking", and "Recycling Rate" became trending with frequencies 4, 4, and 5. In 2015, the keyword "lithium alloys" became a trending topic with a total frequency of 7. In 2018, the keywords "vehicle", "hybrid vehicles", and "nickel metal hydride batteries". Each frequency of 28, 23, and 12 becomes a trending topic. Keywords "secondary batteries", "electric vehicles", and "electrodes" with frequencies 118, 45, and 29 became a trending topic in 2019. In 2020, the keywords "automotive batteries", "waste management", and "battery management systems". Each frequency of 63, 54, and 46 became a trending topic. Keywords "recycling", "electronic waste", and "lithium-ion batteries" became a trending topic in 2021, with a total frequency of 678, 603, and 381. And in 2022, the keywords "battery recycling", "cathodes", and "leaching". With a total frequency of 113, 104, and 90, they became trending topics. Furthermore, in mid of 2023, the keywords "industrial ecology", "recycled materials", and "closed loop". With a total frequency of 6, 6, and 7, they became trending topics. For more details see Figure 11.

Table 5. Trending Topic Based on Authors' Keyword.

Years	Frequent Word	Term Frequency
2010	Vehicle Technology	4
2014	Integrated Approach	4
	Regenerative Braking	4
	Recycling Rate	5
2015	Hybrid Energy Storage Systems (Hess)	3
	Leaching Solution	3
	Lithium Alloys	7
2016	Nickel Metal Hydride	4
	Policy Makers	4
2017	Plug-in Hybrid Vehicles	15
	Recovery	18
	Electric Batteries	32

Table 5. *Cont.*

Years	Frequent Word	Term Frequency
2018	Nickel Metal Hydride Batteries Hybrid Vehicles Vehicles	12 23 28
2019	Electrodes Electric Vehicles (EVS) Secondary Batteries	29 45 118
2020	Battery Management Systems Waste Management Automotive Batteries	46 54 63
2021	Lithium-ion Batteries Electronic Waste Recycling	381 603 678
2022	Spent Lithium-ion Batteries Cathodes Battery Recycling	90 104 113
2023	Industrial Ecology Recycled Materials Closed-Loop	6 6 7

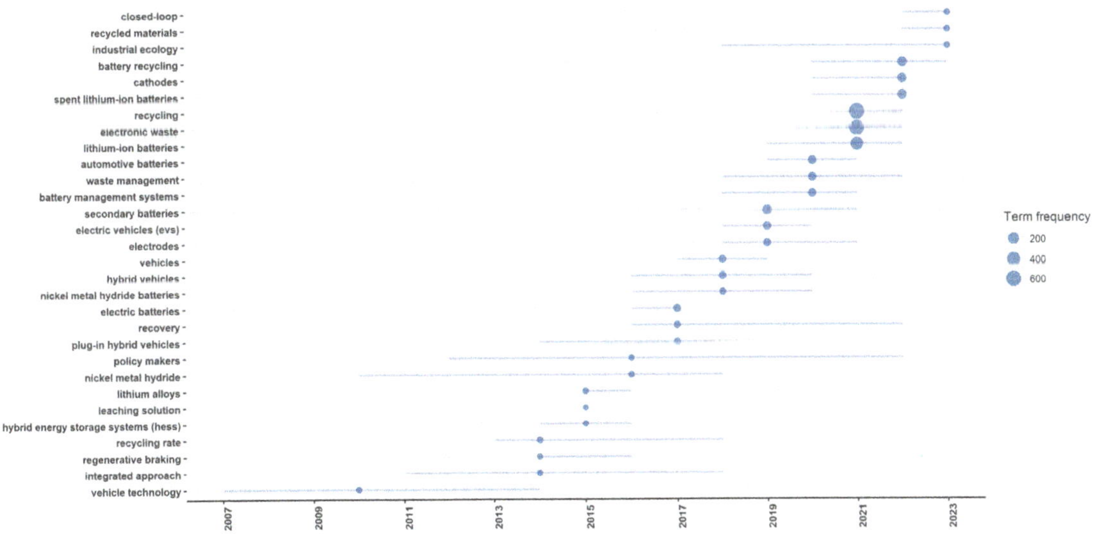

Figure 11. Trending Topics based on Authors' Keywords.

3.13. *TreeMaps*

The treemap depicts words frequently used as keywords in research on electric vehicles and electronic waste. For example, the graph shows that recycling ranks first at 14% with 681 occurrences, electronic waste second at 12% with 603 occurrences, lithium-ion batteries at 8% with 381 occurrences, and electric vehicles and ions at 4% each with 209 and 199 occurrences. TreeMaps can be observed in greater detail in Figure 12.

Figure 12. TreeMap Words.

3.14. Bibliometric VOSviewer

The result is divided into seven clusters. Cluster 1 (red) contains 136 keywords: lithium-ion batteries, lithium compounds, recycle, leaching, and hydrometallurgy. Cluster 2 (green) includes 131 keywords, such as lithium, nickel, metal recovery, and renewable energies. Cluster 3 (blue) contains 120 keywords: electric vehicles, energy storage, energy efficiency, economic analysis, and end-of-life batteries. Cluster 4 (yellow) includes 112 keywords: battery recycling, secondary batteries, sustainable development, and solid waste. Cluster 5 (Violet) contains 80 keywords: life cycle, life cycle analysis, life cycle assessment, environmental impact, and sensitivity analysis. Cluster 6 (green turquoise) has 73 keywords: electronic waste, recycling process, battery pack, and charging batteries. Finally, cluster 7 (orange) contains 72 keywords: waste management, electronic equipment, circular economy, and sustainability. According to Figure 13, the terms with the highest frequency include electronic waste, electric vehicles, recycling, lithium-ion batteries, battery recycling, and others.

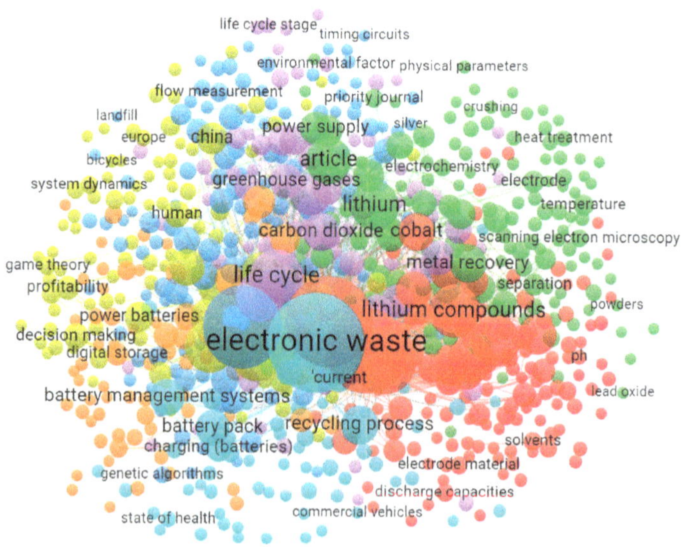

Figure 13. Bibliometric VOSviewer.

4. Discussion

Research in many diverse disciplines has begun adopting bibliometric analysis [68–70]. This study technique can show how articles published in databases are distributed throughout various topics, industries, organizations, and nations. Moreover, many databases, like Scopus, Web of Science, etc., can be utilized to retrieve the raw data for the bibliometric study. We took the source information for this research out of the Scopus Core Collection. The documents were published between 1995 and 20 September 2023, and came from 234 sources (books, journals, etc.).

There were 593 documents overall, of which 363 were articles. A total of 1944 authors have contributed to this field of study. The fundamental data about the documents were extracted using the Biblioshiny Bibliometrix sub-tool and VOSviewer. We observed a significant annual growth rate of 17.71% in the number of articles. Based on the Scopus database, the first publication related to waste electronic components for electric vehicles was found in 1995, with one document published. None of the documents from 1996, 1997, or 1999 were published. Two documents were contributed to the literature in 1998, and one was added in 2000. Throughout the following nine years, not a single new document was released. In 2010, there were two contributing documents. In 2011 and 2012, there were 3 and 2 documents, respectively. Subsequently, there was a consistent increase in the number of documents in the following years. In 2013, there were 4 documents. In 2014, there were 9 documents. In 2015, there were 16 documents. In 2016, there was a decrease to 7 documents. In 2017, there was a significant rise to 27 documents. In 2018, there were 48 documents. In 2019, there were 58 documents, and in 2020, there were 83 documents. The number of documents added to the literature increased in 2022 with a total of 118, then in 2021 with 117. Each year, the number of documents published has been steadily rising. A total of 593 documents were released in 2023.

Few publications were published at the start of the research period, and the research literature was declining. Only 11 citations were provided in 1995; from 2001 to 2009, none were accessible. There were exponential growth in the number of citations in 2013. Citations significantly increased in the year 2020. The most prolific university overall, Tsinghua University, has published 79 publications. China is the leading country regarding corresponding authors, with 37 papers published in multiple countries (MCP) and 147 documents published in a single country (SCP). The USA ranks second, with five MCP documents and 49 SCP documents. With 822 citations, Wang Y. was the author with the most significant influence. From 1998 until now, the word "recycling" has been used the most frequently in the literature, leading to a total count of 678.

Trending topics are analyzed based on the author's keywords. From the last three years in 2020, the keywords "automotive batteries", "waste management", and "battery management systems". Each frequency of 63, 54, and 46 became a trending topic. Keywords "recycling", "electronic waste", and "lithium-ion batteries" became trending topics in 2021, with a total frequency of 678, 603, and 381. And in 2022, the keywords "battery recycling", "cathodes", and "leaching". With a total frequency of 113, 104, and 90, they became trending topics. Furthermore, in mid-2023, the keywords "industrial ecology", "recycled materials", and "closed loop".

5. Conclusions

Municipal electronic waste issues are gaining momentum, and several studies have thoroughly studied the field of connected subjects. This paper used the open-source R language and the bibliometrix package to do a bibliometric analysis based on the data from 593 scientific publications received from Scopus.

An analysis of scientific output reveals that the research area of recycling electronic waste from electric vehicles is still experiencing annual sales growth that is accelerating rapidly. The findings indicate that China and the United States have the most publications, whereas Tsinghua University and Central South University produce the most work. However, when using the number of citations to indicate academic influence, China, with

6177 citations, surpasses the United States's 3334 citations and the United Kingdom's 1415 citations. With 856 citations and 18 published documents, Wang Y was the author with the most significant influence. Additionally, over the past three years, in 2021, the keywords "recycling", "electronic waste", and "lithium-ion batteries" all attained a frequency of 678, 603, and 381, respectively. "Battery recycling", "cathodes", and "leaching" are other significant terms for 2022. The subjects with the highest frequency were 113, 104, and 90. The keywords "industrial ecology", "recycled materials", and "closed loop" will also be used in the middle of 2023. Each issue becomes a trending topic six times in total. We can determine the terms frequently used in studying electric vehicles and electronic trash using TreeMaps. Recycling is mentioned 681 times (14% of all usage), electronic waste 603 times (12%), lithium-ion batteries 381 times (8%), electric vehicles 209 times (4%), ions 199 times (4%), etc. It should be mentioned that this research has significant limitations despite the in-depth investigation. For instance, this analysis is based on data from Scopus, which is unquestionably one of the most reliable and accurate sources of information; however, the trend may be different when other search engines or databases are added, as well as when manuscripts from outside core collections are taken into account. Therefore, more in-depth research in this area is required.

Author Contributions: Writing—original draft, A.N., R.N. and A.S.P. All authors contributed substantially and equally to the writing and original draft preparation. All authors have read and agreed to the published version of the manuscript.

Funding: This research received funding from Hibah Publikasi Terindeks Internasional (PUTI) Q2 Tahun Anggaran 2022-2023, Nomor NKB-707/UN2.RST/HKP.05.00/2022.

Data Availability Statement: As this paper has the character of a bibliometric, no new data were created or analyzed in this study. Data sharing is not applicable to this article.

Conflicts of Interest: The authors declare no conflict of interest.

References

1. Siqi, Z.; Guangming, L.; Wenzhi, H.; Juwen, H.; Haochen, Z. Recovery methods and regulation status of waste lithium-ion batteries in China: A mini review. *Waste Manag. Res.* **2019**, *37*, 1142–1152. [CrossRef] [PubMed]
2. EPA. *Global Greenhouse Gas Emissions Data*; EPA: Washington, DC, USA, 2022.
3. Giannakis, E.; Serghides, D.; Dimitriou, S.; Zittis, G. Land transport CO_2 emissions and climate change: Evidence from Cyprus. *Int. J. Sustain. Energy* **2020**, *39*, 634–647. [CrossRef]
4. Chen, Y.; Wu, G.; Sun, R.; Dubey, A.; Laszka, A.; Pugliese, P. A review and outlook of energy consumption estimation models for electric vehicles. *arXiv* **2020**, arXiv:2003.12873.
5. Verma, S.; Dwivedi, G.; Verma, P. Life cycle assessment of electric vehicles in comparison to combustion engine vehicles: A review. *Mater. Today Proc.* **2021**, *49*, 217–222. [CrossRef]
6. Viola, F. Electric vehicles and psychology. *Sustainability* **2021**, *13*, 719. [CrossRef]
7. Elwert, T.; Goldmann, D.; Römer, F.; Buchert, M.; Merz, C.; Schueler, D.; Sutter, J. Current developments and challenges in the recycling of key components of (hybrid) electric vehicles. *Recycling* **2015**, *1*, 25–60. [CrossRef]
8. Li, H.; Wang, Y.; Fan, F.; Yu, H.; Chu, J. Sustainable Plant Layout Design for End of Life Vehicle Recycling and Disassembly Industry Based on SLP Method, a Typical Case in China. *IEEE Access* **2021**, *9*, 81913–81925. [CrossRef]
9. Blömeke, S.; Scheller, C.; Cerdas, F.; Thies, C.; Hachenberger, R.; Gonter, M.; Herrmann, C.; Spengler, T.S. Material and energy flow analysis for environmental and economic impact assessment of industrial recycling routes for lithium-ion traction batteries. *J. Clean. Prod.* **2022**, *377*, 134344. [CrossRef]
10. Halepoto, H.; Gong, T.; Noor, S.; Memon, H. Bibliometric Analysis of Artificial Intelligence in Textiles. *Materials* **2022**, *15*, 2910. [CrossRef]
11. Donthu, N.; Kumar, S.; Mukherjee, D.; Pandey, N.; Lim, W.M. How to conduct a bibliometric analysis: An overview and guidelines. *J. Bus. Res.* **2021**, *133*, 285–296. [CrossRef]
12. Cabeza, L.F.; Chàfer, M.; Mata, É. Comparative analysis of web of science and scopus on the energy efficiency and climate impact of buildings. *Energies* **2020**, *13*, 409. [CrossRef]
13. Borri, E.; Tafone, A.; Zsembinszki, G.; Comodi, G.; Romagnoli, A.; Cabeza, L.F. Recent trends on liquid air energy storage: A bibliometric analysis. *Appl. Sci.* **2020**, *10*, 2773. [CrossRef]

14. Choi, W.; Kim, J.; Lee, S.; Park, E. Smart home and internet of things: A bibliometric study. *J. Clean. Prod.* **2021**, *301*, 126908. [CrossRef]
15. Tseng, M.-L.; Chang, C.-H.; Lin, C.-W.R.; Wu, K.-J.; Chen, Q.; Xia, L.; Xue, B. Future trends and guidance for the triple bottom line and sustainability: A data driven bibliometric analysis. *Environ. Sci. Pollut. Res.* **2020**, *27*, 33543–33567. [CrossRef] [PubMed]
16. Ejaz, H.; Zeeshan, H.M.; Ahmad, F.; Bukhari, S.N.A.; Anwar, N.; Alanazi, A.; Sadiq, A.; Junaid, K.; Atif, M.; Abosalif, K.O.A. Bibliometric analysis of publications on the omicron variant from 2020 to 2022 in the Scopus database using R and VOSviewer. *Int. J. Environ. Res. Public Health* **2022**, *19*, 12407. [CrossRef]
17. Aria, M.; Cuccurullo, C. Bibliometrix: An R-tool for comprehensive science mapping analysis. *J. Informetr.* **2017**, *11*, 959–975. [CrossRef]
18. Olivetti, E.A.; Ceder, G.; Gaustad, G.G.; Fu, X. Lithium-ion battery supply chain considerations: Analysis of potential bottlenecks in critical metals. *Joule* **2017**, *1*, 229–243. [CrossRef]
19. Fan, E.; Li, L.; Wang, Z.; Lin, J.; Huang, Y.; Yao, Y.; Chen, R.; Wu, F. Sustainable recycling technology for Li-ion batteries and beyond: Challenges and future prospects. *Chem. Rev.* **2020**, *120*, 7020–7063. [CrossRef]
20. Hannan, M.A.; Hoque, M.M.; Hussain, A.; Yusof, Y.; Ker, P.J. State-of-the-art and energy management system of lithium-ion batteries in electric vehicle applications: Issues and recommendations. *IEEE Access* **2018**, *6*, 19362–19378. [CrossRef]
21. Chen, M.; Ma, X.; Chen, B.; Arsenault, R.; Karlson, P.; Simon, N.; Wang, Y. Recycling end-of-life electric vehicle lithium-ion batteries. *Joule* **2019**, *3*, 2622–2646. [CrossRef]
22. Zheng, X.; Zhu, Z.; Lin, X.; Zhang, Y.; He, Y.; Cao, H.; Sun, Z. A mini-review on metal recycling from spent lithium ion batteries. *Engineering* **2018**, *4*, 361–370. [CrossRef]
23. Zackrisson, M.; Avellán, L.; Orlenius, J. Life cycle assessment of lithium-ion batteries for plug-in hybrid electric vehicles—Critical issues. *J. Clean. Prod.* **2010**, *18*, 1519–1529. [CrossRef]
24. Liu, C.; Lin, J.; Cao, H.; Zhang, Y.; Sun, Z. Recycling of spent lithium-ion batteries in view of lithium recovery: A critical review. *J. Clean. Prod.* **2019**, *228*, 801–813. [CrossRef]
25. Miao, Y.; Hynan, P.; Von Jouanne, A.; Yokochi, A. Current Li-ion battery technologies in electric vehicles and opportunities for advancements. *Energies* **2019**, *12*, 1074. [CrossRef]
26. Zeng, X.; Li, J.; Shen, B. Novel approach to recover cobalt and lithium from spent lithium-ion battery using oxalic acid. *J. Hazard. Mater.* **2015**, *295*, 112–118. [CrossRef]
27. Ciez, R.E.; Whitacre, J. Examining different recycling processes for lithium-ion batteries. *Nat. Sustain.* **2019**, *2*, 148–156. [CrossRef]
28. Armand, M.; Axmann, P.; Bresser, D.; Copley, M.; Edstrom, K.; Ekberg, C.; Guyomard, D.; Lestriez, B.; Novák, P.; Petranikova, M. Lithium-ion batteries—Current state of the art and anticipated developments. *J. Power Sources* **2020**, *479*, 228708. [CrossRef]
29. Winslow, K.M.; Laux, S.J.; Townsend, T.G. A review on the growing concern and potential management strategies of waste lithium-ion batteries. *Resour. Conserv. Recycl.* **2018**, *129*, 263–277. [CrossRef]
30. Richa, K.; Babbitt, C.W.; Gaustad, G.; Wang, X. A future perspective on lithium-ion battery waste flows from electric vehicles. *Resour. Conserv. Recycl.* **2014**, *83*, 63–76. [CrossRef]
31. Makuza, B.; Tian, Q.; Guo, X.; Chattopadhyay, K.; Yu, D. Pyrometallurgical options for recycling spent lithium-ion batteries: A comprehensive review. *J. Power Sources* **2021**, *491*, 229622. [CrossRef]
32. Zeng, X.; Li, J.; Liu, L. Solving spent lithium-ion battery problems in China: Opportunities and challenges. *Renew. Sustain. Energy Rev.* **2015**, *52*, 1759–1767. [CrossRef]
33. Xu, C., Dai, Q.; Gaines, L.; Hu, M.; Tukker, A.; Steubing, B. Future material demand for automotive lithium-based batteries. *Commun. Mater.* **2020**, *1*, 99. [CrossRef]
34. Velázquez-Martínez, O.; Valio, J.; Santasalo-Aarnio, A.; Reuter, M.; Serna-Guerrero, R. A critical review of lithium-ion battery recycling processes from a circular economy perspective. *Batteries* **2019**, *5*, 68. [CrossRef]
35. Baars, J.; Domenech, T.; Bleischwitz, R.; Melin, H.E.; Heidrich, O. Circular economy strategies for electric vehicle batteries reduce reliance on raw materials. *Nat. Sustain.* **2021**, *4*, 71–79. [CrossRef]
36. Işıldar, A.; van Hullebusch, E.D.; Lenz, M.; Du Laing, G.; Marra, A.; Cesaro, A.; Panda, S.; Akcil, A.; Kucuker, M.A.; Kuchta, K. Biotechnological strategies for the recovery of valuable and critical raw materials from waste electrical and electronic equipment (WEEE)—A review. *J. Hazard. Mater.* **2019**, *362*, 467–481. [CrossRef]
37. Wang, Y.; An, N.; Wen, L.; Wang, L.; Jiang, X.; Hou, F.; Yin, Y.; Liang, J. Recent progress on the recycling technology of Li-ion batteries. *J. Energy Chem.* **2021**, *55*, 391–419. [CrossRef]
38. Yun, L.; Linh, D.; Shui, L.; Peng, X.; Garg, A.; Le, M.L.P.; Asghari, S.; Sandoval, J. Metallurgical and mechanical methods for recycling of lithium-ion battery pack for electric vehicles. *Resour. Conserv. Recycl.* **2018**, *136*, 198–208. [CrossRef]
39. Cusenza, M.A.; Bobba, S.; Ardente, F.; Cellura, M.; Di Persio, F. Energy and environmental assessment of a traction lithium-ion battery pack for plug-in hybrid electric vehicles. *J. Clean. Prod.* **2019**, *215*, 634–649. [CrossRef]
40. Li, L.; Zhang, X.; Li, M.; Chen, R.; Wu, F.; Amine, K.; Lu, J. The recycling of spent lithium-ion batteries: A review of current processes and technologies. *Electrochem. Energy Rev.* **2018**, *1*, 461–482. [CrossRef]
41. Wang, W.; Wu, Y. An overview of recycling and treatment of spent $LiFePO_4$ batteries in China. *Resour. Conserv. Recycl.* **2017**, *127*, 233–243. [CrossRef]

42. Larouche, F.; Tedjar, F.; Amouzegar, K.; Houlachi, G.; Bouchard, P.; Demopoulos, G.P.; Zaghib, K. Progress and status of hydrometallurgical and direct recycling of Li-ion batteries and beyond. *Materials* **2020**, *13*, 801. [CrossRef] [PubMed]
43. De Souza, L.L.P.; Lora, E.E.S.; Palacio, J.C.E.; Rocha, M.H.; Renó, M.L.G.; Venturini, O.J. Comparative environmental life cycle assessment of conventional vehicles with different fuel options, plug-in hybrid and electric vehicles for a sustainable transportation system in Brazil. *J. Clean. Prod.* **2018**, *203*, 444–468. [CrossRef]
44. Costa, C.M.; Barbosa, J.C.; Gonçalves, R.; Castro, H.; Del Campo, F.; Lanceros-Méndez, S. Recycling and environmental issues of lithium-ion batteries: Advances, challenges and opportunities. *Energy Storage Mater.* **2021**, *37*, 433–465. [CrossRef]
45. Thompson, D.L.; Hartley, J.M.; Lambert, S.M.; Shiref, M.; Harper, G.D.; Kendrick, E.; Anderson, P.; Ryder, K.S.; Gaines, L.; Abbott, A.P. The importance of design in lithium ion battery recycling—A critical review. *Green Chem.* **2020**, *22*, 7585–7603. [CrossRef]
46. Richa, K.; Babbitt, C.W.; Gaustad, G. Eco-efficiency analysis of a lithium-ion battery waste hierarchy inspired by circular economy. *J. Ind. Ecol.* **2017**, *21*, 715–730. [CrossRef]
47. Rothermel, S.; Evertz, M.; Kasnatscheew, J.; Qi, X.; Grützke, M.; Winter, M.; Nowak, S. Graphite recycling from spent lithium-ion batteries. *ChemSusChem* **2016**, *9*, 3473–3484. [CrossRef]
48. Gaines, L.; Singh, M. *Energy and Environmental Impacts of Electric Vehicle Battery Production and Recycling*; Argonne National Lab. (ANL): Argonne, IL, USA, 1995.
49. Biradar, S.; Patil, R.; Ullegaddi, M. Energy storage system in electric vehicle. In Proceedings of the Power Quality '98, Hyderabad, India, 18–18 June 1998; pp. 247–255.
50. Gaines, L.; Cuenca, R. Life-cycle costs of lithium-ion vehicle batteries. *J. Passeng. Cars Mech. Syst. J.* **2000**, *109*, 1920–1931.
51. Messagie, M.; Boureima, F.; Matheys, J.; Sergeant, N.; Timmermans, J.-M.; Macharis, C.; Van Mierlo, J. Environmental performance of a battery electric vehicle: A descriptive Life Cycle Assessment approach. *World Electr. Veh. J.* **2010**, *4*, 782–786. [CrossRef]
52. Hoyer, C.; Kieckhäfer, K.; Spengler, T.S. A strategic framework for the design of recycling networks for lithium-ion batteries from electric vehicles. In *Glocalized Solutions for Sustainability in Manufacturing, Proceedings of the 18th CIRP International Conference on Life Cycle Engineering, Braunschweig, Germany, 2–4 May 2011*; Technische Universität Braunschweig: Braunschweig, Germany, 2011; pp. 79–84.
53. Howes, J. Battery Recycling Interaction with Washington. *SAE Int. J. Mater. Manuf.* **2012**, *5*, 150–159. [CrossRef]
54. Tytgat, J. The Recycling Efficiency of Li-ion EV batteries according to the European Commission Regulation, and the relation with the End-of-Life Vehicles Directive recycling rate. In Proceedings of the 2013 World Electric Vehicle Symposium and Exhibition (EVS27), Barcelona, Spain, 17–20 November 2013; pp. 1–9.
55. Hoyer, C.; Kieckhäfer, K.; Spengler, T.S. Impact of mandatory rates on the recycling of lithium-ion batteries from electric vehicles in Germany. In *Re-Engineering Manufacturing for Sustainability, Proceedings of the 20th CIRP International Conference on Life Cycle Engineering, Singapore, 17–19 April 2013*; Springer Science & Business Media: Berlin, Germany, 2013; pp. 543–548.
56. Ramoni, M.O.; Zhang, H.-C. End-of-life (EOL) issues and options for electric vehicle batteries. *Clean Technol. Environ. Policy* **2013**, *15*, 881–891. [CrossRef]
57. Miedema, J.H.; Moll, H.C. Lithium availability in the EU27 for battery-driven vehicles: The impact of recycling and substitution on the confrontation between supply and demand until 2050. *Resour. Policy* **2013**, *38*, 204–211. [CrossRef]
58. Zeng, X.; Li, J. Spent rechargeable lithium batteries in e-waste: Composition and its implications. *Front. Environ. Sci. Eng.* **2014**, *8*, 792–796. [CrossRef]
59. Hendrickson, T.P.; Kavvada, O.; Shah, N.; Sathre, R.; Scown, C.D. Life-cycle implications and supply chain logistics of electric vehicle battery recycling in California. *Environ. Res. Lett.* **2015**, *10*, 014011. [CrossRef]
60. Träger, T.; Friedrich, B.; Weyhe, R. Recovery concept of value metals from automotive lithium-ion batteries. *Chem. Ing. Tech.* **2015**, *87*, 1550–1557. [CrossRef]
61. Oliveira, L.; Messagie, M.; Rangaraju, S.; Sanfelix, J.; Rivas, M.H.; Van Mierlo, J. Key issues of lithium-ion batteries–from resource depletion to environmental performance indicators. *J. Clean. Prod.* **2015**, *108*, 354–362. [CrossRef]
62. Qiao, Q.; Zhao, F.; Liu, Z.; Jiang, S.; Hao, H. Cradle-to-gate greenhouse gas emissions of battery electric and internal combustion engine vehicles in China. *Appl. Energy* **2017**, *204*, 1399–1411. [CrossRef]
63. Li, J.; Barwood, M.; Rahimifard, S. Robotic disassembly for increased recovery of strategically important materials from electrical vehicles. *Robot. Comput. Integr. Manuf.* **2018**, *50*, 203–212. [CrossRef]
64. Bulach, W.; Schüler, D.; Sellin, G.; Elwert, T.; Schmid, D.; Goldmann, D.; Buchert, M.; Kammer, U. Electric vehicle recycling 2020: Key component power electronics. *Waste Manag. Res.* **2018**, *36*, 311–320. [CrossRef] [PubMed]
65. Diehl, O.; Schönfeldt, M.; Brouwer, E.; Dirks, A.; Rachut, K.; Gassmann, J.; Güth, K.; Buckow, A.; Gauß, R.; Stauber, R. Towards an alloy recycling of Nd–Fe–B permanent magnets in a circular economy. *J. Sustain. Metall.* **2018**, *4*, 163–175. [CrossRef]
66. Han, X.; Liang, Y.; Ai, Y.; Li, J. Economic evaluation of a PV combined energy storage charging station based on cost estimation of second-use batteries. *Energy* **2018**, *165*, 326–339. [CrossRef]
67. Lai, C.-M.; Li, Y.-H.; Cheng, Y.-H.; Teh, J. A high-gain reflex-based bidirectional DC charger with efficient energy recycling for low-voltage battery charging-discharging power control. *Energies* **2018**, *11*, 623. [CrossRef]
68. Borregan-Alvarado, J.; Alvarez-Meaza, I.; Cilleruelo-Carrasco, E.; Garechana-Anacabe, G. A bibliometric analysis in industry 4.0 and advanced manufacturing: What about the sustainable supply chain? *Sustainability* **2020**, *12*, 7840. [CrossRef]

69. Khan, M.A.; Pattnaik, D.; Ashraf, R.; Ali, I.; Kumar, S.; Donthu, N. Value of special issues in the journal of business research: A bibliometric analysis. *J. Bus. Res.* **2021**, *125*, 295–313. [CrossRef]
70. Yu, Y.; Li, Y.; Zhang, Z.; Gu, Z.; Zhong, H.; Zha, Q.; Yang, L.; Zhu, C.; Chen, E. A bibliometric analysis using VOSviewer of publications on COVID-19. *Ann. Transl. Med.* **2020**, *8*, 816. [CrossRef]

World Electric Vehicle Journal

MDPI

Article

Vehicle-Integrated Photovoltaics—A Case Study for Berlin

Philipp Hoth, Ludger Heide *, Alexander Grahle and Dietmar Göhlich

Methods of Product Development and Mechatronics, Technische Universität Berlin, Straße des 17. Juni 135, 10623 Berlin, Germany; philipp.hoth@campus.tu-berlin.de (P.H.); alexander.grahle@tu-berlin.de (A.G.); dietmar.goehlich@tu-berlin.de (D.G.)

* Correspondence: ludger.heide@tu-berlin.de

Abstract: Recent developments in vehicle-integrated photovoltaics (VIPV) offer prospects for enhancing electric vehicle range, lowering operating costs, and supporting carbon-neutral transport, particularly in urban settings. This study evaluates the solar energy potential of parking spaces in Berlin, considering challenges like building and tree shading using digital surface models and weather data for solar simulations. Utilizing open datasets and software, the analysis covered 48,827 parking spaces, revealing that VIPV could extend vehicle range by 7 to 14 km per day, equating to a median annual increase of 2527 km. The findings suggest median yearly cost savings of 164 euros from reduced grid charging. However, the environmental benefits of solar vehicle charging were found to be less than those of traditional grid-connected photovoltaic systems. The study introduces a method to pinpoint parking spaces that are most suitable for solar charging.

Keywords: vehicle-integrated photovoltaics; VIPV; urban environment; solar simulation; trees; GIS; open source; open data

Citation: Hoth, P.; Heide, L.; Grahle, A.; Göhlich, D. Vehicl-Integrated Photovoltaics—A Case Study for Berlin. *World Electr. Veh. J.* **2024**, *13*, 113. https://doi.org/10.3390/wevj15030113

Academic Editors: Wenbin Yu and Guang Zeng

Received: 20 February 2024
Revised: 11 March 2024
Accepted: 12 March 2024
Published: 15 March 2024

1. Introduction

With climate change being a monumental challenge of the 21st century, the European Union (EU) has been implementing policies to reduce greenhouse gas (GHG) emissions. In the 2030 Climate Target plan, the European Commission proposed to cut GHG emissions by at least 55% until 2030—compared to 1990—and pursue climate neutrality by 2050 [1]. The transport sector, responsible for about 25% of GHG emissions in Europe, is a primary contributor in reaching those climate goals.

In the 2022 "REPower EU" plan, the EU adopted a solar energy strategy that aims to accelerate the deployment of solar energy to reduce GHG emissions and the dependence on fossil fuels [2]. Besides stationary applications of photovoltaics (PV), new approaches on utilizing PV in the transport sector can become beneficial in helping reduce carbon emissions.

The market share of Electric Vehicles (EV) has been on the rise over the last decade, and 17 countries have already set targets to phase out the internal combustion engine within the next few decades ([3], p. 15f). For the city of Berlin, the electrification of all private cars would lead to an additional daily electricity demand of more than 6000 MWh [4]. To reduce this demand, producing energy onboard with the use of solar panels has been suggested.

Recently, new developments of vehicle-integrated photovoltaics have appeared. A small number of Original Equipment Manufacturers (OEMs) already offer an optional solar roof with modest energy outputs, and there have been promising developments with fully solar-oriented concepts that are trying to enter the market.

This study will focus on the solar potential of vehicles with integrated photovoltaics in the urban environment of Berlin-Neukölln, Germany, which represents a densely populated urban center in Germany well and has a dataset of all parking spots available (Section 3.1). A digital surface model-based solar simulation was conducted. The model factored in the effect of shading in urban surroundings throughout the day in a year. This model was then evaluated using a dataset of 48,827 parking sites to identify the suitability for the solar charging of parked photvoltaic-equipped EVs.

Additionally, this research will generate a methodological approach to assess solar irradiation at an urban street level, taking seasonal vegetation influence into account. A distinct feature of the method is the open data approach that enables reproducibility for any desired location, with freely available data and free, open-source software.

2. State of the Art

This section provides a literature review of research on vehicle-integrated photovoltaics (VIPV), with a focus on studies that assess vehicle PV potentials in urban environments. Further, the scope of this work is defined, and potential research gaps are identified. In the end of Section 2.5, the research gap is identified and research questions are formulated.

2.1. Commercial Availability of the Technology

A review of current VIPV products, lightweight PV cells, and module technologies was conducted by Commault et al. [5]. It was concluded that VIPV products and developments for various applications are on a steady rise, especially car-based vehicle integration.

Currently available car models with integrated solar roofs are the Toyota Prius PHV [6], the Hyundai Sonata [7], and the Hyundai Ioniq 5. Meanwhile, there are current developments to integrate PV into mass-produced private passenger cars: Sono Motors Sion [8], Lightyear, Mercedes EQXX Concept, Fisker Ocean Fisker, Inc., Manhattan Beach, CA, USA [9], Squad Solar City, Breda, The Nederland [10], and Aptera Sol Aptera Motors Corp., Carlsbad, CA, USA [11].

2.2. Previous Studies on Achievable Energy

Earlier studies about VIPV have mainly focused on supplying additional power to auxiliary systems in applications such as refrigerating, heating, or air conditioning in commercial trucks [12], buses [13], or ambulances [14]. These systems are heavily reliant on battery power when the internal combustion engine is not running. The extra energy supplied from solar panels can extend the capabilities of the onboard batteries and save fuel when the internal combustion engine is running. In the context of this study, PV systems on vehicles are exclusively used to charge the high-voltage (HV) battery of EVs and therefore contribute to the driving energy, which would typically be obtained by charging from the electricity grid.

Yamaguchi et al. [15] assumed a relatively high solar module efficiency of 35%, which enables over 30 km/day of solar driving under an average irradiance of 4 $\text{kWh}/(\text{m}^2 \cdot \text{day})$. This study provides cost targets for different module efficiencies and highlights the importance of low costs for the solar panels.

Lee and Park [16] analyzed electrical scooters in India and found that using a foldable 40 Wp solar panel, up to 45% of the energy demand could be satisfied using solar energy. They did not include building shading in their considerations.

Numerous studies that focus on PV potentials do not include shading characteristics of a real-world environment or set rather simple assumptions about the influence of shading. These studies focus primarily on vehicle parameters, including the PV systems, vehicle geometry, as well as battery and charging management systems. Heinrich et al. [17] have shown that with PV on a typical sedan car roof (1.7–2 m^2; 20% panel efficiency), an additional annual driving distance of 1900–3400 km can be achieved in Freiburg, southern Germany. This amounts to 13–23% of the yearly mean driving distance of cars in Germany (15,000 km). This would also result in approximately one fewer charging stop per month in the summer and an increased range on sunny days. This scenario did not account for losses due to shading. In a second 'at work charging' scenario, a car parked in the sun from 9 a.m. to 5 p.m. from Monday to Friday (assuming no sun exposure the rest of the time) resulted in an annual solar range of 1100–2030 km, which is ~60% of the range from the first 'always unshaded' scenario. The main challenges stated are transformation losses from the panel to the HV battery, as well as possible battery self-consumption. Another issue addressed is the curvature of car roofs, which can lead to a significant irradiance mismatch, resulting in decreased panel efficiency. Open parking/driving during the day is crucial to generate substantial energy yields. Also, the panel technology must be durable and provide an aesthetic appearance for customers. For an improvement of energy yields, other surfaces, such as the hood and/or sides, could also be equipped with solar modules.

Kutter et al. [18] have assessed the yield potential of VIPV on commercial trucks and vans. One of the five scenarios across Stockholm, Freiburg, and Seville was the use-case of a parcel delivery van, which can achieve an annual solar range of 6637 to 11,450 km, which covers 35% to 60% of the annual energy demand. The panel efficiency was assumed at 21%, and shading was not considered in this study. They also found that self-consumption can be a critical factor in the feasibility of VIPV and that it needs to be minimized. The study conducted a break-even analysis, mainly determined by irradiation, electricity prices, and vehicle charging efficiency.

2.3. Weather and Shading

To analyze the real-world potential of VIPV, it is essential to assess the influence of environmental shading. Shadowing is more prominent in urban areas where a high building and vegetation density reduce solar energy yields. There are several studies that used test vehicles to measure shadow cast in real street environments.

Carr et al. [19] measured the solar irradiance on a test vehicle in the Netherlands over a time span of 20 weeks. With extrapolated driving patterns and shadowing events, the authors showed that in good conditions, VIPV could provide up to 41% of the required driving energy and up to 5617 km/year for the reference vehicle. Another study that measured irradiance in a real-world scenario with a test vehicle is the work of Araki et al. [20]. Onboard pyranometer data measurements in Japan showed that shading impacts and a curved car roof decreased the practical solar resources by 25%. It is of note that the vehicle measurements in the Netherlands and Japan were acquired in a mainly non-urban environment. An overview of various case studies and solar irradiance measurements was composed by Araki et al. [21] that showed how different research vehicles perform in different regions. For PHEVs (Plug-in Hybrid Electric Vehicle) an average power generation of 2.1 kWh/day over 100 days was achieved on a Toyota Prius equipped with 800 Wp solar panels [22]. Under the real environment conditions in Japan, they achieved a GHG reduction of 63%. Further measurements under real driving conditions were conducted by Oh et al. [23], Ota et al. [24] in South Korea and Japan. Ota et al. [24] concluded that the PV performance under intermittent shading conditions in an urban setting still achieves a relatively high performance ratio of 0.87. Oh et al. [23] combined irradiance measurements on a solar bus route with irradiance modeling and concluded that there is little economic feasibility with the current technology used in the study. But with better utilization of roof space, it has high potential in the future. Another study of a bus route in New Zealand found that installed solar panels can offset ~8.5% of the electricity demand [25].

Many studies have assessed the solar potential of VIPV with solar irradiance datasets that provide an accurate estimation, considering weather conditions. Most solar simulations in an urban environment assess the solar potentials of rooftops. However, real-world street-level urban shading influences based on digital surface model (DSM) data have been sparsely investigated. DSM-based solar simulations are an attractive tool to assess solar yields with complex shading characteristics over a larger urban area.

The study most closely related to the approach and methodology of this article is the work of Brito et al. [26], who have assessed solar potential in the urban area of Lisbon, Portugal. It is based on the irradiance modeling methodology by Santos et al. [27]. The authors used a Digital Surface Model on which a solar simulation was conducted to include the effect of shadowing in the urban environment. They measured the solar irradiation on parking sites and roads to estimate the solar yield for

VIPV. The results showed that the annual loss due to shadowing can reach 25% for roads and over 50% for urban parking spaces. Despite the shading losses, the average solar range extension was between 10 and 18 km/(day), assuming that there is the installation of a photovoltaics system of 1 kW peak power (kWp), which can significantly reduce charging needs in the summer, especially. Since the study performed by Brito et al. [26] was found to have a very similar approach to this study, it is further discussed in the methodology, Section 3. A further study by Said-Romdhane et al. [28] investigated shading losses by city obstacles and concluded that urban shading is a serious problem for maximizing the photovoltaic potential.

2.4. Life Cycle Assessment

Kanz et al. [29] presented a life cycle assessment (LCA) of a VIPV system on a light utility electric vehicle, the StreetScooter Work L. The study compared the on-board PV electricity generation with grid charging by calculating the CO_2 equivalents per kWh for the location of Cologne, Germany. With a PV array of 930 Wp, an operation time of 8 years, and an average shadowing factor of 30%, an emission factor of 0.357 kg CO_2-eq/kWh could be achieved. The average grid charging emissions amount to 0.435 kg CO_2-eq/kWh. However, if the shadowing factor exceeds 40%, the VIPV emissions would surpass the grid charging emissions. Kanz et al. further consider scenarios of a longer vehicle lifetime of 12 years and a "green" electricity manufacturing process, both of which decrease the emissions of the VIPV solution. On the other hand, the grid charging emissions are also expected to decrease in the future with the further deployment of renewable energies in Germany.

2.5. Research Gap and Research Questions

The current literature review on the potential of VIPV has shown that there are opportunities for further research on spatial solar simulations that take shadowing into account. Specifically, a gap was identified in the research of irradiance modeling for VIPV potentials that accounts for the influence of vegetation. In previous studies, tree shading was assumed to be constant throughout the year, affecting results by up to 50% (see Section 3.4). Furthermore, there are no published studies that have assessed the street-level solar potential in Central Europe and no approaches utilizing only open data and free software.

The scope of this work includes a solar assessment for VIPV at the street level, while considering the influences of building and vegetation shading. The focus lies on the characteristics of urban solar irradiation and the resulting potentials of parked SEVs, assuming a flat horizontal solar panel on the car roof. It is summarized in the following three research questions:

- What solar yield can be achieved by using VIPV on parked vehicles in urban areas, taking into account shading and partial shading of vegetation and buildings?
- What are the potential solar range benefits and charging cost savings for private parked cars with the use of VIPV?
- From an environmental perspective, how does the implementation of VIPV on parked vehicles compare to traditional grid-connected PV systems?

The result quantifies the desired irradiance, solar yield, and range values, while identifying parking site locations that can maximize the solar potential of VIPV. This includes a novel consideration of the impact of seasonally varying vegetation.

3. Methodology

The approach to calculate the solar yield potential of Solar Electric Vehicles (SEVs) is based on a Digital Surface Model (DSM), to which a monthly solar radiation simulation is applied. The solar simulation is enriched by long-term weather data as well as vegetation influence. A detailed database of the parking points enables the assessment of received solar energy on vehicles parked in the study area of Berlin-Neukölln.

All datasets used in this study are publicly available as open data to enable reproducibility. The free geographic information systems QGIS 3.22.1 [30] and GRASS (Geographic Resources Analysis Support System) 7.8 [31] were used to develop the solar yield model and visualizations. OpenStreetMap (OSM) [32] was used as the geographic database.

DSMs are elevation raster maps generated by aerial LiDAR (Light Detection And Ranging) measurements or image-based digital surface models. They provide highly detailed altimetry datasets of cities or other desired regions. DSMs are the foundation for conducting solar simulations to generate solar irradiation maps that provide spatial information about solar energy potential, as well as shading influence by buildings, terrain, and vegetation throughout the day and year [33].

The methodology chosen in this study was found to closely resemble the works of Brito et al. [26], who assessed the "Urban solar potential for vehicle integrated photovoltaics" for the city of Lisbon. Their main concept also consisted of conducting solar simulation models based on DSMs to analyze the solar potential of vehicles in an urban area. Aside from the main idea of the solar assessment, there are also differences in the approaches, as seen in Table 1. A comparison of the study results follows in Section 5.

Table 1. Methodology comparison to Brito et al., 2021 [26].

Topic	Centeno Brito et al. [26]	This Study
Location	Lisbon	Berlin
GIS and Simulation tool	ArcGIS/Area Solar Radiation numeric model	QGIS/GRASS r.sun
Vegetation	Parking sites under trees are assumed to be totally shaded at all times	Tree shading and transmittance of trees is taken into account
Street area	Parking Sites and road network	Parking Sites only
Ground inclination (incline/tilt of roads)	Ground inclination is considered	Ground inclination is neglected (i.e., vehicles are leveled)
Vehicle heights	Vehicle heights are neglected	Vehicle height assumed at 2 m
Life Cycle CO_2 emissions	Out of scope	Considered

One of the major differences of this study compared to Brito et al. [26] is that the Lisbon study assumed that parking spots under trees were being shaded at all times. In this study, parking points under trees are also treated differently. However, the seasonal vegetation transmittance is being taken into account to be able to make accurate solar yield predications about all of the available public parking sites of the study area.

The methodology of this study is presented in the following subchapters, starting from the input parking point data (Section 3.1), the development of the custom Digital Surface Model (Section 3.2), the solar simulation and solar maps (Section 3.3), integration of vegetation (Section 3.4), and the assumptions about the solar yield model and vehicle solar system (Section 3.5). A condensed overview of the methodology is shown in Figure 1.

3.1. Parking Sites in Berlin-Neukölln

In the district (in this case, 'district' refers to the German definition of 'Ortsteil' instead of the 'Bezirk' Neukölln) of Berlin-Neukölln, the location of 48,827 individual public parking sites have been catalogued by the "OpenStreetMap Parking Space Project" by Seidel [34]. Currently, this dataset is the only publicly available high-detail dataset of parking sites in Berlin. The parking sites are represented by a point, their respective map coordinates, and other useful information, such as street name. The dataset extends over the district limits of Neukölln.

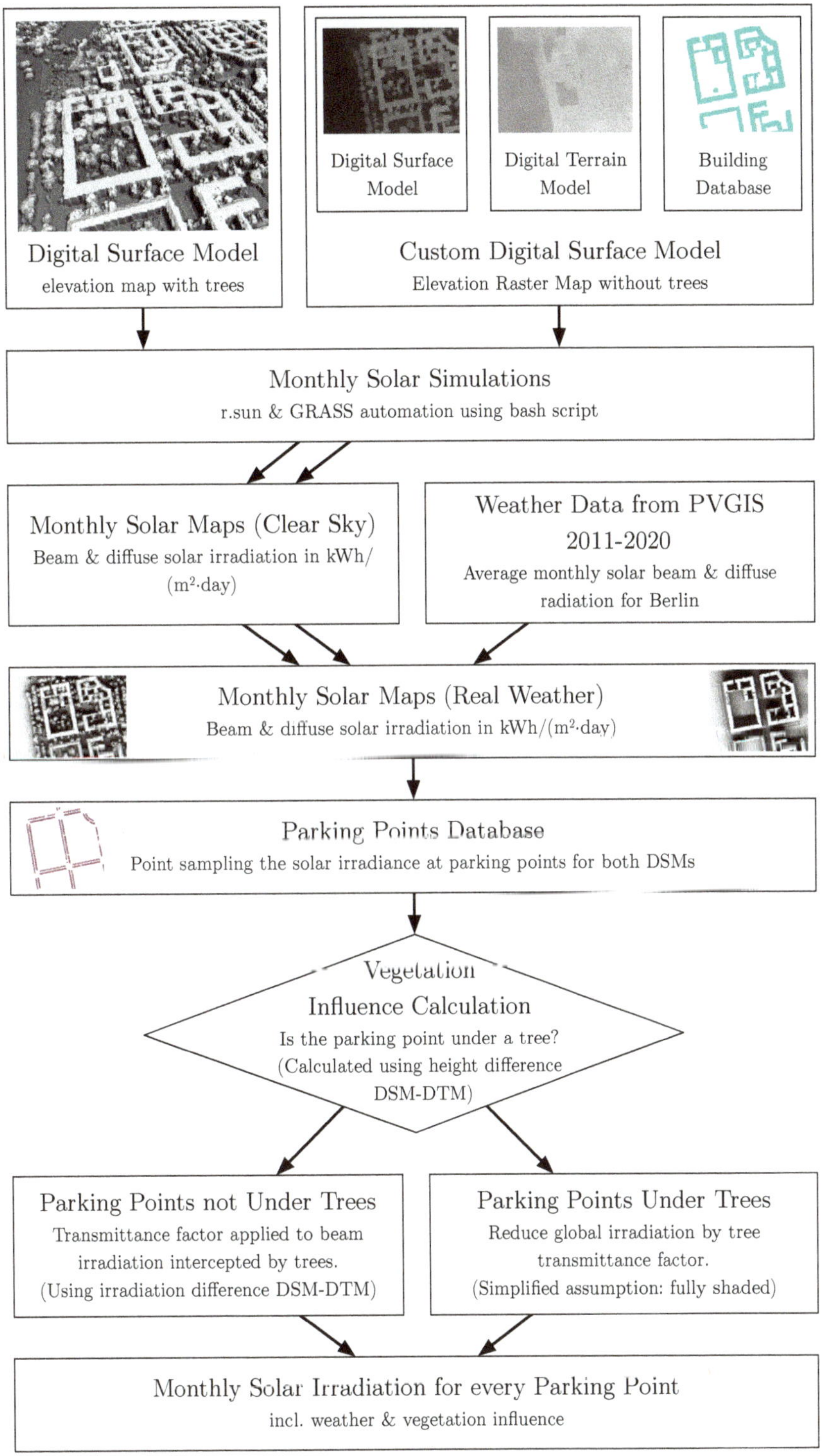

Figure 1. Methodology of the solar irradiation assessment on parking sites.

On average, private vehicles in Germany are parked 23 h of the day [35]; hence, the solar energy received while driving is not taken into account in this study. Therefore, the resulting data are representative of vehicles being parked outside the whole day. While driving, the solar energy yield can vary greatly, depending on the amount of direct solar exposure, time of day, cloudiness, etc.

This study assesses outdoor public parking areas only. According to Seidel et al. in the study area, about 19% of parking spaces are in enclosed or underground parking garages, which will not be considered in this study [34]. Also, the original parking point dataset was purged of 37 unsuitable points (19 of which had no coordinates, and 18 points were located under over-street buildings).

3.2. Development of the Custom Digital Surface Model

In order to calculate the solar potential of vehicles in a dense urban environment, a high-resolution DSM (Digital Surface Model) is needed to simulate sun shading caused by buildings, terrain, and vegetation throughout the day. In 2020, new image-based DSMs of Berlin were published by the Berlin Senate Administration of Urban Development, Construction and Living [36]. In this study, three datasets were used to develop a Custom DSM (from now on called DSM-nt, with nt meaning 'no trees') that enabled the integration of vegetation influence in the solar yield model.

The first dataset is an image-based DSM (bDOM—"Digitales bildbasiertes Oberflächenmodell") that contains the surface elevation of the entire city, including buildings and vegetation. The measurements were taken in the summer, when tree crowns are fully developed. This DSM will be used for the solar simulation, while it also provides the building elevation footprints for the custom DSM-nt, which is purged of trees.

To generate the custom DSM-nt, a second dataset in the form of a DTM (Digital Terrain Model; (DGM—"Digitales Geländemodell")) was used. This DTM contains information about the ground elevation. Buildings, vegetation, and other objects are not displayed. Both raster map datasets have a resolution of 1 m.

The third dataset, the WFS ALKIS Berlin building database supplied in QGIS, was used to crop the building footprints from the DSM to place them on top of the DTM ground. This 'building footprint' approach has been explained and used by Santos et al. [27] to analyze the potential of roof photovoltaics.

The DSM-nt along with the original DSM was then used to conduct the solar simulations discussed in Section 3.3. By simulating both DSMs, the shading influence of trees can be analyzed, as described in Section 3.4.

The DSM elevation models of the study area measure 8 × 6 km and therefore cover 48 km^2 of Berlin's urban area. The final step prior to the solar simulation was the integration of vehicle heights. To properly reflect the car roof's elevation, a vehicle height of 2 m was implemented into the DSM-nt raster map for each parking point.

3.3. Solar Simulation to Generate Monthly Solar Maps

For this study the *r.sun* GIS-based solar radiation model of Hofierka and Šúri [37] was chosen. It calculates the solar irradiance $\left(\mathrm{W/m^2}\right)$ of elevation raster maps, in this case the DSM with trees and the custom DSM-nt of the study area Berlin-Neukölln. The proposed model simulates the sun position and radiation values for any given location, time, and date of the year. It casts shadows caused by elevation, such as terrain, buildings, and vegetation. The solar radiation is being split up into the beam component (direct sunlight) and the diffuse component (scattered light by the atmosphere). The *r.sun* solar radiation model is embedded in the open-source environment GRASS.

The *r.sun* model computes the solar irradiance $\left(\mathrm{W/m^2}\right)$ over a given raster height map for a certain point in time. To simulate an entire day the *r.sun* simulation was done in 15-min steps. These solar irradiance snapshots were then summed up to calculate the solar irradiation values for the day $\left(\mathrm{Wh/\left(m^2 \cdot day\right)}\right)$. The resulting daily solar irradiation maps are called solar maps which contain the irradiation values for every 1×1 m pixel of the raster map.

The automation was accomplished by programming a *bash* script that automatically simulates each timestep and sums up the values for an entire day. To overcome computing time limitations only the representative day for each month was simulated. The representative day of a month in this context is the day with the mean daylength of the respective month. The solar irradiation values of the representative day were then assumed be present for the whole month.

The r.sun radiation model in its basic configuration calculates the solar irradiance under clear-sky conditions. To account for weather influences, in this case cloud coverage, the historical meteorological data from 2011 to 2020 of the European Photovoltaic Geographical Information System (PVGIS) was used to calculate the mean direct and diffuse solar irradiation for each month [38]. The data provided originate from the SARAH2 database, which is derived from satellite images and is validated from measurements of commercial PV modules at the European Joint Research Center.

The clear-sky solar maps were then scaled with the real solar radiation data from the weather database to create solar maps that include real-world weather influences. In this process, the weather station elevation of 39 m was also accounted for since higher elevations result in higher solar irradiation values. Ground albedo as well as diffuse reflections from buildings were not taken into account. Ground albedo can be neglected since it is mainly notable in snow-covered regions. Diffuse or direct reflections from buildings are not within the capabilities of the r.sun solar radiation model.

The solar simulation generated 12 solar maps (one for each month) that contain the spatial distribution of solar irradiation $\left(\mathrm{Wh}/\left(\mathrm{m}^2 \cdot \mathrm{day}\right)\right)$, including weather influences as well as shading areas caused by buildings. A solar map for the whole year was generated as well by combining the monthly solar maps. To extract the solar irradiation values from the solar maps, the QGIS *Point Sampling Tool* was used for the dataset of the 48,827 parking points. The result generated vector point layers (GeoPackages) that contain values of the beam and diffuse solar irradiation for each parking point for every month. The geopackages were exported as .csv/.xslx files to conduct the further vegetation influence and solar yield calculations.

Since the r.sun radiation model calculates the diffuse fraction of the solar irradiation for a flat horizontal plane, the sky view factor (SVF) was integrated into the model. The SVF is important in determining the surface radiation balance in obstructed areas [39]. The SVF is defined as the proportion of visible sky above. In an urban environment, a significant proportion of the visible sky is blocked by buildings and trees, which reduces the diffuse irradiation. The diffuse irradiation is therefore multiplied with the sky view factor (a value between 0 and 1, while 1 represents an unobstructed sky) [40,41]. In this study, the SVF was calculated with the QGIS-integrated SAGA Tool *saga:skyviewfactor* with the cell size method and 16 sectors. The operation was performed on the custom DSM-nt, which means that the SVF for this study is representative for buildings only. A sky view analysis on the DSM with trees was not performed because the resulting values very close to objects (in this case, trees, vehicles, and other objects) were highly inaccurate.

3.4. Integration of Shading by Urban Vegetation

The resulting solar maps of the custom DSM-nt do not contain any information about vegetation, which was purged from the DSM with trees earlier. In this step, the vegetation is reintegrated into the model to simulate the real-world shading effect of urban trees. To evaluate the influence of trees on the solar yield of parked vehicles directly underneath trees, there were three attributes assessed:

The first was whether a parking point is located under a tree or not. This is determined by calculating the difference between the original Digital Surface Model (DSM) and the Digital Terrain Model (DTM). Since the DSM includes trees and the DTM does not, it can be measured if there is a tree above a certain parking point. In this study, it was assumed that if the difference of the DSM and DTM at a parking point is higher than 4 m, then this parking point is under a tree. Since the maximum height for trucks in Germany is 4 m, it is assumed that objects in the DSM that are lower than 4 m are trucks or other vehicles.

The second attribute to evaluate solar yield directly underneath trees is to quantify the transmittance (also referred to as transmissivity) of tree crowns. This topic was investigated by Konarska et al. [42] in the study "Transmissivity of solar radiation through crowns of single urban trees". The transmittance factor for several deciduous tree species was measured both in the summer, when trees are foliated, and winter, when trees are defoliated. Since the majority of trees in the study area of Neukölln are *Tilia* (Linden) and *Acer* (Maple) [43], which shed their leaves in winter, the transmittance factors by Konarska et al. [42] were applied. The mean transmittance factor of tree crowns was found to be 54% in winter and 10% in summer. This transmittance factor affects the received global irradiation (sum of the direct and beam components) under a tree. Similar transmittance factors were found by Oshio and Asawa [44], who measured an average transmittance of 0.1 during a summer afternoon on a tree-shaded sidewalk.

The third attribute required to calculate the shading influence of trees is the vegetation periods throughout the year. From the study on tree shadowing by Rosskopf et al. [45], it has been derived that deciduous trees are leafless from November until March. April to June exhibits the growth period, from July to September the tree crowns are fully developed, and in October, the leaves are shed. The transitions of the vegetation seasons were assumed to be linear. These attributes form the basis of a monthly transmittance factor, as shown in Table 2.

Table 2. Monthly transmittance factor of deciduous trees for global and direct solar irradiation.

Month	Jan.	Feb.	Mar.	Apr.	May	June	July	Aug.	Sept.	Oct.	Nov.	Dec.
Condition		leafless			growing			full crown		shed	leafless	
Transmittance global		54%		43%	32%	21%		10%		32%	54%	
Transmittance direct		48%		36%	25%	14%		3%		25%	48%	

Parking points that are not vertically located under a tree can still receive shading from trees nearby. In this model, the shading influence of trees for parking points not located under trees will only affect the beam component. (As explained in Section 3.3, the diffuse component in this model is only reduced by the sky view factor of the buildings.) To quantify the beam irradiation through tree crowns, the tree transmittance factor for direct irradiation also measured by Konarska et al. was applied (see Table 2).

The transmittance factors were then applied to the solar irradiation values of the affected parking points. The resulting solar irradiation values with weather and tree shading influence are presented in the following chapter.

A limitation of this method can be observed when the sun is at a lower angle in the sky and the irradiance hits locations under a tree. This limitation cannot be avoided when using raster maps. Shadow cast or sunshine under a tree cannot be simulated since the DSM only contains the maximum measured elevation for each 1 m × 1 m pixel. Therefore, in this study, the influence of trees was simplified and categorized in parking points under trees and parking points not under trees.

Even given these limitations, our approach improves on the state of the art in considering tree shading influence in an urban street environment. Other publications have mostly excluded the influence of vegetation in their research.

3.5. Solar Yield Calculation and Vehicle Assumptions

The final step to assess the solar potential of parked vehicles is to calculate the solar yield. The solar yield is the energy of the incoming Global Horizontal Irradiation that is being converted into electrical energy, which is stored in the vehicle battery.

For this study, it is assumed that a Solar Electric Vehicle can hold 1 kWp of installed photovoltaic power, for comparability with Commault et al. [5]. This would be achieved by a panel size of approximately 5 m² and a 20% panel efficiency (also, other combinations are possible, such as a panel with 25% efficiency on a 4 m² surface or a future high efficiency panel with 33.3% efficiency on a 3 m² surface). The solar panel is assumed to be a flat horizontal surface.

The energy conversion efficiency from the solar panels into the battery is assumed at a conservative 75%. The 25% of conversion losses are caused by the DC–DC converter, ohmic losses in cables, and other conversion losses, such as the mismatch of modules (partial shading), temperature influence, and soiling [18]. The average energy consumption of state-of-the-art electric vehicles is 13 kWh/100 km [17].

The electricity price for public AC charging was assumed at a moderate 0.50 €/kWh after comparing several charging energy suppliers from Germany [46]. A monthly base fee was not comprised in the pricing. It is expected that charging prices will increase in the future, which will improve the cost savings of onboard solar panels.

As stated in Section 3.1, the mean private car in Germany is parked for 23 h of the day; hence, the energy yields are calculated for a car being parked the whole day. The full battery effect (solar power not being utilized because the battery is already full) is not taken into account within this study. The assumptions to calculate the solar yield and extended driving range were also chosen to match with the works of Brito et al. [26] to enable a comparison of the study results.

4. Results

4.1. Solar Irradiation in the Urban Environment

The mean yearly solar irradiation in Berlin from 2011 to 2020 is 1132 kWh/m² on a free-standing horizontal plane [38]. The diffuse and the direct sunlight components each contribute to about half of the yearly energy amount. Throughout the year, there are big differences, with December having the lowest irradiation at 17.5 kWh/m² and June having the highest irradiation at 171.5 kWh/m². On the street level, in a dense urban environment like Berlin, these values are often significantly reduced due to shading, especially in winter, as seen in Figure 2. The solar maps shown are the sum of the solar irradiance in time intervals of 15 min throughout the representative day of the month. The dark areas receive the lowest amount of irradiation, while the bright areas receive the highest amount of irradiation, with high proportions of direct sunlight throughout the day. However, even in the shade and under cloudy weather conditions, solar panels can still produce energy via diffuse light from the atmosphere.

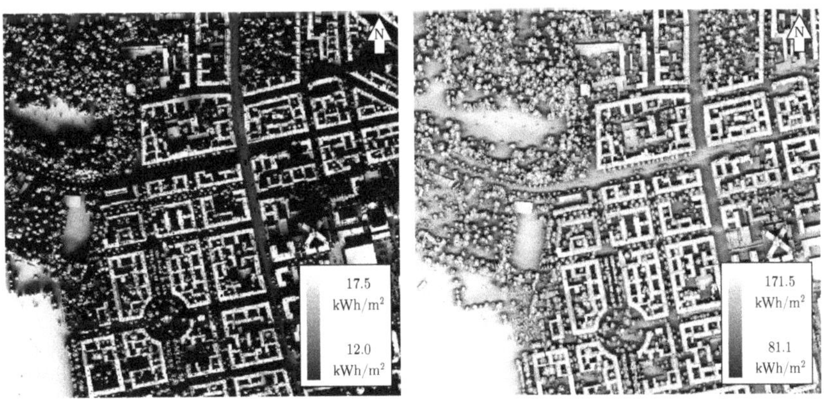

Figure 2. Solar maps for the months of December (**left**) and June (**right**). Exemplary cutout of Schillerkiez. The dark regions correspond to shaded areas. The white areas receive the highest amount of solar irradiation.

It must be pointed out that the minimum values (fully shaded areas) correspond to the diffuse fraction of the solar irradiation. In the r.sun simulation, the diffuse irradiation is assumed to be uniform across the examined area. Therefore, when calculating the solar irradiation at the parking points, the sky view factor (SVF) was taken into account in order to improve the accuracy of the diffuse irradiation component, as described in Section 3.3. Additionally, the influence of trees was thoroughly examined in Section 3.4. The solar simulation was performed for each month and then integrated into the solar yield model to quantify the solar energy being received at each parking point.

4.2. Solar Irradiation at Parking Locations

The results of the solar assessment of parked vehicles over a complete year in Berlin-Neukölln are shown in Figure 3. It can be concluded that the urban environment causes substantial solar energy losses at the street level due to building and vegetation shading. Across the city and throughout the year, there is high variability in solar irradiation. While the best (unshaded) parking point of the dataset receives a yearly irradiation of 1120 kWh/m^2, the mean parking point receives 493 kWh/m^2. A favorable parking point above the 90th percentile still receives 871 kWh/m^2 per year; the median parking point only receives 438 kWh/m^2 per year. On average, more than half of the solar irradiation is lost due to shading.

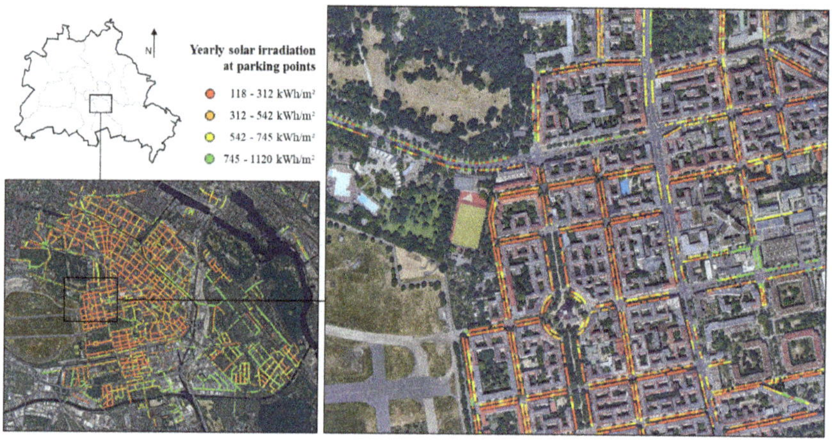

Figure 3. Yearly solar irradiation at parking points in the study area Berlin-Neukölln (**left**). A clos-eup shows the Schillerkiez area (**right**). For visualization purposes, the red points symbolize the lower 40% of parking points, while the other colors represent 20% of parking points each, according to the legend.

The model used in this study confirms that the influence of trees on solar irradiation is very high. In the summer, parking points directly under trees only receive 10% of the solar irradiation compared to unshaded parking points. Also, it becomes apparent that higher buildings south of parking areas significantly decrease the solar irradiation due to shading.

Dense residential areas like Schillerkiez encounter high amounts of shading, while commercial and industrial regions like Grenzallee have more open parking areas. These locations are more suitable for vehicle photovoltaics because people that work in the area park their car here during the day.

Figure 4 shows a histogram of the parking points by their yearly received irradiation. It is noticeable that there is a high concentration of data below 350 kWh/m^2. Nearly all of these rather unfavorable parking points are located directly under a tree and/or are heavily shaded throughout the year.

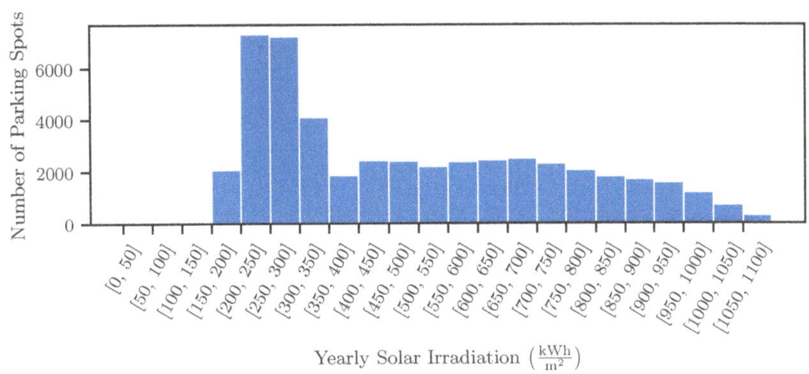

Figure 4. Histogram of the yearly solar irradiation for all parking points.

In Figure 5, a detailed image shows how the developed model can accurately distinguish irradiation values influenced by buildings and trees. Out of the 48,827 parking points, there are 18,223 located vertically under the radius of a tree crown, which amounts to 37% of the parking points. These parking points show a significant shading loss in this model because of the tree transmittance factor described in Section 3.4, especially in the summer months.

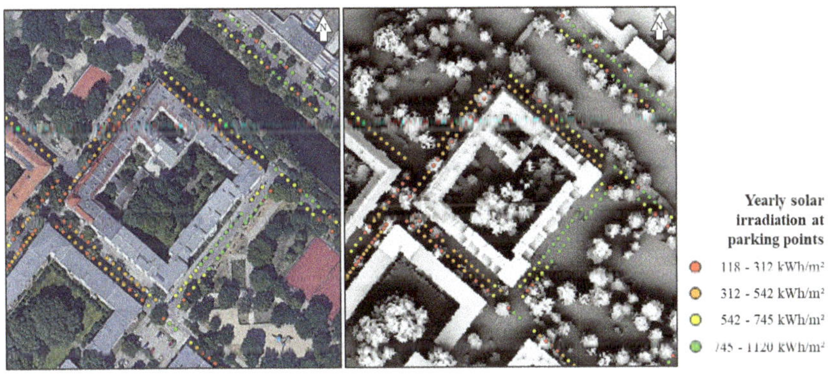

Figure 5. Detailed visualization displaying the influence of buildings and trees on the yearly solar irradiation. On the left, a GoogleMaps image is underlaid, while the right image shows the solar map of the full year. Location: Trusepark.

In Figure 6, the monthly solar irradiation is shown with the characteristic Northern Hemisphere irradiation peak in the summer months. It should be pointed out that there are very few parking points that are unshaded throughout the year. The upper quartile (75th percentile) is subjected to 38% of shading losses, while the median receives 61% of shading losses. The lower quartile (25th percentile) losses amount to a substantial 76%, which is also caused by a heavy dip of irradiation during the vegetation period from June to September, when tree crowns are fully developed.

The beam component of sunlight is crucial for maximizing solar energy yields. Since there is a high amount of shading present in cities, the diffuse fraction of solar irradiance provides a higher proportion of the solar energy yield compared to unshaded conditions. When comparing solar panels on parked vehicles in the study area with unshaded horizontal panels, the irradiation losses can be split up in beam and diffuse irradiation losses, as seen in Figure 7. While the diffuse losses stay relatively constant throughout the year, the beam losses are significantly higher in the summer months due to tree foliage.

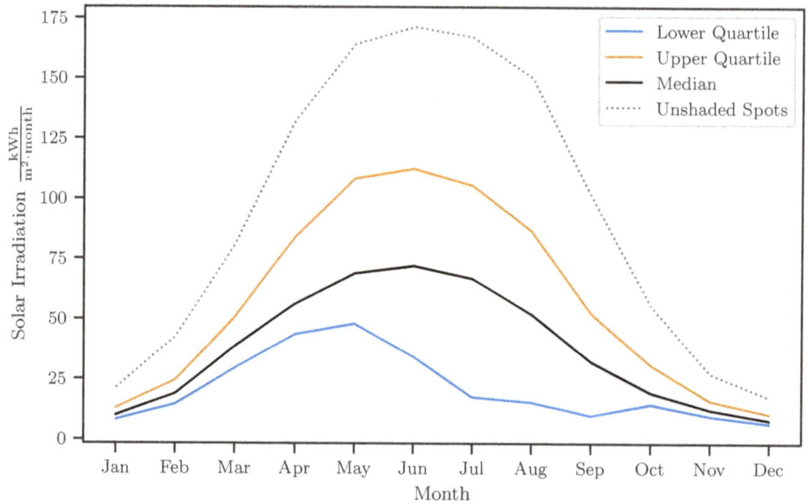

Figure 6. Monthly solar irradiation at parking points throughout the year.

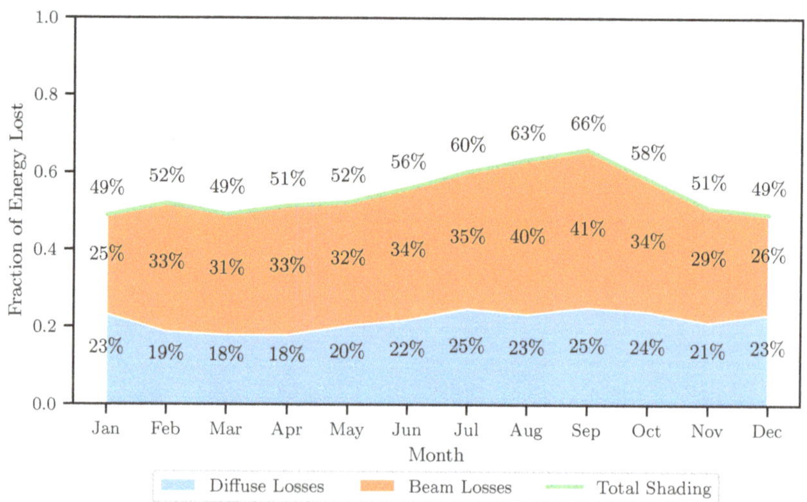

Figure 7. Monthly beam and diffuse irradiation losses as percentages of the Global Horizontal Irradiation.

4.3. Solar Yield Potential, Extended Driving Range, and Cost Savings

With the results of the solar irradiation assessment, the solar yield per vehicle on parking points is now calculated. As discussed in Section 3.5, the solar panel is assumed at 5 m^2, with 20% panel efficiency and further 25% losses between the panel and the battery.

For the analyzed parking points, the median energy capture is about 0.90 kWh/day. For a favorable parking spot (90th percentile) it is 1.79 kWh/day. Meanwhile, an unshaded parking spot provides 2.30 kWh/day.

The seasonal differences of the captured energy between winter (December to February) and summer (June to August) is quite pronounced. The median energy in winter is only 0.31 kWh/day, while in summer, it is 1.57 kWh/day. The difference of these values when compared to a favorable parking point at the 90th percentile is quite high. In a favorable location in winter, 0.51 kWh/day can be captured, while in summer, a value of

3.13 kWh/day can be achieved. An unshaded parking point in the summer provides a benchmark of 3.95 kWh/day.

In Figure 8, the energy captured per month is shown. In winter, the energy is low for all of the parking points, while in summer, there is a wide spread between favorable sunny parking points and heavily shaded points. This correlates to a greater variability of irradiation throughout the city in the summer.

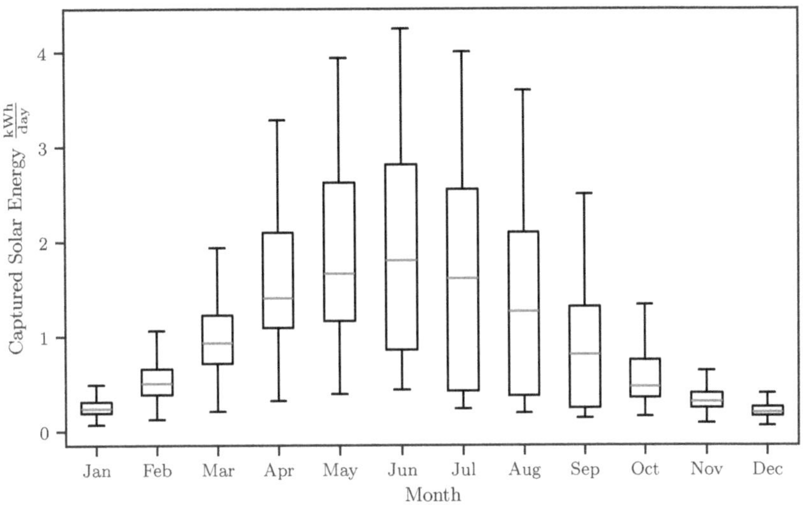

Figure 8. Captured solar energy for a 1 kWp solar array at 75% solar panel-to-battery conversion efficiency.

Assuming an energy consumption of 13 kWh/100 km (an optimistic assumption about the vehicle's efficiency, according to Heinrich et al. [17]; kept by us for compatibility with Brito et al. [26]), the extended driving range (EDR) per day and month can be calculated. Table 3 shows the daily and monthly driving range gained by the solar panels for the median and favorable parking point at the 90th percentile. Additionally, potential cost savings from reduced grid charging were calculated with an assumed electricity price of 0.50 €/kWh.

The average vehicle kilometers traveled (VKT) in Berlin is 25.5 km/day [47]. This implies that at a median parking point, about 27% of the annual kilometers traveled could be powered by energy that is generated from the solar panels. For a favorable parking point, this fraction can be as high as 54%, which can lead to a significant reduction of charging needs. Solar charging a vehicle on a summer day at a favorable parking point could cover 100% of the average daily VKT. However, since the energy delivery from VIPV varies seasonally, the charging infrastructure must be sized to fulfill the worst case charging demand, so the infrastructure savings are limited by the energy savings in the darkest months.

With an assumed PV system price of 2500 € (2.5 €/Wp), the break-even point for a median parking point would be reached after 15.2 years, which exceeds the average lifetime of private vehicles. At a favorable parking point, the break-even point would be reached after 7.7 years. With a lower proposed system cost of 1 €/Wp [18], the break-even point for a median parking point is reached after 6.1 years, and for a favorable parking point, it is reached after 3.1 years.

Table 3. Added solar range and cost savings for the median and favorable parking point.

	Median Parking Point			Favorable Point (90th Percentile)		
Month	**Daily Added Range (EDR)**	**Monthly Added Range (EMR)**	**Monthly Cost Savings**	**Daily Added Range (EDR)**	**Monthly Added Range (EMR)**	**Monthly Cost Savings**
All	6.9 km	2527 km	164.29 €	13.8 km	5027 km	326.74 €
January	1.9 km	57 km	3.73 €	3.0 km	94 km	6.11 €
February	3.9 km	109 km	7.07 €	6.5 km	182 km	11.83 €
March	7.2 km	222 km	14.46 €	11.8 km	367 km	23.83 €
April	10.8 km	325 km	21.12 €	20.4 km	611 km	39.71 €
May	12.8 km	397 km	25.81 €	24.9 km	771 km	50.10 €
June	13.9 km	416 km	27.02 €	26.6 km	799 km	51.92 €
July	12.4 km	385 km	25.02 €	24.6 km	763 km	49.58 €
August	9.7 km	300 km	19.52 €	21.2 km	657 km	42.71 €
September	6.2 km	186 km	12.11 €	13.8 km	413 km	26.82 €
October	3.6 km	112 km	7.26 €	7.3 km	227 km	14.74 €
November	2.4 km	71 km	4.63 €	3.9 km	118 km	7.69 €
December	1.5 km	47 km	3.03 €	2.5 km	76 km	4.96 €

5. Discussion

The results presented in the previous chapter show that significant shading losses are inevitable in an urban environment. Yet, the yearly added range of over 5000 km for a favorable parking location of the 90th percentile should not be neglected. It can be concluded that significant range increases—or charging energy decreases—are possible for vehicles at favorable parking spots under favorable conditions. However, the fixed parts of the transportation system, such as power grid infrastructure, charging infrastructure, and vehicle battery size, can only be reduced by a very small amount (less than 5%) when VIPV is widely implemented, as these components need to be designed for unfavorable times and places, where the energy gain is minimal. Another positive impact of VIPV could be the occurrence of fewer deep discharge events of vehicle batteries [48]. Fewer deep discharging events can result in higher battery lifetime and therefore less battery waste.

5.1. Comparison to Previous Research

When comparing the results to the findings of Brito et al. [26], the outcomes are similar. While Brito et al. estimated an annual average EDR for the median parking location to be about 10 km/(day · kWp), this study found an annual median EDR of 6.9 km/(day · kWp). This difference is most likely attributed to Lisbon being located further south and having more favorable weather conditions with lower cloud cover. The yearly unshaded solar irradiation in Berlin is 1117 kWh/m^2, whereas in Lisbon, it amounts to 1764 kWh/m^2 [38], which is 1.58 times higher than in Berlin. Brito et al. [26] also concluded that the solar potential on roads is higher than on parking sites because the distance to buildings and trees is higher in the middle of the road. This can also be applied to the results of this research, whereas it can presumed as well that the solar yield while driving in Berlin is higher on average than in a parking location. This obviously depends greatly on the choice of the parking location.

Further, Brito et al. stated that for the median urban parking space, about 33% of the annual typical driving distance could be supplied by energy generated by PV. In Berlin, this value was estimated to be 27%. It is worth mentioning that the typical driving distance used by Brito et al. [26] was 11,000 km/a, while in Berlin, a lower distance of 9323 km/a [47] was assumed, which increases the relative annual solar added range. A notable difference between the studies are the monthly irradiation losses. Brito et al. [26] found irradiation losses in winter to be much higher than in the summer, while in the study of Berlin, irradiation losses were relatively constant throughout the year. This is most likely due to the integration of seasonal vegetation shading in this study. Since trees are blocking about 90% of the solar irradiance in the summer, the expected irradiation gains due to higher insolation angles of the sun might be mitigated by the tree foliage. Even though the results of the studies of Lisbon and Berlin are comparable, they are yet to be verified with experimental measurement data.

5.2. Limitations and Methodological Implications

The model assumptions and methodological implications have various impacts on the model results. The vehicle assumptions were chosen to match the assumptions by Brito et al. Whether a PV car roof panel with 1 kWp is feasible with current car geometries and state-of-the-art technology could be argued. Also, the assumption of the roof being flat and horizontal is a simplification since car roofs are usually curved surfaces. However, this aspect was taken into account by assuming conservative 25% conversion losses from the panel into the vehicle battery.

Another aspect that was not considered (congruent with Brito et al. [26]) is the battery saturation effect. There might be instances when the vehicle battery is fully charged and therefore cannot receive any more energy. This leads to a potential overestimation of solar energy yields within this study [49,50]. Other neglected aspects (for comparability purposes) were temperature influences and battery self-consumption [17].

Further, the energy output of VIPV on a daily or even hourly basis is highly dependent on the actual weather. In this study, weather data were averaged over time; hence, the daily solar yield and resulting added ranges can vary greatly from day to day, depending on the weather conditions and solar irradiance. The solar simulation time step interval of 15-min steps could also be decreased to increase the model's precision.

Another limitation is the assumption of the parking spots/vehicles being points. This implies that a location is either in direct sunlight or not. In a real-world scenario, there are frequent instances where the solar panels would be partially shaded, which entails a technological challenge for module manufacturers to utilize partial shading conditions [51].

The solar simulation also incorporates several limitations. One is that albedo and direct/diffuse reflections by buildings are not considered. Also, the diffuse irradiation was reduced by a sky view factor of the DSM-nt (without trees), as described in Section 3.3. This leads to an overestimation of the diffuse component of the sunlight for vehicles not parked under trees but near trees. The introduced tree transmittance factors also introduce inaccuracies because they do not distinguish the exact location of a vehicle under a tree. A tree is a highly complex and partially transmittant 3D object that alters the solar irradiance. Regarding this problem, raster DSM elevation maps are subjected to limitations. Even though DSMs are excellent tools for spatial analyses, they also induce certain constraints regarding the 3D nature of objects, especially trees. An example of this problem is a vehicle under a tree receiving sunshine from an angle. Hence, the approach of vertical shading was used in this study to approximate the shadow influence by trees at parking points directly underneath. This could lead to an underestimation of solar yields under trees.

The parking point classification (whether the point is located under a tree or not) highly depends on the accuracy of the DSM and DTM. The tree threshold of 4 m can lead to errors, i.e., smaller trees, trucks, or objects that are taller than 4 m can lead to an incorrect classification. Also, parking points under bridges or buildings can be misinterpreted (manual purging required).

5.3. Sustainability

When applying the assumptions of Kanz et al. [29] to the results within this study, it becomes apparent that the shadowing factor (shading losses) for the average parking point exceeds 50%. Therefore, VIPV would not be more sustainable than grid charging when taking all parking points into account. This means that to make VIPV a more sustainable option than grid charging, the consistent use of favorable unshaded parking locations is necessary. Kanz et al. [29] also conclude that the outcomes of VIPV LCA studies strongly depend on the location of use of the vehicle, the annual irradiation, and the carbon footprint of the grid-mix in that location.

A different approach in evaluating the sustainability of VIPV is the comparison to stationary photovoltaics, such as a rooftop or utility-scale PV. The combined annual solar yield of all 48,827 parking points in the study area is a hypothetical 18 GWh. If this photovoltaic capacity was installed in an unshaded place like a roof, it could produce 41 GWh. These roof panels could also be sloped at a 40° angle and advantageous azimuth to generate an even greater energy yield of 48.6 GWh [38]. While VIPV can improve the charging autonomy of vehicles, they are subjected to higher shading and an unfavorable orientation compared to stationary solar panels.

In an unshaded scenario, the efficiency of VIPV onboard charging is higher than the efficiency of PV grid charging. The VIPV system only requires a DC–DC voltage conversion to charge the battery, while PV grid charging undergoes a DC–AC conversion, grid transmission losses, and, finally, AC–DC conversion within the onboard charger. The conversion losses of grid PV charging, therefore, are greater than VIPV onboard charging. However, in shaded conditions, the efficiency advantage of VIPV rapidly disappears, and it becomes more efficient to charge the vehicle via grid PV.

When comparing VIPV to rooftop solar installations from an environmental perspective, the lifetime of each system plays a very important role, as most negative environmental consequences are caused by the manufacturing process, which can be "amortized" by the energy produced during the service life. For rooftop PV installations, a lifetime of 25 years is usually assumed [52]. For the lifetime of electric vehicles, no good data set exists yet, as the wide adoption of electric vehicles is still in progress. The average fleet age in Germany is 11.3 years [53]. However, due to vehicles leaving the fleet both through scrapping and export, as well as political and economical factors affecting fleet composition, the lifetime of future VIPV-equipped cars is uncertain. A higher lifetime or highly efficient recycling program for VIPV equipment can significantly improve their ecological footprint.

6. Conclusions and Outlook

Photovoltaics are one of the future pillars of sustainable energy generation. Next to stationary PV on roofs or dedicated PV farms, VIPV could be used in scenarios where available space for stationary PV is limited and unshaded parking is an option.

The results of this study imply that VIPV will not solve the energy supply challenges for urban electric vehicles. VIPV, rather, can unfold its potential in certain use-cases, where it is assured that the solar yield is being maximized by avoiding shaded areas. Only then can VIPV become a sustainable contributor to future transportation. However, in higher latitudes, this potential is mainly restricted to the spring and summer months.

Our research shows that up to 5000 km annually can be generated using VIPV on passenger cars in ideal unshaded parking locations in Berlin, Germany. However, this applies to 10% of the observed parking locations in the study area.

The utilization of solar energy in stationary applications is more favorable because it (ideally) is unshaded, the panels can be aligned at an ideal angle and azimuth, and the conditions of use are less challenging than PV integration into a vehicle.

Photovoltaics in world regions south of Berlin (51° latitude) could significantly improve solar performance. Heinrich et al. concluded that in an unshaded scenario, the yearly solar range could be increased by 50% in Spain or even 60% in California, USA, due to higher irradiation levels [17].

Urban areas are the most challenging environments for VIPV. In rural regions and road networks, the solar yields are expected to be significantly higher. To ensure a better utilization of VIPV, designated solar parking areas could be established by communities. Mobile map applications could provide valuable information to inform drivers about favorable parking locations for solar charging. Conversely, the results of our study could also be used to identify well-shaded parking spots, where vehicles without solar panels heat up the least.

A further assessment of the potential of commercial vehicles and trucks is highly suggested. Large flat roof areas are suitable for PV, and the driving profiles are often located outside of urban areas.

The approach of this study is exclusively based on open data and free software to enable a barrier-free reproducibility for other world regions. Our methodology can be applied to different cities in a further study, analyzing the solar potential for different kinds of buildings and road layouts, tree coverage, and parking strategies and developing a generic prediction for the feasibility of VIPV in different types of cities. Further research can also be conducted on solar irradiance modeling of VIPV potential while driving. The consideration of trees can be further improved by using 3D photogrammetry.

Author Contributions: Conceptualization: L.H.; Data curation: L.H.; Funding acquisition: D.G.; Investigation: P.H., L.H. and A.G.; Methodology: P.H., L.H. and A.G.; Project administration: D.G.; Software: P.H. and L.H.; Supervision: L.H. and A.G.; Validation: L.H. and A.G.; Visualization: P.H. and L.H.; Writing—original draft: P.H.; Writing—review and editing: P.H., L.H., A.G. and D.G. All authors have read and agreed to the published version of the manuscript.

Funding: This work was supported by the German Federal Ministry for Education and Research (BMBF) as part of the "Research Campus Mobility2Grid" project, under grant number 03SF0674A. The authors acknowledge the financial support and resources provided for the development of the code and research findings reported in this paper. We acknowledge support by the German Research Foundation and the Open Access Publication Fund of TU Berlin.

Data Availability Statement: These data were derived from the following resources available in the public domain:ALKIS Berlin (https://daten.berlin.de/datensaetze/alkis-berlin-geb%C3%A4ude-wfs); DGM Berlin (https://www.berlin.de/sen/sbw/stadtdaten/geoportal/landesvermessung/geotopographie-atkis/dgm-digitale-gelaendemodelle/); bDOM Berlin (https://www.berlin.de/sen/sbw/stadtdaten/geoportal/landesvermessung/geotopographie-atkis/bdom-digitales-bildbasiertes-oberflaechenmodell/) as well as the Neukölln Parking Space Dataset (https://parkraum.osm-verkehrswende.org/project-prototype-neukoelln/report/, all accessed on 25 September 2023).

Conflicts of Interest: The authors declare no conflicts of interest.

References

1. European Commission. Delivering the European Green Deal. Available online: https://commission.europa.eu/strategy-and-policy/priorities-2019-2024/european-green-deal/delivering-european-green-deal_en (accessed on 25 September 2023).
2. European Commission. Solar Energy. Available online: https://energy.ec.europa.eu/topics/renewable-energy/solar-energy_en (accessed on 25 September 2023).
3. International Energy Agency. Global EV Outlook 2020. Available online: https://iea.blob.core.windows.net/assets/af46e012-18c2-44d6-becd-bad21fa844fd/Global_EV_Outlook_2020.pdf (accessed on 16 November 2023).
4. Straub, F.; Göhlich, D. Modelling and Simulation of a fully Electrified Urban Private Transport—A case Study for Berlin, Germany. In Proceedings of the 2023 IEEE Transportation Electrification Conference & Expo (ITEC), Detroit, MI, USA, 21–23 June 2023; pp. 1–8. [CrossRef]
5. Commault, B.; Duigou, T.; Maneval, V.; Gaume, J.; Chabuel, F.; Voroshazi, E. Overview and Perspectives for Vehicle-Integrated Photovoltaics. *Appl. Sci.* **2021**, *11*, 11598. [CrossRef]
6. Miyoshi, T. Solar Charging System for Prius PHV. *J. Jpn. Soc. Appl. Electromagn. Mech.* **2017**, *25*, 379–382. [CrossRef]
7. Hyundai Motor Group. Everything About the Sonata Hybrid's Solar Roof. Available online: https://www.hyundaimotorgroup.com/story/CONT0000000000091508 (accessed on 25 September 2023).
8. Sono Motors. Sion. Available online: https://web.archive.org/web/20221113201804/https://sonomotors.com/en/sion/ (accessed on 25 September 2023).
9. Fisker, Inc. Fisker Ocean. Available online: https://www.fiskerinc.com/ocean (accessed on 25 September 2023).
10. Squad Mobility BV. Squad Soar City Car. Available online: https://www.squadmobility.com/squad#specs (accessed on 25 September 2023).
11. Aptera Motors Corp. Driven by the Sun. Available online: https://aptera.us/ (accessed on 25 September 2023).
12. Kühnel, M.; Hanke, B.; Geißendörfer, S.; von Maydell, K.; Agert, C. Energy forecast for mobile photovoltaic systems with focus on trucks for cooling applications. *Prog. Photovoltaics Res. Appl.* **2017**, *25*, 525–532. [CrossRef]
13. Wendeker, M.; Gęca, M.J.; Grabowski, Ł.; Pietrykowski, K.; Kasianantham, N. Measurements and analysis of a solar-assisted city bus with a diesel engine. *Appl. Energy* **2022**, *309*, 118439. [CrossRef]

14. Almonacid, G.; Muñoz, F.J.; de la Casa, J.; Aguilar, J.D. Integration of PV systems on health emergency vehicles. The FIVE project. *Prog. Photovoltaics Res. Appl.* **2004**, *12*, 609–621. [CrossRef]
15. Yamaguchi, M.; Nakamura, K.; Ozaki, R.; Kojima, N.; Ohshita, Y.; Masuda, T.; Okumura, K.; Satou, A.; Nakado, T.; Yamada, K.; et al. Analysis for the Potential of High-Efficiency and Low-Cost Vehicle-Integrated Photovoltaics. *Solar RRL* **2022**, *7*, 2200556. [CrossRef]
16. Lee, K.Y.; Park, S. Reducing Charging Burden of Light Electric Vehicles by Integrated Photovoltaic Modules. In Proceedings of the 2022 IEEE Vehicle Power and Propulsion Conference (VPPC), Merced, CA, USA, 1–4 November 2022; pp. 1–6. [CrossRef]
17. Heinrich, M.; Kutter, C.; Basler, F.; Mittag, M.; Alanis, L.; Eberlein, D.; Schmid, A.; Reise, C.; Kroyer, T.; Neuhaus, D.; et al. Potential and Challenges of Vehicle Integrated Photovoltaics for Passenger Cars. In Proceedings of the 37th European Photovoltaic Solar Energy Conference and Exhibition, Virtual, 7–11 September 2020; pp. 1695–1700. [CrossRef]
18. Kutter, C.; Alanis, L.E.; Neuhaus, D.H.; Heinrich, M. Yield potential of vehicle integrated photovoltaics on commercial trucks and vans. In Proceedings of the 38th European PV Solar Energy Conference and Exhibition, Virtual, 6–10 September 2021.
19. Carr, A.; van den Tillaart, E.; Burgers, A.; Köhler, T.; Newman, B. Vehicle integrated photovoltaics: Evaluation of the energy yield potential through monitoring and modelling. In Proceedings of the 37th EUPVSEC European PV Solar Energy Conference and Exhibition, Lisbon, Portugal, 7–11 September 2020; WIP GmbH & Co Planungs-KG: München, Germany, 2020.
20. Araki, K.; Ota, Y.; Yamaguchi, M. Measurement and Modeling of 3D Solar Irradiance for Vehicle-Integrated Photovoltaic. *Appl. Sci.* **2020**, *10*, 872. [CrossRef]
21. Araki, K.; Carr, A.; Chabuel, F.; Commault, B.; Derks, R.; Ding, K.; Duigou, T.; Ekins-Daukes, N.; Gaume, J.; Hirota, T.; et al. State-of-the-Art and Expected Benefits of PV-Powered Vehicles. Available online: https://iea-pvps.org/wp-content/uploads/2021/07/IEA_PVPS_T17_State-of-theart-and-expected-benefits-of-VIPV_report.pdf (accessed on 11 March 2024).
22. Masuda, T.; Araki, K.; Okumura, K.; Urabe, S.; Kudo, Y.; Kimura, K.; Nakado, T.; Sato, A.; Yamaguchi, M. Next environment-friendly cars: Application of solar power as automobile energy source. In Proceedings of the 2016 IEEE 43rd Photovoltaic Specialists Conference (PVSC), Portland, OR, USA, 5–10 June 2016; IEEE: Piscataway, NJ, USA, 2016. [CrossRef]
23. Oh, M.; Kim, S.M.; Park, H.D. Estimation of photovoltaic potential of solar bus in an urban area: Case study in Gwanak, Seoul, Korea. *Renew. Energy* **2020**, *160*, 1335–1348. [CrossRef]
24. Ota, Y.; Araki, K.; Nagaoka, A.; Nishioka, K. Evaluating the Output of a Car-Mounted Photovoltaic Module Under Driving Conditions. *IEEE J. Photovoltaics* **2021**, *11*, 1299–1304. [CrossRef]
25. ur Rehman, N.; Hijazi, M.; Uzair, M. Solar potential assessment of public bus routes for solar buses. *Renew. Energy* **2020**, *156*, 193–200. [CrossRef]
26. Brito, M.C.; Santos, T.; Moura, F.; Pera, D.; Rocha, J. Urban solar potential for vehicle integrated photovoltaics. *Transp. Res. Part D Transp. Environ.* **2021**, *94*, 102810. [CrossRef]
27. Santos, T.; Gomes, N.; Freire, S.; Brito, M.; Santos, L.; Tenedório, J. Applications of solar mapping in the urban environment. *Appl. Geogr.* **2014**, *51*, 48–57. [CrossRef]
28. Said-Romdhane, M.B.; Skander-Mustapha, S.; Slama-Belkhodja, I. Analysis study of city obstacles shading impact on solar PV vehicle. In Proceedings of the 2021 4th International Symposium on Advanced Electrical and Communication Technologies (ISAECT), Alkhobar, Saudi Arabia, 6–8 December 2021; pp. 1–6. [CrossRef]
29. Kanz, O.; Reinders, A.; May, J.; Ding, K. Environmental Impacts of Integrated Photovoltaic Modules in Light Utility Electric Vehicles. *Energies* **2020**, *13*, 5120. [CrossRef]
30. QGIS Association. QGIS Geographic Information System. Available online: http://www.qgis.org (accessed on 11 March 2024).
31. GRASS Development Team. *Geographic Resources Analysis Support System (GRASS GIS) Software, Version 8.2*; Open Source Geospatial Foundation: Dover, DE, USA, 2022. [CrossRef]
32. OpenStreetMap Contributors. OpenStreeMap Dataset. Available online: https://www.openstreetmap.org (accessed on March 13 2024).
33. Freitas, S.; Catita, C.; Redweik, P.; Brito, M. Modelling solar potential in the urban environment: State-of-the-art review. *Renew. Sustain. Energy Rev.* **2015**, *41*, 915–931. [CrossRef]
34. Seidel, A. Parkraumanalyse für den Berliner Ortsteil Neukölln—Methoden- und Ergebnisbericht. Available online: https://parkraum.osm-verkehrswende.org/project-prototype-neukoelln/report (accessed on March 13 2024).
35. Kuhnimhof, T.; Nobis, C. Ergebnisbericht. Available online: https://bmdv.bund.de/SharedDocs/DE/Anlage/G/mid-ergebnisbericht.pdf?__blob=publicationFile (accessed on March 13 2024).
36. Senatsverwaltung für Stadtentwicklung, Bauen und Wohnen. Geotopographie/ATKIS. Available online: https://www.berlin.de/sen/sbw/stadtdaten/geoportal/landesvermessung/geotopographie-atkis/ (accessed on 25 September 2023).
37. Hofierka, J.; Šúri, M. The solar radiation model for Open source GIS: Implementation and applications. In Proceedings of the Open Source GIS—GRASS Users Conference 2002, Trento, Italy, 11–13 September 2002.
38. European Commission, Joint Research Centre Energy Efficiency and Renewables Unit. PVGIS-5 Geo-Temporal Irradiation Database. Available online: https://re.jrc.ec.europa.eu/pvg_tools/en/ (accessed on March 13 2024).
39. Dirksen, M.; Ronda, R.; Theeuwes, N.; Pagani, G. Sky view factor calculations and its application in urban heat island studies. *Urban Clim.* **2019**, *30*, 100498. [CrossRef]
40. Böhner, J.; Antonić, O. Chapter 8 Land-Surface Parameters Specific to Topo-Climatology. In *Developments in Soil Science*; Elsevier: Amsterdam, The Netherlands, 2009; pp. 195–226. [CrossRef]

41. Häntzschel, J.; Goldberg, V.; Bernhofer, C. GIS-based regionalisation of radiation, temperature and coupling measures in complex terrain for low mountain ranges. *Meteorol. Appl.* **2005**, *12*, 33–42. [CrossRef]
42. Konarska, J.; Lindberg, F.; Larsson, A.; Thorsson, S.; Holmer, B. Transmissivity of solar radiation through crowns of single urban trees—Application for outdoor thermal comfort modelling. *Theor. Appl. Climatol.* **2013**, *117*, 363–376. [CrossRef]
43. Referat Freiraumplanung und Stadtgrün. Straßenbäume in Berlin. Available online: https://www.berlin.de/sen/uvk/natur-und-gruen/stadtgruen/daten-und-fakten/stadtbaeume/ (accessed on March 13 2024).
44. Oshio, H.; Asawa, T. Estimating the Solar Transmittance of Urban Trees Using Airborne LiDAR and Radiative Transfer Simulation. *IEEE Trans. Geosci. Remote Sens.* **2016**, *54*, 5483–5492. [CrossRef]
45. Rosskopf, E.; Morhart, C.; Nahm, M. Modelling Shadow Using 3D Tree Models in High Spatial and Temporal Resolution. *Remote Sens.* **2017**, *9*, 719. [CrossRef]
46. Verivox GmbH. Ladetarife für E-Autos. Übersicht Ladetarife: Anbieter und Kosten. Available online: https://www.verivox.de/elektromobilitaet/ladetarife/ (accessed on 3 February 2023).
47. Kirk, E. In Deutschland sind Kfz-Halter*innen im Schnitt 11.085 Kilometer p. a. Unterwegs. Available online: https://www.check24.de/unternehmen/presse/pressemitteilungen/in-deutschland-sind-kfz-halter*innen-im-schnitt-11.085-kilometer-p.-a.-unterwegs-2059/ (accessed on March 13 2024).
48. Kouzelis, A. Model-Based Impact Analysis of Vehicle-Integrated Photovoltaics and Vehicle-to-Grid on Electric Vehicle Battery Life. Available online: http://resolver.tudelft.nl/uuid:f3f716b3-f57e-486b-b5b9-8d5edf643d41 (accessed on 25 October 2023).
49. Lodi, C.; Gil-Sayas, S.; Currò, D.; Serra, S.; Drossinos, Y. Full-battery effect during on-board solar charging of conventional vehicles. *Transp. Res. Part D Transp. Environ.* **2021**, *96*, 102862. [CrossRef]
50. Birnie, D.P. Analysis of energy capture by vehicle solar roofs in conjunction with workplace plug-in charging. *Sol. Energy* **2016**, *125*, 219–226. [CrossRef]
51. Götz, D.; Hahn, D.; Gottschalg, R.; Dassler, D.; Schindler, S.; Hanifi, H. Evaluation of shading tolerance of PV modules with different module designs for mobile applications by simulation, indoor and outdoor measurements. In Proceedings of the 36th European Photovoltaic Solar Energy Conference and Exhibition, Marseille, France, 9–13 September 2019; pp. 1–6.
52. Martinopoulos, G. Are rooftop photovoltaic systems a sustainable solution for Europe? A life cycle impact assessment and cost analysis. *Appl. Energy* **2020**, *257*, 114035. [CrossRef]
53. Kraftfahrt-Bundesamt. Durchschnittsalter der Personenkraftwagen Wächst. Available online: https://www.kba.de/DE/Statistik/Fahrzeuge/Bestand/Fahrzeugalter/2021/2021_b_kurzbericht_fz_alter_pdf.pdf;jsessionid=A280D8FA3BFB96F7C153FA41124F3002.live11292?__blob=publicationFile&v=2 (accessed on 18 November 2023).

energies

MDPI

Article

Investigation on Traffic Carbon Emission Factor Based on Sensitivity and Uncertainty Analysis

Jianan Chen [1], Hao Yu [2], Haocheng Xu [2], Qiang Lv [2], Zongqiang Zhu [2], Hao Chen [1], Feiyang Zhao [1,*] and Wenbin Yu [1,*]

1 School of Energy and Power Engineering, Shandong University, Jinan 250061, China; chenjianan@mail.sdu.edu.cn (J.C.); 202314499@mail.sdu.edu.cn (H.C.)
2 China Automotive Engineering Research Institute Co., Ltd., Chongqing 401122, China; yuhao1@caeri.com.cn (H.Y.); xuhaocheng@caeri.com.cn (H.X.); lvqiang@caeri.com.cn (Q.L.); zhuzongqiang@caeri.com.cn (Z.Z.)
* Correspondence: fyzhao@sdu.edu.cn (F.Z.); wbyu@sdu.edu.cn (W.Y.)

Abstract: The premise for formulating effective emission control strategies is to accurately and reasonably evaluate the actual emission level of vehicles. Firstly, the active subspace method is applied to set up a low-dimensional model of the relationship between CO_2 emission and multivariate vehicle driving data, in which the vehicle specific power (*VSP*) is identified as the most significant factor on the CO_2 emission factor, followed by speed. Additionally, acceleration and exhaust temperature had the least impact. It is inferred that the changes in data sampling transform the establishment of subspace matrices, affecting the calculation of eigenvector components and the fitting of the final quadratic response surface, so that the emission sensitivity and final fitting accuracy are impressionable by the data distribution form. For the *VSP*, the best fitting result can be obtained when the *VSP* conforms to a uniform distribution. Moreover, the Bayesian linear regression method accounts for fitting parameters between the *VSP* and CO_2 emission factor with uncertainties derived from heteroscedastic measurement errors, and the values and distributions of the intercept and slope α and β are obtained. In general, the high-resolution inventory of the carbon emission factor of the tested vehicle is set up via systematically analyzing it, which brings a bright view of data processing in further counting the carbon footprint.

Keywords: CO_2 emission; vehicle specific power; active subspaces; multivariate analysis; uncertainty analysis

Citation: Chen, J.; Yu, H.; Xu, H.; Lv, Q.; Zhu, Z.; Chen, H.; Zhao, F.; Yu, W. Investigation on Traffic Carbon Emission Factor Based on Sensitivity and Uncertainty Analysis. *Energies* **2024**, *17*, 1774. https://doi.org/10.3390/en17071774

Academic Editor: Anastassios M. Stamatelos

Received: 13 March 2024
Revised: 5 April 2024
Accepted: 6 April 2024
Published: 8 April 2024

1. Introduction

The transportation sector is the third largest CO_2 emission source after power and industry. Carbon emissions from transportation can cause significant harm to the climate and environment [1]. Therefore, how to reduce vehicle exhaust pollution and greenhouse gas emissions has become an undeniable challenge [2]. Roughly, China emitted 35% of the world's total CO_2 emissions in 2023 [3]. Road transportation contributes to 70~80% of the CO_2 emissions in the transportation sector of China, which still dominates as one of the largest parts of total carbon emissions [4,5]. An action plan of decarbonizing China's road transport is becoming among the stated national policies [6]; meanwhile, the technology roadmap for energy saving and new energy vehicles is being undertaken by many researchers and institutes, with the aim to achieve carbon neutrality by 2060.

How to accurately and reasonably evaluate the actual emission level of vehicles is a prerequisite for formulating effective emission control strategies. Many scholars have conducted research on this issue. When characterizing the emission characteristics of different operating conditions, instantaneous operating parameters such as speed [7,8], acceleration [9,10], fuel consumption [11,12], and driving distance [13] are used while the vehicle specific power (*VSP*) is a newly proposed alternative parameter in correlation with speed and acceleration. Jimenez et al. [14] found that the *VSP* could capture the majority of the dependence of light vehicle emissions on driving conditions when combined with traffic flow data, thus improving the accuracy of the predicted emissions. Song et al. [15] compared the estimated fuel consumption with actual data and demonstrated that the *VSP* bin distribution model is reliable and accurate in estimating fuel consumption. When the real-time speed data are available, the speed-specific *VSP* bin distribution model may help monitor dynamic transportation, to facilitate quantifying the relationship between *VSP* distributions and vehicle fuel consumption or emissions. Forcetto et al. [16] applied the *VSP* as an additional parameter for improving the evaluation of the vehicle dynamics, which can add a better comprehension of the real drive emission (RDE) dynamic, complementing the regulatory parameters. By combining fuel consumption rates with *VSP* distributions, Zhang, L. et al. [17] proposed an improved method for evaluating eco-driving behavior depending on the *VSP* distribution at a specific speed, providing a potential way to evaluate fuel consumption and emission combined with traffic conditions.

Vehicle CO_2 emission is typically influenced by a combination of multiple factors, so it is necessary to reveal the impacts of simultaneous multi-factor variables on emission factors, in order to summarize the methods for constructing a high-resolution database of the carbon emission factor. At present, existing multi-factor variable analysis models, such as COPERT [18,19], MOBILE [20,21], IVE [22], etc., are all based on static factors such as vehicle type, vehicle mileage, fuel quality, and fuel volatility to study the vehicle emission factors of motor vehicles. The traditional one-at-a-time (OAT) sensitivity analysis applied in the sensitivity analysis can only predict the relative sensitivity of a single input parameter at a time, so that it precludes the synergistic effects between multiple input parameters. A multivariate analysis is able to reflect the combined effects of multiple variables on output, which displays a more accurate and flexible performance than an OAT sensitivity analysis. However, a multi-factor analysis of emission factors based on dynamic driving conditions is rarely reported.

The active subspace (AS) is an emerging dimensionality reduction method for multivariate analysis, which can obtain the low-dimensional structure of the multivariate function by transforming the high-dimensional space [23,24]. At the same time, the component values of the active direction vector can provide global sensitivity information of the target quantity relative to the input parameters [25]. In the field of engineering, the active subspace method has been applied in sensitivity analysis studies of turbomachinery [26,27] and combustion [28–30], but, up to now, few reports of AS application in transportation emissions analysis had been presented.

Moreover, as the data size increases, the uncertainties driven by the error distribution from the sampling data play an important role in the accounting accuracy of carbon emissions under varied transportation scenarios. Thus, uncertainty quantification is necessary when formulating emission factor models based on real-driving cycles.

Bayesian inference methods have been used widely for model calibration and uncertainty analysis [31,32]. The goal of the Bayesian regression method is to characterize the parameter distribution consistent with the given experimental dataset, instead of finding the best fit for regression model parameters [33,34]. Li et al. [35] used Bayesian approaches to explicitly accommodate the uncertainty of the model predictions. Mudgal, A. et al. [36] modeled the speed profiles of drivers by using a Bayesian inference methodology, and then estimated the vehicular emissions using past experimental data. Martin et al. [37] presented a new Bayesian methodology called the Cambridge Automotive Research Modelling Application (CARma), which was able to categorize the sources of uncertainty and calibrate uncertain parameters to present the results as probability distribution functions.

The active subspace method and Bayesian linear fitting method are used in this study to combine multivariate analysis with uncertainty analysis, explore the sensitivity of vehicle emissions to operating parameters, and improve the accuracy of emission prediction by considering the distribution of operating points. This method comprehensively studies the impact of vehicle driving parameters on emission factors, laying the foundation for constructing a high-resolution database of the carbon emission factor. The study is structured as follows: In Section 2, the methodologies of this study are introduced, including the data processing and the principles of data analysis methods, such as multivariate analysis and uncertainty analysis. Then, Section 3 describes the details analysis of the sensitivity and the uncertainty of the carbon emission factor. Finally, some concluding remarks are given in Section 4.

2. Materials and Methods

2.1. Data Source

In China, light gasoline vehicles account for a large proportion of road traffic. Meanwhile, China has fully implemented the national VI motor vehicle emission standards. Therefore, in this study, the multi-purpose gasoline passenger vehicle that meets the national VI motor vehicle emission standards was chosen. The source of data used in this study is the RDE test conducted by China Automotive Engineering Research Institute Co., Ltd. (Chongqing, China).

The RDE test was conducted once and employed by using on-board emission measurement system OBS-ONE GS Unit from Japan's HORIBA company to measure the vehicle's emissions per second. The test route is shown in Figure 1. The test route is located in Chongqing, China. The total length of this route is 82.096 km, and the maximum altitude and minimum altitude are 499.4 m and 240.9 m, respectively. The test data are divided into three working conditions based on the vehicle's driving speed, urban ($v \leq 60$ km/h), suburban (60 km/h < $v \leq 90$ km/h), and highway ($v > 90$ km/h), as shown in Figure 2, and the distribution of driving conditions is shown in Table 1.

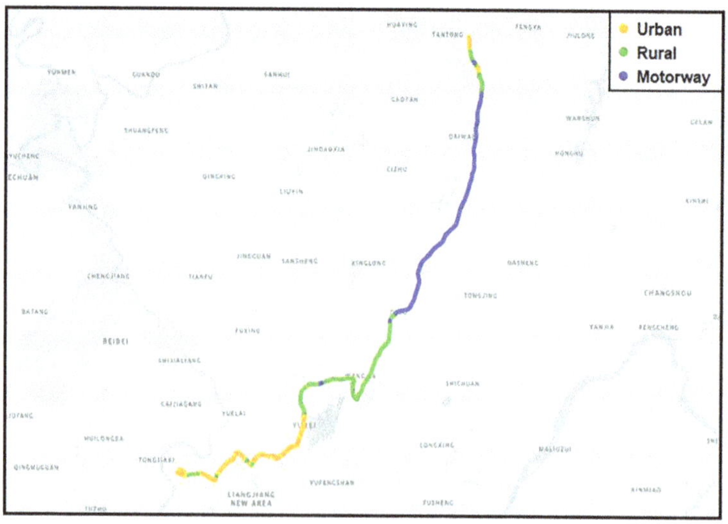

Figure 1. Testing route.

Table 1. Distribution of vehicle driving conditions.

	Urban	Rural	Motorway	Total
Average Speed (km/h)	28.2	77.6	106.0	
Distance (km)	27.2	26.7	28.3	82.096
Duration (h:min:s)	0:57:52	0:20:36	0:16:00	1:34:28

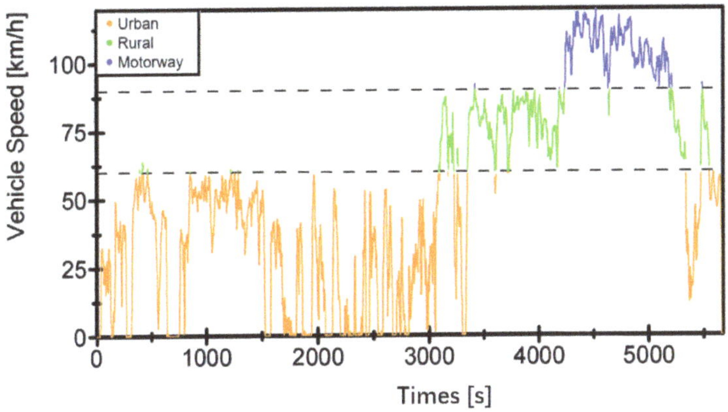

Figure 2. The speed structure of vehicle driving.

2.2. VSP Binning Method

VSP is a dynamic proxy variable of the vehicle with vector nature, defined as the power output per unit mass of motor vehicle towed by the engine (kw/t) or m^2/s^3 [14]. For light gasoline vehicles, the formula for *VSP* is computed as [38]:

$$VSP = v \times (1.1a + 0.132) + 0000302 \times v^3 \quad (1)$$

where v is the driving speed of the tested vehicle in a unit of m/s, and a is the driving acceleration of the tested vehicle in a unit of m/s^2.

The values of *VSP* can be one-to-one corresponded and classified with cluster bins, thus using the overall *VSP* distribution as a function of average vehicle speed for statistical analysis. Since there is a certain degree of discreteness in RDE data, this study conducted cluster analysis on *VSP* [38], when *VSP* bin = k, and $VSP \in [k - 0.5, k + 0.5]$. The data are mainly concentrated in *VSP* bin $\in [-20, 20]$, accounting for 96.93%. Therefore, this study mainly focuses on the data in this range, and the *VSP* bin distribution is shown in Figure 3.

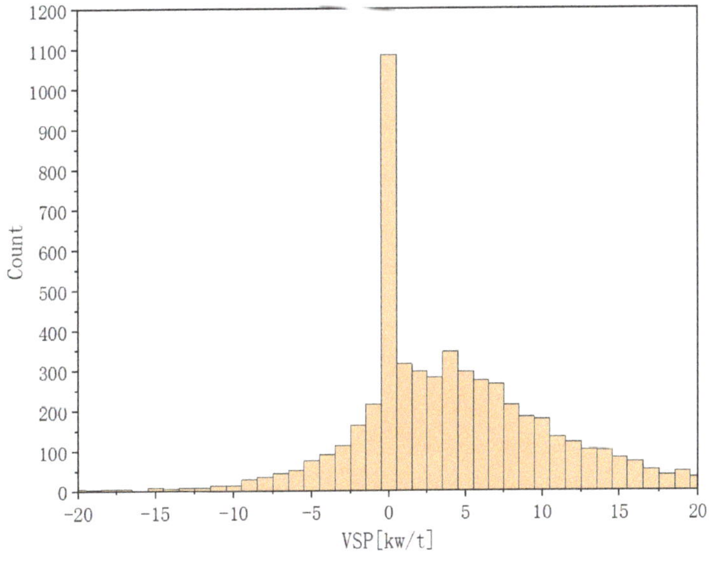

Figure 3. *VSP* bin distribution of the RDE data.

The traffic carbon emission factor generally refers to the carbon emissions generated per unit workload, in order to evaluate carbon emission efficiency. This study applies the speed-specific CO_2 emission factor based on driving distance to characterize the mass of CO_2 emitted by motor vehicles per unit distance (g/km), which can be obtained by Equation (2):

$$EF^k = \frac{\sum ER^k}{\sum v^k} \times 3600 \quad (2)$$

where EF^k, ER^k, and v^k, respectively, represent the CO_2 emission factor (g/km), CO_2 emission rate (g/s), and driving speed (km/h) of the tested vehicle when *VSP* bin = k.

2.3. *Active Subspace Method*

The active subspace is a type of low-dimensional structure in a function of several variables. It can achieve dimensionality reduction through transformations in high-dimensional spaces, which essentially involves important directions with higher sensitivity in high-dimensional spaces through linear space transformations. In the present study, a one-dimensional active subspace is observed to figure out the key contributors to the variability in the emission factor response. The input–output response diagram is approximated by a linear model. This method is applicable when both conditions are satisfied that the activity subspace is one-dimensional and the relationship between the quantity of interest and its input parameters is approximately monotonic. The sufficient summary plot shows the relationship between the linear combination of the quantity of interest and input parameters (i.e., active variables), with the weights of the linear combination being the components of the normalized gradient of the linear model. Each point on the plot represents a set of inputs and corresponding outputs of the model. The specific flowchart of performing the active subspace method for sensitivity analysis is shown in Figure 4. In the current study, speed, acceleration, *VSP*, CO_2 emission rate, and exhaust temperature are selected as input variables, and corresponding CO_2 emission factor as the quantity of interest.

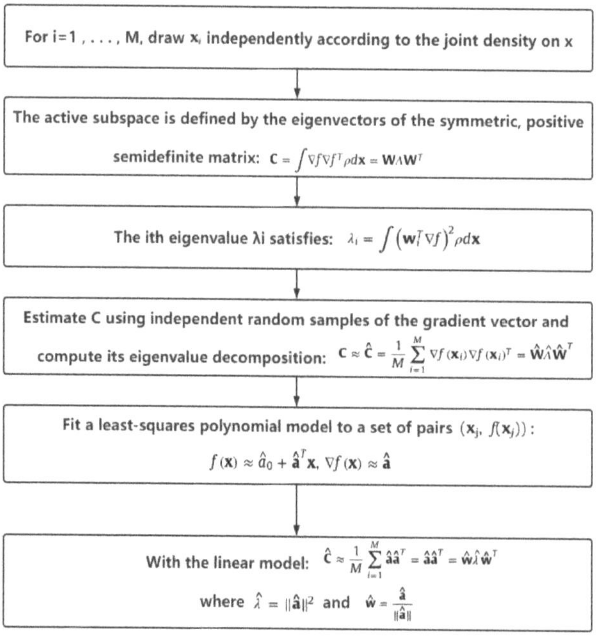

Figure 4. Workflow of active subspace method.

The active subspace is defined by the eigenvectors of a symmetric positive semidefinite matrix:

$$C = \int \nabla f \nabla f^{T} \rho dx = W\Lambda W^{T} \tag{3}$$

where W is the orthogonal matrix of the eigenvector, and Λ is the diagonal matrix of non-negative eigenvalues arranged in descending order.

To estimate the eigenvectors and eigenvalues, independent random samples of gradient vector are used to estimate C and calculate its eigenvalue decomposition [39]:

$$C \approx \hat{C} = \frac{1}{M}\sum_{i=1}^{M} \nabla f(x_i)\nabla f(x_i)^{T} = \hat{W}\hat{\Lambda}\hat{W}^{T} \tag{4}$$

where x_i are based on $\rho(x)$ drawing randomly and independently.

In this study, the gradient of polynomial approximation is worked out and the least-squares polynomial model is fitted to a pair of $(x_j, f(x_j))$. When the polynomial approximation is a linear function of x, the calculation amount of $\hat{W}$ decreases sharply. The gradient of the global linear model is constant for all x:

$$f(x) \approx \hat{a}_0 + \hat{a}^{T} x \tag{5}$$

$$\nabla f(x) \approx \hat{a} \tag{6}$$

Thus, $\hat{C}$ can be expressed as:

$$\hat{C} \approx \frac{1}{M}\sum_{i-1}^{M} \hat{a}\hat{a}^{T} = \hat{a}\hat{a}^{T} = \hat{w}\hat{\Lambda}\hat{w}^{T} \tag{7}$$

where $\hat{w} = \hat{a}/\|\hat{a}\|$ is the normalized gradient of the linear model. In a linear model, only one-dimensional active subspace can be identified.

2.4. Bayesian Linear Regression

Bayesian linear regression is a method that uses probability distribution rather than point estimation to construct linear regression. The response variable y is not a single value to be estimated, but a probability distribution assumed to be extracted from a normal distribution. The posterior probability distribution of model parameters $P(\beta \mid y,X)$ is conditional on the inputs and output of the training, as calculated in Equation (8):

$$P(\beta|y, X) = \frac{P(y|\beta, X)P(\beta|X)}{P(y|X)} \tag{8}$$

which is equal to the likelihood $P(y \mid \beta,X)$ multiplied by the prior probability distribution $P(\beta \mid X)$ of the parameter β of the given input and divided by the normalization constant.

In this study, the data with heteroscedastic measurement errors (errors with different variances) in both variables are regression-fitted through Bayesian theory. It is assumed that the independent variable ξ and the dependent variable η follow a Gaussian distribution, and that ξ is a random vector of n data points extracted from a certain probability distribution [40,41]. According to the usual additive model, the dependent variable η depends on ξ:

$$\eta_i = \alpha + \beta\xi_i + \varepsilon_i \tag{9}$$

where ε_i is a random variable which represents the intrinsic scatter in η_i about the regression relationship and (α, β) are the regression coefficients. The mean of ε_i is assumed to be zero and the variance σ^2 is constant. The values (x, y) measured with errors are observed instead of the actual values of (ξ, η). The measured values are assumed to be related to the actual values as:

$$x_i = \xi_i + \varepsilon_{x,i} \tag{10}$$

$$y_i = \eta_i + \varepsilon_{y,i} \tag{11}$$

where $\varepsilon_{x,i}$ and $\varepsilon_{y,i}$ are, respectively, the random measurement errors on x_i and y_i, which are normally distributed with known variances $\sigma^2_{x,i}$ and $\sigma^2_{y,i}$ and covariance $\sigma_{xy,i}$. The variances and covariance are assumed to be the same for each data point in this study.

Due to the complexity of parameter distribution, the Markov chain Monte Carlo (MCMC) method is introduced to conduct efficient sampling and promote the final convergence to the target distribution. By using MCMC, the mathematical expectation of the posterior distribution inferred by Bayesian inference is obtained as the estimated value of the parameter.

3. Results and Discussion

3.1. Multivariate Sensitivity Analysis of CO_2 Emission Factor

The components of the eigenvector are plotted in Figure 5; the single-column active variable weights quantify the sensitivity of the output to input parameters. The larger the weights, the greater the changes in the CO_2 emission factor caused by the corresponding parameters. The CO_2 emission rate (CO_2/[g/s]) is definitely the most influential input parameter for the CO_2 emission factor, and the *VSP* (vsp/[kw/t]) and speed (v/[km/h]) are coming next, with the *VSP* having a greater weight than speed. In addition, the influence of acceleration (a/[m/s^2]) and exhaust temperature (ExhaustTemp/[degC]) is minimal.

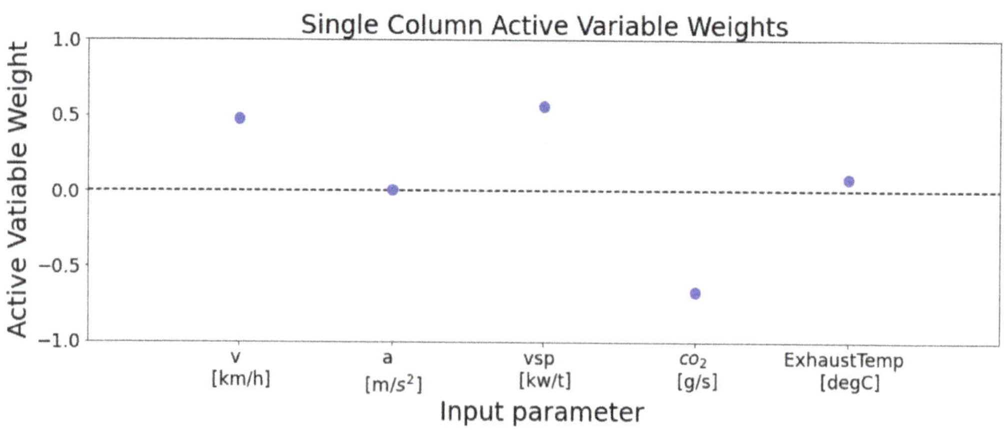

Figure 5. Input parameter weights calculated using all operating points.

The symbol of the input parameter weight can represent the positive and negative ratio relationship between the output parameter and input parameters. Meanwhile, the trend of the change can also be quantitatively described by the quadratic response surface curves and the influence of the eigenvector components on the activity variables (as seen in Figure 6). When all other input parameters suppose fixed values, the CO_2 emission factor decreased by 3.333% when v is increased from 5 to 10, and continually decreased by 3.330% when v reached 15. In general, the CO_2 emission factor will increase with a decrease in speed. This result may be caused by the definition of the emission factor, which is based on the driving distance (Equation (2)). When the speed decreases, $\sum v^k$ decreases, and the amplitude of the $\sum ER^k$ change is much smaller than $\sum v^k$, hence resulting in an increase in EF^k.

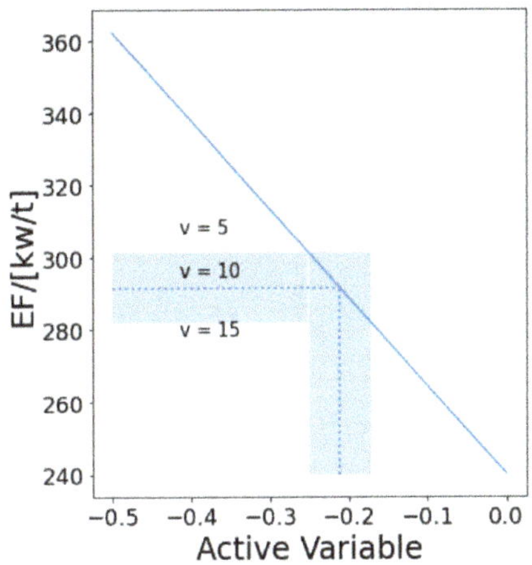

Figure 6. Estimation of output change with respect to input parameter values.

3.2. Inference of CO_2 Emission Factor by Bayesian Regression

The relationship of the CO_2 emission factor with the *VSP* is plotted in Figure 7, with the error bars indicating a 2σ- uncertainty of the data points. The test data are first fitted by the method of ordinary least squares without considering errors, and the fitting function is:

$$\eta = 5.243\,\xi + 142.6 \tag{12}$$

where η is the value of the CO_2 emission factor, and ξ is the *VSP* bin value.

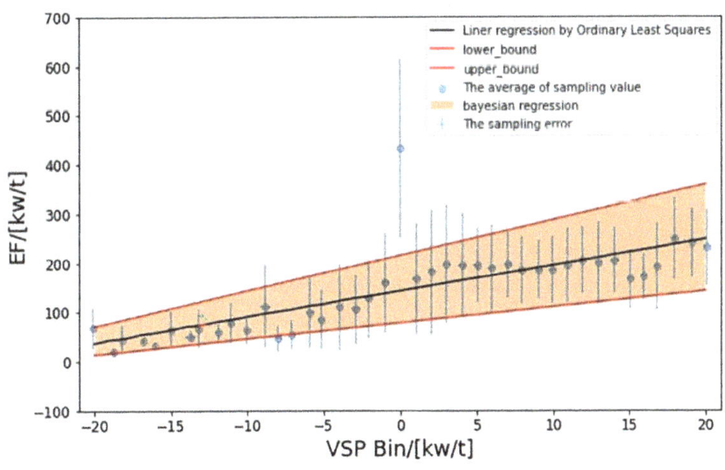

Figure 7. Regression results for *VSP* and CO_2 emission factor.

The test data are not a fixed value but a distribution within a certain range; the CO_2 emission factor should also be taken within a certain range. When considering sampling errors, the uncertainty of the CO_2 emission factor is plotted within the shaded area bounded by the upper and lower 95% confidence interval (95% CI) (in Figure 7). Meanwhile, the posterior probability distributions of parameters α and β are shown in Figure 8, as well as MCMC trace-plotting the samples of α and β under the Bayesian framework. The mean posterior distribution of α is 146.5362, while the mean posterior distribution of β is 5.1986. The progressions of the samples plotted in the trajectories of α and β seem to converge well without significant drift.

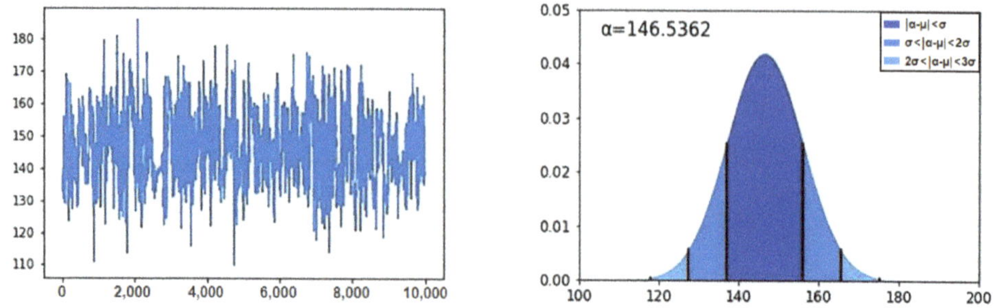

(a)

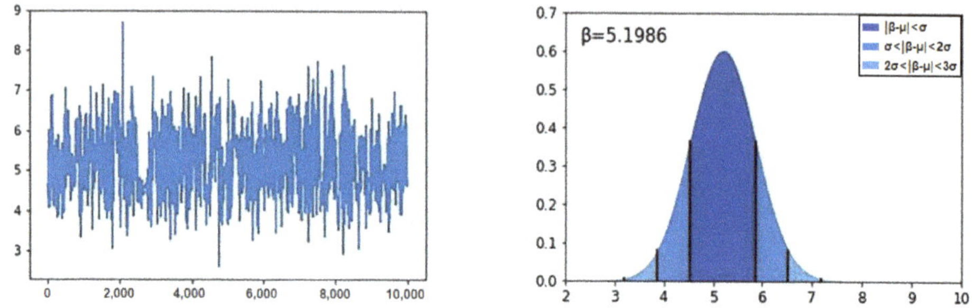

(b)

Figure 8. Tracing plots and posterior probability distribution of parameter (α, β) by MCMC under Bayesian inference: (**a**) tracing plots and posterior probability distribution of parameter α by MCMC; (**b**) tracing plots and posterior probability distribution of parameter β by MCMC.

For the surge of the CO_2 emission factor when *VSP* bin = 0 (Figure 7), the reason may be related to the lower vehicle speed. The CO_2 emission factor increases with the decrease in vehicle driving speed [42], and approximately 86.67% of the ultra-low speed operating points (v < 10 km/h) are accumulated in the *VSP* bin = 0 cluster. Meanwhile, the nosedive in vehicle driving speed in the range of *VSP* bin = 0 may result in sustained carbon dioxide emissions without the mileage being increased, causing a rising change shape in the CO_2 emission factor.

3.3. The Influence of Distribution Functions on Multivariate Analysis

Since the surge of the CO_2 emission factor when *VSP* bin = 0 in Section 3.2, only the cases with the *VSP* bin not being equal to zero are considered in this section, to mitigate the accounting uncertainties. The *VSP* distribution is determined to be approximately the Rayleigh distribution using the marginal distribution plot and Q–Q plot (Figure 9), with the corresponding *p*-value being less than 0.05. It can be seen from Figure 9a that, when $VSP > 5$ and $VSP < -5$, the marginal distribution of the *VSP* is in good agreement with the Rayleigh distribution; however, when $-5 < VSP < 5$, there is a certain deviation. For the Q–Q plot (Figure 9b), the quantile of the probability distribution of the *VSP* is basically fit to the Rayleigh distribution.

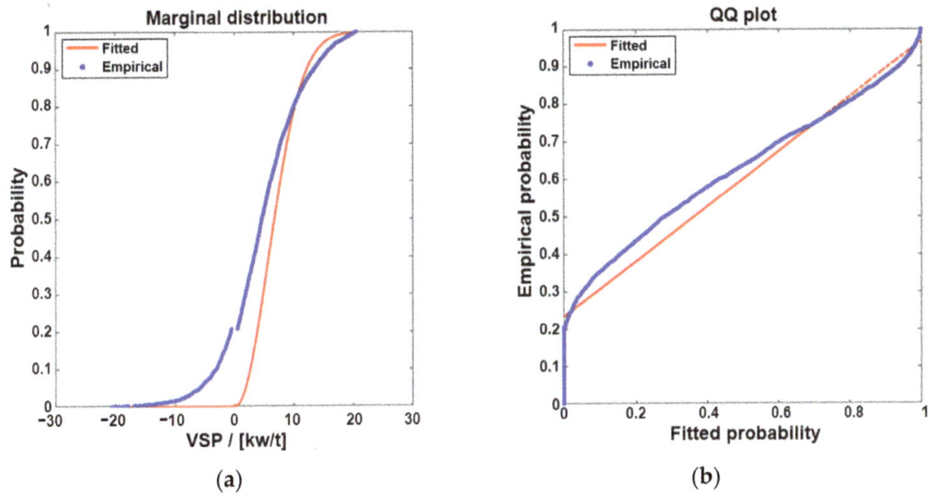

Figure 9. The marginal distribution of input parameters and Rayleigh distribution and the Quantile–Quantile plots: (**a**) the marginal distribution; (**b**) Q–Q plot.

Then, the *VSP* data are set to the uniform distribution and Rayleigh distribution, respectively, when the distribution functions of other input variables remain unchanged. All input variables are normalized and brought into the active subspace for a sensitivity analysis. The linear combinations of the CO_2 emission factor relative to the inputs (i.e., the active variables) exhibit a univariate trend as given in Figure 10, and the relationship between the output and active variables can be represented via a quadratic polynomial:

$$g(y) \approx C_0 + C_1 y + C_2 y^2 \approx C_0 + C_1(\hat{w}^T x) + C_2(\hat{w}^T x)^2 \tag{13}$$

The function $g(y)$ outputs the CO_2 emission factor (g/km) where y represents the active variable, which is the weighted sum of the input parameters after scaling.

The CO_2 emission can be understood through a powerful information combination of the parameter weights and sufficient summary plots. The positive and negative weights of the input parameters can be used to predict the trend of the changes in the active variables affected by the input parameters, thereby further predicting the trend of the changes in the output parameters. For example, as plotted in Figure 10a, since the *VSP* index weight is negative for the EF, a greater *VSP* index decreases the active variable, which results in the decrease in EF. The opposite trend of influence compared to Bayesian fitting is led by the comprehensive influence of other input parameters in multivariate analysis.

It is inferred that the changes in data sampling transform the establishment of subspace matrices, thus affecting the calculation of eigenvector components and the fitting of the final quadratic response surface. As a result, the acceleration and speed, along with the *VSP*, are combined to become the most influenced factor of the CO_2 emission factor when the *VSP* is sampled according to the Rayleigh distribution. The fitting coefficients C_0, C_1, and C_2 in Equation (13) and the corresponding coefficient of determination (R-squared) values for each of the *VSP* distribution functions are tabulated in Table 2. Overall, the result of the R-squared from the *VSP* sampled according to uniform distributions shows a better performance on the goodness of fit. It also suggests that the data shown in Figure 10 are more in line with a quadratic polynomial. Hence, the input parameter distribution function determines the relationship between the combination of inputs and the output of interest to a certain extent.

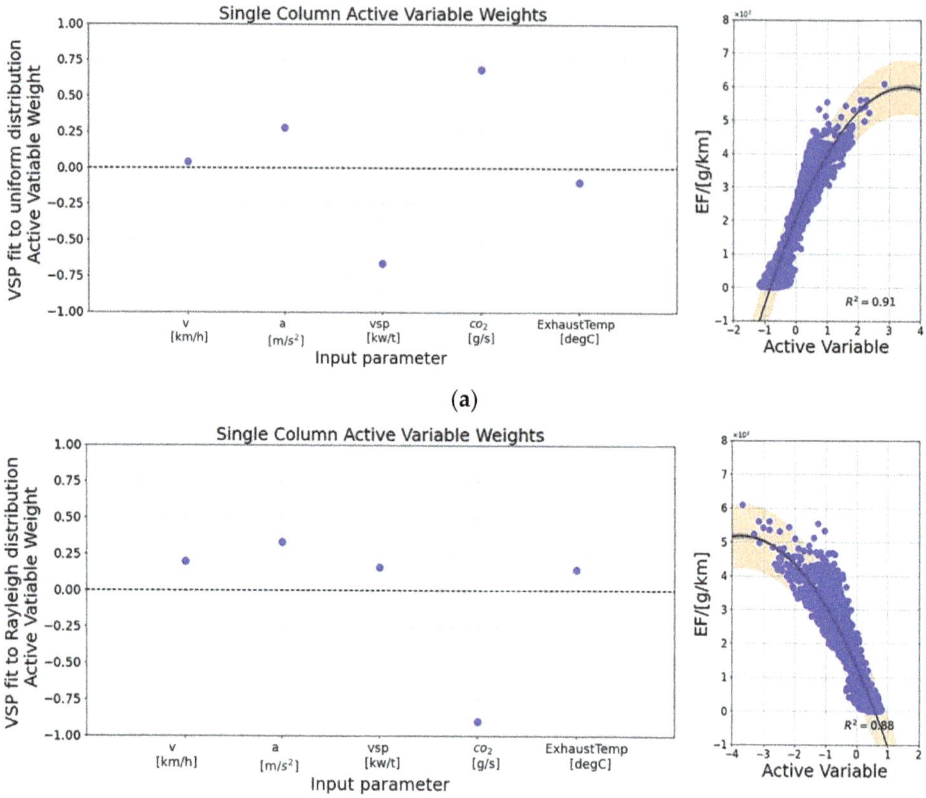

(**a**)

(**b**)

Figure 10. Input parameter weights and sufficient summary plots calculated when *VSP* bin = 0 is excluded from data sampling: (**a**) input parameter weights and sufficient summary plots when *VSP* fit to uniform distribution; (**b**) input parameter weights and sufficient summary plots when *VSP* fit to Rayleigh distribution.

Table 2. Summary of quadratic model coefficients.

Distribution	C_0	C_1	C_2	R^2
Rayleigh	122.59	−124.90	−29.29	0.88
Uniform	202.63	224.32	−31.76	0.91

So far, comprehensive studies on the impact of vehicle driving parameters on emission factors by multivariate sensitivity analysis and uncertainty analysis have been carried out. It is highlighted with high-resolution database processing prepared for the further modelling of the carbon footprint in transportation.

4. Conclusions

In the present study, dynamic vehicle data contributing to traffic carbon emissions are comprehensive studied in the aspects of data sensitivity and uncertainty. The active subspace method can identify which input parameters are the most important through magnitudes of the input parameter weights, while exploring how the combination of inputs is related to the output of interest, without the expense of multiple simulations. It is concluded that the CO_2 emission factor is most sensitive to the *VSP*. The method has great potential to readily derive the relationship between the combination of inputs and outputs in a complex domain without the expense of multiple simulations. And the relationship between the input parameters (i.e., the active variables) and CO_2 emission factor is able to be formulated using a quadratic function. Moreover, two domains with varied *VSP* data distributions are set up to evaluate the sampling diversity on the emission sensitivity. At the same time, a conclusion similar to that in Section 3.3 can be obtained by changing the distribution function of other vehicle operating parameters. But, due to the study only focusing on the RDE data of one light gasoline vehicle, it cannot be guaranteed that the sensitivity analysis of diesel vehicles or non-light gasoline vehicles is consistent with this study.

Concluded from a sensitivity analysis, the relationship between the *VSP* and CO_2 emission factor is explored via the Bayesian linear regression method with the sampling uncertainties considered. The uncertainty of the CO_2 emission factor within the upper and lower 95% CI is determined when considering the sampling errors. The uncertainty quantifications for the calibrating parameters α and β are well demonstrated by using the heteroscedasticity measurement errors of both variables. In addition, the reason for the surge of the EF in the *VSP* bin = 0 cluster is discussed by combining it with the active subspace method. Compared with traditional linear or nonlinear methods, the proposed method takes into account the uncertainty of the parameter distribution and improves the fitting accuracy.

Generally, this article comprehensively studies the impact of vehicle driving parameters on emission factors via multivariate sensitivity analysis and uncertainty quantification analysis. The input parameters that have a significant impact on emission factors are obtained, and the influence of the error distribution and distribution function of the sampled data on the fitting results is explored. On this basis, a new perspective for modelling traffic carbon emission with a high-resolution database is proposed. And it is conducive to improving the accuracy of carbon counting under varied transportation scenarios, laying the foundation for constructing a high-resolution database of the carbon emission factor.

Author Contributions: Conceptualization, F.Z.; methodology, J.C. and F.Z., validation, J.C.; formal analysis, J.C.; investigation, J.C., H.Y., H.X., Q.L., Z.Z. and H.C.; data curation, H.Y., H.X., Q.L. and Z.Z.; visualization, H.C.; writing—original draft, J.C.; writing—review and editing, F.Z.; funding acquisition, F.Z. and W.Y.; resources, H.Y., H.X., Q.L. and Z.Z.; supervision, F.Z. and W.Y.; project administration, F.Z. and W.Y. All authors have read and agreed to the published version of the manuscript.

Funding: This research was funded by the Open Funds of Chongqing Key Laboratory of Vehicle Emission and Economizing Energy, grant number No. PFJN-09, and the Shandong Provincial Natural Science Foundation, grant number No. 2022HWYQ-061.

Data Availability Statement: Restrictions apply to the availability of these data. Data were obtained from China Automotive Engineering Research Institute Co., Ltd. and are available Feiyang Zhao with the permission of China Automotive Engineering Research Institute Co., Ltd.

Conflicts of Interest: Authors Hao Yu, Haocheng Xu, Qiang Lv and Zongqiang Zhu were employed by the company China Automotive Engineering Research Institute Co., Ltd. The remaining authors declare that the research was conducted in the absence of any commercial or financial relationships that could be construed as a potential conflict of interest.

References

1. Hu, G.; Zeng, W.; Yao, R.; Xie, Y.; Liang, S. An integrated assessment system for the carrying capacity of the water environment based on system dynamics. *J. Environ. Manag.* **2021**, *295*, 113045. [CrossRef] [PubMed]
2. Chen, Z.; Li, B.; Jia, S.; Ye, X. Modeling and simulation analysis of vehicle pollution and carbon reduction management model based on system dynamics. *Environ. Sci. Pollut. Res. Int.* **2023**, *30*, 14745–14759. [CrossRef] [PubMed]
3. IEA. *CO2 Emissions in 2023*; IEA: Paris, France, 2024. Available online: https://www.iea.org/reports/co2-emissions-in-2023 (accessed on 27 April 2023).
4. Li, Y.; Dong, H.; Lu, S. Research on application of a hybrid heuristic algorithm in transportation carbon emission. *Environ. Sci. Pollut. Res.* **2021**, *28*, 48610–48627. [CrossRef] [PubMed]

5. Wang, L.; Zhao, Y.; Wang, J.; Liu, J. Regional inequality of total factor CO_2 emission performance and its geographical detection in the China's transportation industry. *Environ. Sci. Pollut. Res. Int.* **2022**, *29*, 3037. [CrossRef] [PubMed]
6. IEA. *Energy Technology Perspectives 2015*; OECD Publishing: Paris, France, 2015.
7. Yao, Z.; Cao, H.; Cui, Z.; Wang, Y.; Huang, N. Research on Urban Distribution Routes Considering the Impact of Vehicle Speed on Carbon Emissions. *Sustainability* **2022**, *14*, 15827. [CrossRef]
8. Dong, Y.; Xu, J.; Ni, J. Carbon emission model of vehicles driving at fluctuating speed on highway. *Environ. Sci. Pollut. Res. Int.* **2023**, *30*, 18064–18077. [CrossRef] [PubMed]
9. Rakha, H.; Ahn, K.; Trani, A. Development of VT-Micro model for estimating hot stabilized light duty vehicle and truck emissions. Transportation research. *Part D Transp. Environ.* **2004**, *9*, 49–74. [CrossRef]
10. Kim, W.; Kim, C.; Lee, J.; Kim, J.; Yun, C.; Yook, S. Fine particle emission characteristics of a light-duty diesel vehicle according to vehicle acceleration and road grade. *Transp. Res. Part D Transp. Environ.* **2017**, *53*, 428–439. [CrossRef]
11. Zhao, F.; Liu, F.; Liu, Z.; Hao, H. The correlated impacts of fuel consumption improvements and vehicle electrification on vehicle greenhouse gas emissions in China. *J. Clean. Prod.* **2019**, *207*, 702–716. [CrossRef]
12. Li, X.; Yu, B. Peaking CO_2 emissions for China's urban passenger transport sector. *Energy Policy* **2019**, *133*, 110913. [CrossRef]
13. Pu, X.; Lu, X.; Han, G. An improved optimization algorithm for a multi-depot vehicle routing problem considering carbon emissions. *Environ. Sci. Pollut. Res. Int.* **2022**, *29*, 54940–54955. [CrossRef] [PubMed]
14. Jimenez-Palacios, J.L. *Understanding and Quantifying Motor Vehicle Emissions with Vehicle Specific Power and TILDAS Remote Sensing*; Massachusetts Institute of Technology: Cambridge, MA, USA, 1999.
15. Song, G.; Lei, Y. Characteristics of low-speed vehicle-specific power distributions on urban restricted-access roadways in Beijing. *Transp. Res. Rec.* **2011**, *2233*, 90–98. [CrossRef]
16. Forcetto, A.L.S.; de Salvo Junior, O.; Maciel Filho, F.F.; de Fátima Andrade, M.; de Almeida, F.; Flávio Guilherme, V. Improving the assessment of RDE dynamics through vehicle-specific power analysis. *Environ. Sci. Pollut. Res. Int.* **2022**, *29*, 59561–59574. [CrossRef] [PubMed]
17. Zhang, L.; Zhu, Z.; Zhang, Z.; Song, G.; Zhai, Z.; Yu, L. An improved method for evaluating eco-driving behavior based-on speed-specific vehicle-specific power distributions. *Transp. Res. Part D Transp. Environ.* **2022**, *113*, 103476. [CrossRef]
18. Amoatey, P.; Omidvarborna, H.; Baawain, M.S.; Al-Mamun, A. Evaluation of vehicular pollution levels using line source model for hot spots in Muscat, Oman. *Environ. Sci. Pollut. Res. Int.* **2020**, *27*, 31184–31201. [CrossRef] [PubMed]
19. Smit, R.; Awadallah, M.; Bagheri, S.; Surawski, N.C. Real-world emission factors for SUVs using on-board emission testing and geo-computation. *Transp. Res. Part D Transp. Environ.* **2022**, *107*, 103286. [CrossRef]
20. EPA. *User's Guide to Mobile4 (Mobile Source Emission Factor Model)*; Office of Mobile Sources U.S. Environmental Protection Agency Ann Arbor: Washington, DC, USA, 1989.
21. EPA. User's Guide to MOBILE6.1 and MOBILE6.2: Mobile Source Emission Factor Model [CP/OL]. 2008. Available online: http://www.epa.gov/oms/m6.htm,2008/04/05 (accessed on 27 April 2023).
22. Khazini, L.; Kalajahi, M.J.; Blond, N. An analysis of emission reduction strategy for light and heavy-duty vehicles pollutions in high spatial–temporal resolution and emission. *Environ. Sci. Pollut. Res. Int.* **2022**, *29*, 23419–23435. [CrossRef]
23. Charles, R.; Crawford. *Active Subspaces: Emerging Ideas for Dimension Reduction in Parameter Studies*; SIAM: Philadelphia, PA, USA, 2016.
24. Constantine, P.G.; Emory, M.; Larsson, J.; Iaccarino, G. Exploiting active subspaces to quantify uncertainty in the numerical simulation of the HyShot II scramjet. *J. Comput. Phys.* **2015**, *302*, 1–20. [CrossRef]
25. Constantine, P.G.; Diaz, P. Global sensitivity metrics from active subspaces. *Reliab. Eng. Syst. Saf.* **2017**, *162*, 1–13. [CrossRef]
26. Seshadri, P.; Shahpar, S.; Constantine, P.; Parks, G.; Adams, M. Turbomachinery Active Subspace Performance Maps. *J. Turbomach.* **2018**, *140*, 041003. [CrossRef]
27. Bahamonde, S.; Pini, M.; De Servi, C.; Schiffmann, J.; Colonna, P. Corrigendum to "Active subspaces for the optimal meanline design of unconventional turbomachinery" [*Appl. Therm. Eng.* **2017**, *127*, 1108–1118]. *Appl. Therm. Eng.* **2019**, *150*, 1353–1355. [CrossRef]
28. Ji, W.; Ren, Z.; Marzouk, Y.; Law, C.K. Quantifying kinetic uncertainty in turbulent combustion simulations using active subspaces. *Proc. Combust. Inst.* **2019**, *37*, 2175–2182. [CrossRef]
29. Zhang, L.; Wang, N.; Wei, J.; Ren, Z. Exploring active subspace for neural network prediction of oscillating combustion. *Combust. Theory Model.* **2021**, *25*, 570–587. [CrossRef]
30. Lin, K.; Zhou, Z.; Wang, Y.; Law, C.K.; Yang, B. Using active subspace-based similarity analysis for design of combustion experiments. *Proc. Combust. Inst.* **2023**, *39*, 5177–5186. [CrossRef]
31. Chen, D.; Dahlgren, R.A.; Shen, Y.; Lu, J. A Bayesian approach for calculating variable total maximum daily loads and uncertainty assessment. *Sci. Total Environ.* **2012**, *430*, 59–67. [CrossRef] [PubMed]
32. Rajamand, S.; Caglar, R. Control of voltage and frequency based on uncertainty analysis using Bayesian method and effective power flow control of storage role in electrical vehicle charging station. *Sustain. Energy Grids Netw.* **2022**, *32*, 100837. [CrossRef]
33. Elster, C.; Toman, B. Bayesian uncertainty analysis for a regression model versus application of GUM Supplement 1 to the least-squares estimate. *Metrologia* **2011**, *48*, 233–240. [CrossRef]
34. Lu, D.; Ye, M.; Hill, M.C. Analysis of regression confidence intervals and bayesian credible intervals for uncertainty quantification. *Water Resour. Res.* **2012**, *48*, W09521.1–W09521.20. [CrossRef]

35. Li, X.; Feng, J.; Wellen, C.; Wang, Y. A Bayesian approach of high impaired river reaches identification and total nitrogen load estimation in a sparsely monitored basin. *Environ. Sci. Pollut. Res. Int.* **2017**, *24*, 987–996. [CrossRef]
36. Mudgal, A.; Hallmark, S.; Carriquiry, A.; Gkritza, K. Driving behavior at a roundabout: A hierarchical Bayesian regression analysis. *Transp. Res. Part D Transp. Environ.* **2014**, *26*, 20–26. [CrossRef]
37. Martin, N.P.D.; Bishop, J.D.K.; Choudhary, R.; Boies, A.M. Can UK passenger vehicles be designed to meet 2020 emissions targets? A novel methodology to forecast fuel consumption with uncertainty analysis. *Appl. Energy* **2015**, *157*, 929–939. [CrossRef]
38. EPA. *Methodology for Developing Modal Emission Rates for EPA's Multi-Scale Motor Vehicle and Equipment Emission System*; EPA: Washington, DC, USA, 2002.
39. Jefferson, J.L.; Gilbert, J.M.; Constantine, P.G.; Maxwell, R.M. Active subspaces for sensitivity analysis and dimension reduction of an integrated hydrologic model. *Comput. Geosci.* **2015**, *83*, 127–138. [CrossRef]
40. Yu, W.; Zhao, F.; Yang, W.; Zhu, Q. Uncertainty quantifications of calibrating laser-induced incandescence intensity on sooting propensity in a wick-fed diffusion flame burner. *Fuel* **2021**, *289*, 119921. [CrossRef]
41. Kelly, B.C. Some Aspects of Measurement Error in Linear Regression of Astronomical Data. *Astrophys. J.* **2007**, *665*, 1489–1506. [CrossRef]
42. Shuai, M.; Zhihui, H.; Liang, J.; Haiguang, Z.; Tian, M. CO_2 and NOx emission characteristics from a heavy-duty China VI diesel truck based on portable emission measurement system. *Acta Sci. Circumstantiae* **2022**, *42*, 341–350. [CrossRef]

energies

MDPI

Article

Low-Friction and -Knocking Diesel Engine Cylindrical-Tapered Bore Profile Design

Junhong Zhang [1,2], Ning Wang [1], Jian Wang [1,3,*], Hui Wang [1,3], Xueling Zhang [2], Huwei Dai [1] and Jiewei Lin [1,*]

1 State Key Laboratory of Engines, Tianjin University, Tianjin 300354, China; zhangjh@tju.edu.cn (J.Z.); 2021201385@tju.edu.cn (N.W.); dhwmail@tju.edu.cn (H.D.)
2 Department of Mechanical Engineering, Tianjin Renai College, Tianjin 301636, China; zhangxueling0@163.com
3 Weichai Power Co., Ltd., Weifang 261000, China
* Correspondence: wangjian@weichai.com (J.W.); linjiewei@tju.edu.cn (J.L.); Tel.: +86-13884712670 (J.W.); +86-13752356026 (J.L.)

Abstract: To reduce the friction loss and the piston-knocking noise from the perspective of the design of the cylinder bore profile, the piston-ring cylinder bore (PRCB) dynamic model of an L6 diesel engine was developed using AVL-Excite-Piston & Rings. Based on the full-scale test method, the effects of bore taper, starting height of tapered profile, and ellipticity on the friction power and knocking energy of the PRCB system were investigated, and the optimization of the design of the bore profile was carried out with the objectives of minimizing the system's friction power and the peak knocking kinetic energy. The results showed that the taper of the cylinder bore has the greatest influence on the system's friction power and the peak knocking kinetic energy, followed by the starting height of the conical profile. For the peak knocking kinetic energy of the piston, there was an obvious interaction between the taper and the starting height of the conical profile. When the taper was 35 μm and 45 μm, the peak knocking kinetic energy showed a decreasing and then increasing trend with the increase in the starting height of the profile, and when the taper was 55 μm the peak knocking kinetic energy monotonically was decreased with the increase in the starting height of the conical profile. The optimization results showed that the system's friction power was decreased by 15.05% and the peak knocking kinetic energy was decreased by 21.41% for a taper degree of 55 μm, a tapered profile starting height of 31 mm, and an ellipticity of 50 μm compared to the initial cylindrical cylinder bore.

Keywords: diesel engine; piston-ring cylinder bore; cylinder hole profile; friction power; knocking kinetic energy

Citation: Zhang, J.; Wang, N.; Wang, J.; Wang, H.; Zhang, X.; Dai, H.; Lin, J. Low-Friction and -Knocking Diesel Engine Cylindrical-Tapered Bore Profile Design. *Energies* **2024**, *17*, 2042. https://doi.org/10.3390/en17092042

Academic Editor: Pietro Zunino

Received: 5 March 2024
Revised: 23 April 2024
Accepted: 24 April 2024
Published: 25 April 2024

1. Introduction

With the enhancement of diesel engines, the thermal and mechanical load of their cylinder bore has increased accordingly, and the deformation of the cylinder bore has also increased [1–3]. Studies [4,5] have shown that an increase in the degree of irregular cylinder-bore deformation would lead to an increase in friction loss between the piston and cylinder bore. The friction between the piston assembly and cylinder bore accounts for half of the total friction loss of the engine [6,7]. Research has shown that for every 10% increase in engine friction loss, fuel consumption would increase by 3% to 5%, and the engine's effective output power would also decrease accordingly. The irregular deformation of the cylinder bore will lead to an increase in cylinder clearance, which will cause an increase in piston-knocking noise [8,9]. Therefore, rationalizing the design of the cylinder bore pattern, reducing the friction power and the peak knocking kinetic energy of the PRCB system, is of great significance to improving the economy and NVH performance of diesel engines. Most scholars have chosen to optimize the piston structure [10–12] and rationally design the surface texture of the friction pair [13–15] to reduce the friction loss and the knocking kinetic energy of the PRCB. Few scholars have improved the friction and lubrication characteristics and knocking characteristics of the PRCB system by rationally designing the macrostructure of the cylinder bore. At present, from the perspective of designing the cylinder bore profile, the use of a non-cylindrical cylinder liner design in the cold state has become another effective method to improve the friction performance of the ring group-cylinder liner system.

The average friction-effective pressure of the piston was reduced [16,17] by expanding the bottom diameter of the cylinder bore of the gasoline engine through the honing process in the experimental test. Numerical calculations were utilized in the literature to compare the deformation of the conical and elliptical cross-section cylinder liner with that of the traditional cylindrical cylinder liner of a certain gasoline engine, and it was found [18,19] that the radial deformation of the elliptical–cone combination cylinder liner was reduced by 70%~80% under the thermal load condition. Alshwawra et al. [20] studied three initial bore shapes of a certain gasoline engine—elliptical, conical–elliptical, and reverse free-form—and found that the reverse free-form bore was closest to a "round" shape at each height after applying thermal and mechanical loads to the bore, and the oil film contact area between the piston ring and the cylinder wall was increased, which provided a new reference for the bore pattern to reduce the friction power of the piston ring group. Halbhuber et al. [21] compared the cumulative friction power of cylindrical bores under the consideration of piston cooling conditions and found that the cumulative friction power of the piston group under cylindrical-tapered compensated bores was 4.4% lower than that using uncompensated bores.

Due to the complexity and high cost of the reverse free-form process, a tapered cylindrical bore is used to improve the geometric performance of the gasoline engine in the cold state, and the tapered bore reduces the friction loss by increasing the distribution of the oil film in the friction gap; however, too large a gap between the piston and the cylinder wall can lead to an increase in the piston-knocking noise, while too small a gap can lead to an increase in the friction power, and the design parameters of the cylinder bore pattern have different degrees of influence on the friction and knocking performance of the piston. Moreover, the design parameters of the cylinder bore pattern have different influences on the piston friction and knocking performance, and there is an interaction between the parameters. At present, there is little literature on the influence of the design parameters of the cylinder bore pattern on the piston friction and knocking characteristics, and there is no systematic design method for the cylinder bore pattern for engineering applications.

Therefore, the influence of the cold design parameters of the cylinder bore pattern on the friction and knocking energy of the piston-ring cylinder bore system in an L6 diesel engine was studied. With the objective function of minimizing the friction power and peak knocking kinetic energy, the parameters of the cylinder bore pattern were optimized, and the optimal design of the cylinder bore pattern was obtained. This provided some theoretical guidance for reducing friction loss and reducing piston-knocking noise from the perspective of cylinder bore design.

2. Numerical Modeling

2.1. The Piston's Dynamic Modeling

In the motion plane of the diesel engine's crank-connecting rod mechanism, the forces on the piston are shown in Figure 1, where F_{rx} and F_{ry} are the normal and tangential forces of the piston ring on the piston, respectively, F_{cx} and F_{cy} are the normal and tangential forces of the cylinder bore on the piston, respectively, F_p is the force on the piston pin, F_{rod} is the force of the linkage system at the con rod's small-end bearing, F_{gc} is the piston gravity, F_{gas} is the gas force, mc is the mass of the piston, M_{gc} is the gravity-induced piston moment around the pin axis, M_{pc} is the friction moment at the piston pin, and β is the crankshaft angle.

After analyzing the force on the piston, the dynamic equations of the piston in the reciprocation, radial motion, and rotation around the pin were expressed as Equations (1)–(3), respectively.

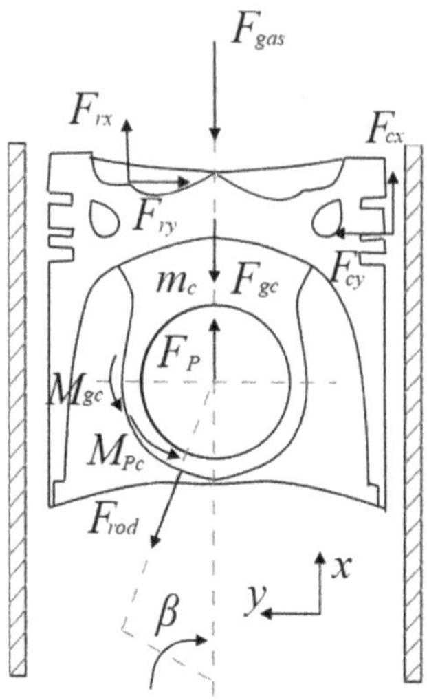

Figure 1. Schematic diagram of piston movement force analysis under calibration conditions.

Forces in the x-direction:

$$m_c \times \ddot{x}_c = F_{cx} + F_{rx} - F_{gcx} - F_{gas} + F_{px} - F_{rodx} \tag{1}$$

Forces in the y-direction:

$$m_c \times \ddot{y}_c = F_{cy} - F_{ry} - F_{gcy} - F_{py} + F_{rody} \tag{2}$$

Rotation around the pin:

$$\Theta_c \times \ddot{w}_c = M_c + M_r + M_{gc} + M_{pc} \tag{3}$$

where $\ddot{x}_c$ and $\ddot{y}_c$ are the acceleration of the piston motion in the x-direction and y-direction, respectively, F_{gcx} and F_{gcy} are the component force of piston gravity along the x-direction and y-direction, respectively, F_{px} and F_{py} are the load on the piston pin along the x-direction and the force in the y-direction, respectively, F_{rodx} and F_{rody} are the force of the linkage system at the con rod's small-end bearing along the x-direction and the force in the y-direction, respectively, Θ_c is the piston's moment of inertia around the pin, $\ddot{w}_c$ is the angular acceleration, Mc is the moment due to piston–liner contact, and M_r is the moment due to ring axial and radial force.

During the working process of the diesel engine, the piston not only reciprocates along the cylinder bore in the height direction, it also includes transverse movement and rotation around the pin, which has a great influence on the piston's knocking on the inner wall of the cylinder bore to produce knocking noise. Therefore, the piston-knocking kinetic energy was used to measure the degree of piston knocking on the inner wall of the cylinder bore; the specific formula is given as follows:

$$E = E_v + E_W = \frac{1}{2} m_c \dot{x}_c^2 + \frac{1}{2} \Psi \dot{w}_c^2 \tag{4}$$

where E_v and E_W are the kinetic energy of radial motion and rotational kinetic energy respectively, m_c is the mass of the piston, $\dot{x}_c$ is the radial velocity, Ψ is the rotational moment of inertia, and $\dot{w}_c$ is the rotational angular velocity.

2.2. The Piston Ring Group Dynamic Model

The force analysis of the piston ring is shown in Figure 2, where m_R is the mass of the piston ring, $\ddot{x}$ is the acceleration of the piston ring in the x-direction, s is the center of gravity, M is the piston ring cross-section center position, P_{above} and P_{below} are the pressure distributions on the upper and lower surfaces of the piston ring, respectively, P_{behind} is the pressure distribution on the inside of the piston ring, $F_{fric,ax}$ is the axial friction between the cylinder bore and piston ring, $F_{gas,ax}$ is the gas axial force, ξ is the piston ring torsion angle, $F_{gas,rad}$ is the gas radial force, $F_{fric,rad}$ is the radial friction between the moving surface of the piston ring and the cylinder bore, and $F_{hydr,ax}$ and $F_{hydr,rad}$ are the axial and radial hydrodynamic friction, respectively.

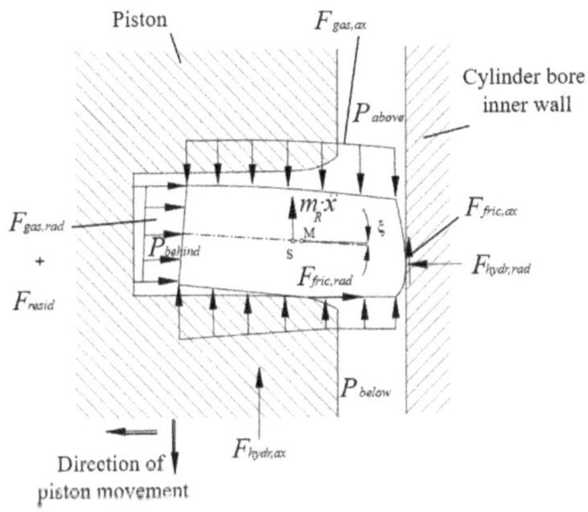

Figure 2. Schematic diagram of piston ring force under calibration conditions.

When the ring is floating (the ring is not in contact with the ring groove), the force equilibrium equations between the ring and the ring groove and the dynamic force equilibrium equations can be expressed as follows:

$$F_{contact} = F_{mass,x} + F_{fric,ax} + F_{gas,ax} + F_{bend} \tag{5}$$

$$m_{ring} \cdot \ddot{x}_{ring} = \Sigma F = F_{damp,ax} + F_{fric,ax} + F_{gas,ax} + F_{bend} \tag{6}$$

where $F_{contact}$ is the contact force between the piston ring and ring groove, $F_{mass,x}$ is the ring's gravity force in the axial direction, F_{bend} is the ring's bending force; m_{ring} is the ring's mass, $\ddot{x}_{ring}$ is the piston ring's axial acceleration, $F_{damp,ax}$ is the axial damping force, and $F_{tension}$ is the tangential elasticity of the piston rings.

The piston ring's radial force equation can be expressed as follows:

$$F_{contact,rad} = F_{fric,rad} + F_{gas,rad} + F_{tension} \tag{7}$$

The equilibrium equation for the bending moment of the ring section around the form center M is given as follows:

$$\sum M = \sum (F_i \cdot h_i) + M_{pre-twist} = M_{elatic} \cdot \xi \tag{8}$$

where F_i is the radial and axial forces on the piston rings, h_i is the distance from each component force to the center of the piston ring's cross-section, $M_{pre\text{-}twist}$ is the moment on the piston ring, and M_{elatic} is the bending moment of the piston ring.

2.3. Friction Force Modeling

The state of the oil film between the piston ring group/skirt and the bore liner has an important influence on the dynamics of the PRCB. The pressure distribution of the oil film can be described by the mean Reynolds equation, which is given as follows:

$$\frac{\partial}{\partial x}\left(\Phi_x h^3 \frac{\partial p_h}{\partial x}\right) + \frac{\partial}{\partial y}\left(\Phi_y h^3 \frac{\partial p_h}{\partial y}\right) = 6\nu U\left(\frac{\partial \bar{h}_T}{\partial y} + \sigma \frac{\partial \Phi_s}{\partial y}\right) + 12\nu \frac{\partial \bar{h}_T}{\partial t} \tag{9}$$

where p_h is the oil film pressure, U is the piston movement speed, h is the nominal oil film thickness, $\bar{h}_T$ is the average value of actual oil film thickness, σ is the combined roughness, ν is the lubricant power viscosity, Φ_x and Φ_y are the pressure flow factors, Φ_s is the shear flow factor, and t is the time.

The calculation assumes that the lubricant film is a Newtonian fluid; the variation in lubricant viscosity with temperature should be described by applying Vogel's formula:

$$\nu(T) = A \times \exp(\frac{B}{T + C})$$

where A, B, and C are the coefficients of the viscosity equation related to the type of diesel engine, which in this paper take values of 2.4×10^{-5}, 1641, and 170, respectively, and T is the temperature.

During the reciprocating motion of the piston, the oil film's thickness and pressure are in a state of dynamic change. When the piston moment reaches a certain level, the thickness of the oil film decreases rapidly, which leads to the direct contact of the piston skirt or the piston ring with the micro-convexity on the cylinder bore's surface, and the normal pressure of the contact surface at this time is determined by the contact model of the micro-convexity of the GREENWOOD.

The ratio of the friction pair to the combined roughness of the friction pair is defined as the minimum film thickness ratio, which is calculated as follows:

$$H_{min} = \frac{h_{min}}{\sigma} \tag{10}$$

where h_{min} is the minimum film thickness, H_{min} is the minimum film thickness ratio, and σ is the combined roughness. An H_{min} value of less than 1 indicates boundary lubrication, and H_{min} value of 1 to 4 indicates mixed lubrication, and an H_{min} value greater than 4 indicates hydrodynamic lubrication.

The friction between the piston ring group and the inner wall of the cylinder bore can be derived from Stribeck's formula:

$$F_f = F_n \mu \tag{11}$$

where F_n is the force of pressure between the piston/piston ring and the inner wall of the cylinder bore, and μ is the coefficient of friction.

$$\mu = f(\frac{v}{A \cdot v^2 + B \cdot |v| + C} + D \cdot v) \tag{12}$$

where f is the overall scaling factor; A, B, C, and D are related to the form of the diesel engine; and v is the dynamic viscosity coefficient.

2.4. Cylinder Bore Deformation Calculation

Firstly, one-dimensional simulation on the working process of an inline six-cylinder diesel engine was carried out by AVL BOOST [22], and the main parameters of the engine were as shown in Table 1. The thermal boundary conditions, such as gas temperature and convective heat transfer coefficient, were obtained from one-dimensional simulations, as shown in Figure 3, which shows the variation in the convective heat transfer coefficient and temperature in the combustion chamber with the crankshaft angle, and it can be seen that both of them reach their maximum value near the upper stop (90CA), due to the huge heat generated by the gas explosion in the cylinder at this time.

Table 1. Main engine parameters.

Parameters	Value
Maximum power	235 kW
Maximum moment	1300 N·m
Number of cylinders	6
Diameter of a cylinder	107 mm
Stroke	126 mm
Engine capacity	6.8 L
Compression ratio	19.5

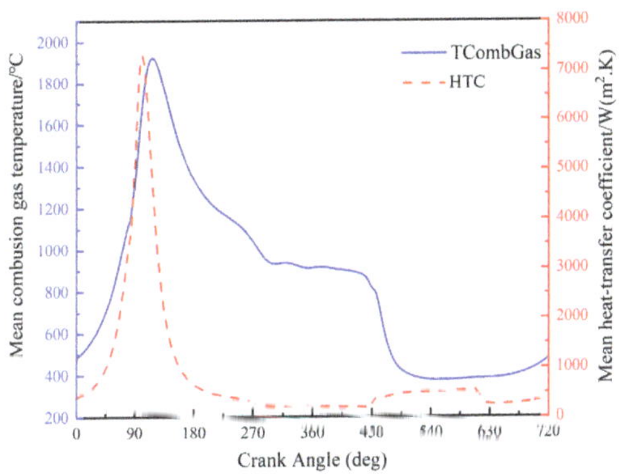

Figure 3. Gas temperature and convective heat transfer coefficient.

In the actual working process of the diesel engine, due to the thermal inertia of the body material, the effect of the heat load of the high-temperature gas on the body is limited to a thin layer on the surface of the material. The temperature of the vast majority of the region is approximate to stabilize; that is, the wall surface of the combustion chamber can be formed without a time-varying steady-state temperature field [23].

The time-averaged convective heat transfer coefficient and temperature of the gas were applied to the finite element model in the form of a third-type boundary condition. In the simulation process, the deformation of the cylinder bore is mainly determined by the inhomogeneous thermal load, and since it is assumed that the thermal load is steady-state and the fuel of each cylinder is consistent in the calculation process, the radial deformation of each cylinder due only to the thermal load is roughly the same, but the deformation of cylinders 2, 3, and 4 is affected by the adjacent two cylinders at the same time, while cylinders 1 and 6 are affected by only one of the neighboring cylinders; thus, in a comprehensive view, the deformation of cylinders 1 and 6 is symmetrically similar, while the deformation of cylinders 2, 3, and 4 is similar, so the calculation process can be simplified to cylinders 1, 2, and 6 in order to ensure the accuracy of the premise, greatly reducing the computing time. The deformation of the engine body was calculated in ABAQUS, and the whole deformation of the engine was as shown in Figure 4. The radial deformation of the bore surface of cylinder 2 was also extracted as a boundary condition for the subsequent dynamic calculations.

Figure 5 shows the radial deformation of the cross-section of the inner wall at different heights in the thermal state of the cylindrical cylinder bore. Since the 0° anti-thrust side (ATS) and 180° thrust side (TS) were subjected to thermal loads in addition to piston-knocking loads, the deformation in the 0° and 180° directions was larger than that in the 90° and 270° directions at different heights, and the overall deformation showed an "oval shape". Figure 6 shows the radial deformation of the cylinder bore surface on the

ATS and TS. Due to the influence of the bolt preload, the deformation at the top of the cylinder bore is small, and the deformation on the ATS and TS is 49.35 μm and 49.45 μm, respectively. Due to the large thermal load on the top half of the cylinder bore, the middle and upper portion of the cylinder bore show more prominent expansion deformation, and the maximum deformation occurs on the TS, with an amplitude of 122.93 μm; the lower part of the cylinder bore shows a smaller expansion deformation relative to the upper part, and the maximum deformation occurs on the TS, with an amplitude of 108.04 μm.

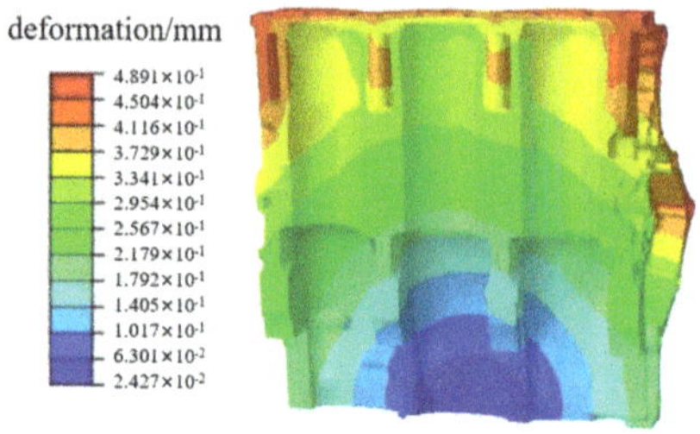

Figure 4. Schematic diagram of cylinder bore deformation.

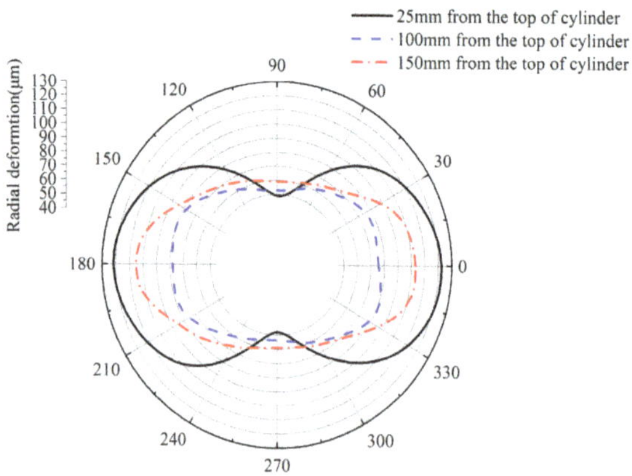

Figure 5. Radial deformation at different heights of the cylinder bore.

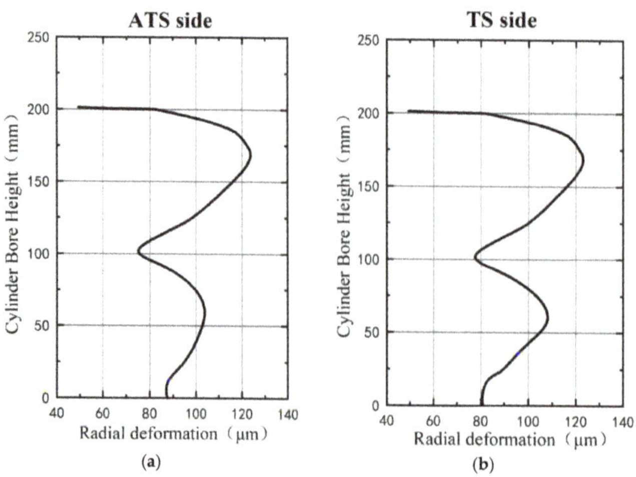

Figure 6. Radial deformation of the ATS and TS: (**a**) ATS radial deformation; (**b**) TS radial deformation.

2.5. The Piston-Ring Cylinder Bore (PRCB) Dynamic Model

The working condition of the engine is complicated; many factors can affect the normal operation of piston friction pairs. The dynamic model of the piston-ring cylinder bore was established by AVL Piston & Rings software R2020.1 [24], which can fully account for the lubrication of the piston group. The modeling process included the following assumptions: (1) Only the motion of the piston in the plane of the TS and ATS was considered. (2) The crankshaft speed was even. (3) The piston was considered to be an elastic body; the cylinder liner, connecting rod, and crankshaft were defined as rigid bodies; and the hinge clearance was zero.

In the modeling process, the cylinder bore deformation profile obtained from the cylinder bore deformation calculation was input as the geometric boundary condition of the calculation.

2.6. Dynamic Model Calibration

The piston leakage is directly affected by the structure of the piston ring group and the deformation of the cylinder bore's inner wall. The accuracy of the PRCB dynamic model established in the previous section was verified by comparing the piston leakage results in the numerical model with that in the experiment. The test process was as follows: block all channels of crankcase interaction with the outside world; keep the crankcase sealing normal (add 0.2 kPa pressure in the crankcase; the amount of air leakage is no more than 5 L/min), after which the throttle is fully open; adjust the engine speed from 1300 r/min to 2100 r/min and record the average leakage of air at each speed point in the process. Figure 7 shows the comparison of the average leakage of a single cylinder in a cycle of the diesel engine (the ratio of total air leakage of the diesel engine to the number of cylinders) in the experiment and the numerical model. It can be seen that the measured and simulated values of piston air leakage under each rotational speed were of the same order of magnitude, and the maximum error was 5.52%, verifying the accuracy of the PRCB dynamic model.

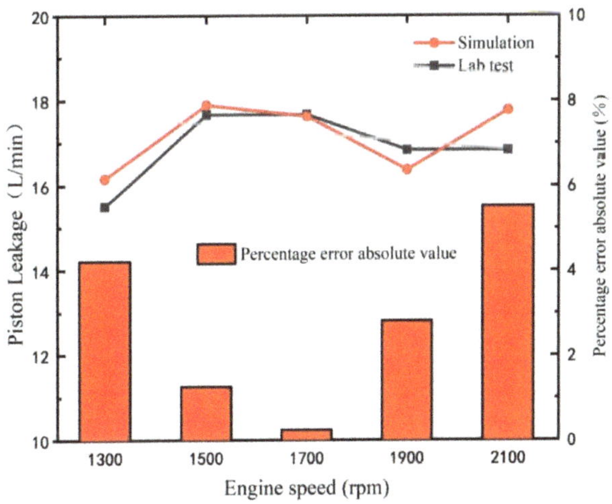

Figure 7. Comparison between measured and simulated average air leakage of a single piston cylinder.

2.7. Design of Cylinder Bore Profiles

A comprehensive experimental method was used to study the influence of the bore pattern parameters on the friction and knocking characteristics of the piston, and the optimal design of the bore pattern was derived to minimize the friction power of the PRCB system and minimize the peak knocking kinetic energy.

A combination of cylindrical taper and elliptical cross-section was adopted for the cylinder bore's cold-state profile, as shown in Figure 8, where A is the taper, and the friction power between the piston and bore was reduced through the optimization of the taper; B is the starting height of the conical profile, and the second-order movement of the piston was improved through the optimization of B; C is the degree of ellipticity, and the ellipse of the difference between the long and short axes of half of the difference between the C is a positive number indicating the different heights that can be attenuated under the cylinder bore's radial deformation to reduce the friction power; and L is the cylinder bore diameter. Subsequent calculations directly superimposed the cold and hot profiles of the cylinder bore as boundary conditions for the dynamic calculations.

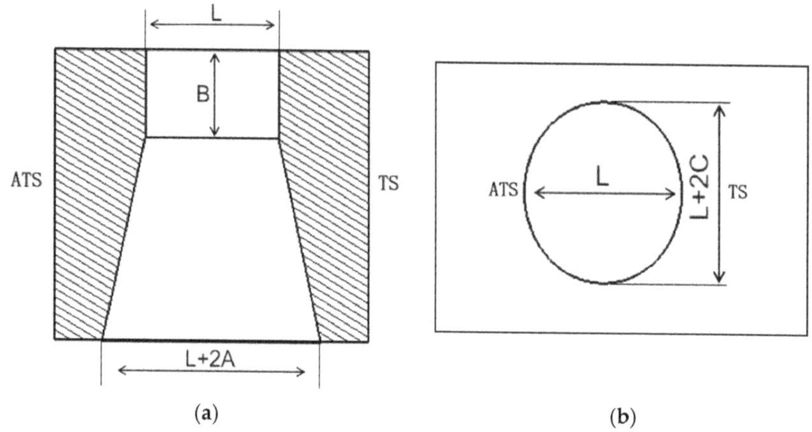

Figure 8. Cold profile of cylinder bore: (**a**) front cutaway view; (**b**) top view.

3. Effect of Cylindrical-Tapered Cylinder Bore Shape Parameters on Friction and Knocking

The control variable method was used to analyze the effects of A, B, and C on the piston friction and knocking characteristics, where the piston friction characteristics were evaluated by the average friction power consumption per cycle, and the knocking characteristics were evaluated by the peak kinetic energy of the piston knocking per cycle.

The effects of the cylindrical-tapered bore profile parameters taper (A), tapered profile start height (B), and ellipticity (C) on the piston skirt friction consumption, piston ring friction consumption, and peak knocking kinetic energy are shown in Figure 9. Overall, the cylindrical-tapered bore has improved friction and knocking characteristics compared to the cylindrical bore.

As shown in Figure 9a, at B = 62 mm, C = 40 μm, and peak kinetic energy was decreased with the increase in taper (A); the skirt friction power was decreased from 232 J to 216 J, the ring group friction power was decreased from 77.87 J to 69.32 J, and the peak kinetic energy was decreased from 256.12 N·mm to 191.97 N·mm. Among them, the rate of decrease in skirt friction power was from fast to slow, while the speed of the decrease in ring group friction power and peak knocking kinetic energy tended to be constant. All three decreased because with the increase in taper (A), the thickness of the oil film between the piston ring's moving surface bore wall and the piston skirt bore wall increases, the area of solid–solid contact between the friction partners decreases, and the dry friction decreases, so the skirt's friction work and the piston ring's friction work are both reduced. At the same time, with the increase in oil film thickness, the viscous shear force of the oil film also increases, and the rotation speed of the piston's knocking moment is reduced, so the peak kinetic energy of the piston knocking is reduced.

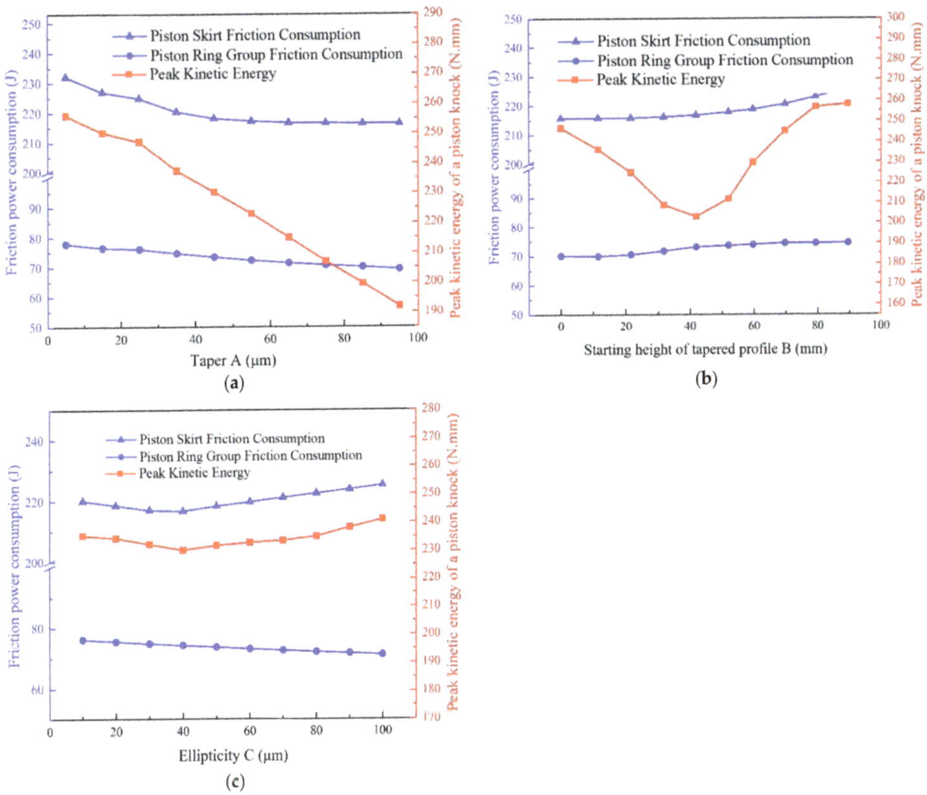

Figure 9. Cylinder bore profile parameters' effects on the piston friction lubrication characteristics and knocking characteristics by (**a**) taper (B = 62 mm, C = 40 μm); (**b**) starting height of tapered profile (A = 45 μm, C = 40 μm); and (**c**) ellipticity (A = 45 μm, B = 62 mm).

As shown in Figure 9b, when B was increased at A = 45 μm and C = 40 μm, the piston skirt friction power and ring set friction power increased from 215 J to 225 J and from 70.15 J to 74.54 J, respectively. The peak piston-knocking kinetic energy decreased and then increased with the increase in B, with a minimum value of 202.79 N at B = 41 mm. The reason for the change in these three parameters is that with the increase in the starting height (B) of the profile, the range of tapered compensation along the height gradually decreases, so the piston skirt and the piston ring moving surface and the cylinder wall in the piston are gradually reduced within a cycle of the average clearance, resulting in a gradual decrease in the average thickness of the lubricant film, and the friction in the solid–solid contact area increases, resulting in an increase in dry friction, so the skirt friction and the friction of the piston ring work increase. The peak knocking kinetic energy is determined by the piston's radial speed and rotation speed. The reduction in the skirt's oil film thickness makes the oil film pressure increase, and the piston's radial speed is weakened, the oil film's viscous shear force decreases, and the piston's angular speed increases, but the piston's radial speed decreases significantly, so the peak knocking kinetic energy as B changes from 0 to 41 mm follows a downward trend, and then as B continues to increase, the angular velocity of the piston's rotation increases more than the radial velocity decreases, so the peak knocking kinetic energy shows an upward trend.

As shown in Figure 9c, at A = 45 μm and B = 62 mm, with the increase in C, both the friction power of the piston skirt and the peak knocking kinetic energy show a decreasing and then increasing trend, achieving minimum values of 216.74 J and 230.27 N·mm at C = 40 μm. The friction power of the piston ring group shows a decreasing trend from

76.25 J to 71.11 J with the increase in C. The reasons for these changes in the three parameters are as follows: with the increase in ellipticity (C), the second-order deformation amplitude of the cylinder bore is gradually weakened, and each cross-section of different heights of the cylinder bore is closer to the "round", while the lubricating oil film around the piston skirt and the piston ring is more uniformly distributed, so the friction power consumption of the piston skirt and the friction power consumption of the piston ring group are gradually reduced. At the same time, because of the increase in ellipticity (C), the gap between the TS and ATS of the piston skirt does not change significantly, so the peak value of knocking kinetic energy does not change significantly.

3.1. Factorial Analysis

The cold-profile parameter factor levels for the cylinder bore under the full-scale test are shown in Table 2. Three parameters related to the design of the cylinder bore's cold profile—namely, taper (A), starting height of tapered profile (B), and ellipticity (C)—were selected as the test factors, and three levels were taken for each factor. The level sizes were selected based on the results of the previous calculations.

Table 2. Factor level values of type line parameters under the full-scale test.

Level	A/μm	B/mm	C/μm
1	35	31	30
2	45	41	40
3	55	51	50

The test scheme and calculation results are shown in Table 3. The effects of the three parameters (A, B, and C) on the piston skirt friction power, the piston ring group friction power, the total friction power (the sum of the ring group friction power and the skirt friction power), and the peak knocking kinetic energy were investigated, where Ai, Bi, and Ci denote the taper, starting height of the tapered profile, and ellipticity at their corresponding i_{th}-level values, respectively.

Table 3. Full-scale test scenarios and calculations.

Programmatic	Skirt Friction Power/J	Ring Group Friction Power/J	Total Friction-Power/J	Peak Knocking Kinetic Energy/(N·mm)
A1B1C1	215.62	71.86	287.48	219
A1B1C2	215.62	71.36	286.98	220
A1B1C3	215.79	70.89	286.68	220
A1B2C1	216.09	72.98	289.04	212
A1B2C2	216.26	72.42	288.69	212
A1B2C3	216.18	71.95	288.13	212
A1B3C1	217.53	73.37	290.90	219
A1B3C2	217.42	72.75	290.19	219
A1B3C3	217.36	72.25	289.60	219
A2B1C1	214.61	69.83	284.41	205
A2B1C2	214.58	69.33	283.91	205
A2B1C3	214.52	68.88	283.41	205
A2B2C1	215.08	71.45	286.53	202
A2B2C2	215.08	70.95	286.03	202
A2B2C3	215.06	70.48	285.56	203
A2B3C1	215.85	72.19	288.04	204
A2B3C2	215.88	71.63	287.48	204
A2B3C3	215.85	71.12	286.98	204
A3B1C1	214.44	68.00	282.43	203
A3B1C2	214.44	67.56	281.99	203
A3B1C3	214.41	67.11	281.55	203
A3B2C1	214.49	70.06	284.56	199
A3B2C2	214.52	69.56	284.11	200
A3B2C3	214.49	69.12	283.61	200
A3B3C1	215.11	71.15	286.30	198
A3B3C2	215.08	70.59	285.71	198
A3B3C3	215.17	70.12	285.29	198

3.2. Analysis of Results

The data obtained from the calculations were analyzed using the range method, and the range results reflected the influence degree of each factor on the calculation results. The factor with the largest range has the highest influence on the results and is the main factor. Range is defined as follows:

$$\text{Range} = \max\left|\overline{K}_i - \overline{K}_j\right| (i \neq j) \tag{13}$$

where $K_{i(j)}(i(j) = 1,2,3)$ is equal to the sum of certain indicators of level i in the sum column of Table 3, $\overline{K}_{i(j)} = K_{i(j)}/s$ is the arithmetic mean of K_i, and s is the number of factor levels.

The influencing factors are taper (A), taper start height (B), ellipticity (C), and the interaction between the three (A × B, A × C, B × C,), and the calculated indices are the skirt friction power, the piston ring group friction power, the total friction power (the sum of the friction power of the skirt and the friction power of the ring group), and the peak of the kinetic energy of the piston knocking. Tables 4 and 5 show the range method tables.

Table 4. Range method of friction characteristics due to a single parameter.

Parameter Factor Level	Skirt Friction Power/J	Range	Piston Ring Group Friction Power/J	Range	Total Friction Power/J	Range	Peak Knocking Kinetic Energy/N·mm	Range
A1	216.41		72.22		288.628		217	
A2	215.17	1.71	70.65	3.30	285.8255	4.69	203	16
A3	214.70		68.91		283.9375		200	
B1	214.88		69.41		284.321		209	
B2	215.26	1.27	71.01	2.27	286.2385	3.51	204	4
B3	216.15		71.69		287.8315		207	
C1	215.41		71.21		286.622		206	
C2	215.44	0.02	70.68	1.00	286.1205	0.97	207	1
C3	215.41		70.21		285.6485		207	

Table 5. Range method of friction characteristics due to interactions between profile parameters.

Norm	Interactions	Range
Skirt friction power	A × B	0.56
	A × C	0.03
	B × C	0.06
Piston ring friction power	A × B	0.86
	A × C	0.06
	B × C	0.09
Total friction power	A × B	0.38
	A × C	0.06
	B × C	0.12
Peak knocking kinetic energy (physics)	A × B	11.80
	A × C	2.95
	B × C	2.95

3.2.1. Effect of Taper (A) with Respect to Different Values of Ellipticity (C)

The influence of taper (A) on the friction and knocking characteristics under different values of ellipticity (C) is shown in Figure 10. The starting height of the conical profile (B) remained unchanged at 41 mm, the taper (A) was varied from 35 μm to 55 μm, and different colors indicate different values of ellipticity (C). The dotted lines in Figure 10 represent the values corresponding to the cylindrical cylinder bores (not to be repeated later). For all of the values of C studied in this work, the friction power of the skirt and the ring group, along with the peak knocking kinetic energy, decreased with the increase in A. The optimal parameter set was obtained as A = 55 μm, C = 50 μm, and B = 41 mm; as shown in Figure 10d, under this parameter set, compared to cylindrical cylinder bores, the friction power of the skirt was reduced by 10.63%, the friction power of the ring group was reduced by 24.38%, and the peak knocking kinetic energy was reduced by 22.64%.

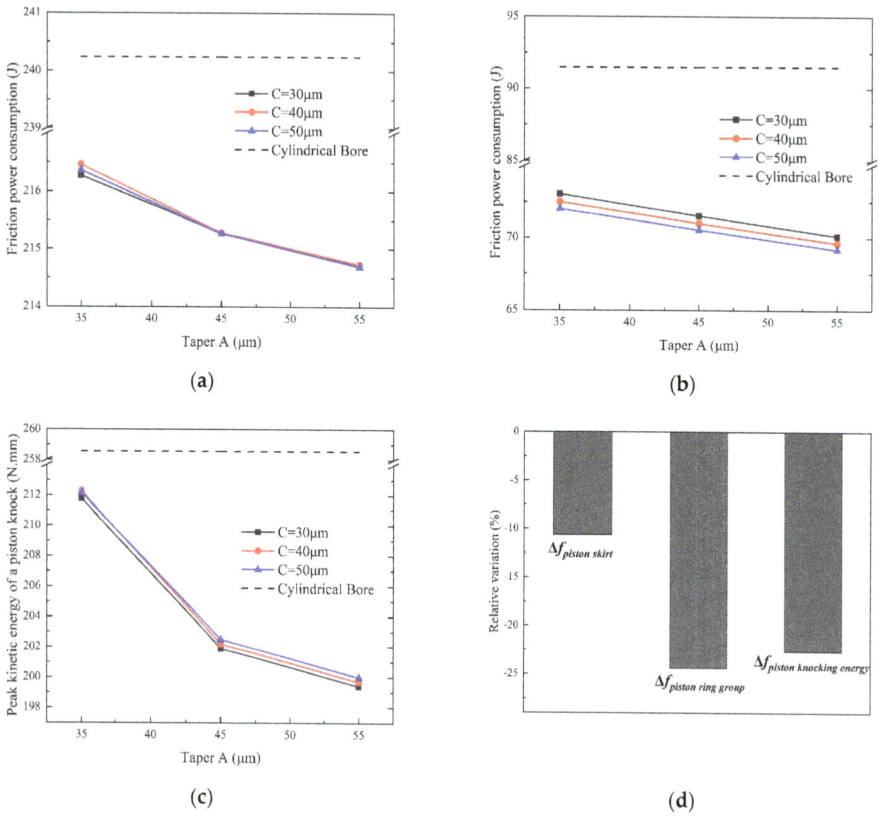

Figure 10. Effect of taper (A) on the piston friction lubrication characteristics and knocking characteristics with different values of ellipticity (C): (**a**) piston skirt friction power; (**b**) piston ring group friction power; (**c**) peak kinetic energy of piston knocking; (**d**) relative changes brought by the optimal parameter group (A = 55 mm, C = 50 μm, B = 41 mm).

3.2.2. Effect of Starting Height of Tapered Profile (B) with Respect to Different Values of Ellipticity (C)

Figure 11 shows the influence of the starting height of the conical profile (B) on the friction and knocking characteristics under different values of ellipticity (C). The taper (A) was kept unchanged at 55 μm, and the starting height of the conical profile (B) was varied from 31 mm to 51 mm. The friction power of the skirt and ring group increased with B regardless of the C value, and the peak knocking kinetic energy decreased with the increase in B. The optimal parameter set was obtained as A = 55 μm and C = 50 μm, and the peak knocking kinetic energy decreased with the increase in B. The optimal parameter set was obtained as A = 55 μm, C = 50 μm, and B = 31 mm; as shown in Figure 11d, under this parameter set, compared to cylindrical cylinder bores, the skirt friction power was reduced by 10.35%, the ring group friction power was reduced by 23.28%, and the peak knocking kinetic energy was reduced by 23.30%.

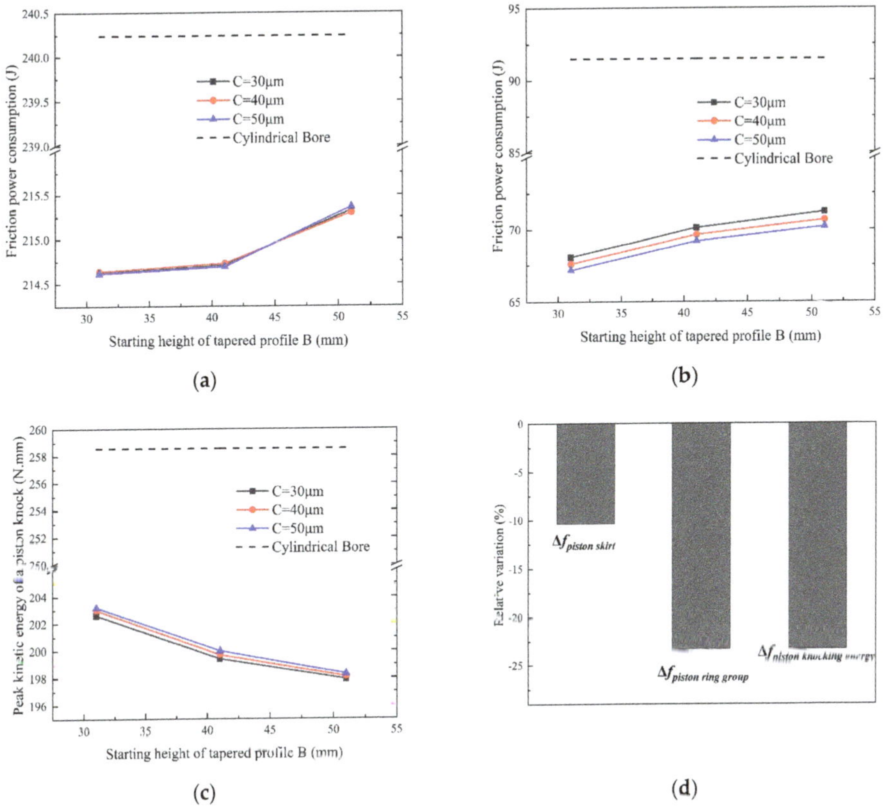

Figure 11. Effect of starting height of tapered profile (B) on the piston friction lubrication characteristics and knocking characteristics with different values of ellipticity (C): (**a**) piston skirt friction power; (**b**) piston ring group friction power; (**c**) peak kinetic energy of piston knocking; (**d**) relative changes brought by the optimal parameter group (B = 31 mm, C = 50 μm, A = 55 μm).

3.2.3. Effect of Starting Height of Tapered Profile (B) with Respect to Different Values of Taper (A)

The influence of the starting height of the conical profile (B) on the friction characteristics and knocking characteristics under different values of taper (A) is shown in Figure 12. The ellipticity (C) remained unchanged at 50 μm, and the starting height of the tapered profile (B) was varied from 31 μm to 51 μm. The friction power of the skirt and the ring group increased with B regardless of the value of A. However, this growth rate was decreased with increasing B. For the peak knocking kinetic energy, there was an interaction between A and B. When A was taken as 35 μm and 45 μm, the peak piston-knocking kinetic energy showed a decreasing and then increasing trend with the increase in B. When the taper (A) was 55 μm, the peak piston-knocking kinetic energy monotonically decreased with the increase in B. The optimal parameter set was taken as A = 55 μm, C = 50 μm, and B = 31 mm. As shown in Figure 12d, under this parameter set, compared to cylindrical cylinder bores, the friction power of the skirt was reduced by 10.66%, the friction power of the ring group was reduced by 26.59%, and the peak knocking kinetic energy was reduced by 21.41%.

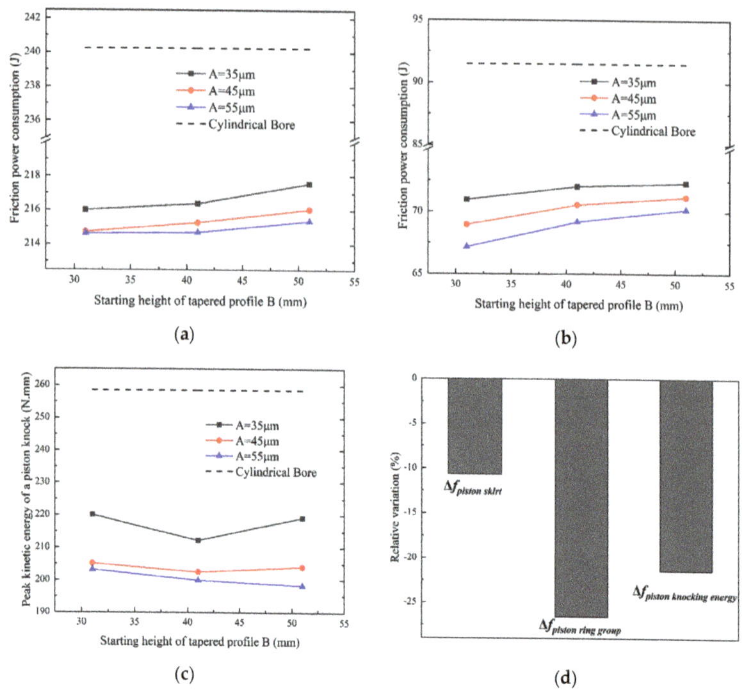

Figure 12. Effect of starting height of tapered profile (B) on the piston friction lubrication characteristics and knocking characteristics with different values of taper (A): (**a**) piston skirt friction power; (**b**) piston ring group friction power; (**c**) peak kinetic energy of piston knocking; (**d**) relative changes brought by the optimal parameter group (A = 55 μm, B = 31 mm, C = 50 μm).

3.3. Comparison between Optimal and Baseline Design

Compared with the peak kinetic energy of the piston knocking, taking the friction consumption of the piston system as the main consideration, A = 55 μm, B = 31 mm, and C = 50 μm were selected as the optimized cylinder bore profile parameter combination; the variation in friction characteristics and knocking characteristics with the crankshaft rotation angle for the cylindrical bore and optimized cylindrical-tapered bore is shown in Figure 13.

After profile parameter optimization, the total friction power of the piston group was reduced by 49.95 J, which mean a decrement of 15.05%. The decrement of friction power was beneficial for reduced fuel consumption. The peak piston knock kinetic energy was reduced by 55.35 N·mm, representing a decrement of 21.41%. The reduction in knocking kinetic energy can reduce the risk of engine cavitation and improve engine NVH performance.

Figure 14 shows the comparison of the minimum film thickness ratio and film pressure on the friction pair. Only the minimum film thickness ratio of the first is shown in Figure 14b, since the first compressor ring accounts for the majority of the friction power in the ring group. Compared with that in the cylindrical bore, the minimum film thickness ratio between the piston skirt and the piston ring moving surface of the cylindrical-tapered bore was improved, as shown in Figure 14a, so the latter resulted in a smaller friction power. The distribution of the oil film pressure on the skirt at the moment of piston knocking is shown in Figure 14c,d. The peak oil film pressure was smaller in the cylindrical-tapered bore and the area of the oil-film-bearing area was larger, which effectively absorbed the kinetic energy of the piston's second-order motion, so the peak knocking kinetic energy was reduced in the latter case.

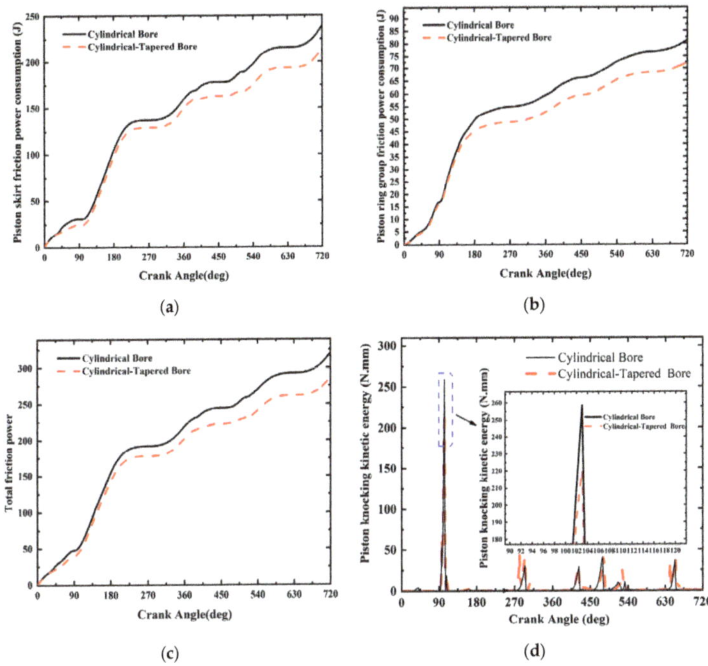

Figure 13. Comparison of piston friction characteristics and knocking characteristics before and after optimization. (**a**) skirt friction power change; (**b**) ring group friction power change; (**c**) total friction power change; (**d**) knocking kinetic energy change.

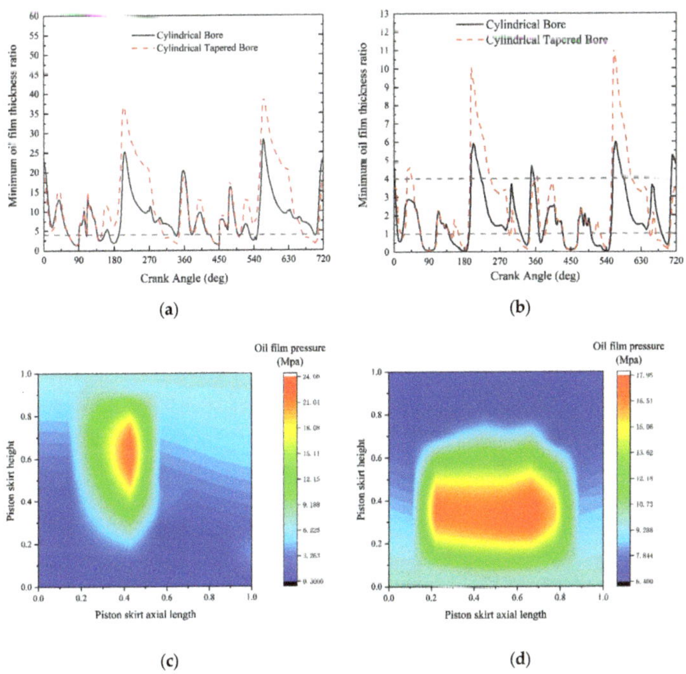

Figure 14. Comparison of oil film thickness and oil film pressure distribution in the cylinder bore before and after optimization: (**a**) minimum oil film thickness ratio change in the skirt; (**b**) minimum

oil film thickness ratio change for the 1st ring; (**c**) oil film pressure distribution in the skirt at the moment of piston knocking (pre-optimization); (**d**) oil film pressure distribution in the skirt at the moment of piston knocking (post-optimization).

4. Conclusions

(1) The taper has the greatest effect on the total friction power and the peak piston-knocking kinetic energy, followed by the starting height of the conical profile, while the ellipticity has the smallest effect; for the peak piston-knocking kinetic energy, there was a significant interaction between the taper and the starting height of the conical profile.

(2) Compared with the cylindrical bore, the optimized skirt friction work of the cylindrical-tapered bore decreased from 240.24 J to 214.62 J, with a decrease of 10.66%; the piston ring friction work decreased from 91.5 J to 67.17 J, with a decrease of 26.59%; and the peak knocking kinetic energy decreased from 258.55 N·mm to 203.2 N·mm, with a decrease of 21.41%.

(3) Compared with the cylindrical bore, the minimum oil film thickness on the moving surface of the piston skirt and the piston ring was increased in the cylindrical-tapered bore, which reduced the contact time of dry friction and increased the contact time of hydrodynamic friction at the piston-knocking moment. The cylindrical-tapered bore provided a larger contact area and oil film thickness, which absorbed some of the energy of the second-order motion of the piston and attenuated the impact.

Author Contributions: Conceptualization, H.D. and N.W.; methodology, J.L.; software, N.W.; validation, N.W., J.W. and X.Z.; formal analysis, N.W.; investigation, H.W.; resources, J.W.; data curation, H.W.; writing—original draft preparation, N.W.; writing—review and editing, H.D.; visualization, X.Z.; supervision, J.L.; project administration, J.Z.; funding acquisition, J.Z. All authors have read and agreed to the published version of the manuscript.

Funding: This research was funded by the National Natural Science Foundation of China (Grant No. 52205166).

Data Availability Statement: The original contributions presented in the study are included in the article, further inquiries can be directed to the corresponding authors.

Conflicts of Interest: Authors Jian Wang and Hui Wang were employed by the Weichai Power Co., Ltd. The remaining authors declare that the research was conducted in the absence of any commercial or financial relationships that could be construed as a potential conflict of interest.

References

1. Bi, Y.; Wang, P.; Luo, L.; Wang, H.; Xin, Q.; Lei, J.; Shen, L. Analysis of out-of-round deformation of a dry cylinder liner of a non-road high-pressure common-rail diesel engine based on multi-field coupling. *J. Braz. Soc. Mech. Sci.* **2021**, *43*, 50. [CrossRef]
2. Bi, Y.; Wang, P.; Xiang, R.; Wen, J.; Lei, J.; Shen, L.; Xin, Q. Numerical investigation on the operating characteristics of the cylinder liners of a turbocharged diesel engine. *Sādhanā* **2021**, *46*, 150. [CrossRef]
3. Li, G.; Gu, F.; Wang, T.; Lu, X.; Zhang, R.; Ball, A. A dynamic deformation based lubrication model between the piston rings and cylinder liner. In Proceedings of the 2017 23rd International Conference on Automation and Computing (ICAC), Huddersfield, UK, 7–8 September 2017; pp. 1–6. [CrossRef]
4. Delprete, C.; Razavykia, A. Piston dynamics, lubrication and tribological performance evaluation: A review. *Int. J. Engine Res.* **2020**, *21*, 725–741. [CrossRef]
5. Sato, K.; Fujii, K.; Ito, M.; Koda, S. *Application to Engine Development of Friction Analysis by Piston Secondary Motion Simulation in Consideration of Cylinder Block Bore Distortion*; 0148-7191; Technical Paper; SAE: Washington, DC, USA, 2006. [CrossRef]
6. Styles, G.; Rahmani, R.; Rahnejat, H.; Fitzsimons, B. In-cycle and life-time friction transience in piston ring–liner conjunction under mixed regime of lubrication. *Int. J. Engine Res.* **2014**, *15*, 862–876. [CrossRef]
7. Wang, Y.; Ma, X.; Li, T.; Lu, X.; Li, W. Influence of thermal effect in piston skirt lubrication considering thermal deformation of piston and cylinder liner. *Int. J. Engine Res.* **2023**, *24*, 14680874231155571. [CrossRef]
8. Li, G.; Gu, F.; Wang, T.; Lu, X.; Zhang, L.; Zhang, C.; Ball, A. An improved lubrication model between piston rings and cylinder liners with consideration of liner dynamic deformations. *Energies* **2017**, *10*, 2122. [CrossRef]

9. Lu, Y.; Li, S.; Wang, P.; Liu, C.; Zhang, Y.; Müller, N. The analysis of secondary motion and lubrication performance of piston considering the piston skirt profile. *Shock. Vib.* **2018**, *2018*, 3240469. [CrossRef]
10. Zhang, J.; Piao, Z.; Liu, S. Influence of skirt profile structure of gasoline engine piston on the friction and wear characteristics under standard conditions. *J. Tribol.* **2018**, *140*, 021703. [CrossRef]
11. Totaro, P.P.; Westerfield, Z.; Tian, T. *Introducing a New Piston Skirt Profile to Reduce Engine Friction*; 0148-7191; Technical Paper; SAE: Washington, DC, USA, 2016. [CrossRef]
12. Pawlus, P.; Reizer, R. Functional importance of honed cylinder liner surface texture: A review. *Tribol. Int.* **2022**, *167*, 107409. [CrossRef]
13. Guo, Z.; Yuan, C.; Liu, P.; Peng, Z.; Yan, X. Study on influence of cylinder liner surface texture on lubrication performance for cylinder liner–piston ring components. *Tribol. Lett.* **2013**, *51*, 9–23. [CrossRef]
14. Grabon, W.; Pawlus, P.; Wos, S.; Koszela, W.; Wieczorowski, M. Evolutions of cylinder liner surface texture and tribological performance of piston ring-liner assembly. *Tribol. Int.* **2018**, *127*, 545–556. [CrossRef]
15. Mezghani, S.; Demirci, I.; Yousfi, M.; El Mansori, M. Mutual influence of crosshatch angle and superficial roughness of honed surfaces on friction in ring-pack tribo-system. *Tribol. Int.* **2013**, *66*, 54–59. [CrossRef]
16. Hu, Y.; Meng, X.; Xie, Y.; Fan, J. Mutual influence of plateau roughness and groove texture of honed surface on frictional performance of piston ring–liner system. *Proc. Inst. Mech. Eng. Part J. J. Eng. Tribol.* **2017**, *231*, 838–859. [CrossRef]
17. Edtmayer, J.; Lösch, S.; Hick, H.; Walch, S. Comparative study on the friction behaviour of piston/bore interface technologies. *Automot. Engine Technol.* **2019**, *4*, 101–109. [CrossRef]
18. Alshwawra, A.; Pohlmann-Tasche, F.; Stelljes, F.; Dinkelacker, F. Enhancing the Geometrical Performance Using Initially Conical Cylinder Liner in Internal Combustion Engines—A Numerical Study. *Appl. Sci.* **2020**, *10*, 3705. [CrossRef]
19. Alshwawra, A.; Pasligh, H.; Hansen, H.; Dinkelacker, F. Increasing the roundness of deformed cylinder liner in internal combustion engines by using a non-circular liner profile. *Int. J. Engine Res.* **2021**, *22*, 1214–1221. [CrossRef]
20. Alshwawra, A.; Pohlmann-Tasche, F.; Stelljes, F.; Dinkelacker, F. Effect of freeform honing on the geometrical performance of the cylinder liner—Numerical study. *SAE Int. J. Engines* **2022**, *16*, 463–486. [CrossRef]
21. Halbhuber, J.; Wachtmeister, G. *Effect of Form Honing on Piston Assembly Friction*; 0148-7191; Technical Paper; SAE: Washington, DC, USA, 2020. [CrossRef]
22. AVL LIST GmbH. *BOOST Rev 2017.1 Users Guide*; AVL LIST GmbH: Steiermark, Austria, 2017.
23. Liang, X.; Wang, Y.; Huang, S.; Yang, G.; Tang, L.; Cui, G. *Investigation on Cylinder Bore Deformation under Static Condition Based on Fourier Decomposition*; 0148-7191; Technical Paper; SAE: Washington, DC, USA, 2017. [CrossRef]
24. AVL LIST GmbH. *Excite Piston & Rings Rev 2017.1 Users Guide*; LIST GmbH: Steiermark, Austria, 2017.

designs

MDPI

Article

An Innovative Mechanical Approach to Mitigating Torque Fluctuations in IC Engines during Idle Operation

Daniel Silva Cardoso [1], Paulo Oliveira Fael [1], Pedro Dinis Gaspar [1,*] and António Espírito-Santo [2]

[1] C-MAST—Center for Mechanical and Aerospace Science and Technologies, Calçada Fonte do Lameiro, 6201-001 Covilhã, Portugal; silva.cardoso@ubi.pt (D.S.C.); pfael@ubi.pt (P.O.F.)
[2] IT—Institute of Telecommunications, Calçada Fonte do Lameiro, 6201-001 Covilhã, Portugal; aes@ubi.pt
* Correspondence: dinis@ubi.pt

Abstract: Internal combustion engines have been a major contributor to air pollution. Replacing these engines with electric propulsion systems presents significant challenges due to different countries' needs and limitations. An active, purely mechanical solution to the problem of irregular torque production in an alternative internal combustion engine is proposed. This solution uses an actuator built on a camshaft and a spring, which stores and returns energy during the engine operating cycle, allowing torque production to be normalized, avoiding heavy flywheels. Designed for control throughout the engine's duty cycle, this system incorporates a cam profile and a spring mechanism. The spring captures energy during the expansion stroke, which is then released to the engine during the intake and compression strokes. Simple, lightweight, and efficient, this system ensures smoother and more consistent engine operations. It presents a viable alternative to the heavy and problematic dual-mass flywheels that were introduced in the 1980s and are still in use. This innovative approach could significantly enhance the performance and reliability of alternative internal combustion engines without notable energy losses.

Keywords: internal combustion engine; driving torque; resistive torque; mass inertia torque; torque fluctuations; energy storage mechanism; camshaft; spring

Citation: Cardoso, D.S.; Fael, P.O.; Gaspar, P.D.; Espírito-Santo, A. An Innovative Mechanical Approach to Mitigating Torque Fluctuations in IC Engines during Idle Operation. *Designs* **2024**, *8*, 47. https://doi.org/10.3390/designs8030047

Academic Editors: Wenbin Yu, Guang Zeng and Junnian Wang

Received: 4 March 2024
Revised: 8 May 2024
Accepted: 15 May 2024
Published: 17 May 2024

1. Introduction

1.1. Background

The internal combustion engine (ICE) exhibits inherent characteristics of torque fluctuations throughout its operational cycle. These fluctuations are a consequence of the engine's mechanism, where the energy from fuel is converted into mechanical energy by discrete combustion events. Each combustion stroke generates a distinct torque impulse, contributing to an uneven torque output over the engine cycle. This phenomenon impacts the smoothness of engine operation and has broader implications for engine efficiency and longevity [1–5].

Historically, the engine industry has continually evolved to address these torque inconsistencies, driven by the pursuit of enhanced engine performance, smoothness, and efficiency. The traditional approach to mitigating torque variations has largely centered on passive mechanical solutions, such as the utilization of conventional flywheels and dual mass flywheels (DMF) [6]. However, while effective in smoothing out rotational speeds, the flywheel adds to the engine's mass and is incapable of adjusting itself to the engine operation variations. On the other hand, the mechanism proposed in this study seeks to offer a relatively lightweight solution that harmonizes the produced torque without affecting the engine's capability for acceleration or deceleration. This initiative is part of a broader research endeavor aimed at studying and developing a mechanism capable of functioning across all engine speed ranges and under various loads. Currently, the focus is on enhancing performance during idle and low-speed operations, where the demand for such systems is most critical [2,7].

Recent advancements employing mechanical, electromechanical, and control strategies have significantly improved torque harmonization and reduced engine vibrations and rotational speed variations. This serves as a clear indication of both the opportunity and the need for further advancements in this domain, emphasizing the shift away from traditional, bulkier, and less efficient solutions towards more accurate and responsive technologies. Moreover, this transition facilitates the integration of engines into hybrid vehicles, range extenders, and microcogeneration systems [8–13].

This study builds upon approaches introduced by Lin et al. [14] and Arakelian et al. [15]. Although the first study focuses on reducing torque variations in the camshaft and the second emphasizes inertia effects in slider-crank mechanisms, both studies commonly utilize a cam to control the mitigating mechanism and a spring to act as energy storage. Another approach, conducted by Cardoso and Fael [16], has demonstrated the feasibility of using switched reluctance machines to replace traditional flywheels, offering a method to correct engine output torque and speed variations, albeit at the cost of a complex system. Despite these advancements, a gap remains in seamlessly integrating these technologies into existing engine designs without complex control systems or significant modifications to the engine architecture. This research aims to fill this gap by developing a mechanically focused solution that maintains engine performance, particularly at low rotational speeds.

1.2. Problem Statement

A series of multifaceted challenges underscore the quest to enhance torque management in internal combustion engines. At the core of these challenges lies the inherent nature of alternative ICE operations, where the intermittent combustion process leads to nonuniform torque output. This irregularity manifests as torque ripple or fluctuations, which not only compromises the smoothness and responsiveness of the engine but also adversely affects overall efficiency and increases vibrations and noise [17,18].

One of the primary issues in contemporary torque management is the limitation of traditional mechanical solutions. While these methods, such as the utilization of flywheels, have been fundamental in maintaining engine operation at low speeds, they fall short of addressing the demands of modern engines. The added mass and inertia of traditional solutions often impede the engine's dynamic response, failing to align with the growing emphasis on lightweight and compact engine designs. Moreover, these approaches do not actively adapt to varying operational conditions, thus limiting their effectiveness in dynamic and diverse operation scenarios [19–21].

Another critical aspect is the evolution of engine technology itself, with advancements in engine design, including the trend towards more compact, lightweight engine designs, different cylinder arrangements, unusual cylinder number engines (this can lead to extreme torque production irregularities), cylinder deactivation technologies, and the advent of hybrid and electric vehicles, highlighting the need for systems that offer greater precision and adaptability [22–27].

1.3. Objectives

Study of Torque Decomposition in a Slider Crank Engine: The objective pretends to study the decomposition of output torque in an alternative engine design. This involves analyzing how different torque types (such as driving torque, resistive torque, and mass inertia torque) contribute to the overall output torque in this kind of engine.

Development of a Camshaft: The main objective of this research is to design and develop a camshaft, which is the main component of the actuator implemented to smooth the torque output. The key challenge is in creating a cam profile that can precisely mimic the torque output profile and can act on the spring, compressing it to store mechanical energy during the expansion stroke and subsequently receiving energy from the spring to return it to the engine during strokes when the engine is not producing positive work.

Development of an Integrated Camshaft-Actuator System: With this objective, an integrated system controlled by the developed cam is intended to be designed and developed. In this process, some considerations must be taken, with some of the most important parts to study and develop being the spring stiffness and preload. The spring needs to be studied to store as much energy as possible without interfering with the engine's natural movement and performance.

Evaluation and Demonstration of System Adaptability and Performance: Lastly, the evaluation of the performance of the developed system during idle operation is aimed at.

1.4. Study Contributions

The major contributions of this work are divided between the development of a mechanical mechanism and advancements in the field of internal combustion engines. The specific contributions of this study are outlined as follows:

Advanced Torque Decomposition Analysis: A comprehensive analysis of torque decomposition in alternative engines is provided. This approach enhances understanding of how different types of torque—driving, resistive, and mass inertia—interact and contribute to overall engine performance. This insight is useful for future work, for example, focusing on studying engine torque output smoothness and examining how geometric and construction parameters influence torque production.

Camshaft Development Methodology: This work includes the design and development of a camshaft that accurately replicates the engine's torque output profile. The design process encounters challenges due to the irregularities of the engine's torque output and the numerous parameters influencing the cam's profile. This paper presents a methodology for cam development.

Development of a Torque Mitigating Mechanism: This study introduces the development of a fully mechanical torque mitigating mechanism. The mechanism is designed to absorb excess energy from the engine system during periods of high output and strategically release it during phases where additional energy is required. This cyclic absorption and redistribution are tailored to operate seamlessly across different periods of time and varying operational conditions.

Integration of a Camshaft-Actuator with a single-cylinder engine: Beyond individual components, this study focuses on integrating the developed mechanism, which includes a camshaft for control and a spring for energy storage, with a single-cylinder engine. This integration aims to enhance torque smoothness and ensure that speed variations during the engine cycle remain closely aligned with the engine's average rotational speed.

2. Engine Torque Analysis

In considering the output torque at the crankshaft of ICEs, as illustrated in Figure 1, torque fluctuation arises mainly from three sources. Firstly, the driving torque generated mostly during the expansion stroke in ICEs leads to irregular driving actuation. This fluctuation is consistent across different crankshaft speeds, depending solely on crankshaft angle (φ) and pressure force ($F_{pressure}$). Secondly, fluctuation occurs due to the inertia force from the acceleration (α) of the imbalanced mass. This type of fluctuation is more pronounced at higher speeds as it is proportional to the square of the engine's angular speed (ω). Lastly, fluctuating resistive loads, such as those in engine valve systems, impart irregular torque to the camshaft, independent of engine angular speed and related to crankshaft angle [14], [28,29].

To mitigate these fluctuations, passive and active techniques have been employed. Flywheels, as referred to previously, for example, can store and release kinetic energy to smooth out speed fluctuations but are less effective at low speeds due to their large inertia. This affects the engine performance and its capacity to change the rotational speed when solicited and the use of this represents an expressive mass increment. Torsional vibration dampers offer a solution across various speeds, but this involves dissipation of energy. Active control techniques, in contrast, counteract torque fluctuations by providing an opposing force, but at the cost of complex control techniques or input of external energy [30,31].

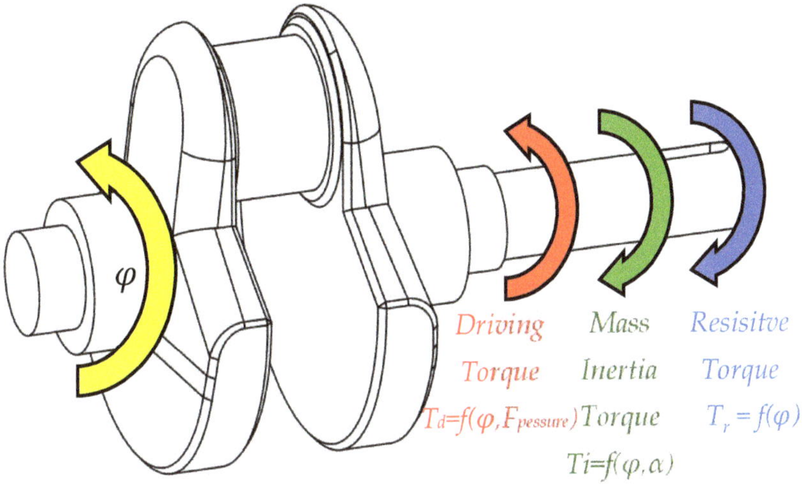

Figure 1. Representation of how the three types of torque mentioned above act on the crankshaft.

For a clearer understanding of the torque production cycle in internal combustion engines, the analysis is often conducted using a single-cylinder engine model. This approach effectively isolates three types of torque fluctuations, which are more difficult to discern in multicylinder engines due to the overlapping of multiple-stroke events.

These events significantly contribute to torque fluctuations and can be categorized into two groups: those that generate positive torque and those that produce negative torque. Only the expansion that represents a quarter (on Otto) of the strokes and is part of the first group is directly useful for the engine's primary purpose. However, the magnitude of this useful stroke at engine full load is about fifteen times greater than the average of the other three strokes. However, despite their negative torque, strokes such as intake, compression, and exhaust are essential for the engine's cycle.

The torque generated by an engine is a composite of the three types of torque previously described, as outlined in Equation (1). The capability of alternative engines to generate torque stems from the slider–crank mechanism, depicted in Figure 2. Through dynamic analysis of this mechanism, both the inertia torque and driving torque of the engine can be accurately determined.

$$T_t = T_d + T_i - T_r \tag{1}$$

T_t—*total engine torque* [Nm]
T_d—*driving torque* [Nm]
T_i—*mass inertia torque* [Nm]
T_r—*resistive torque* [Nm]

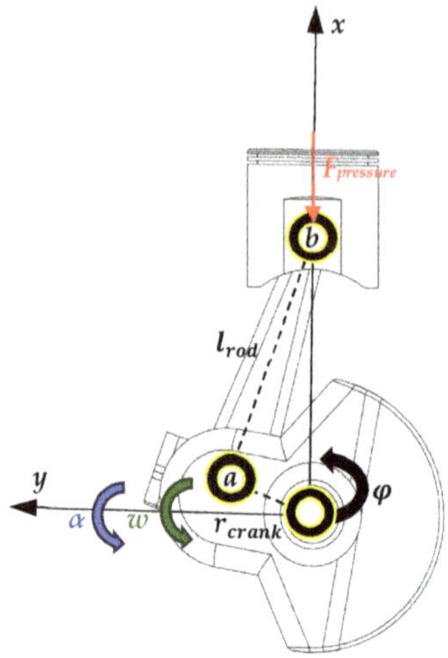

Figure 2. Slider−crank mechanism from a single-cylinder engine.

2.1. Driving Torque

ICEs are volumetric machines designed to convert the chemical energy in fuel into mechanical energy, which is then transferred to the crankshaft. They utilize a slider−crank mechanism to convert pressure variations from fuel combustion in the combustion chamber into mechanical movement. This is due to the design of the alternative engine that only allows the gases in the cylinder to expand in the piston direction.

The force resulting from these pressure variations can be expressed as the product of the cylinder's internal pressure by the piston area, as shown in Equation (2). This equation clearly illustrates the generation of driving force in ICEs due to pressure variation, calculated from Equation (3), determined using the first law of thermodynamics.

For a comparative analysis, Figure 3 demonstrates how the values of pressure exerted in the piston can act in the piston when the engine is running at idle, i.e., when the engine only produces the torque needed to keep it running and when the presence of the flywheel is most necessary to store energy from the expansion stroke for the rest of the strokes. In contrast, Figure 4, derived from an analysis based on Figure 2, culminating in Equation (4), depicts the generation of driving torque throughout a 720° crankshaft rotation in an Otto engine under identical operational conditions to those shown in Figure 3 [32].

$$F_{pressure} = (p_{cyl} - p_{atm})\frac{\pi B^2}{2} \tag{2}$$

$$p_{cyl} = \frac{2\left(\delta Q_f - \delta Q - \delta Q_{vap}\right) + p_{cylprev}\left(\frac{\gamma+1}{\gamma-1}V_{cylprev} - V_{cyl}\right)}{\frac{\gamma+1}{\gamma-1}V_{cyl} - V_{cylprev}} \tag{3}$$

$F_{pressure}$—*pressure force acting on the piston* [N]
B—*bore* [m]
p_{cyl}—*cylinder pressure* [Pa]
p_{atm}—*atmosferic pressure* [Pa]
δQ_f—*energy released by the fuel* [J]
δQ—*nenergy transferred in the form of heat* [J]
δQ_{vap}—*aenergy released by the vaporization of the fuel inside the cylinder* [J]
$p_{cylprev}$—*pressure in the cylinder at the previous iteration* [Pa]
γ—*expansion coefficient*

$V_{cylprev}$—*volume in the cylinder at the previous iteration* $\left[m^3\right]$
V_{cyl}—*volume in the cylinder at the current iteration* $\left[m^3\right]$

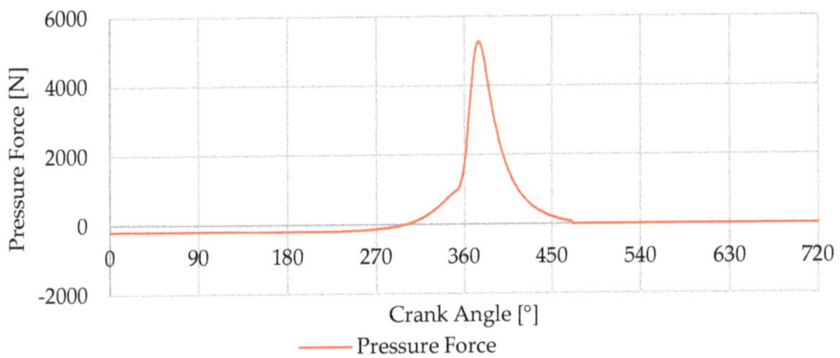

Figure 3. $F_{pressure}$ in the piston from Otto strokes.

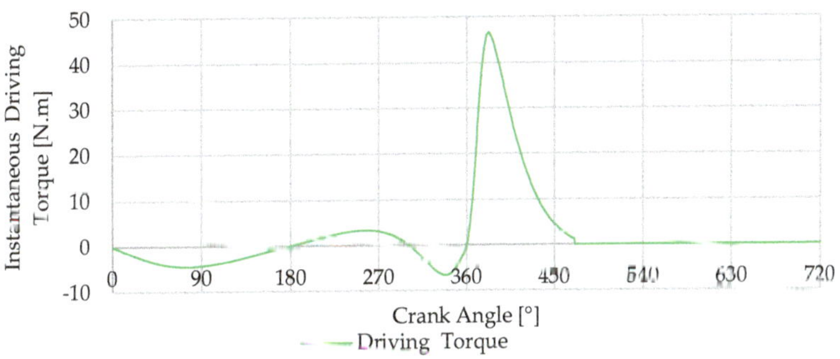

Figure 4. Driving torque from the $F_{pressure}$ on the slider–crank mechanism.

$$T_d - F_{pressure} sin\varphi(1 + \frac{r_{crank}}{l_{rod}} cos\,\varphi) - T_r + T_i \tag{4}$$

T_d—*driving engine torque* [Nm]
r_{crank}—*crank radius* [m]
l_{rod}—*connecting rod length* [m]
φ—*crankshaft rotation angle* [rad]

2.2. Mass Inertia Torque

Mass inertia torque, hereinafter referred to as inertia torque, arises from the moving parts, particularly during changes in their speed. In the context of an ICE, this involves components like the crankshaft, connecting rods, pistons, and other moving parts. The inertia of these components plays a significant role in the engine's responsiveness, vibrations, and overall performance. Inertia torque is influenced by several factors, including the mass and distribution of the engine's moving parts and the speed. The faster the components rotate, the more significant the inertia torque becomes, as it takes more force to change the speed of these rapidly moving parts. This phenomenon is especially noticeable during rapid acceleration or deceleration, where the inertia torque can have a substantial impact on engine performance. The inertia forces can be broken down into components along the piston's movement line (x-axis), which induce forces on the crankshaft and thereby contribute to its torque, presented in Equation (5), and forces acting perpendicularly to the former (y-axis), which do not directly affect the torque generated by the engine, presented in Equation (6). The mass responsible for inducing inertia forces includes the mass of the piston group and the connecting rod. For simplification purposes, this mass is divided between two distinct points: point a, located on the crankshaft pin (as shown in Figure 2), where, according to literature, we can consider as comprising 2/3 of the connecting rod mass, and point b, located on the piston pin, where the piston group mass and 1/3 of the connecting rod mass are assumed, as outlined in Equations (7) and (8) [33].

$$F^x_{inertia} = (m_a + m_b) r_{crank} \omega^2 \cos\varphi + \left(m_b \frac{r_{crank}}{l_{rod}}\right) r_{crank} \omega^2 \cos 2\varphi \tag{5}$$

$$F^y_{inertia} = m_a r_{crank} \omega^2 \sin\varphi \tag{6}$$

$$m_a = \frac{2}{3} m_{cr} \tag{7}$$

$$m_b = \frac{1}{3} m_{cr} + m_{pg} \tag{8}$$

$F^x_{inertia}$—$inertia\ forces\ on\ x - axis$ [N]
m_a—$mass\ considered\ at\ point\ a,\ crankshaft\ pin$ [kg]
m_b—$mass\ considered\ at\ point\ b,\ piston\ pin$ [kg]
ω—$angular\ speed$ [rad/s]
$F^y_{inertia}$—$inertia\ forces\ on\ y - axis$ [N]
m_{pg}—$mass\ of\ piston\ group$ [kg]
m_{cr}—$mass\ of\ connecting\ rod$ [kg]

To calculate the torque induced by inertia forces on the crankshaft, as detailed in Equation (9), consideration is given only to the masses previously divided and located at point b. This approach is based on the premise that the crankshaft is properly balanced, factoring in the masses on the crankshaft pin and the equivalent connecting rod mass at the point a, ensuring that the rotational balance point of this mass aligns with the crankshaft's center of rotation.

$$T_i = \frac{m_b}{2} {r_{crank}}^2 \omega^2 \left(\frac{r_{crank}}{2 l_{rod}} \sin\varphi - \sin 2\varphi - \frac{3 r_{crank}}{2 l_{rod}} \sin 3\varphi\right) \tag{9}$$

T_i—$mass\ inertia\ torque$ [Nm]

Figure 5 represents the forces that lead to the torque profile obtained in Figure 6. For clarity, green vertical lines indicate the top dead center and blue vertical lines denote the bottom dead center. Figure 6 illustrates the magnitude of the inertia torque produced by the movement of the internal components in a single-cylinder slider crank engine model at 800 RPM without charge, representing idling operation.

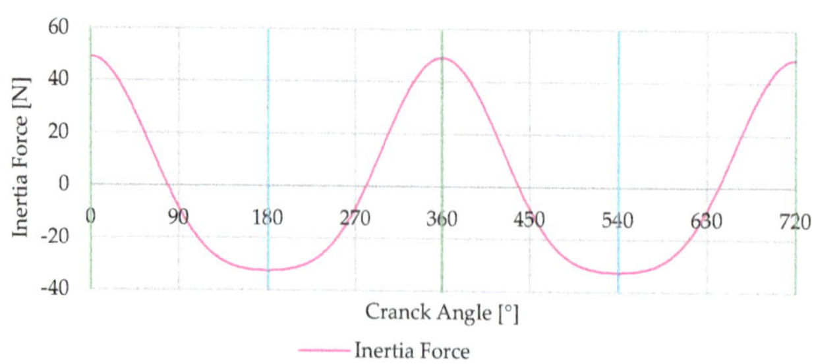

Figure 5. Inertia forces on the crankshaft at 800 RPM from the piston group and connecting rod.

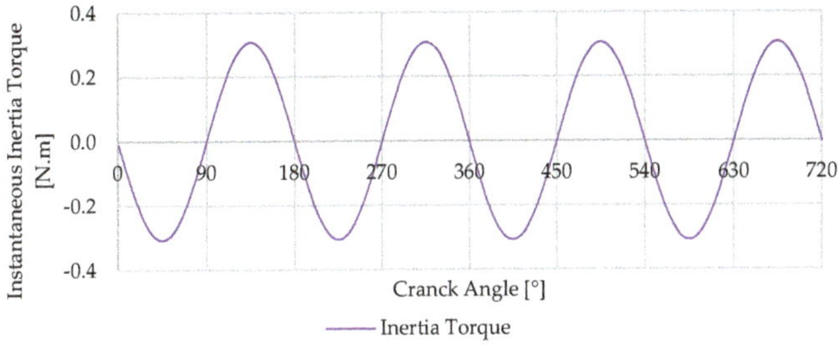

Figure 6. Instantaneous inertia torque on the crankshaft from internal components in a single-cylinder slider–crank engine.

2.3. Resistive Torque

The resistive torque comes essentially from camshafts, oil pumps, or other engine auxiliary elements. This torque type is primarily due to resistive forces that the engine components must overcome during operation. For the purposes of this work, only the resistive torque from camshafts will be considered. In the context of camshafts, this type of torque fluctuates based on the engine cycle, particularly in the valve operation phase, where the opening and closing of valves require varying levels of force. In the camshaft context, the resistive torque is a consequence of the need to compress (negative torque) and then release (resulting in positive torque) the valve springs. The magnitude of this torque fluctuation can vary significantly based on the springs' stiffness and the valve mechanism's design. At lower to medium speeds, the resistive torque fluctuation tends to be more substantial than the inertia torque fluctuation. To quantitatively analyze resistive torque in ICEs, one must consider the spring stiffness, the preload force, the rocker mechanism, and the displacement of valves (see Figures 7 and 8). The spring force from the valve mechanism can be expressed by Equation (10) [34].

$$F = k\delta + P \tag{10}$$

k—$spring\ stiffness$ [N/m]
δ—$valve\ displacement$ [m]
P—$preload\ of\ the\ spring$ [N]

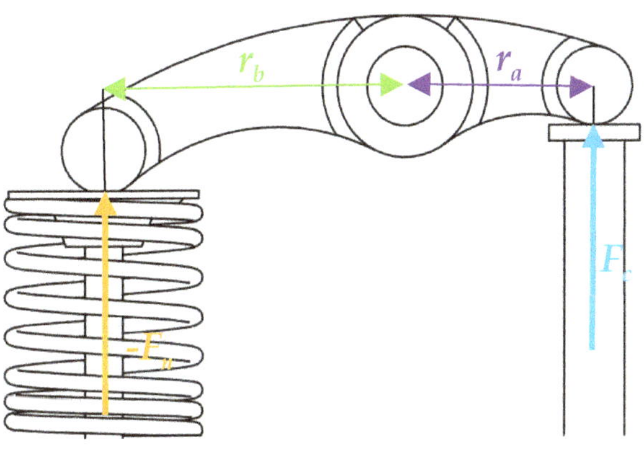

Figure 7. Rocker arm diagram from the valve train.

$$r_a F_c + r_b(-F_n) = 0 \tag{11}$$

r_a—*distance from the push rod to the rocker centre* [m]
r_b—*distance from the valve center to the rocker centre* [m]
F_n—*force from the rocker on the valve* [N]
F_c—*force from the cam on the rocker* [N]

$$\vec{T}_c + \vec{r}_c \times \left(-\vec{F}_c\right) = 0 \quad (12)$$

$\vec{T}_c$—*torque on the camshaft* [Nm]
$\vec{r}_c$—*vector from the center of the cam to the contact point with the follower* [m]

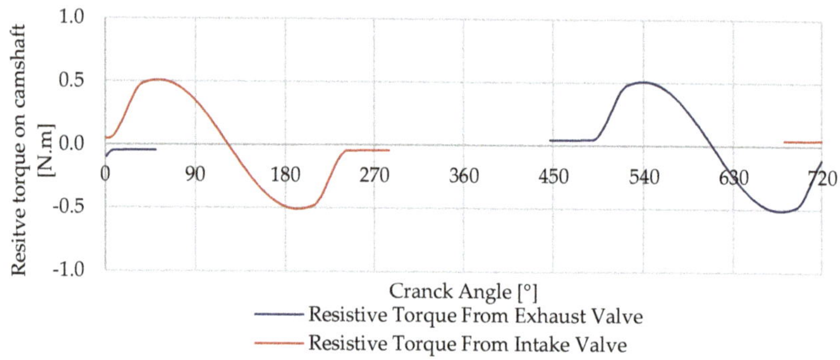

Figure 8. Resistive torque from intake and exhaust valve mechanism.

3. Design of the Torque Correction Mechanism

3.1. Design Concept

To enhance the output torque's smoothness, a specifically engineered mechanism is proposed to create a counteracting torque. This design pretends to minimize fluctuations in the engine's output torque. Key components of this system include a spring for energy storage and a cam tailored to align with the engine's output torque characteristics. A CAD model of this mechanism is shown in Figure 9. The mechanism is structured to generate a counteracting torque at any rotation angle θ, matching the output torque in magnitude but opposite in direction.

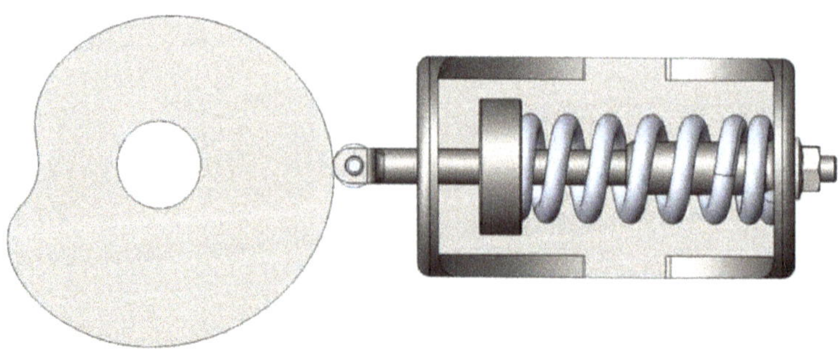

Figure 9. CAD model from the mechanism, with the cam designed to mimic the profile torque desire, the roller follower, and the spring responsible for storing the energy.

To achieve the desired mechanism response, the specially designed cam must replicate the engine's opposing torque. It can be installed on the camshaft or with a 2:1 transmission ratio, similar to the engine's distribution system, which may increase friction and consequently, resistive torque, leading to a more complex system. This is due to the torque contributing events being dispersed over the 720° crankshaft rotation in a 4-stroke cycle. The direct integration of this mechanism into the crankshaft would simplify implementation but is impractical for addressing resistive and driving torque irregularities, though it would be effective for inertia torque, as shown in Figure 6 with its 360° periodicity. Figures 10–12 illustrate a potential implementation in a single-cylinder engine using the existing distribution system and camshaft to simplify the system and minimize friction. This implementation assumes available space on the camshaft for the cam and that the camshaft shaft and transmission ratio can support the counteracting torque with these premises being valid for a demonstrative implementation in this specific engine construction.

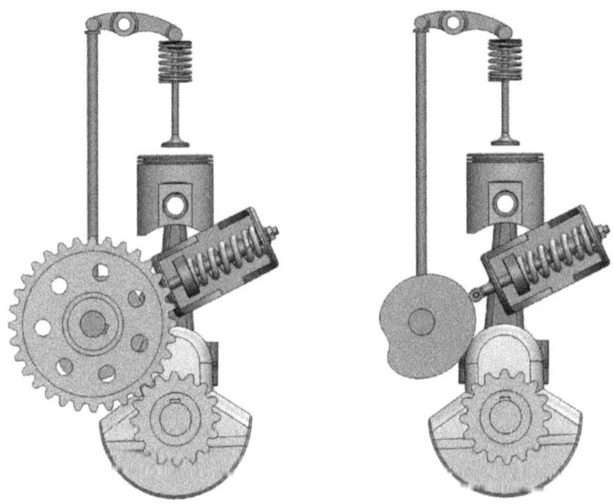

(a) (b)

Figure 10. Front views of the CAD model from the (**a**) mechanism integrated on the single-cylinder engine, on image (**b**) the camshaft gear was removed to show the positioning of the balancing cam.

(a) (b)

Figure 11. Perspective views—left side—of the CAD model from the (**a**) mechanism integrated on the single-cylinder engine, in image (**b**), the camshaft gear was removed to show the positioning of the balancing cam.

(a) (b)

Figure 12. Perspective views—right side—of the CAD model from the (**a**) mechanism integrated on the single-cylinder engine, in image (**b**), the camshaft gear was removed to show the positioning of the balancing cam.

3.2. Determination of the Balancing Cam Profile

The cam's profile is essential for controlling the actuator in response to the engine's needs. This cam was developed by considering the torque acting upon it, which is then transformed and stored in the spring. To design the cam profile, it is required to determine the cam radius in the function of the crankshaft rotation angle φ, for a previously defined torque profile. Figure 13 is a diagram of the forces acting on the cam [14,33,34].

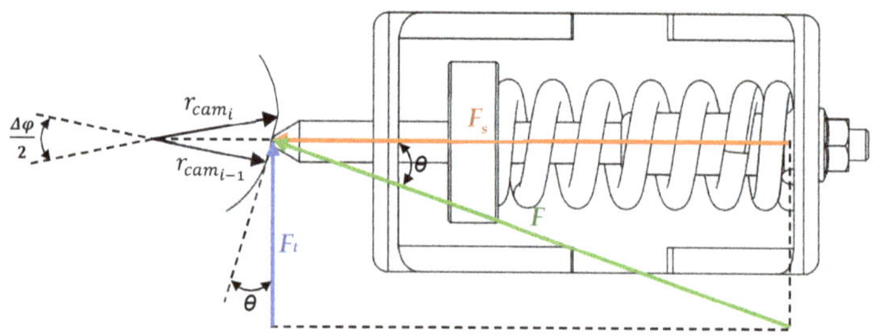

Figure 13. Diagram of the mechanism used to deduce the equations that allow the balancing cam to be obtained that mimics the desired torque profile.

The torque on the mechanism camshaft T_{cam} as the product of the tangential force F_t by cam radius:

$$T_{cam} = F_t r_{cam_i} \tag{13}$$

T_{cam}—*torque on mechanism camshaft* [Nm]
r_{cam_i}—*cam radius* [m]
F_t—*tangential force* [N]

From Figure 13 F_t is the product from the F_s by the $tan\theta$ as presented in Equation (14), and F_t can be directly substituted into Equation (13), which gives the torque as a function of F_s as presented in Equation (15), F_s depends on the spring conditions and the displacement induced by the cam on the spring, as shown by Equation (16).

$$F_t = F_s tan\theta \tag{14}$$

F_s—*spring force* [N]

$$T_{cam} = F_s tan\theta \, r_{cam} \tag{15}$$

$$F_s = k(r_{cam} - r_{initial}) + P \tag{16}$$

$r_{initial}$—*cam initial radius* [m]

Based on the defined torque for the cam T_{cam}, $tan\theta$ can be determined using Equation (17). Then, following Equation (18), which arises from the analysis of the diagram in Figure 13, the radius of the cam can be calculated as a function of the crankshaft angle.

$$tan\theta = \frac{T_{cam}}{r_{cam_i}(k(r_{cam_i} - r_{initial}) + P)} \tag{17}$$

$$r_{cam_i} = r_{cam_{i-1}} + \left(\frac{\Delta\varphi}{2}\right)\left(r_{cam_{i-1}} tan\theta\right) \tag{18}$$

To convert the previously obtained polar coordinates into cartesian coordinates for better visualization and plotting of the cam, Equations (19) and (20) are used.

$$x = r_{cam} cos\left(\frac{\varphi}{2}\right) \tag{19}$$

$$y = r_{cam} sin\left(\frac{\varphi}{2}\right) \tag{20}$$

Finally, when integrating a roller follower, including the roller's radius in the cam profile calculation is necessary. This is done using Equations (21) and (22), ensuring the cam profile meets all specified conditions for the roller follower. This crucial step guarantees that the cam design is effectively customized for the specific follower type in the engine mechanism.

$$x_r = x - r_{roller} cos\left(\left(\frac{\varphi}{2}\right) - \theta\right) \tag{21}$$

$$y_r = y - r_{roller} sin\left(\left(\frac{\varphi}{2}\right) - \theta\right) \tag{22}$$

r_{roller}—*roller radius* [m]

4. Modelling of Torque Correction Mechanism

4.1. Engine Modelling and Simulation

The development of the engine simulation, performed using MATLAB® R2024a (MathWorks, Natick, MA, USA), is based on a modeling approach that analyzes the engine's instantaneous torque for each degree of crankshaft rotation. It was specifically constructed to dissect engine torque into three distinct components.

The simulation inputs geometric parameters to model the slider–crank mechanism, iterating through one-degree increments of crankshaft rotation. It spans 720 degrees of rotation to simulate, among other things, torque values and speed variations at each degree, covering a full cycle of the 4-stroke engine under study.

As shown in Section 2 through various graphs, this modeling framework successfully segments engine torque into three main components: the inertia torque driven by mass acceleration, as specified in Equation (9); torque due to pressure differences within the engine cylinder, detailed in Equation (3); and the resistive torque produced by the valve mechanism, outlined in Equation (12).

The engine parameters are outlined in Table 1. The simulation framework is organized around valve timing, divided into four segments that correspond to each stroke of the Otto cycle, thereby covering all necessary conditions to derive operational parameters at each crankshaft angle, as will be analyzed in Section 4.3.

Table 1. Engine parameters used to perform the simulation used to present the idle torque profile.

Slider–Crank	Valve Train
$r = 0.022$ m	$k = 588,000$ N/m
$l = 0.073$ m	$P = 47$ N
$B = 0.06$ m	$r_a = 0.022$ m
$m_{pg} = 0.144$ kg	$r_b = 0.022$ m
$m_{cr} = 0.123$ kg	$IVO = 0°$
	$IVC = 220°$
$\omega = 83.8$ rad/s	$EVO = 470°$
	$EVC = 720°$

This simulation incorporates elements that yield instantaneous torque curves closely resembling those obtained through experimental means. Figure 14 showcases the results from prior experimental work conducted to validate the curves derived from the implemented methodology [3].

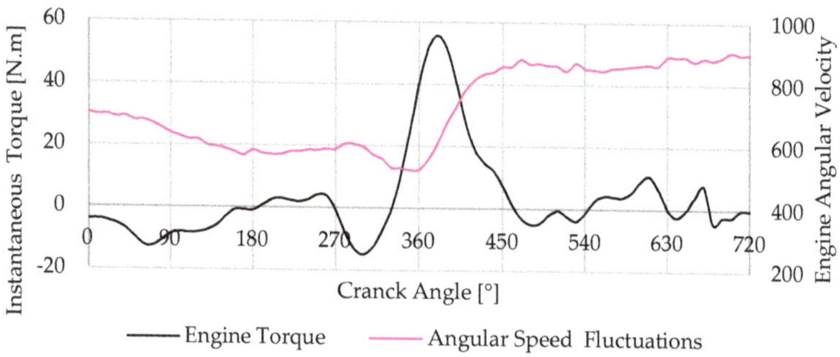

Figure 14. Experimental output torque from the single-cylinder engine at idle, 800 RPM.

4.2. Balancing Cam Mechanism Modelling

For the torque correction mechanism's development, a torque profile established from the process described earlier is used. The mechanism and cam, crafted based on equations detailed in Section 3.2, aim to correct the torque profile illustrated in Figure 15, which represents an engine idle at 800 RPM. In defining the cam profile, in addition to the torque profile, the input parameters are the spring's stiffness k, spring preload P, initial cam radius $r_{initial}$, and if the mechanism uses a roller follower like the one shown in the CAD representations of the mechanism r_{roller}; These parameters are outlined in Table 2.

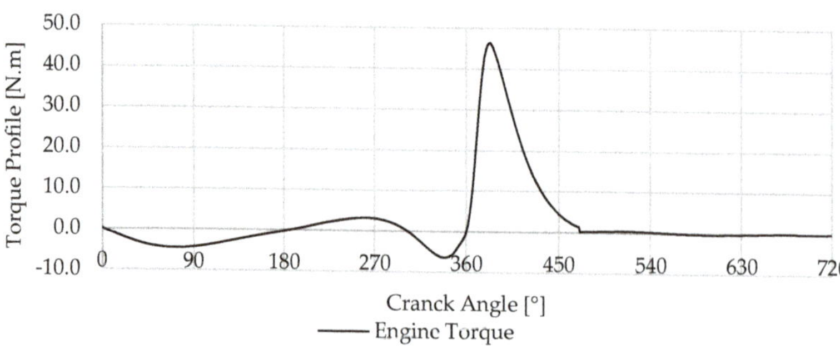

Figure 15. Output torque from single-cylinder engine during idle at 800 RPM, used to design the cam profile.

Table 2. Balancing mechanism parameters.

Cam	Spring
$r_{initial} = 0.05$ m $r_{roller} = 0.08$ m	$k = 100,000$ N/m $P = 100$ N

Based on the torque profile outlined, the cam shown in Figure 15 was developed through an iterative process involving the parameters in Table 2 and the torque profile in Figure 15, which shows two cam profiles, one for a follower with and one without a roller. The desired cam profile aimed to maintain a radius not exceeding 50 mm for integration purposes in the specified engine. In Figure 16, the cam analysis shows three distinct zones. Section 1, in the first quadrant, shows the beginning of a cam profile depression, signifying energy release from the spring to the crankshaft, countering the resistive torque during valve opening, and the driving torque of the air-fuel mixture intake. Section 2 highlights a marked decrease in cam radius, associated with compensating the compression process of the air–fuel mixture, with a sharp radius increase towards the end of compression and the start of the expansion. This radius increase continues during the third quadrant, storing energy in the spring from the expansion to utilize during the other strokes. Section 3, located in the fourth quadrant, mirrors the analysis of Section 1, about the exhaust process. Additionally, in Figure 16, a perfect circumference is depicted by a red dotted line, allowing for a comparison between the cam profile variations and a perfect circle profile.

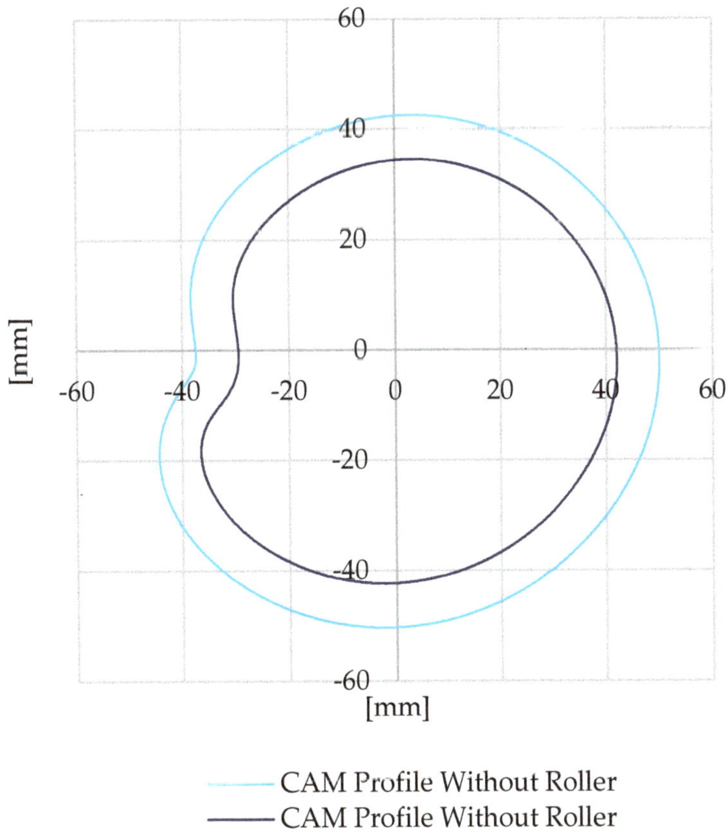

Figure 16. Representation of the determined CAM profile if the follower is with roller or without, considering the torque profile presented previously.

Building upon the previously described cam and maintaining the parameters used for its design, the torque returned by the mechanism to the engine is illustrated in Figure 17. Figure 18 shows the torque delivered from the mechanism into the crankshaft.

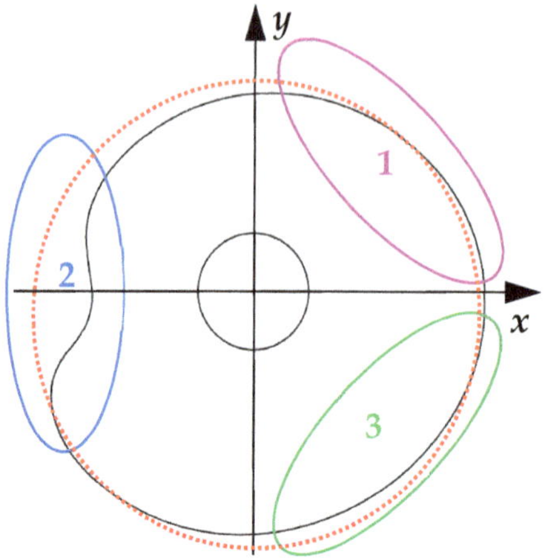

Figure 17. Esquematization of the determined CAM profile, considering the torque profile presented previously.

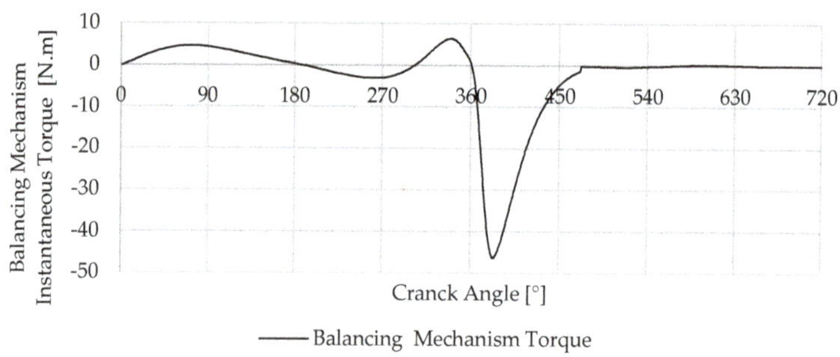

Figure 18. Torque delivered from the mechanism into the crankshaft.

4.3. Balancing Cam Mechanism Integration with the Engine

The integration of the balancing cam mechanism with the engine simulation, as established in Sections 4.1 and 4.2, is detailed in the processes outlined in Figure 19, the flowchart representing the simulation steps. The simulation begins by setting the engine's geometric parameters, initial conditions, and the cam profile for the balancing cam mechanism. The flowchart in Figure 19 tracks the engine cycle through each crank angle φ from 0 to 720 degrees, covering one complete cycle of a four-stroke engine. This encompasses all phases of engine operation: intake, compression, combustion, expansion, and exhaust. During each phase, the simulation calculates various parameters, such as temperatures and pressures within each stroke. These values are crucial for determining the torque and angular speed in each phase. The torque and angular speed calculated by the simulation are then refined by the balancing cam mechanism. This mechanism uses a predefined cam profile and spring parameters to compute the output torque from the mechanism and adjust the engine's simulated torque based on the established initial conditions.

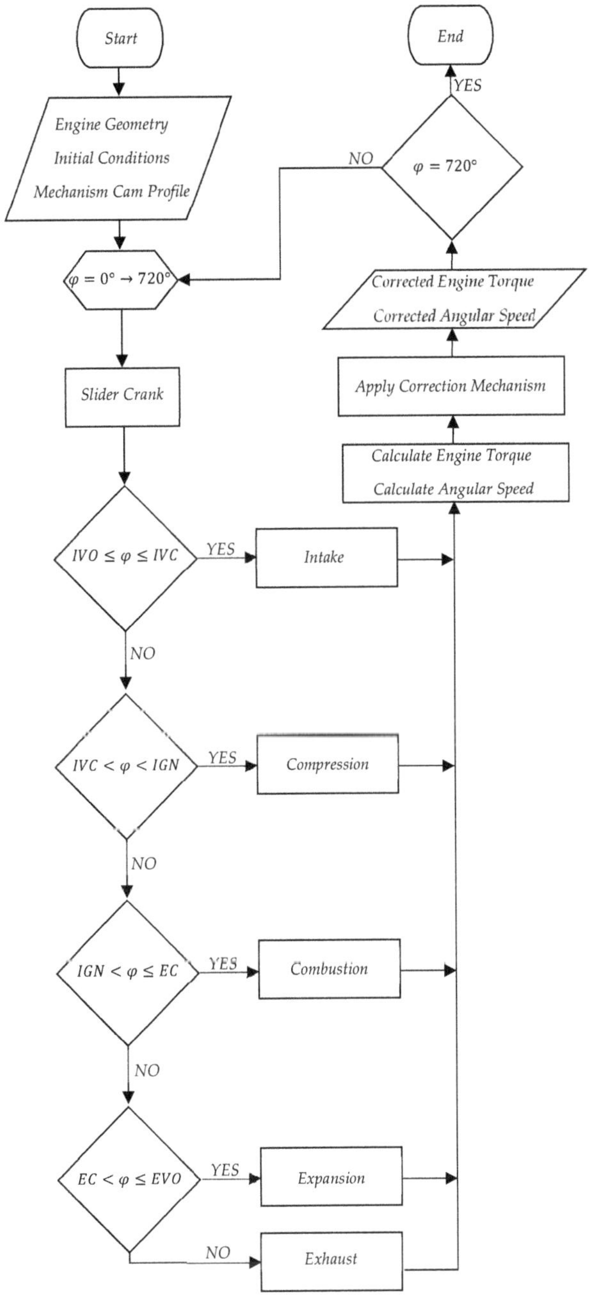

Figure 19. Simulation Flowchart.

The corrections made to the engine torque and angular speed by this integration represent the outputs the simulation aims to achieve. The simulation cycle is documented until the crank angle φ reaches 720 degrees, marking the end of one full engine cycle.

4.4. Simulation Results

The simulations were performed to analyze and determine the distinct characteristics of driving torque, resistive torque, and inertia torque within an internal combustion engine. These types of torque, which have been previously discussed individually, are now collectively examined in Figure 20. This Figure illustrates the engine's output torque profile under the simultaneous influence of driving, resistive, and inertia torque. This comprehensive analysis is crucial as it reveals how the combined effects of these different torques impact the overall performance of the engine. Particularly noteworthy in this idle scenario is that the three types, the torque resulting from the valve mechanism, the inertia of the engine's internal components, and driving torque during intake and exhaust, contribute to the same magnitude for the final engine torque. The variations in engine torque in these conditions are observed during the moments when the piston reaches its extreme positions—the bottom dead center and the top dead center. At these junctures, the piston experiences significant accelerations and decelerations, which in turn influence the directional reversal of its motion. Other variations outcomes from the operation of opening and closing the intake valve approximability from 0 to 180° and the exhaust valve from 540 to 720°, the most pronounced fluctuations from to 180 540° come from the compression and expansion process.

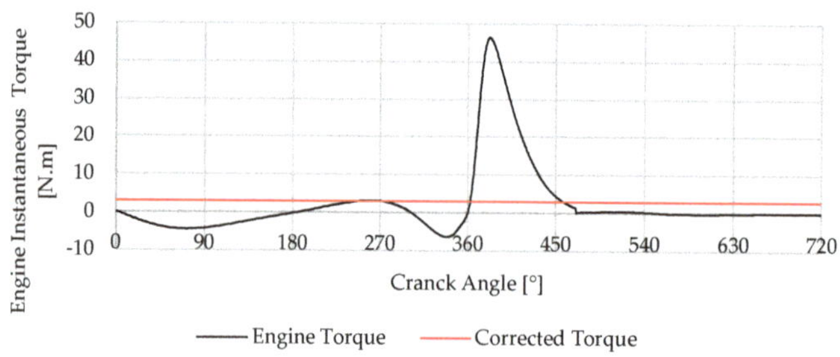

Figure 20. Profile of the engine output torque with the mechanism in red and without the mechanism in black.

In Figure 20, it is possible to observe the overlay of two torque profiles—one from the engine and the other from the mechanism. The computational results assume a loss-free system, indicating that the torque resulting from the system implementation is a consistent profile. This supports the thesis presented, validating the effectiveness of the implemented system in maintaining a stable torque during idle operation.

Torque variations result in corresponding changes in engine speed, which are more noticeable in idle regimes due to the lack of load masking these variations. Figure 21 demonstrates the engine speed variations over a cycle, consistent with the previously analyzed torque variations. As the mechanism suppresses torque fluctuations, it directly impacts reducing speed variations, consequently decreasing vibrations and noise.

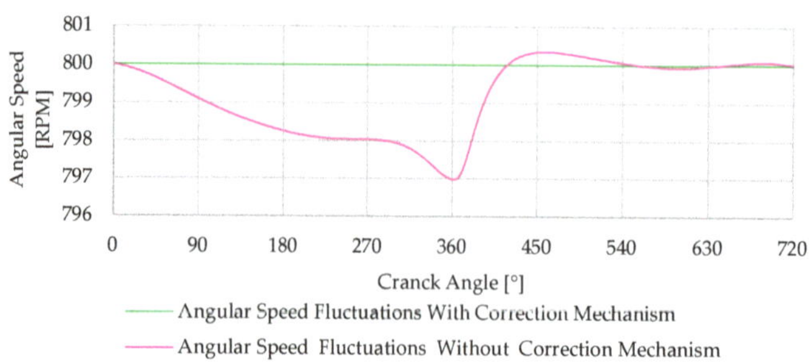

Figure 21. Angular speed variation over the cycle.

4.5. Limitations and Future Work

The need for an energy storage system in ICEs, such as the traditionally used flywheel or the system presented in this work, is more pronounced during low-speed operations or idling. The system developed and designed in this research is specifically tailored to function when the engine is idling. However, as the load imposed on the engine or its rotational speed increases, the influence of this mechanism is lower. To address this limitation, ongoing research efforts are focused on designing a system that can adapt the mechanism to the engine by varying the spring preload based on the engine load and rotation. This adaptation aims to make the system more influential in smoothing the engine's output torque. The development of this system faces several challenges. One significant challenge is that as the engine's rotation speed or load increases, variations primarily occur in the driving torque and inertia torque; consequently, increasing the spring's preload to accommodate these changes can inadvertently amplify the resistive torque correction component from the mechanism, which does not vary under the conditions. This increase in resistive torque, due to the higher preload of the spring, may introduce irregularities in the engine's output torque. Future work will focus on overcoming these challenges by refining the developed cam and the mechanism to balance the varying torque components effectively.

5. Conclusions

This paper introduced a balancing mechanism designed to smooth the output torque of the engine. The torque model proposed for the engine revealed that output torque fluctuations result from a combination of multiple factors, each contributing differently to the final torque produced by the engine. By employing a balancing mechanism with a specifically designed cam profile, the engine operation can be smoothed during idle periods without relying on traditional and heavy flywheels.

The simulations conducted in Section 4 allowed us to study the influence of various factors on engine torque and their actual impact on engine torque. Variations in torque production reflect speed changes throughout the engine cycle, which are responsible for generating vibrations. By developing a system capable of addressing these vibrations, engine operation cannot only be stabilized but also reduce engine mass and the need for energy absorbing dampers.

Although this system does not significantly influence high-speed engine operation, it does not introduce irregularities in the engine's normal functioning. It proves to be an effective solution for correcting engine operation during idling and opens opportunities for developing a system capable of adapting to different engine operating regimes.

Evaluating the performance of our novel mechanism alongside traditional systems like heavy flywheels, we observed notable differences. While flywheels serve well in specific conditions, they do not offer the necessary adaptability and tend to increase the overall weight of the engine system, and inclusively change the engine behavior. On the other hand, our mechanism has the potential to adjust dynamically to meet operational demands and efficiently smooth torque fluctuations with much less impact on engine weight and behavior.

The preliminary results obtained align with those presented in similar studies, albeit for different purposes, such as those by Lin et al. [14] and Arakelian et al. [15], which employ the use of cam-based mechanisms for torque correction. These findings reinforce the validity of our approach and highlight the versatility of cam mechanisms in various engineering applications beyond their traditional uses. By leveraging cam-based designs, we can achieve a consistent correction of torque fluctuations, which is critical for enhancing the operational stability and efficiency of internal combustion engines; this convergence of results across different studies underscores the potential of our proposed mechanism.

This study introduces a unique mechanism for torque harmonization that meets the current and emerging needs of modern engines, emphasizing efficiency and adaptability. Such advancements are instrumental in the development of new technologies, including range extenders, microcogeneration systems, and innovative engine configurations, representing a substantial progression in engine technology.

6. Patents

This and subsequent works resulted in a patent application submitted to the Portuguese Institute of Industrial Property (*INPI*) under application number 119346, with the title "Método Implementado por Computador para a Conceção de um Atuador de Equilíbrio para um Motor, Atuador de Equilíbrio, Programa de Computador e Meio de Leitura Associados".

Author Contributions: Conceptualization, D.S.C. and P.O.F.; methodology, D.S.C. and P.O.F.; software, D.S.C. and P.O.F.; validation, P.O.F., P.D.G. and A.E.-S.; formal analysis, P.O.F., P.D.G. and A.E.-S.; investigation, D.S.C. and P.O.F.; data curation, P.O.F.; writing—original draft preparation, D.S.C.; writing—review and editing, D.S.C., P.O.F. and P.D.G.; supervision, P.O.F., P.D.G. and A.E.-S. All authors have read and agreed to the published version of the manuscript.

Funding: This research was funded in part by the Fundação para a Ciência e Tecnologia (FCT) and C-MAST (Centre for Mechanical and Aerospace Science and Technologies) for their support in the form of funding under the project UIDB/00151/2020 (https://doi.org/10.54499/UIDB/00151/2020; https://doi.org/10.54499/UIDP/00151/2020, accessed on 3 January 2024).

Data Availability Statement: Data are contained within the article.

Acknowledgments: We would like to express our gratitude to the Entrepreneurship, Careers, and Alumni Office (GESPA) at the University of Beira Interior for facilitating our collaboration with a company specializing in patent development. We are also thankful to Susana Rodrigues from Inventa, who expertly guided us through the process of writing the patent application.

Conflicts of Interest: The authors declare no conflicts of interest.

References

1. Solmaz, H.; Karabulut, H. A mathematical model to investigate the effects of misfire and cyclic variations on crankshaft speed fluctuations in internal combustion engines. *J. Mech. Sci. Technol.* **2015**, *29*, 1493–1500. [CrossRef]
2. Babagiray, M.; Solmaz, H.; İpci, D.; Aksoy, F. Modeling and validation of crankshaft speed fluctuations of a single-cylinder four-stroke diesel engine. *Proc. Inst. Mech. Eng. Part D J. Automob. Eng.* **2022**, *236*, 553–568. [CrossRef]
3. Cardoso, D.S.; Fael, P.O.; Espírito-Santo, A. Instantaneous angular velocity and torque on Otto single-cylinder engine: A theoretical and experimental analysis. *Energy Rep.* **2020**, *6*, 43–48. [CrossRef]
4. Filipi, Z.S.; Assanis, D.N. A nonlinear, transient, single-cylinder diesel engine simulation for predictions of instantaneous engine speed and torque. *J. Eng. Gas Turbines Power* **2001**, *123*, 951–959. [CrossRef]
5. Antonopoulos, A.K.; Hountalas, D.T. Effect of instantaneous rotational speed on the analysis of measured diesel engine cylinder pressure data. *Energy Convers. Manag.* **2012**, *60*, 87–95. [CrossRef]
6. Munde, K.H.; Mehtre, V.K.; Ware, D.S.; Kamble, D.P. Review on Performance of Dual Mass Flywheel over Conventional Flywheel. *Math. Stat. Eng. Appl.* **2022**, *71*, 496–505. [CrossRef]
7. Dawange, S.V.; Kadlag, V.L. A Review Paper on Vibration Analysis of DI Engine. *Int. J. Sci. Res. IJSR* **2015**, *4*, 759–761.
8. Cardoso, D.; Nunes, D.; Faria, J.; Fael, P.; Gaspar, P.D. Intelligent Micro-Cogeneration Systems for Residential Grids: A Sustainable Solution for Efficient Energy Management. *Energies* **2023**, *16*, 5215. [CrossRef]
9. Mittal, V.; Shah, R.; Przyborowski, A. Analyzing the Usage of Wankel Engine Technology in Future Automotive Powertrains. *SAE Int. J. Sustain. Transp. Energy Environ. Policy* **2023**, *5*, 1–13. [CrossRef]
10. Schaper, U.; Sawodny, O.; Mahl, T.; Blessing, U. Modeling and torque estimation of an automotive Dual Mass Flywheel. In Proceedings of the 2009 American Control Conference, St. Louis, MO, USA, 10–12 June 2009; pp. 1207–1212. [CrossRef]
11. Ayana, E.; Plahn, P.; Wejrzanowski, K.; Mohan, N. Active torque cancellation for transmitted vibration reduction of low cylinder count engines. *IEEE Trans. Veh. Technol.* **2011**, *60*, 2971–2977. [CrossRef]
12. Anjum, R.; Yar, A.; Ahmed, Q.; Bhatti, A. Model-Based Unified Framework for Detection and Mitigation of Cyclic Torque Imbalance in a Gasoline Engine. *J. Eng. Gas Turbines Power* **2021**, *143*, 071013. [CrossRef]
13. Zhang, X.; Liu, H.; Zhan, Z.; Wu, Y.; Zhang, W.; Taha, M.; Yan, P. Modelling and Active Damping of Engine Torque Ripple in a Power-Split Hybrid Electric Vehicle. *Control Eng. Pract.* **2020**, *104*, 104634. [CrossRef]
14. Lin, D.Y.; Hou, B.J.; Lan, C.C. A balancing cam mechanism for minimizing the torque fluctuation of engine camshafts. *Mech. Mach. Theory* **2017**, *108*, 160–175. [CrossRef]
15. Arakelian, V.V.; Briot, S. Simultaneous inertia force/moment balancing and torque compensation of slider-crank mechanisms. *Mech. Res. Commun.* **2010**, *37*, 265–269. [CrossRef]

16. Cardoso, D.S.; Fael, P.O. Simulation and analysis of a switched reluctance machine for flywheel replacement. In *Lecture Notes in Engineering and Computer Science, Proceedings of the World Congress on Engineering 2018, London, UK, 4–6 July 2018*; Springer: Singapore, 2019; pp. 819–823. [CrossRef]
17. Kim, G.W.; Shin, S.C. Research on the torque transmissibility of the passive torsional vibration isolator in an automotive clutch damper. *Proc. Inst. Mech. Eng. Part D J. Automob. Eng.* **2015**, *229*, 1840–1847. [CrossRef]
18. Vehicular Technology Society; Institute of Electrical and Electronics Engineers. Active Torque Ripple Damping in Direct Drive Range Extender Applications: A Comparison and an Original Proposal. In Proceedings of the 2015 IEEE Vehicle Power and Propulsion Conference (VPPC), Montreal, QC, Canada, 19–22 October 2015. [CrossRef]
19. Jianguo, B.; Xudong, L.; Ming, Z.; Kaixiong, L. Design and Optimization for the Main Dimension of Flywheel Motor Based on Torque Density. In Proceedings of the 2018 IEEE 3rd Advanced Information Technology, Electronic and Automation Control Conference (IAEAC), Chongqing, China, 12–14 October 2018; IEEE: New York, NY, USA, 2018; pp. 2156–2162. [CrossRef]
20. Galvagno, E.; Velardocchia, M.; Vigliani, A.; Tota, A. Experimental Analysis and Model Validation of a Dual Mass Flywheel for Passenger Cars. In Proceedings of the SAE 2015 World Congress & Exhibition, Detroit, MI, USA, 21–23 April 2015. [CrossRef]
21. Zhang, Y.; Zhang, X.; Qian, T.; Hu, R. Modeling and simulation of a passive variable inertia flywheel for diesel generator. *Energy Rep.* **2020**, *6*, 58–68. [CrossRef]
22. Fraser, N.; Blaxill, H.; Lumsden, G.; Bassett, M. Challenges for Increased Efficiency through Gasoline Engine Downsizing. *SAE Int. J. Engines* **2009**, *2*, 991–1008. [CrossRef]
23. Hannan, M.A.; Azidin, F.A.; Mohamed, A. Hybrid electric vehicles and their challenges: A review. *Renew. Sustain. Energy Rev.* **2014**, *29*, 135–150. [CrossRef]
24. Cardoso, D.S.; Fael, P.O.; Espírito-Santo, A. A review of micro and mild hybrid systems. *Energy Rep.* **2020**, *6*, 385–390. [CrossRef]
25. Mastrangelo, G.; Micelli, D.; Sacco, D. Extreme Downsizing by the two-cylinder gasoline engine from Fiat. *ATZautotechnology* **2011**, *11*, 18–25. [CrossRef]
26. Omanovic, A.; Zsiga, N.; Soltic, P.; Onder, C. Increased internal combustion engine efficiency with optimized valve timings in extended stroke operation. *Energies* **2021**, *14*, 2750. [CrossRef]
27. Bech, A.; Shayler, P.J.; McGhee, M. The Effects of Cylinder Deactivation on the Thermal Behaviour and Performance of a Three Cylinder Spark Ignition Engine. *SAE Int. J. Engines* **2016**, *9*, 1999–2009. [CrossRef]
28. Kim, C.J.; Kang, Y.J.; Lee, B.H.; Ahn, H.J. Determination of optimal position for both support bearing and unbalance mass of balance shaft. *Mech. Mach. Theory* **2012**, *50*, 150–158. [CrossRef]
29. Guo, J.; Zhang, W.; Zou, D. Investigation of dynamic characteristics of a valve train system. *Mech. Mach. Theory* **2011**, *46*, 1950–1969. [CrossRef]
30. Pfabe, M.; Woernle, C. Reducing torsional vibrations by means of a kinematically driven flywheel—Theory and experiment. *Mech. Mach. Theory* **2016**, *102*, 217–228. [CrossRef]
31. Fan, H.; Jing, M.; Wang, R.; Liu, H.; Zhi, J. New electromagnetic ring balancer for active imbalance compensation of rotating machinery. *J. Sound Vib.* **2014**, *333*, 3837–3858. [CrossRef]
32. Blair, G.P. *Design and Simulation of Four-Stroke Engines*; SAE International: Warrendale, PA, USA, 1999; ISBN 978-0-7680-0440-3.
33. Uicker, J.J., Jr.; Pennock, G.R.; Shigley, J.E. *Theory of Machines and Mechanisms*, 5th ed.; Oxford University Press: New York, NY, USA, 2016; ISBN 9780190264482.
34. Budynas, R.G.; Nisbett, K.J. *Shigley's Mechanical Engineering Design*, 9th ed.; McGraw-Hill: New York, NY, USA, 2010; ISBN 978-0-07-352928-8.

energies

MDPI

Article

Combustion Mechanism of Gasoline Detonation Tube and Coupling of Engine Turbocharging Cycle

Diyun Huang, Jiayong Wang, Minshuo Shi, Puze Yang and Binyang Wu *

State Key Laboratory of Engines, Tianjin University, Tianjin 300072, China; hdytj@tju.edu.cn (D.H.); 1023201121@tju.edu.cn (J.W.); sms_1072@tju.edu.cn (M.S.); yangpz1@tju.edu.cn (P.Y.)
* Correspondence: binyang.wu@tju.edu.cn

Abstract: Traditional exhaust-gas turbocharging exhibits hysteresis under variable working conditions. To achieve rapid-intake supercharging, this study investigates the synergistic coupling process between the detonation and diesel cycles using gasoline as fuel. A numerical simulation model is constructed to analyze the detonation characteristics of a pulse-detonation combustor (PDC), followed by experimental verification. The comprehensive process of the flame's deflagration-to-detonation transition (DDT) and the formation of the detonation wave are discussed in detail. The airflow velocity, DDT time, and peak pressure of detonation tubes with five different blockage ratios (BR) are analyzed, with the results imported into a one-dimensional GT-POWER engine model. The results indicate that the generation of detonation waves is influenced by flame and compression wave interactions. Increasing the airflow does not shorten the DDT time, whereas increasing the BR causes the DDT time to decrease and then increase. Large BRs affect the initiation speed of detonation in the tube, while small BRs impact the DDT distance and peak pressure. Upon connection to the PDC, the transient response rate of the engine is slightly improved. These results can provide useful guidance for improving the transient response characteristics of engines.

Keywords: pulse detonation; deflagration to detonation; blockage ratio; turbine; numerical simulation

Citation: Huang, D.; Wang, J.; Shi, M.; Yang, P.; Wu, B. Combustion Mechanism of Gasoline Detonation Tube and Coupling of Engine Turbocharging Cycle. *Energies* **2024**, 17, 2466. https://doi.org/10.3390/en17112466

Academic Editors: Wenbin Yu and Guang Zeng

Received: 9 April 2024
Revised: 9 May 2024
Accepted: 18 May 2024
Published: 22 May 2024

1. Introduction

When an engine steps from a low load to a high load, because of the hysteresis of the response to the exhaust gas turbocharger, the intake charge becomes far too low to produce the required acceleration, which seriously worsens the engine's performance and may cause the engine to stall [1]. Therefore, improving the turbocharger's transient response capability is the key to improving the acceleration performance of the engine. Pulse detonation combustion approximates constant volume combustion and is characterized by a fast combustion rate and a small entropy increase [2–4]. It can obtain high-temperature and high-pressure working fluid in an extremely short time, provide additional energy for the exhaust gas turbocharger, promote the rapid rotation of the turbine to supply sufficient fresh air for the engine, and enable the engine to switch to the steady operation mode quickly and smoothly through the valve, injection, and other synergistic control.

The pulse-detonation combustor (PDC) is the core component for generating pulse detonation waves. Optimizing the structure of the combustor can effectively enhance the intensity of detonation combustion, thus improving the transient response rate of the engine. The critical technologies of pulse detonation combustion mainly include air intake, detonation, and gas exchange, wherein the fuel atomization and mixing of fuel, the geometry of the detonation combustor, and the composition of the oxidant and the ignition energy each have an important influence on ignition.

Compared with conventional combustion, the conditions for the formation of detonation combustion are more stringent, especially for the control of the temporal and spatial distribution of the mixture and the initial ignition. Therefore, early research on detonation mainly focused on gaseous fuels such as hydrogen [5–7], methane [8,9], and ethylene [10,11]. However, liquid fuels are typically used in practical applications because gaseous materials are more difficult to store and transport. The structure of liquid-fueled detonation is more complex and much more difficult to control than that of gas-fueled detonation in terms of aspects including but not limited to the initial state of the mixtures (pressure and temperature), the particle parameters (e.g., particle shape, size, and distribution), the complicated multiphysical processes such as droplet breakup, atomization, and evaporation, as well as incomplete mixing of fuel and air [12,13].

Some studies have shown that the activity of liquid fuels is low; therefore, it is necessary to add a certain percentage of high-activity gaseous fuels to liquid fuels [14,15] or to adopt oxygen-enriched [16–18] and preheating [19–21] methods to accelerate the evaporation and atomization of the liquid fuels. Adopting atomizers [22–24] is another common method to improve fuel atomization and mixing. Frolov et al. [25] fabricated a one-meter-long pre-detonator and an air-assisted atomizer to initiate the detonation in the main detonation tube. Wolanski et al. [26] developed a special liquid fuel atomization system that uses preheated air to assist the early evaporation and mixing of liquid fuel to produce a homogeneous fuel–air mixture. Wu et al. [27] used an ultrasonic acoustic atomizer to finely atomize liquid fuel to increase the pre-evaporation degree of liquid fuel.

The installation of obstacles in the detonation tube also plays a key role in the generation of detonation waves. Common forms of obstacles include orifice plates, Shchelkin spirals, and grooves [28,29]. These obstacles can induce strong disturbances during flame propagation, which can enhance the evaporation and mixing rates of fuel, thereby accelerating combustion and heat release [30]. Research has shown that channel length and height, obstacle spacing, and blockage ratio (BR) all have varying degrees of influence on the deflagration-to-detonation transition (DDT) process. Heilbronn et al. [31] and Goodwin et al. [32] investigated the effects of the BR on the flame acceleration process using experimental and numerical methods, and the results showed that a BR between 0.3 and 0.6 will accelerate the DDT process efficiently, and a BR exceeding this interval will cause the DDT distance to increase drastically. Ciccarelli et al. [33] conducted experimental studies on the effects of obstacle spacing and BR on the DDT process, finding that obstacle spacing has a huge effect on the DDT distance at high BR and acquires the shortest DDT distance when obstacle spacing is roughly equal to the pipe diameter. Saeid et al. [34] studied the spatial height ratio (S/H) of the obstacle spacing to the pipe diameter; they found that in inhomogeneous mixtures, detonation could not be triggered in the tube when the S/H was equal to 10 and that the DDT time was considerably reduced when the S/H was reduced to 2.5. Wang et al. [35] investigated the effect of the number of jet obstacles on the DDT process, speculated that there existed an optimal number of jet obstacles, and concluded that the optimal number of jet obstacles may vary with changes in channel size and obstacle spacing. Coates et al. [36] compared the influences of obstacles of different shapes on flame acceleration and found that rectangular obstacles induced the largest flame acceleration. The aforementioned studies suggest that suitable obstacles would accelerate the combustion reaction rate, which in turn would increase the flame speed and pressure, causing the DDT to occur more easily in the PDC.

Recent studies on pulse detonation combustion mainly focused on aero engines and rocket engines [37–39]; therefore, they have adopted the detonation cycle to replace the Brayton cycle to achieve high thermal efficiency and power performance. The main approach used to optimize the DDT process has been optimizing the ignition and the PDC structure. Most of the research objects have been gaseous fuels, and when liquid fuel has been the research object, it has typically been aviation kerosene [40–42]. A series of complex methods have been devised to accelerate the evaporation and atomization process of fuels [43,44]. In this study, we rely on the engine fuel supply system to inject high-pressure fuel directly into the detonation tube so that the fuel is rapidly broken and atomized into tiny particles (particle size < 8.5 μm), which substantially simplifies the multiphysical process of liquid fuel in the volatilization process. For numerical simulation, most studies have utilized two-dimensional (2D) simulation to reduce the computation time [45–47]; however, in the study of two-phase detonation, the mixing, atomization, and evaporation process of liquid fuel is the key to the formation of combustible mixtures. Computational fluid dynamics (CFD) can simulate the whole process of fuel from injection to mixing with the oxidant, which can provide a comprehensive understanding of how the three-dimensional (3D) detonation structure evolves so as to better optimize the detonation and combustion process. The idea of this study is to provide additional energy to the diesel engine turbine through the high-temperature and high-pressure working fluid discharged by the PDC so that the turbine can obtain enough power to increase the speed of the compressor and exhibit a fast intake response to improve the transient response performance of the heavy-duty diesel engine. At present, research on the synergistic coupling of the pulse detonation cycle and the diesel cycle to improve the transient response of the turbocharger is still scarce.

In this study, based on relevant literature, a numerical simulation is used to optimize the design of the straight tube detonation combustion process, and an implementation scheme for the detonation combustor is proposed. The numerical simulation results are experimentally verified and then imported into the one-dimensional coupling model of the engine equipped with the detonation afterburning module. The model is intended to serve as a design reference to facilitate the improvement in the transient response performance of heavy-duty diesel engines.

2. Materials and Methods

2.1. Computational Models

Figure 1 presents the configuration and meshing of the computational model, in which the lengths of the injection section, premixing section, ignition section, and main detonation section are 135, 104, 131, and 1030 mm, respectively, and the total length of the tube is 1400 mm. The main detonation section of the tube has a diameter of D = 50 mm, eight groups of symmetrically distributed solid obstacles are arranged inside the tube, the gap between the upper and lower obstacles is d = 34 mm, the spaces between obstacles 1–5 are 80 mm each, and the space between obstacles 6 and 7 is 160 mm; moreover, BR is defined as $BR = 1 - d^2/D^2$. For the suppression of flame and pressure wave back-propagation, a conical thrust wall is arranged behind the premixing section. Because the main detonation section is long, the injectors are placed in two different positions to better organize the distribution of the mixture in the tube. Injector 1 is installed horizontally in the front of the injection section, while injector 2 is installed vertically at around the one-third point along the length of the main detonation section. The two injectors are of the same model, and both use an eight-hole fuel injector with a nozzle orifice diameter of 0.169 mm.

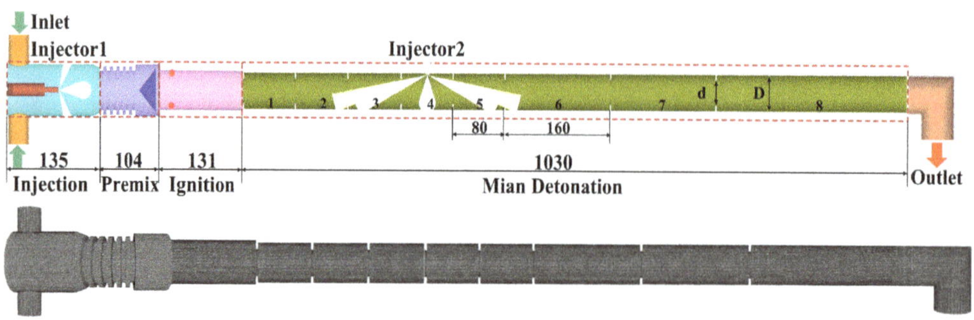

Figure 1. The configuration and meshing of the computational model (unit: mm).

2.2. Computational Numerical Simulation Methods and Boundary Models

The various physical and chemical models used in the computational model are shown in Table 1; the governing equations for all models are solved using CONVERGE. The characterization fuel of gasoline is isooctane (C_8H_{18}), and the chemical reaction mechanism is a 42-component, 142-step simplified model. In the simulation, the basic grid size is set to 4 mm and is supplemented by an adaptive mesh refinement technique; the minimum grid size is 0.5 mm, and the maximum number of grids is about 1 million.

Table 1. Physical and chemical models of simulation.

Process	Model
Turbulence	RNG k–ε
Spray break	KH-RT
Spray-wall interaction	Rebound/slide
Fuel collision	NTC
Drop evaporation	Frossling
Wall heat transfer	Han and Reitz
Combustion	SAGE chemical reaction solver and simplify chemical reaction mechanism

To ensure that the detonation tube can be filled with fresh air smoothly, the inlet pressure is set to 0.15 MPa, the temperature is 325 K, the outlet pressure is set to 0.1 MPa, and the total temperature is 300 K. Two symmetrically distributed spherical ignition sources, each with a radius of 1 mm, are set up at 26 mm from the left side of the main detonation tube; the ignition energy is 1 J, and the ignition pulse width is 5 ms. The spray cone angle of the injector is 149°, and the injection pressure is 100 MPa.

To make the mixture distribution in the detonation tube more uniform, the fuel quantity of injector 1 is set to 150 mg, that of injector 2 is set to 110 mg, and the injection interval between the two injectors is 6 ms. According to the pulse detonation cycle process, the detonation combustion can be divided into five stages:

1. Intake stage: the intake valve is opened, allowing fresh air to flow into the detonation tube to provide oxidants for subsequent detonation combustion, which is usually carried out at the same time as the purge process.
2. Injection stage: the injector is opened, and fuel mixes with the fresh air in the tube, driven by the airflow and compression wave successively dispatched into the ignition section and the main detonation section.
3. Ignition stage: high-energy igniter discharge instantly ignites the mixture near the spark plug, and a compression wave causes the initial flame to travel into the main detonation section.
4. DDT stage: flames form and propagate, and deflagration gradually transitions to detonation in the tube.
5. Purge stage: the intake valve is reopened, and the exhaust gas in the pipe is discharged by fresh air to prepare for the next detonation.

2.3. Experimental Validation

The experimental setup for the pulse detonation system, detailed in Figure 2, is mainly composed of a control system, an acquisition system, an air intake system, a fuel system, and an ignition system. Among them, the detonation tube used in the experiment is designed in reference to the model parameters in Figure 1, with air flowing vertically into the tube from the side of the injection section and an overall equivalence ratio of about 1.06. The liquid fuel used in the test is gasoline with an octane number of 92, the approximate fuel is C_8H_{16}, and the oxidant is air. The Chapman–Jouguet (CJ) detonation pressure and detonation wave speed are 1.88 MPa and 1796.4 m/s [48], respectively, and the gasoline/air equivalence ratio is 1. According to two-phase detonation research [49], there is a certain loss of detonation wave when using liquid-fueled detonation. Hence, when the peak pressure is up to 80% of the CJ detonation pressure (1.5 MPa), and the detonation wave speed is up to 65% of the CJ detonation wave velocity (1168 m/s), the detonation is considered successful.

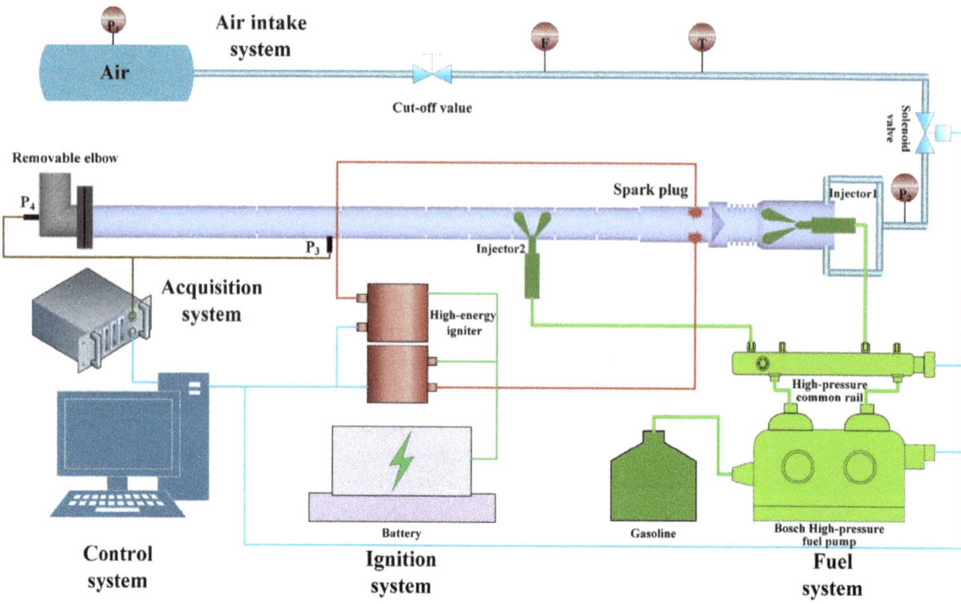

Figure 2. Schematic of the experimental setup for the pulse detonation system.

The control system, developed independently by our research group, can control the whole process of intake, fuel injection, and ignition constantly and accurately. The acquisition system is a self-editing module based on the LabVIEW platform, which allows the number and frequency of acquisition to be freely adjusted according to demand; consequently, the pressure fluctuation in the detonation tube can be accurately collected. To detect whether a detonation wave is generated in the detonation tube, three pressure sensors are set up in the intake section and the main detonation section. A dynamic piezoresistive pressure sensor P_2 with a range of 0–2 MPa is installed at the intake inlet to monitor the intake pressure in the detonation tube. Two Kistler6125C (Kistler, Switzerland) piezoelectric sensors (P_3–P_4), each with a range of 0–30 MPa, are installed on the outer wall of the main detonation section to detect the triggering and development process of the detonation wave in the tube. As shown in Figure 2, P_3 is installed at the head of the seventh section, and the distance to the end of the tube is 450 mm. To measure the total pressure of the detonation tube at the gas outlet, an L-shaped elbow is added to the end of the detonation tube, and the piezoelectric sensor P_4 is placed on the wall facing the burning gases. For the fuel system, the Bosch high-pressure common rail system is utilized to provide high-pressure fuel to both injectors, and the ignition system uses two high-energy igniters driven by 24 V batteries to provide 1 J of ignition energy to each of the spark plugs. All dynamic data of the test are acquired by a NIUSB6259 (National Instruments, Austin, TX, USA) high-speed data acquisition card with thirty-two channels (single channel sampling rate of 1.25 MS/s) and a sampling frequency of 2×10^4 MS/s.

Figure 3 presents a comparison of the trend of detonation pressure curves obtained via a numerical simulation and an experiment under a single detonation with an inlet air pressure of 0.15 MPa. It can be seen from the figure that the peak pressure and change trend of the two curves are basically consistent. The peak pressure of the test is 2.1 MPa, which satisfies the characteristics of detonation. However, there is a certain error in the simulation and test data during the pressure drop stage because the piezoelectric sensor has the inherent characteristics of return hysteresis. In summary, it can be considered that the numerical simulation in this paper can reflect the ignition and detonation process of PDC and can be used to study different parameters.

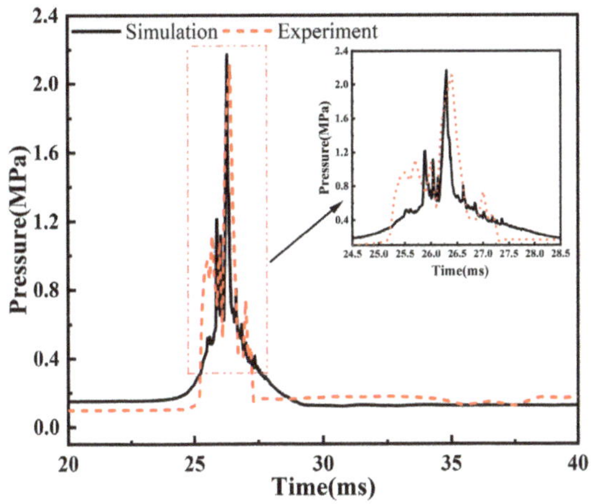

Figure 3. Comparison of detonation pressure between the experiment and numerical simulation.

3. Results

3.1. Process Analysis of the Injection to Ignition Stage

The development process of the mixture is shown in Figure 4. Injector 1 sprays most of the fuel near the wall of the tube. As a result of the impact, a plurality of vortex clusters is formed close to the wall; moreover, the fuel is continuously rolled backward by the airflow and then enters the premixing section. There is weak airflow movement in the small grid near the wall of the premixing section, which can largely prevent the fuel from adhering to the wall. The fuel pauses briefly after contacting the conical body, and the flow of the mixture through the conical body becomes more uniform under the push of the airflow, which can effectively reduce the fuel concentration zone. Subsequently, the mixture quickly passes through the conical body and reaches the ignition point of the spark plug, at which time the mixture exhibits an approximately equivalent distribution.

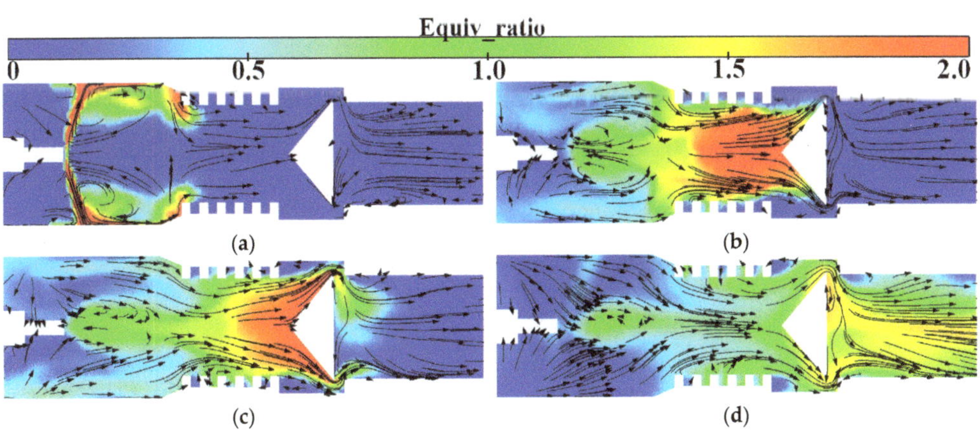

Figure 4. Process of fuel injection and mixing: (**a**) distribution of mixture equivalence ratio and flow field (t = 13 ms); (**b**) distribution of mixture equivalence ratio and flow field (t = 15 ms); (**c**) distribution of mixture equivalence ratio and flow field (t = 18 ms); (**d**) distribution of mixture equivalence ratio and flow field (t = 20 ms).

Figure 5 shows the development of the mixture and flame in the ignition section during the ignition stage (t = 18–23 ms). As shown in Figure 5a, at the initial stage of ignition (t = 18 ms), the mixture in the ignition section exhibits a distribution wherein the middle is concentrated, and the surroundings are dilute because of the faster flow velocity near the gap between the conical body and the wall. In contrast, it can be seen from Figure 5b,c that the airflow movement is relatively slow at the wall near the spark plug, which can favor the stable formation of the initial flame kernel. The development of the flame front is basically consistent with the contour of the chemical equivalent ratio, and it gradually moves from the periphery to the middle. Afterward, because of the airflow movement of the compression wave, the flame kernel is driven forward continually, passing through the first obstacle at t = 20.4 ms and gradually entering the main detonation tube.

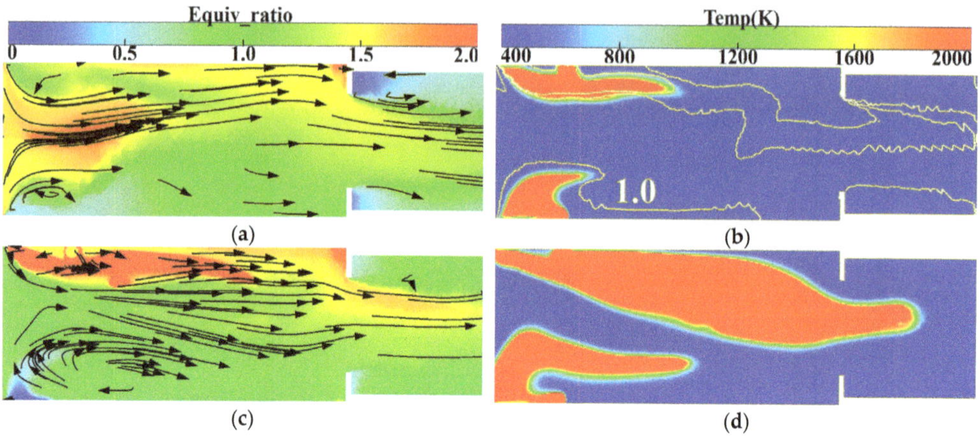

Figure 5. Process of ignition and the initial formation of flame: (**a**) distribution of mixture equivalence ratio and flow field (t = 18 ms); (**b**) distribution of mixture temperature and equivalence ratio contour lines (t = 19 ms); (**c**) distribution of mixture equivalence ratio and flow field (t = 20 ms); (**d**) distribution of mixture temperature (t = 20.4 ms).

3.2. Process Analysis of Flame Acceleration-to-Detonation Transition and Detonation Wave Development

When the flame enters the main explosion section, owing to the presence of obstacles, the high-temperature airflow is often blocked and compressed when it flows around the obstacles, resulting in a sudden rise in the compression wave. As shown in Figure 6a, the initial compression wave is mainly distributed in the tube wall and near the flame front, where the compression wave sweeping through will cause a sudden change in the thermodynamic state of the unburned gas (i.e., the pressure, temperature, and density will all increase), which will intensify the combustion of the flame, increase the heat release, and ultimately enable the flame to propagate at a faster speed. In the meantime, the increase in flame speed will enhance the strength of the compression wave, thus creating a feedback mechanism for flame acceleration. This feedback mechanism makes the distance between the flame front and the compression wave gradually decrease, as shown in Figure 6c, and finally, at t = 24.7 ms, at the end of the sixth section of the main detonation tube, the flame front and the leading shock wave coincide. The state of the compression wave in the tube at this time with the influence of the obstacles is shown in Figure 6e; the trend of airflow in the tube is fast in the middle and slow in the periphery, so its velocity profile is a "bullet" type. The shock wave velocity can reach 718 m/s, which is a form of flame propagation before the detonation wave is triggered. Subsequently, the high-temperature flame surface (T > 2500 K) expands rapidly in the tube, accelerated by the obstacles and coupled with high-temperature, high-pressure reactants. The "detonation wave kernel" is formed at the front end of the obstacle in the seventh section, corresponding to the initial state of detonation wave formation, as depicted in Figure 6d,f. At this point, the velocity of the detonation wave reaches 1283 m/s, with a peak pressure of 0.56 MPa.

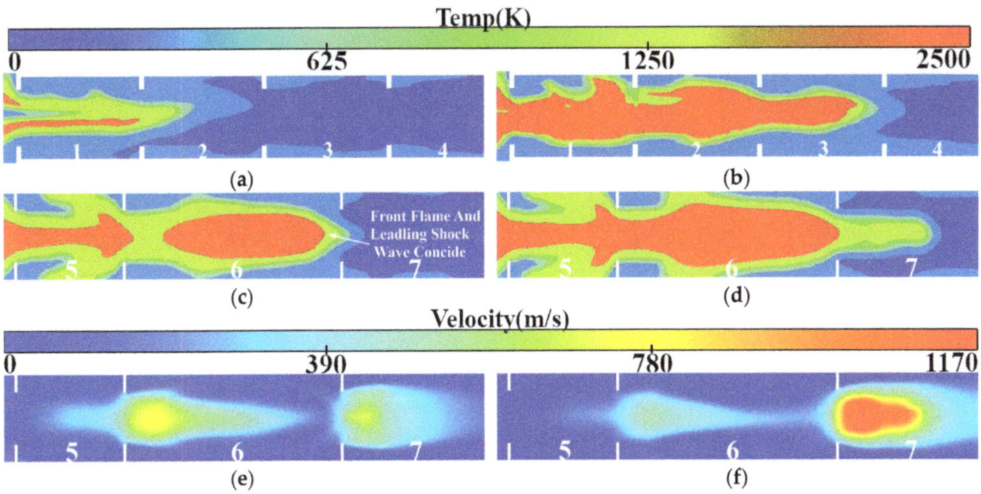

Figure 6. Process of flame acceleration and detonation initiation: (**a**) distribution of mixture temperature (t = 23 ms); (**b**) distribution of mixture temperature (t = 24 ms); (**c**) distribution of mixture temperature (t = 24.7 ms); (**d**) distribution of mixture velocity (t = 24.7 ms); (**e**) distribution of mixture velocity (t = 24.7 ms); (**f**) distribution of mixture velocity (t = 24.8 ms).

Once formed, the detonation shock wave will propagate to both ends. However, owing to the presence of a larger amount of unburned high-temperature and high-pressure mixture downstream, the initial detonation shock wave exhibits a conical shape and expands further downstream. As depicted in Figure 7a, detonation waves are observed in sections 7 and 8 of the tube at t = 25.1 ms. Notably, the trigger mechanisms for these two detonation waves are entirely different. The first detonation wave originates from the progressive acceleration of the flame, leading to the combustion flame's DDT; in contrast, the second detonation wave arises from the direct detonation of high-temperature and high-pressure reactants due to shock wave-wall interactions. Figure 7b shows that the high-temperature and high-pressure reactants at the front end of the eighth section are reignited after touching the leading shock wave near the wall and react rapidly to form a new detonation wave. In comparison to the first shock wave, both the initiation time and distance of the second shock wave are significantly reduced, causing the generation speed of the second detonation wave to be greater than the attenuation speed of the first detonation wave. Consequently, multiple detonations occur within the tube. This phenomenon effectively retards decoupling between shock waves and reaction surfaces during the transmission process while inhibiting downstream back transmission of detonation waves, thereby ensuring stable propagation.

As can be seen from Figure 7c, the decay of the seventh section of the detonation wave results in further augmentation of the velocity and sweep range of the detonation wave in the eighth section, which continues to propagate downstream. Figure 7d demonstrates that at t = 25.3 ms, a high-temperature flame overspreads the detonation tube within the seventh section. At this juncture, the velocity of the detonation wave inside the tube can reach up to 1950 m/s while achieving a maximum burst pressure of 1.98 MPa. Moreover, it is observed that there is a transition in the shape of the detonation wave front from conical to horizontal forward propulsion. The detonation wave reaches the exhaust outlet at t = 25.5 ms, as depicted in Figure 7e. Because of a sharp pressure drop when it reaches the outlet, a smaller range of detonation waves is generated near the outlet of the PDC. This substantial pressure difference leads to a surge of the wave velocity at the outlet, with maximum speeds reaching up to 2033 m/s. Simultaneously, the pressure of the detonation wave in the tube begins to decrease, with a peak pressure of 0.8 MPa. Because there is no combustible mixture at the leading edge of the PDC, the generation of new detonation waves becomes unattainable. Subsequently, after 0.5 ms, the velocity of the detonation wave drops below 1000 m/s and gradually attenuates until it reaches a stable state.

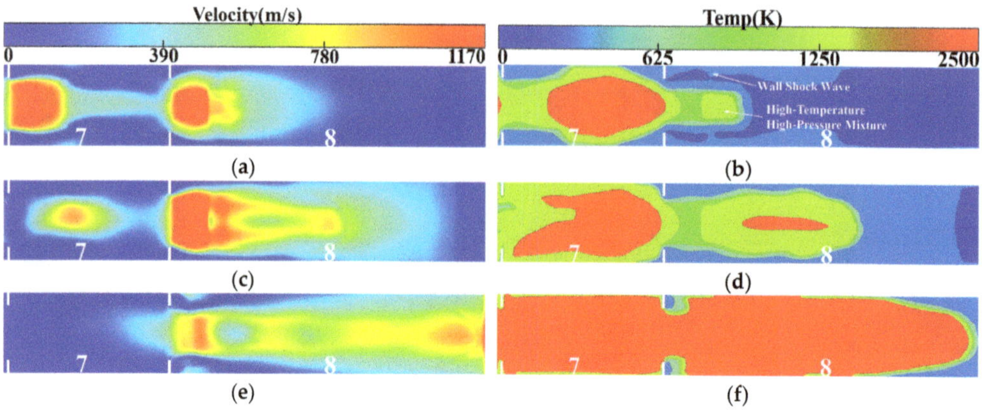

Figure 7. Process of flame and detonation wave propagation: (**a**) distribution of mixture velocity (t = 25.1 ms); (**b**) distribution of mixture temperature (t = 25.1 ms); (**c**) distribution of mixture velocity (t = 25.3 ms); (**d**) distribution of mixture temperature (t = 25.3 ms); (**e**) distribution of mixture velocity (t = 25.5 ms); (**f**) distribution of mixture temperature (t = 25.5 ms).

The aforementioned section provides a comprehensive description of the process involving the formation, ignition, and detonation of the gasoline–air mixture within a tube containing obstacles. To further enhance the optimization of the PDC structure, an analysis is conducted on the BR parameters of the detonation tube based on this foundation.

3.3. *Effect of Different BRs on Detonation and Combustion Characteristics of the Detonation Tube*

The BR is an important parameter that affects the detonation in the detonation tube. Figure 8 illustrates the gas flow velocities and the DDT times of five different detonation tubes with BRs of 0.3 to 0.7 at the moment of ignition (t = 18 ms). Moreover, this figure clearly demonstrates the considerable impact of BR on gas flow velocity within the detonation tube. A higher value of BR corresponds to a slower gas flow rate in the tube. The gas flow velocity greatly affects the diffusion rate of the fuel particles, thereby influencing the distribution of the mixture throughout the tube.

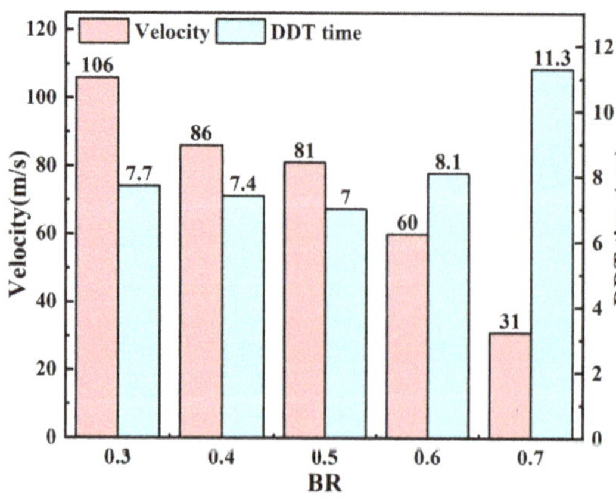

Figure 8. The airflow velocity and DDT time vary with BR.

High BR results in considerable resistance in the tube, which affects the development and step of the compression wave, leading to a substantial increase in the required DDT time. However, the DDT time does not exhibit a monotonic increase as BR decreases; on the contrary, it exhibits a trend of initial decrease and subsequent increase. The minimum DDT time occurs when the BR is approximately 0.5, and the DDT time at BR = 0.7 is 61% higher than that at BR = 0.5. High BR considerably impacts the detonation speed of the detonation tube.

Notably, BR also exerts an influence on the DDT distance of the detonation tube. As depicted in Figure 9a, when the BR ranges from 0.4 to 0.7, the DDT distance remains relatively consistent, with detonation occurring at the onset of the seventh obstacle. Conversely, when BR is 0.3, detonation occurs at the onset of the eighth obstacle and results in the furthest DDT distance, indicating that a low BR has a greater impact on the detonation distance of the DDT.

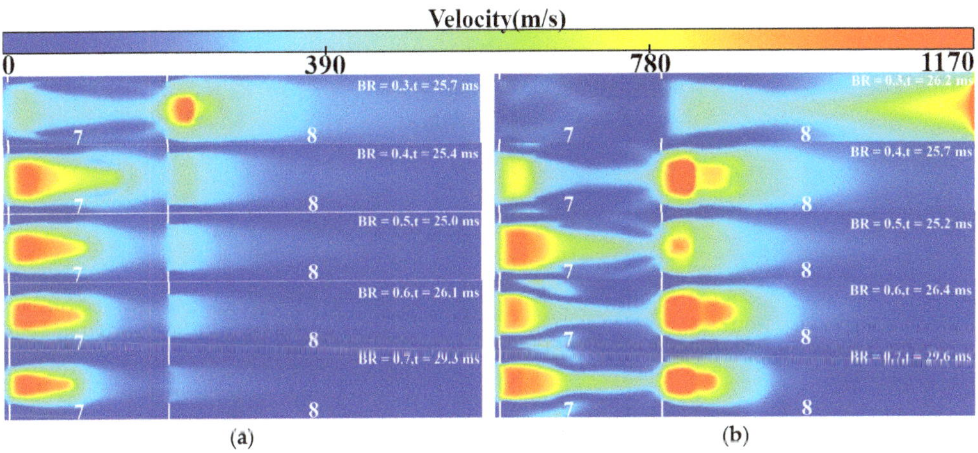

Figure 9. DDT distance and detonation wave distribution vary with BR: (**a**) DDT distance at different BRs; (**b**) distribution of detonation wave at different BRs.

The analysis presented in Figure 9b investigates multiple detonation phenomena in the tube with different BRs. Multiple detonation phenomena are not observed in the tube when the BR ranges from 0.3 to 0.5, and the detonation wave is directly propagated from the beginning of the eighth section to near the outlet when the BR is 0.3. However, for BRs of 0.4 or 0.5, only one detonation wave is generated in the tube; a shock wave is also generated, but its velocity is lower than $0.67V_{CJ}$ (780 m/s), which does not reach the velocity of a detonation wave. When the BR increases to 0.6–0.7, two detonation waves are generated within the tube.

Results demonstrate that the propagation mechanism of detonation waves in the tube with continuous obstacles remains consistent. The detonation wave in the process of traversing obstacles will exhibit attenuation, failure, and re-initiation of the phenomenon. This can be attributed to the fact that the detonation wave is a shock wave coupled with a chemical reaction and that the law of its development through the obstacles is similar to the law of shock wave diffraction in the obstacles. In other words, as airflow passes through the tiny concave corner, compression waves are generated due to the blockage. The magnitude of these compression waves is influenced by both airflow velocity and obstacle shape. Therefore, during the initiation stage of the detonation wave, which involves the interaction between the flame and the shock wave, the turbulence effect does not play a leading role. This also explains why the DDT time does not decrease with the increase in airflow movement. Specifically, the generation and propagation of detonation waves are jointly determined by interactions among flames, the leading shock wave, obstacles, and local geometric conditions.

The peak pressure and instantaneous heat release rate curves of five different detonation tubes with BRs ranging from 0.3 to 0.7 are depicted in Figures 10a and 10b, respectively. It can be observed from Figure 10a that as BR increases, the peak pressure of the detonation tube exhibits an initial increase followed by a decrease. Notably, the detonation tubes with BRs of 0.5 and 0.6 demonstrate relatively high peak pressures. When the BR is 0.3, the fuel particles disperse rapidly due to the high flow velocity in the tube, which is unfavorable to the formation of a stable initial fire core. Moreover, owing to little obstacle blockage, there is weak disturbance and a small area affected by compression waves, resulting in a negligible impact on the unburned gas state and minimal overall heat release. When the BR is 0.7, although the DDT time is prolonged, the peak pressure of the detonation wave remains high. This can be attributed to low airflow velocity within the tube impeding flame propagation speed and causing slow development of the detonation wave during its formative stage.

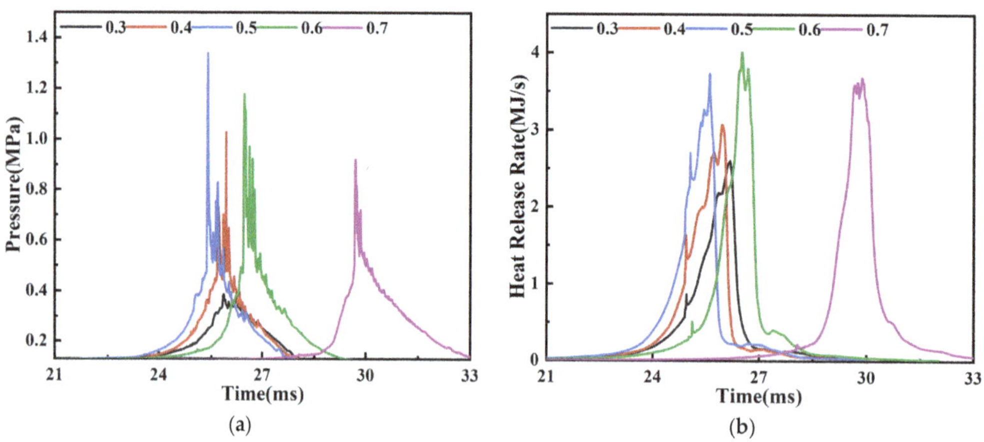

Figure 10. Pressure and heat release rate vary with BR: (**a**) pressure at different BRs; (**b**) heat release rate at different BRs.

The heat release time in the tube becomes longer as the BR decreases, as depicted in Figure 10b. Conversely, as the BR increases, the heat release becomes more concentrated. Although the peak heat release values are higher when BR = 0.6 and 0.7, the maximum pressure is lower than when BR = 0.5. This is because increasing the BR increases the resistance and results in a certain pressure loss within the tube, which emphasizes that detonation wave triggering is a consequence of flame and compression wave interaction. The DDT time is observed to be the shortest and the pressure to be the largest when the BR is 0.5, as depicted in Figure 10a,b. Deviating from this optimal value of BR (in the positive or negative direction) will adversely affect both the fuel particle diffusion speed (making it either excessively slow or fast) and the compression wave magnitude. Even with adjustments made to ignition and injection timing, satisfactory results will not be achieved, and some problems will arise, such as diminished heat release and prolonged detonation cycle time. Through optimization, it has been determined that the optimal DDT time is 6.7 ms, and the peak pressure is 2.2 MPa when the BR of the detonation tube is 0.54.

3.4. Effect of the Detonation Tube Coupled Exhaust Gas Turbine on Diesel Engine Performance

To investigate the impact on the engine performance after accessing the pulse detonation tube, a one-dimensional engine model is constructed based on the GT-POWER platform, which mainly includes cylinders, a supercharging system, an intercooling system, and a detonation afterburning module. The engine's operating parameters are presented in Table 2.

Table 2. GT-POWER simulation engine parameters.

Item	Value	Unit
Cylinder number	6	-
Bore	126	mm
Stroke	155	mm
Speed	700	r/min
Compression rate	17	-
Displacement	11.6	L
Air intake method	Turbocharged, intercooled	-
Nozzle number × diameter	8 × 0.217	mm

The engine employs a variable-geometry turbocharger. The total outlet pressure and pulse flow from a previous acquisition of the detonation tube are imported at the environmental boundary of the afterburning module e300-2 to drive the turbine and compressor together with the exhaust gas discharged from the engine. The combined cycle arrangement according to GT-POWER is illustrated in Figure 11.

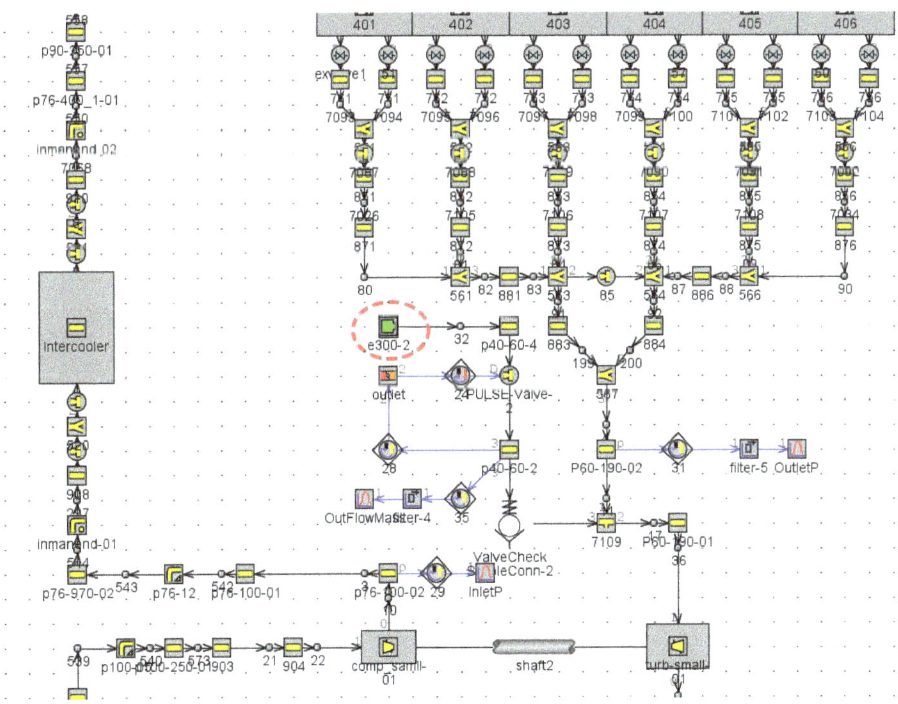

Figure 11. GT-POWER model layout.

The engine's constant speed is set at 700 r/min. From 0 to 5 s, the engine operates in an idle state with a constant fuel quantity of 35 mg. At 5 s, the fuel quantity begins gradually increasing and reaches its maximum value of 212 mg after 0.5 s. At this time, the outlet pressure and pulse flow of the detonation tube at a frequency of 7 Hz are inputted into the e300-2 module of the GT engine model.

Figure 12 presents the transient response abilities of the original engine and the engine connected to the detonation tube, with a variable-geometry turbocharger (VGT) blade opening of 0.55. As depicted in the figure, it takes 13.3 s for the original diesel engine to achieve an intake pressure of 1.4 bar; however, when the PDC is connected to supply exhaust gas to the turbine, this response time is shortened to 8.2 s, a reduction of 38.3% This clearly indicates that connecting the PDC significantly improves the compressor supercharging response and yields outstanding supercharging effects.

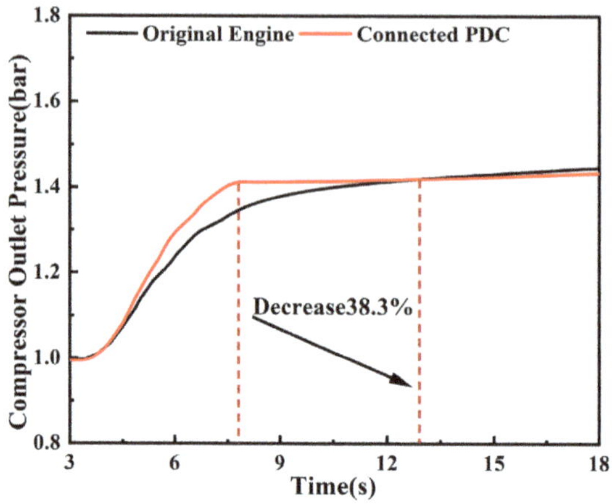

Figure 12. Comparison of compressor outlet pressure between the original engine and connected PDC at a VGT opening of 0.55.

The PDC also demonstrates a certain effect in enhancing the engine's transient response performance. Figure 13 illustrates the impact curve of the PDC outlet's exhaust gas flow on the engine's intake flow and torque at a turbine VGT blade opening of 0.55. It is evident that, under the current VGT setting, connecting the PDC results in a reduction of 1.1 s (approximately 11%) in response time when the intake flow reaches 65 g/s compared to the original engine. Meanwhile, when the engine torque increases to 1600 Nm, it only takes 5.9 s, which is about 11% shorter than the original engine's response time.

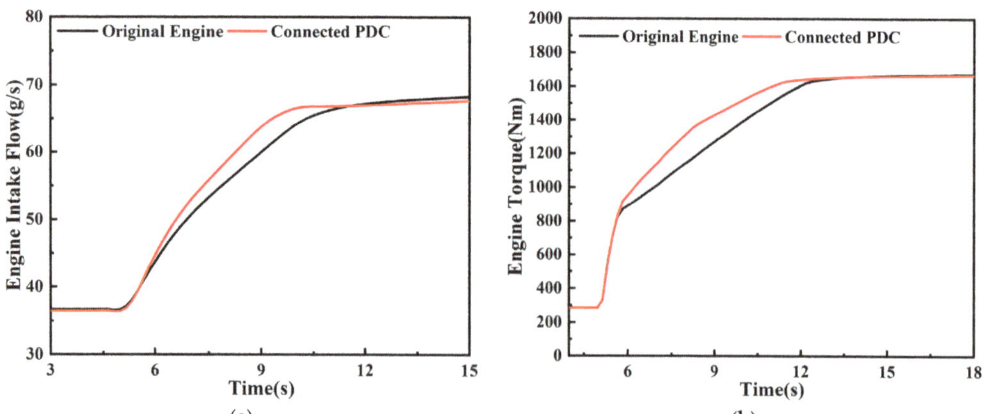

Figure 13. Comparison of transient response performance between the original engine and connected PDC: (**a**) engine intake flow response; (**b**) engine torque response.

From the aforementioned results, it can be concluded that the engine will experience a substantial improvement in the transient response of intake supercharging when the detonation combustion tube is installed. With an appropriate VGT opening, the engine intake supercharging response time can be effectively shortened, which can meet the transient demand of the engine under special working conditions. These results provide insights and numerical references that will be valuable for improving the transient operating efficiencies of engines.

4. Conclusions

In this study, the detonation characteristics of high-pressure direct-injection gasoline–air mixtures in a long straight detonation tube are investigated using numerical simulations, the effects of different BRs on flame acceleration and the DDT process are determined, and an efficient PDC structure is established. The exhaust gas generated by the PDC is further used to promote turbine operation, effectively compensating for the problem of turbocharging delay and quickly realizing the purpose of pressure boost and intake volume adjustment. This study is intended to serve as a design reference to facilitate the improvement in the transient response performance of heavy-duty diesel engines. The key conclusions are as follows:

1. In the ignition stage, the formation of the initial flame kernel is significantly influenced by the local airflow velocity and the mixture equivalent ratio, while flame development predominantly follows the chemical equivalent ratio contour. In the flame acceleration stage, turbulent flow and compression waves synergistically promote flame propagation, establishing a feedback mechanism for flame acceleration. During the detonation wave development stage, detonation is triggered by one of the following two mechanisms: one involves detonation resulting from the coupling between flames and shock waves, and the other entails direct detonation initiated by collision and reflection of compression waves with the wall.
2. The detonation mechanism of a detonation wave remains fundamentally consistent across varying BRs, but as the BR increases, the airflow in the tube slows, and the DDT time decreases and then increases. DDT time is 61% higher when the BR is 0.7 than when the BR is 0.5, indicating that a high BR seriously impacts detonation velocity and results in substantial pressure loss. When the BR is 0.54, the DDT time is the shortest at 6.7 ms. The BR has less effect on the DDT distance, but a low BR will prolong the DDT distance.
3. When BR $\leq$ 0.5, there are not two types of detonation waves in the detonation tube, indicating that an excessively small BR will affect the strength and range of the compression waves. At this time, the blockage in the tube is small, high-strength leading shock waves cannot be formed, and the interaction between shock waves and the wall surface is insufficient to directly initiate detonation. The flames, compression waves, and local geometric conditions collectively determine the generation and propagation of detonation waves, and re-initiation will occur under appropriate conditions.
4. When the PDC exhaust gas flow is connected to the turbine, the VGT blade opening is 0.55, the intake air is supercharged to 1.4 bar, the response time of the compressor is 5.1 s shorter than that of the original engine, and the response rate is increased by 38.3%. Moreover, the time required for the engine to reach its rated torque of 1600 Nm decreases from 7 s to 5.9 s, indicating that the PDC can effectively improve the transient acceleration response process of the diesel engine and provide sufficient intake air charge for the diesel engine in a short time.

Author Contributions: Writing—original draft preparation, D.H.; writing—review and editing, B.W.; validation, J.W.; investigation, M.S. and P.Y. All authors have read and agreed to the published version of the manuscript.

Funding: This research was funded by the National Defense Science and Technology Key Laboratory Foundation Program, grant number 6142212210205.

Data Availability Statement: The data presented in this study are available on request from the corresponding author. The data are not publicly available due to privacy.

Conflicts of Interest: The authors declare no conflicts of interest.

References

1. Shi, Y.; Zheng, G.; Chen, H.; Wang, L. Transient Behavior Study of HD Diesel Engine and the Effects of Turbochargers. In *Proceedings of the FISITA World Automotive Congress;* Lecture Notes in Electrical Engineering; Springer: Berlin/Heidelberg, Germany, 2013; Volume 190. [CrossRef]
2. Roy, G.D.; Frolov, S.M.; Borisov, A.A.; Netzer, D.W. Pulse detonation propulsion: Challenges, current status, and future perspective. *Prog. Energy Combust. Sci.* **2004**, *30*, 545–672. [CrossRef]
3. Liu, J.Y.; Wang, Z.W.; Zhang, Z.X.; Li, J.L.; Qin, W.F.; Huang, J.J. Reheat effect on the improvement in efficiency of the turbine driven by pulse detonation. *Def. Technol.* **2024**, *31*, 200–210. [CrossRef]
4. Anand, V.; Gutmark, E. Rotating detonation combustors and their similarities to rocket instabilities. *Prog. Energy Combust. Sci.* **2019**, *73*, 182–234. [CrossRef]
5. Jin, S.; Qi, L.; Zhao, N.B.; Zheng, H.T.; Meng, Q.Y.; Yang, J.L. Experimental and numerical research on rotating detonation combustor under non-premixed conditions. *Int. J. Hydrogen Energy* **2020**, *45*, 10176–10188. [CrossRef]
6. Wang, Z.W.; Qin, W.F.; Huang, J.J.; Wei, L.S.; Yang, Y.X.; Wang, Y.Q.; Zhang, Y. Numerical investigation of the effect of jet intensity from internal jet tube on detonation initiation characteristics. *Int. J. Hydrogen Energy* **2022**, *47*, 13732–13745. [CrossRef]
7. Zhao, M.J.; Zhang, H.W. Origin and chaotic propagation of multiple rotating detonation waves in hydrogen/air mixtures. *Fuel* **2020**, *275*, 117986. [CrossRef]
8. Luan, Z.Y.; Huang, Y.; Gao, S.J.; You, Y.C. Formation of multiple detonation waves in rotating detonation engines with inhomogeneous methane/oxygen mixtures under different equivalence ratios. *Combust. Flame* **2022**, *241*, 112091. [CrossRef]
9. Peng, H.Y.; Liu, W.D.; Liu, S.J.; Zhang, H.L.; Jiang, L.X. Hydrogen-air, ethylene-air, and methane-air continuous rotating detonation in the hollow chamber. *Energy* **2020**, *211*, 118598. [CrossRef]
10. Wang, G.Y.; Liu, W.D.; Liu, S.J.; Zhang, H.L.; Peng, H.Y.; Zhou, Y.F. Experimental verification of cylindrical air-breathing continuous rotating detonation engine fueled by non-premixed ethylene (vol 189, pg 722, 2021). *Acta Astronaut.* **2022**, *193*, 795–796. [CrossRef]
11. Wang, Y.; Le, J.; Wang, C.; Zheng, Y. A non-premixed rotating detonation engine using ethylene and air. *Appl. Therm. Eng.* **2018**, *137*, 749–757. [CrossRef]
12. Musick, B.J.; Paudel, M.; Ramaprabhu, P.K.; McFarland, J.A. Numerical simulations of droplet evaporation and breakup effects on heterogeneous detonations. *Combust. Flame* **2023**, *257*, 113035. [CrossRef]
13. Wen, H.C.; Fan, W.Q.; Xu, S.; Wang, B. Numerical study on droplet evaporation and propagation stability in normal-temperature two-phase rotating detonation system. *Aerosp. Sci. Technol.* **2023**, *138*, 108324. [CrossRef]
14. Huang, X.X.; Lin, Z.Y. Analysis of coupled-waves structure and propagation characteristics in hydrogen-assisted kerosene-air two-phase rotating detonation wave. *Int. J. Hydrogen Energy* **2022**, *47*, 4868–4884. [CrossRef]
15. Yao, S.B.; Guo, C.H.; Zhang, W.W. Effects of droplet evaporation on the flow field of hydrogen-enhanced rotating detonation engines with liquid kerosene. *Int. J. Hydrogen Energy* **2023**, *48*, 33335–33345. [CrossRef]
16. Ding, C.W.; Wu, Y.W.; Xu, G.; Xia, Y.Q.; Li, Q.; Weng, C.S. Effects of the oxygen mass fraction on the wave propagation modes in a kerosene-fueled rotating detonation combustor. *Acta Astronaut.* **2022**, *195*, 204–214. [CrossRef]
17. Han, X.P.; Huang, Y.K.; Zheng, Q.; Xiao, Q.; Xu, H.; Wang, F.; Wu, Y.W.; Feng, W.K.; Weng, C.S. Study of the characteristics and combustion efficiency of liquid kerosene/oxygen-enriched air rotating detonation wave with different modes. *Fuel* **2024**, *355*, 18. [CrossRef]
18. Zhang, Q.B.; Qiao, X.Q.; Fan, W.; Wang, K.; Tan, F.G.; Wang, J.G. Study on operation and propulsion features of a pulse detonation rocket engine with secondary oxidizer injection. *Appl. Therm. Eng.* **2020**, *180*, 115661. [CrossRef]
19. Ma, Y.; Zhou, S.B.; Ma, H.; Ge, G.Y.; Yu, D.H.; Zou, G.; Liang, Z.T.; Zhang, T.F. Experimental investigation on propagation characteristics of liquid-fuel/preheated-air rotating detonation wave. *Int. J. Hydrogen Energy* **2022**, *47*, 24080–24092. [CrossRef]
20. Qiu, H.; Bai, Q.D.; Han, J.X.; Huang, B.Y.; Liu, Z.Y.; Weng, C.S. Experimental research on self-initiation process of rotating detonation wave by high-temperature ethylene-rich gas. *Fuel* **2024**, *357*, 129795. [CrossRef]
21. Wang, J.S.; Lin, W.; Huang, W.D.; Shi, Q.; Zhao, J.F. Numerical study on atomization and evaporation characteristics of preheated kerosene jet in a rotating detonation scramjet combustor. *Appl. Therm. Eng.* **2022**, *203*, 117920. [CrossRef]
22. Zhao, J.F.; Ren, Y.J.; Tong, Y.H.; Lin, W.; Nie, W.S. Atomization of a liquid jet in supersonic crossflow in a combustion chamber with an expanded section. *Acta Astronaut.* **2021**, *180*, 35–45. [CrossRef]
23. Malik, V.; Salauddin, S.; Hytovick, R.; Bielawski, R.; Raman, V.; Bennewitz, J.; Burr, J.; Paulson, E.; Hargus, W.; Ahmed, K. Detonation wave driven by aerosolized liquid RP-2 spray. *Proc. Combust. Inst.* **2023**, *39*, 2807–2815. [CrossRef]
24. Xu, G.; Wu, Y.W.; Kang, C.H.; Lei, T.; Qiu, Y.M.; Ding, C.W.; Weng, C.S. Propagation behaviors of kerosene-fueled rotating detonation wave with varied atomizer locations. *Aerosp. Sci. Technol.* **2023**, *142*, 108676. [CrossRef]
25. Frolov, S.M. Liquid-fueled, air-breathing pulse detonation engine demonstrator: Operation principles and performance. *J. Propuls. Power* **2006**, *22*, 1162–1169. [CrossRef]
26. Wolanski, P.; Balicki, W.; Perkowski, W.; Bilar, A. Experimental research of liquid-fueled continuously rotating detonation chamber. *Shock Waves* **2021**, *31*, 807–812. [CrossRef]

27. Wu, Y.W.; Han, Q.X.; Yang, G.Y. Effects of an acoustic atomizer upon liquid-fueled detonation initiations in a detonation tube. *Exp. Therm. Fluid Sci.* **2019**, *109*, 109863. [CrossRef]
28. Zhang, Q.B.; Wang, K.; Dong, R.X.; Fan, W.; Lu, W.; Wang, Y.J. Experimental research on propulsive performance of the pulse detonation rocket engine with a fluidic nozzle. *Energy* **2019**, *166*, 1267–1275. [CrossRef]
29. Zhang, B.; Liu, H.; Wang, C. On the detonation propagation behavior in hydrogen-oxygen mixture under the effect of spiral obstacles. *Int. J. Hydrogen Energy* **2017**, *42*, 21392–21402. [CrossRef]
30. Zheng, K.; Jia, Q.H.; Ma, Z.M.; Xing, Z.X.; Hao, Y.M.; Yu, M.G. Experimental and numerical investigation on the premixed methane/air flame propagation in duct with obstacle gradients. *Process Saf. Environ. Protect.* **2023**, *178*, 893–904. [CrossRef]
31. Heilbronn, D.; Barfuss, C.; Sattelmayer, T. Influence of geometry on flame acceleration and DDT in H₂-CO-air mixtures in a partially obstructed channel. *J. Loss Prev. Process Ind.* **2021**, *71*, 104493. [CrossRef]
32. Goodwin, G.B.; Houim, R.W.; Oran, E.S. Effect of decreasing blockage ratio on DDT in small channels with obstacles. *Combust. Flame* **2016**, *173*, 16–26. [CrossRef]
33. Ciccarelli, G.; Fowler, C.J.; Bardon, M. Effect of obstacle size and spacing on the initial stage of flame acceleration in a rough tube. *Shock Waves* **2005**, *14*, 161–166. [CrossRef]
34. Saeid, M.H.S.; Khadem, J.; Emami, S. Numerical investigation of the mechanism behind the deflagration to detonation transition in homogeneous and inhomogeneous mixtures of H2-air in an obstructed channel. *Int. J. Hydrogen Energy* **2021**, *46*, 21657–21671. [CrossRef]
35. Wang, J.B.; Zhao, X.Y.; Fan, L.Y.; Pan, J.F.; Zhu, Y.J. Effects of the quantity and arrangement of reactive jet obstacles on flame acceleration and transition to detonation: A numerical study. *Aerosp. Sci. Technol.* **2023**, *137*, 108269. [CrossRef]
36. Coates, A.M.; Mathias, D.L.; Cantwell, B.J. Numerical investigation of the effect of obstacle shape on deflagration to detonation transition in a hydrogen-air mixture. *Combust. Flame* **2019**, *209*, 278–290. [CrossRef]
37. Zhou, S.B.; Liu, F.; Ning, H.M.; Hu, N. Experimental investigation on a rotating detonation combustor with the pulse operating frequency of 10 Hz. *Acta Astronaut.* **2024**, *215*, 642–652. [CrossRef]
38. Gray, S.; McLoughlin, M.; Ciccarelli, G. Fuel-oxygen mixing and detonation propagation in a linear rotating detonation rocket engine geometry. *Combust. Flame* **2024**, *260*, 113250. [CrossRef]
39. Luo, S.B.; Sun, Y.H.; Song, J.W.; Liu, J. Performance analysis of a hybrid pulse detonation engine using liquid hydrogen as fuel. *Int. J. Hydrogen Energy* **2022**, *47*, 21537–21551. [CrossRef]
40. Feng, W.K.; Zheng, Q.; Xiao, Q.; Meng, H.L.; Han, X.P.; Cao, Q.; Huang, H.L.; Wu, B.W.; Xu, H.; Weng, C.S. Effects of cavity length on operating characteristics of a ramjet rotating detonation engine fueled by liquid kerosene. *Fuel* **2023**, *332*, 126129. [CrossRef]
41. Xu, S.D.; Song, F.L.; Wu, Y.; Zhou, J.P.; Cheng, P.; Yang, X.K.; Chen, X. Experimental investigation on combustion efficiency of a partially premixed kerosene-air rotating detonation combustor. *Fuel* **2022**, *329*, 125418. [CrossRef]
42. Li, X.F.; Li, J.Z.; Qin, Q.Y.; Jin, W.; Yuan, L. Experimental study on detonation characteristics of liquid kerosene/air rotating detonation engine. *Acta Astronaut.* **2024**, *215*, 124–134. [CrossRef]
43. Tunik, Y.V.; Gerasimov, G.Y.; Levashov, V.Y.; Mayorov, V.O. Busemann diffuser for supersonic ramjet on detonation combustion of kerosene vapor. *Acta Astronaut.* **2022**, *198*, 495–501. [CrossRef]
44. Sato, T.; Chacon, F.; Gamba, M.; Raman, V. Mass flow rate effect on a rotating detonation combustor with an axial air injection. *Shock Waves* **2021**, *31*, 741–751. [CrossRef]
45. Zhao, W.D.; Deiterding, R.; Liang, J.H.; Wang, X.X.; Cai, X.D.; Duell, J. Adaptive simulations of flame acceleration and detonation transition in subsonic and supersonic mixtures. *Aerosp. Sci. Technol.* **2023**, *136*, 108205. [CrossRef]
46. Warimani, M.; Azami, M.H.; Khan, S.A.; Ismail, A.F.; Saharin, S.; Ariffin, A.K. Internal flow dynamics and performance of pulse detonation engine with alternative fuels. *Energy* **2021**, *237*, 121719. [CrossRef]
47. Jiang, C.; Pan, J.; Li, J.; Shi, X.; Quaye, E.K. Numerical simulation of detonation re-initiation in a 90° bifurcated channel filled with n-heptane/air mixture. *Acta Astronaut.* **2023**, *202*, 497–510. [CrossRef]
48. Wang, Z.W.; Wang, Y.Q.; Peng, C.X.; Zheng, L.X. Experimental study of pressure back-propagation in a valveless air-breathing pulse detonation engine. *Appl. Therm. Eng.* **2017**, *110*, 62–69. [CrossRef]
49. Qiu, H.; Su, Z.; Xiong, C. Experimental investigation on multi-cycle two-phase spiral pulse detonation tube of two configurations. *Proc. Inst. Mech. Eng. Part G-J. Aerosp. Eng.* **2019**, *233*, 4166–4175. [CrossRef]

energies

MDPI

Article

The Influence of Injector Nozzle Diameter on High-Density and Lean Mixture Combustion in Heavy-Duty Diesel Engines

Yize Liu * and Wanhua Su

State Key Laboratory of Engines, Tianjin University, Tianjin 300072, China
* Correspondence: liuyize@tju.edu.cn

Abstract: In order to improve the fuel economy of heavy-duty diesel engines under high-load conditions, based on the combustion pathway model, it is proposed that the proportion of lean mixture with $0 < \Phi < 1$ is the most important spray characteristic affecting the overall diesel combustion process. Answering the question of how to increase the proportion of lean mixture inside the spray is the key to achieving the efficient and clean combustion of diesel engines. This paper investigated the mechanism of injector nozzle diameter on the in-cylinder air–fuel mixture and combustion process based on a high-density and lean mixture characteristic combustion strategy. The experimental results show that with an increase in nozzle diameter, the peak pressure and instantaneous heat release rate significantly increase, the combustion duration is shortened by about 20%, and the heat release becomes more concentrated. At 1200 rpm and IMEPg~2.3 MPa conditions, the indicated thermal efficiency increases by 1.3%, reaching a maximum of 51.5%. The numerical simulation results show that with the increase in nozzle diameter from 0.169 mm to 0.218 mm, the spray ejection momentum per unit time increases by 30%, the momentum transferred to the air by the spray increases, the oxygen transport process becomes more intense, and the air entrainment mass during the spray free development stage increases by 42%. The proportion of lean mixture inside the spray throughout the entire spray development process increases, resulting in an increase in the heat release rate of the lean mixture, making the overall combustion more intense and concentrated.

Keywords: heavy-duty diesel engine; injector nozzle diameter; lean mixture; spray characteristic; air entrainment; fuel–air mixing; thermal efficiency

Citation: Liu, Y.; Su, W. The Influence of Injector Nozzle Diameter on High-Density and Lean Mixture Combustion in Heavy-Duty Diesel Engines. *Energies* **2024**, *17*, 2549. https://doi.org/10.3390/en17112549

Academic Editor: Dimitrios C. Rakopoulos

Received: 24 April 2024
Revised: 17 May 2024
Accepted: 22 May 2024
Published: 24 May 2024

1. Introduction

In order to address the environmental issues brought about by global climate change, the Paris Agreement aims to reduce greenhouse gas emissions to near zero by the end of this century [1]. Road transport accounts for 72.06% of greenhouse gas emissions from the transport sector as a whole, and the diesel engine is still the most widely used means of road transportation [2,3]. The intensification of energy consumption and the urgent need for environmental protection have prompted the upgrading of traditional diesel engine technology to meet the requirements of efficient and clean combustion.

In order to further improve the thermal efficiency of diesel engines, extensive research has been carried out on combustion modes and various innovations have been made. In terms of combustion theory, the pursuit of in-cylinder homogeneous charge combustion became a key research direction for internal combustion engines following the proposal of the homogeneous charge compression ignition combustion mode [4]. However, this combustion mode still had some difficulties, such as high pressure rise rate and limited operating load [5]. Under heavy-load conditions, diesel engines predominantly use diffusion combustion. The large cycle fuel injection mass and slow heat release rate limit the improvement of thermal efficiency.

Researchers have made significant progress in the development of high mixing rate combustion technology for diesel engines [6,7]. The parameters of the fuel injection system determine the initial fuel injection momentum and the subsequent fragmentation and entrainment process. Zhai [8] et al. investigated the effect of injector nozzle diameter on injection characteristics under different injection pressure conditions. As the injection pressure increases and the injector nozzle diameter decreases, the time for the liquid length to reach a stable stage is shortened, and strong oxidation reactions are simultaneously distributed in both the upstream and downstream regions of the flame. Compared to using a micro-hole diameter (D = 0.07 mm, Pinj = 100 MPa), increasing the injection pressure to 300 MPa (D = 0.133 mm, Pinj = 300 MPa) can more effectively reduce the soot produced per unit of fuel mass and improve the thermal efficiency of diesel engines. The increase in fuel injection pressure enhances the speed of radial air entrainment at the initial stage of fuel injection, which increases the spray angle and significantly improves the fuel–air mixing rate [9,10]. Shi [11] et al. studied the effect of injection pressure on the combustion characteristics of diesel engine low-temperature wall-impinging ignition, finding that increasing the injection pressure is beneficial for optimizing high-temperature combustion but may lead to unstable low-temperature ignition. With an increase in the injection pressure from 40 MPa to 100 MPa, the ignition delay at 840 K ambient temperature initially shortened and then prolonged, with the combustion condition being best at 80 MPa. Reducing the fuel injection mass is detrimental to low-temperature ignition, as the ignition transitions from stable to unstable, and then to misfire, as the injection mass decreases from 35–50 mg to 20–30 mg, and further to 15 mg. Comprehensive analysis reveals that the minimum fuel mass required for stable wall-impinging ignition approximately increases linearly with injection pressure increasing, thus reducing injection pressure and increasing fuel mass help to improve combustion quality and enhance environmental adaptability. Xia [12] et al. conducted experimental analysis on the spray characteristics under different nozzle diameters (100, 300, 350 μm) and injection pressures (60, 100, 140, 180 MPa) within a constant volume bomb to elucidate the effects of key factors such as nozzle diameter, injection pressure, and thermodynamic boundary conditions on fuel spray characteristics and critical characteristics. The 350 μm nozzle, due to its higher injection flow rate, results in a significantly larger spray volume compared to other nozzles, while the spray generated by the 100 μm nozzle is easily affected by the ambient air entrainment movement, thereby enhancing the radial expansion process of the spray. The effect of injection pressure on the liquid-phase fuel length, volume, and spray cone angle is not significant, but it significantly affects the fuel–air mixture process. Even under supercritical conditions, increasing the injection pressure still significantly improves the mixing quality in the downstream region of the spray. In summary, the influence of the injector nozzle diameter on in-cylinder fuel–air mixture and combustion process is not clear, and a universal rule has not yet been summarized. The optimization scheme of injector nozzle diameter needs further improvement. Addressing the above issues, this paper adopts a combined approach of bench experiments and numerical simulation to investigate the mechanism of the injector nozzle diameter on the in-cylinder fuel–air mixture and combustion process, based on high-density and lean mixture characteristic combustion strategy, aiming to provide guidance and suggestions for the selection of injector nozzle diameter in the development of diesel engine combustion systems.

2. Spray Characteristics of High-Efficiency Clean Combustion in Diesel Engines

Traditional diesel engine combustion is a highly complex physicochemical process, where liquid diesel undergoes processes such as droplet formation, collision, breakup, atomization, evaporation diffusion, air entrainment, and mixing after leaving the injector before ignition and combustion can occur. John Dec et al. used the two-dimensional laser Rayleigh scattering (LRS) technique to quantitatively study the concentration fields of gas and liquid phase sprays and the generation of soot in an optical engine, and based on quantitative experimental results, proposed a modern diesel engine combustion process model [13] (Figure 1). After the fuel jet leaves the nozzle, it rapidly atomizes as the spray progresses forward, and high-temperature ambient gas is entrained into the spray, causing the temperature rise and phase change in the liquid-phase fuel, leading to chemical reactions. A high equivalence ratio premixed flame (Φ~4, T = 825 K) appears at the head of the spray, generating heat to heat up the fuel–air mixture to about 1600 K, contributing to approximately 10–15% of the total heat release. In the mid–late stage of spray development, a high-temperature diffusion flame envelope of 2700 K appears around the entire fuel jet, and combustion products rich in fuel from the spray interior such as CO, UHC, PM are continuously transported to this envelope, where they undergo diffusion combustion (including the generation of CO_2 and H_2O, NOx generation, and oxidation of soot). It can be seen that traditional diesel engine combustion is a predictable combustion, closely related to fuel injection, which is easier to control and more suitable for high-load conditions. However, the inherent nature of heterogeneous combustion inevitably results in NOx and soot emissions. Under the flame lift height downstream, the diesel spray forms a premixed reaction zone with an equivalence ratio close to four, and the premixed combustion products will become the "fuel" for the downstream diffusion combustion process of the spray, including the precursor of soot generation. In other words, the distribution characteristics of equivalence ratio inside the spray at the moment of ignition will affect the total amount of harmful emissions such as soot and the heat release rate of the diffusion combustion process.

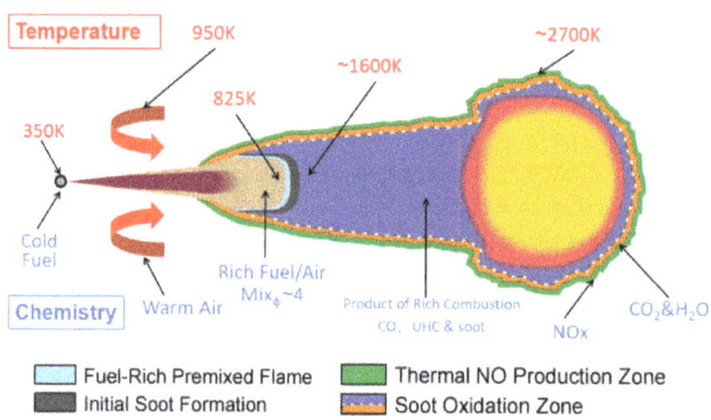

Figure 1. The combustion and emission generation model of traditional diesel engine.

Pickett and Siebers from Sandia National Laboratories in the United States first proposed the concept of leaner lifted flame combustion (LLFC) [14]. By enhancing the air entrainment process within the flame lift height, the equivalence ratio of the mixture in the premixed reaction zone is kept below or equal to 2 (Figure 2), achieving a completely smokeless and relatively low-temperature combustion strategy. The key to achieving the LLFC strategy is to increase the flame lift height and strengthen the spray air entrainment capability. Pickett et al. [14,15] conducted experimental studies in optical constant volume chambers and found that diesel LLFC combustion could be achieved when the ambient temperature and pressure were below specific values. Additionally, reducing the injector nozzle diameter or using oxygen-containing fuels can broaden the operating range of LLFC, as shown in Figure 3. Polonowski et al. [16] found that achieving a continuous smokeless LLFC mode in actual engines is challenging. It requires injection pressures as high as 240 MPa, moderate exhaust gas recirculation rates, and lower intake air temperatures to achieve completely smokeless LLFC with a two-hole injector equipped diesel engine. For injectors with six or more holes, the secondary entrainment of combustion products and the interactions between sprays, making LLFC combustion unsustainable. It is worth noting that the above experimental results were obtained under moderate-load conditions. Achieving LLFC under high-load conditions requires the use of oxygen-containing fuels [17,18], as shown in Figure 4.

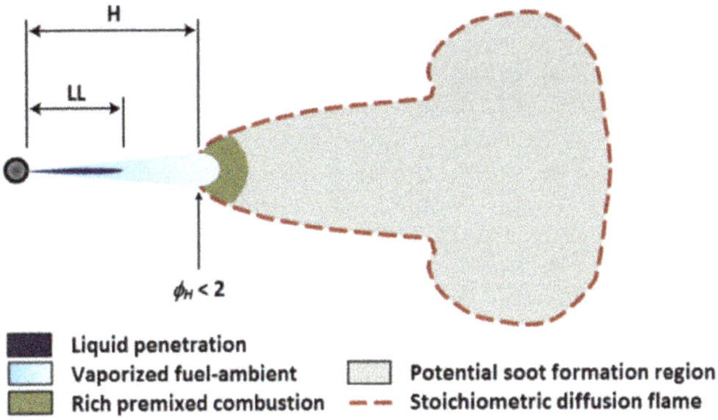

Figure 2. Conceptual model of LLFC combustion strategy.

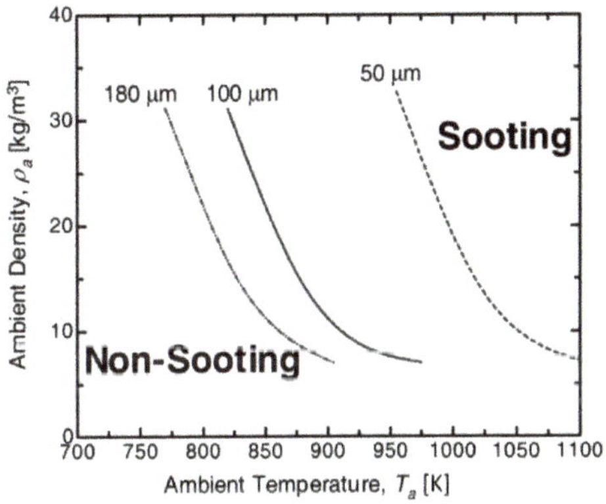

Figure 3. Relationship between soot formation status and injector nozzle diameter in constant volume combustion chamber.

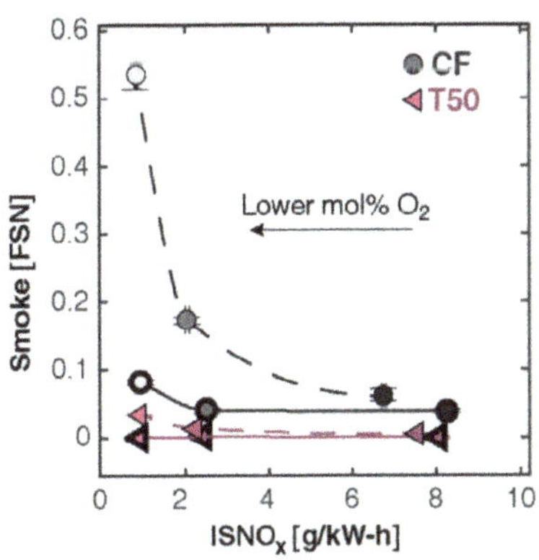

Figure 4. Influence of oxygen-containing fuels on soot and NOx emissions.

In the combustion process of diesel engines, the final amount of soot generated is the result of the combined effects of its formation and oxidation processes. Pursuing combustion processes with zero soot generation is not the only way to address soot emissions. However, in many LLFC test results, excessively low intake air temperatures and oxygen concentrations are used to extend flame lift height, affecting the overall fuel–air mixing process and leading to prolonged combustion duration, sharp increases in HC and CO emissions levels, and significant reductions in combustion efficiency and engine thermal efficiency. Therefore, simply pursuing combustion modes with flame lift height positions where the mixture equivalence ratio is below two does not meet the requirements of current high-efficiency, clean, low-carbon diesel engines. The inherent uneven mixing of diesel spray leads to a complex distribution of equivalence ratios inside the spray. To determine the relevant reaction pathways, the mixture is divided into different regions based on local fuel–oxygen equivalence ratios: $0 < \Phi < 1$, $1 < \Phi < 2$, and $\Phi > 2$. Different mixtures in different equivalence ratio regions follow specific reaction pathways, as shown in Figure 5 In high-temperature environments above 1400 K, lean mixtures with $0 < \Phi < 1$ directly oxidize to CO_2, releasing a large amount of heat instantly. Rich mixtures with $1 < \Phi < 2$ first generate intermediate product CO, which then reacts with oxygen to form CO_2, with the rate of heat release controlled by the fuel–air mixing rate. Overly rich mixtures with $\Phi > 2$ exceed the equivalence ratio boundary for soot generation. Therefore, lean mixtures with $0 < \Phi < 1$ can complete the combustion heat release process in the shortest time scale, while mixtures with $\Phi > 1$ inevitably undergo the mixing processes, with the mixing time scale far exceeding the chemical reaction time scale, resulting in combustion rates limited by the mixing rate.

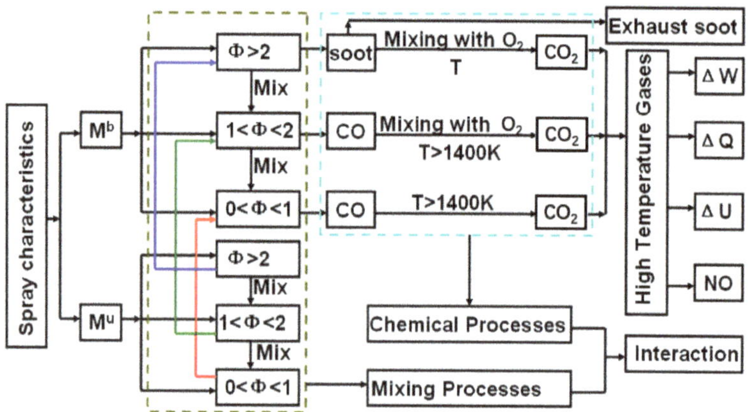

Figure 5. The schematic diagram of spray combustion path of heavy-duty diesel engine.

Diesel spray will form a premixed reaction zone downstream of the flame lift height, where the combustion products in this zone are mainly composed of over-rich mixtures with $\Phi \geq 2$, serving as "fuel" for the downstream diffusion combustion process of the spray. From the analysis of the combustion path, achieving complete combustion of the rich mixture in the premixed reaction zone unavoidably requires a relatively long time scale for the mixing process. It is only through the incomplete combustion product and unburned fuel transport during the fuel–air mixing process that contact with oxygen can achieve complete combustion. At this point, the combustion becomes a mixed-dominated, staged process with a large amount of intermediate product generation. Introducing the Φ–T diagram of CO generation [19] to characterize the combustion efficiency of the diffusion combustion process, it can be seen from Figure 6 that the combustion path curves of diffusion combustion differ significantly at different mixing rates. The smaller the mixing time, that is, the faster the mixing rate, and the further the combustion path is from the CO generation area, indicating that more fuel develops along the lean mixture path throughout the combustion process. At the time scale level that cannot be characterized by the Φ–T diagram, the mixing rate largely affects the duration of diffusion combustion, thereby influencing combustion phase and piston work capacity.

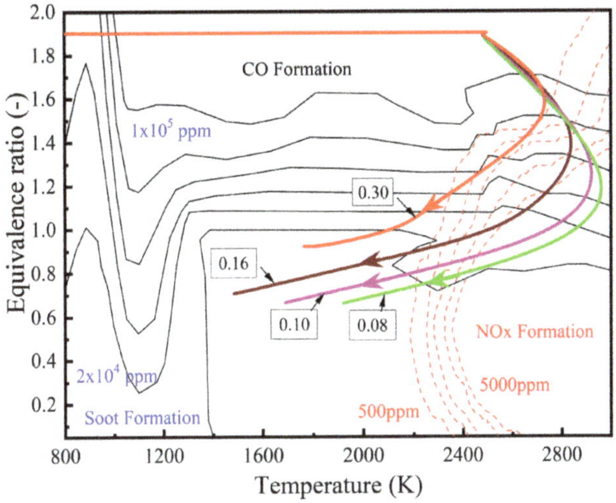

Figure 6. Influence of different mixing rates on the diesel diffusion combustion process.

In summary, the proportion of lean mixtures with $0 < \Phi < 1$ is the most important spray characteristic affecting the overall diesel combustion process. Increasing the proportion of lean mixtures with $0 < \Phi < 1$ inside the spray can enhance the heat release during the premixed combustion stage, and on the other hand, significantly reduce the dependence of the diffusion combustion stage on the mixing rate, making the overall heat release process more concentrated. Ultra-high proportions of lean mixtures are an ideal spray characteristic for efficient and clean combustion. Therefore, increasing the proportion of lean mixtures throughout the ignition timing and the entire spray development process is the key to achieving high-efficiency clean diesel combustion.

3. Test Platform and 3-D Simulation Model

3.1. Single-Cylinder Engine Test Platform

In this study, experiments were conducted on a single-cylinder engine test platform. The original engine parameters are listed in Table 1, and the schematic diagram of a single-cylinder engine test platform is shown in Figure 7. During the experiment, the first cylinder of the original engine was used as the test cylinder. The intake system of the test cylinder used external pressurization combined with intake heating and an independent exhaust gas recirculation system, allowing flexible control of intake pressure, temperature, and EGR rate. In this study, three different injector nozzle diameters were used, with nozzle diameters of 0.169/0.203/0.218 mm. Table 2 shows the parameter settings for the experimental conditions. Table 3 provides specific information on the measurement instruments and their error rates. Single-cylinder exhaust components were detected using a Horiba 7100 exhaust gas analyzer made in Japan, and soot emissions were measured with an AVL 415 smoke meter made in Austria. Cylinder pressure signals were collected using a 6125C cylinder pressure sensor and 5165A charge amplifier by the Kistler company made in Switzerland. The uncertainty of the single-cylinder engine test platform was calculated as follows: Total uncertainty = Square root of {(uncertainty of soot)^2 + (uncertainty of load)^2 + (uncertainty of speed)^2 + (uncertainty of temperature)^2 + (uncertainty of air flow)^2 + (uncertainty of diesel measurement)^2 + (uncertainty of pressure acquisition)^2 + (uncertainty of angle encoder)^2}, and after calculation, the total uncertainty was confirmed to be 1.47%. In addition, cylinder pressure data were measured every 0.5 crank angle degree and the average was taken from 100 consecutive engine cycles. The gaseous emissions and the particulate mass emission were continuously measured for 5 min and the average results were calculated. The steady-state tests were repeated twice to ensure that the results were repeatable, within the experimental uncertainties of the measurements.

Table 1. Main technical parameters of heavy-duty diesel engine.

Parameter	Value
Cylinder Diameter/mm	116
Stroke/mm	150
Displacement/L	9.5
Compression Ratio	18.5
Intake Swirl Ratio	1.1
Combustion Chamber Shape	Stepped-lip type
Intake Mode	Turbocharged and Intercooled
Rated Power (kW)	294 (at 1900 r/min)
Maximum Torque (N·m)	1800 (1000–1400 r/min)
Maximum Cylinder Pressure (MPa)	24

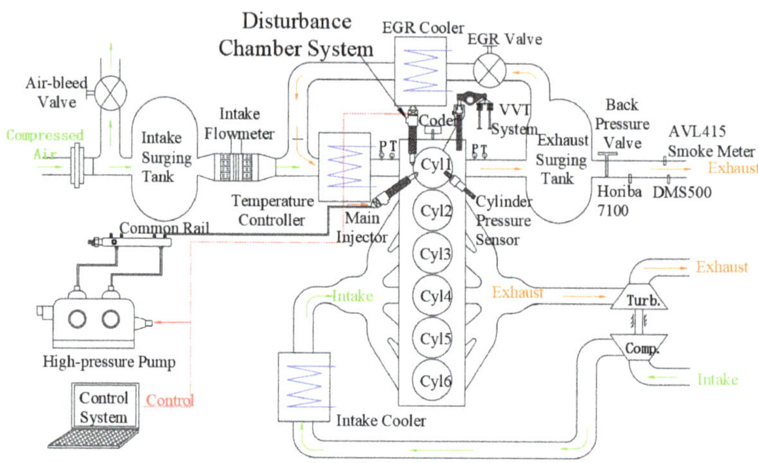

Figure 7. Schematic diagram of single-cylinder engine test platform.

Table 2. Experimental operating conditions parameter.

Parameter	Value	Note
Speed	1200 r/min	Constant
Total Cycle Fuel Mass	180 mg	90% load, constant
Common Rail Pressure	180 MPa	Constant
Fuel Injection Timing	−2~8 deg. ATDC	Variable
Nozzle Diameter	0.169 mm; 0.203 mm; 0.218 mm	Variable
Intake Valve Closing Timing	−100 deg. ATDC (Using RIVCT structure, original engine −150 deg. ATDC)	Constant
Intake Pressure	2.7~4.15 bar	Variable
Intake Temperature	330 ± 3 K	Constant

Table 3. Measuring instruments and error rates.

	Instruments Type	Error
Cylinder pressure sensor	KISTLER 6125C 0–30 MPa	±0.01%F·S
Charge amplifier	5165A	±0.05%F·S
Emission collection equipment	Horiba 7100	±0.1%F·S
Smoke detector	AVL 415	±0.05%F·S
Temperature sensor	PT100 0–700 K	±0.1%F·S
Pressure sensor	0–1 MPa	±0.25%F·S

3.2. Injector Hydraulic System Simulation Model

To simplify the simulation model of the injector hydraulic system, the following basic assumptions are proposed [20]: (1) the rail pressure remains uniformly constant during the injection process; (2) the fuel temperature remains constant during the injection process; (3) the viscosity, density, and elastic modulus of the fuel are considered constant; (4) elastic deformation of the components in the system is not considered; (5) leakage caused by machining issues in flat and conical seals and its effect on chamber pressures are not considered. The following equations form the mathematical model describing the dynamic process of the hydraulic system [21]:

(1) The internal pipes of the injector are treated as concentrated volumes, and fuel flow follows the law of mass conservation

$$Q_{gi} = Q_{go} + \frac{V_g}{E} * \frac{dp_g}{dt}, \tag{1}$$

where Q_{gi} is the flow into the pipe, Q_{go} is the flow out of the pipe, V_g is the pipe volume, E is the elastic modulus of the fuel, p_g is the fuel pressure inside the pipe, and t is time.

(2) The influence of the shape of the control chamber and pressure-bearing groove on the operation of the injector can be neglected, and the volume is the main influencing factor. It can be expressed as a volume chamber model, and fuel flow follows the law of mass conservation

$$\sum Q_i = \frac{V_i}{E} * \frac{dp_i}{dt} + \frac{dV_i}{dt}, \tag{2}$$

where Q_i is the change in flow rate in the i volume chamber, V_i is the volume of the i volume chamber, p_i is the pressure inside the i volume chamber, and t is time.

(3) Treating the control rod and needle valve as a whole, according to Newton's second law, the motion equation of the needle valve is

$$m\frac{^2x}{t} = F_C + F_y + F_z - F_k - k(xo + x), \tag{3}$$

where x is the lift of the needle valve, m is the total mass of the control rod and needle valve, F_c is the force of the fuel on the needle valve at the pressure-bearing groove, F_y is the force of the fuel on the needle valve at the pressure chamber, F_z is the supporting force at the needle valve seat, F_k is the force of the control chamber on the control rod, k is the stiffness of the needle valve spring, and x_0 is the preload of the needle valve spring.

Based on the above mathematical model, a simulation model of the injector hydraulic system is built in the AMESim 2020 simulation platform, as shown in Figure 8. To verify the accuracy of the simulation model, the actual fuel injection rate is measured using the EFS high-pressure common rail test bench. The test was carried out under the conditions of 180 MPa common rail pressure and 1.5 ms pulse width. After filtering and shaping, the measured fuel injection rate matches well with the simulated fuel injection rate under the same conditions, as shown in Figure 9. This indicates that the simulation results can reflect the actual fuel injection pattern, and the simulation model of the high-pressure common rail injector hydraulic system is reliable.

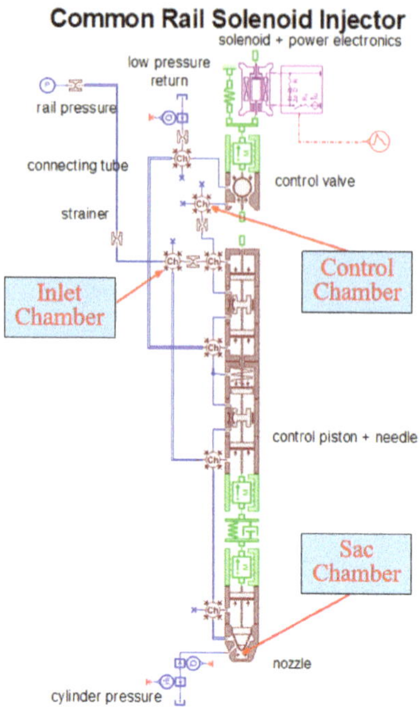

Figure 8. Simulation model of high-pressure common rail injector hydraulic system.

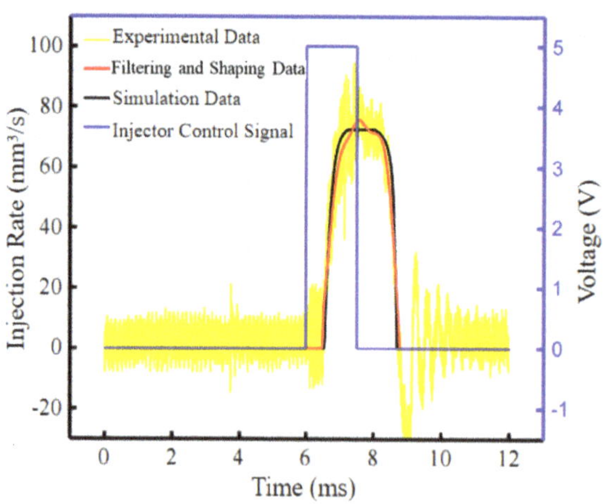

Figure 9. Comparison of fuel injection rate experimental results and simulation results.

3.3. 3-D Combustion Simulation Model

Based on the geometric parameters of the combustion chamber, modeling will be carried out using Solidworks 2023. Since the original engine combustion chamber is a rotationally symmetric regular revolving body, and the eight injector nozzles are evenly distributed circumferentially, a one-eighth combustion chamber will be used for simulation to improve computational speed. Three-dimensional simulation models were established based on the computational fluid dynamics software CONVERGE 2.3 (Figure 10), with a base grid size of 4.0 mm. The important mathematical models used in the simulation calculations are shown in Table 4.

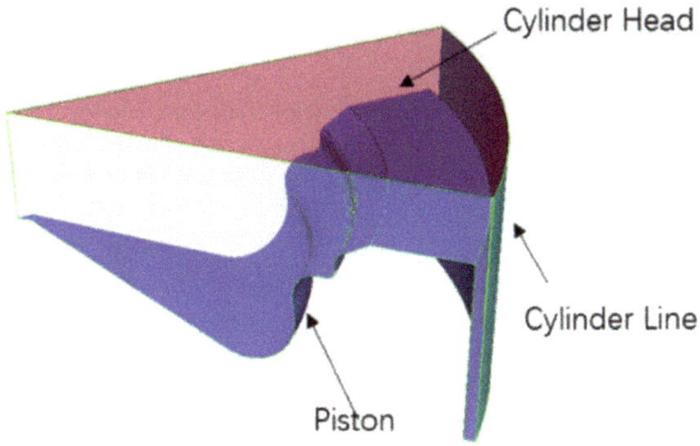

Figure 10. Geometric model of the combustion chamber.

Table 4. Models used in simulation calculations.

Parameter	Value
Turbulence Model	RNG k-ε
Combustion Model	SAGE Chemical Reaction Solver + Simplified Chemical Reaction Kinetics Mechanism
Heat Transfer Model	Han and Reitz
Spray Breakup Model	KH-RT
Spray Wall Impact Model	O'Rourke and Amsden
Droplet Collision Model	NTC collision
Spray Evaporation Model	Frossling
Base Grid Size/mm	1

Based on the obtained experimental data, the simulation model was calibrated and validated. The calibration condition had an intake pressure of 4.15 bar and an injection timing of −2 deg. ATDC, with other test conditions referenced from Table 2. The simulation results for injectors with different nozzle diameters were compared with actual experimental results, as shown in Figure 11. Comparing the simulation results with the experimental results, the peak cylinder pressure error was less than 0.1%, the crank angle error corresponding to the peak cylinder pressure was less than 0.5 deg, and the peak heat release rate error was less than 2%. It can be considered that the simulated data closely matches the experimental data, and the established simulation model can be used to accurately describe the combustion process of the diesel engine.

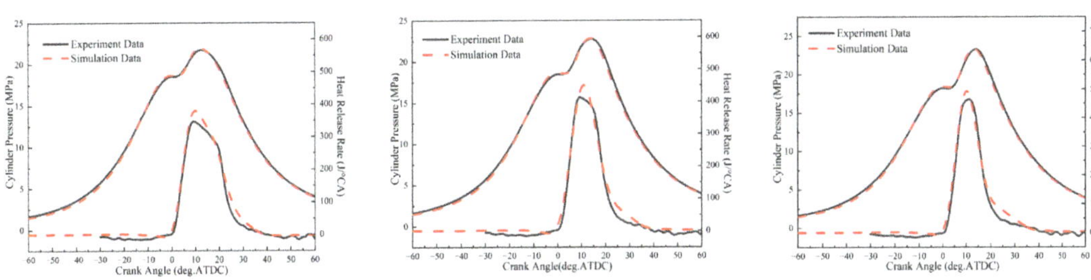

(a) (b) (c)

Figure 11. Comparison of cylinder pressure and heat release rate between simulation and experiment: (**a**) 0.169 mm; (**b**) 0.203 mm; (**c**) 0.218 mm.

4. Results and Discussion

4.1. Experimental Study on the Influence of Injector Nozzle Diameter on Combustion Process

Experimental research on the influence of the injector nozzle diameter on combustion and emissions was conducted based on the aforementioned test platform. Figure 12 presents experimental data including cylinder pressure, instantaneous heat release rate, and average in-cylinder temperature for three injector nozzle diameters at an intake pressure of 4.15 bar and a rail pressure of 180 MPa. The injection strategy was single injection, and the injection start time was −2 deg. ATDC. As the injector nozzle diameter increases from 0.169 mm to 0.218 mm, the maximum cylinder pressure increases, and the combustion heat release process becomes more intense and concentrated. The peak heat release rate increases by 33.9%, and the combustion duration shortens from 26.3 deg to 20.8 deg. Additionally, both the combustion center CA50 and the combustion end CA90 move closer to the top dead center. Furthermore, observation of the average in-cylinder temperature reveals that increasing the injector nozzle diameter leads to a greater temperature rise during the combustion process, with the timing of the temperature peak significantly advancing.

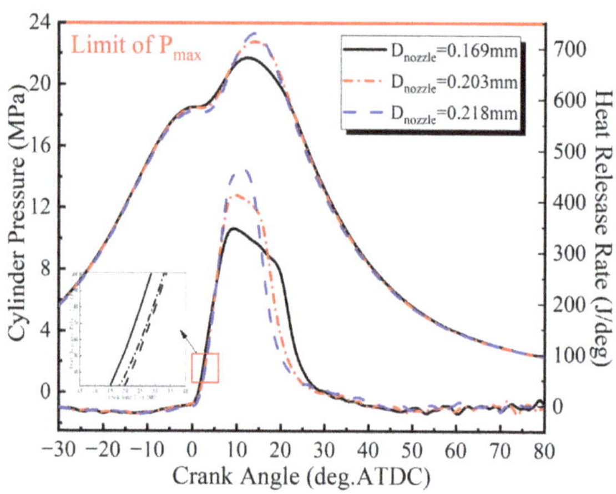

(a)

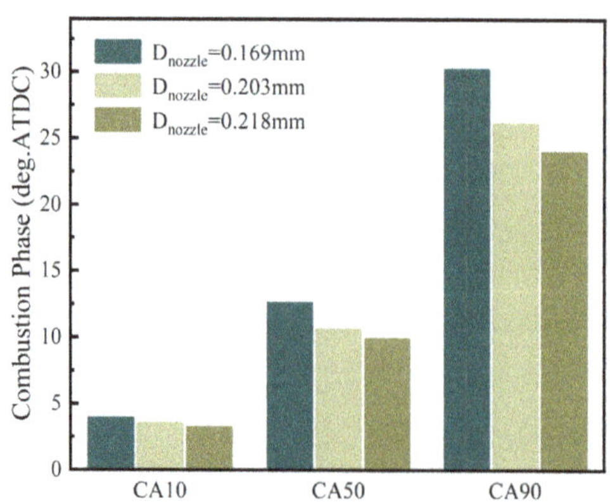

(b)

Figure 12. *Cont.*

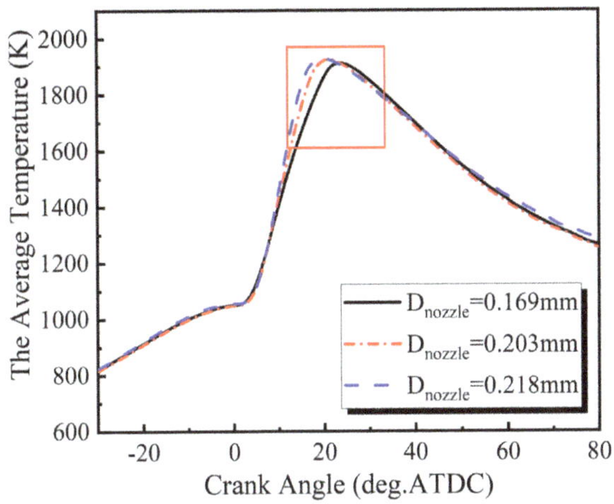

(c)

Figure 12. Combustion characteristic parameters corresponding to three injector nozzle diameters: (**a**) cylinder pressure and heat release rate; (**b**) combustion phase; and (**c**) average temperature.

Figure 13 illustrates the impact of injector nozzle diameter on major harmful emissions. Experimental results indicate that there is little difference in the emissions of CO, soot, and UHC for different nozzle diameters. The original emissions of CO and soot from the engine are much lower than the emission limits corresponding to the WHSC test cycle in the Euro VI emission standard. The original emission of UHC is close to the aforementioned emission limit. When the nozzle diameter increases from 0.169 mm to 0.218 mm, the NOx emissions exhibit a trend of first increasing and then decreasing. This is because increasing the nozzle diameter leads to a greater temperature rise in the cylinder during diesel combustion, resulting in an increase in NOx production. However, when the nozzle diameter increases to 0.218 mm, the overall combustion duration shortens by 20%, and the duration of high-temperature NOx generation elements also decreases accordingly, leading to a reduction in NOx emissions.

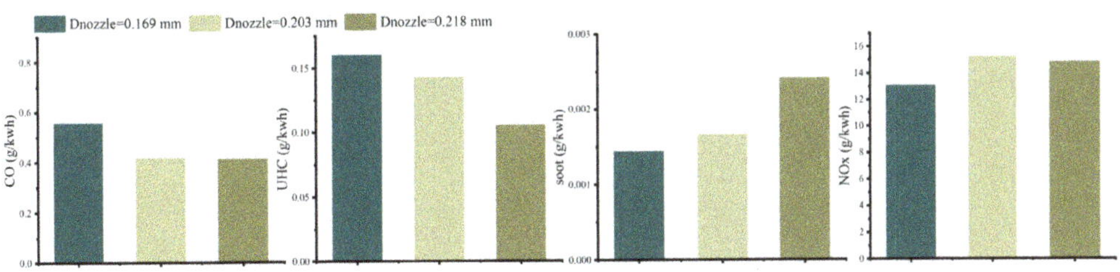

Figure 13. Influence of injector nozzle diameter on major harmful emissions.

Figure 14 shows the influence of the injector nozzle diameter on energy loss distribution. As the nozzle diameter increases, the heat transfer loss initially increases and then decreases. This is mainly due to the increase in the nozzle diameter, which significantly increases the cylinder temperature and average pressure during diesel combustion, affecting the temperature difference and heat transfer coefficient in the wall heat transfer process, resulting in an increase in wall heat transfer rate. However, on the other hand, the 20% reduction in combustion duration reduces the duration of high-temperature in-cylinder, so under the interplay of these two factors, heat transfer loss exhibits a trend of initially increasing and then decreasing with the increase in nozzle diameter. The increase in nozzle diameter increases the proportion of constant volume combustion, allowing for a more thorough expansion of the working fluid and reducing exhaust temperature. Consequently, exhaust loss decreases monotonically with the increase in nozzle diameter. Benefiting from the simultaneous reduction in heat transfer loss and exhaust loss, indicated thermal efficiency increases with the increase in nozzle diameter. Under the condition of 1200 rpm and IMEPg of 2.3 MPa, using a nozzle diameter of 0.218 mm, the indicated thermal efficiency reaches as high as 51.5%.

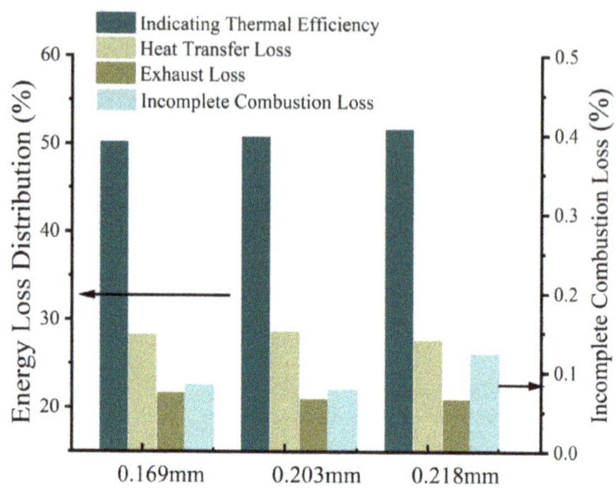

Figure 14. Influence of injector nozzle diameter on energy loss distribution.

Based on the comprehensive analysis of the test results, under the thermodynamic boundary conditions in this study, as the nozzle diameter increases from 0.169 mm to 0.218 mm, the combustion process of diesel becomes more rapid, with a 20% reduction in the duration of combustion. Furthermore, the combustion is sustained for a shorter period, and the proportion of constant volume combustion increases. Indicated thermal efficiency at high-load conditions also rises from 49.8% to 51.5%. While the experimental results provide insights at a macroscopic level into the impact of nozzle diameter on heat release rate, thermal efficiency, and emissions, a microscopic understanding of the variation in nozzle diameter on in-cylinder fuel–air mixture and combustion processes is lacking. The following sections will utilize numerical simulation methods to analyze in-cylinder spray air entrainment motion and the distribution of mixed gas within the spray, aiming to elucidate the influence mechanism of nozzle diameter on the combustion process from the perspective of fuel–air mixture.

4.2. Influence of Nozzle Diameter on Spray Internal Mixing Quality

Changes in nozzle diameter affect the pressure fluctuation characteristics inside the injector, but the range of pressure fluctuation is mainly influenced by the effective injection pressure. Therefore, it is necessary to focus on the impact of nozzle diameter on the pressure loss characteristics inside the injector. Three injectors used in the experiments were simulated in the AMESim model to analyze the effects of nozzle diameter on effective injection pressure and actual fuel injection rate. It should be noted that the three injectors used in this study are all conical hole nozzles, with a nozzle coefficient K of 0.5. The coefficient K is defined as $(D_{inlet} - D_{outlet})/10$. Previous studies [22] have shown that when $K > 0$, it indicates that the nozzle along the flow direction is contracting, which will significantly inhibit the geometrically induced cavitation process inside the nozzle. Therefore, this study does not consider the influence of cavitation process inside the nozzle.

Friction in the internal pipeline, pipeline diameter switching, connection points, and the opening position of the needle valve of the fuel injection system all contribute to fuel pressure loss, ultimately leading to an actual fuel injection pressure much lower than the pressure inside the common rail. Figure 15a shows the pressure distribution and the form of pressure loss at various positions inside the injector when the needle valve is fully open. Along-the-way loss refers to the energy loss caused by fuel flow friction with the pipe wall in a straight pipe, mainly occurring in the external high-pressure fuel pipe and the internal high-pressure fuel passage. There are also local pressure losses inside the injector, including sudden expansion and sudden contraction of the pipeline, especially at the needle valve. Changes in the nozzle diameter affect the local resistance coefficient between the sac chamber and the nozzle, but the greater influence on effective injection pressure is its effect on the flow rate of the entire high-pressure fuel passage. The flow velocity of fuel in the entire fuel injection system increases, and both the along-the-way loss and local loss in the internal fuel injection system significantly increase, leading to a significant decrease in effective injection pressure, as shown in Figure 15. For the three nozzle diameter injectors, under the common rail pressure of 180 MPa and the cycle fuel mass quantity of 180 mg, the average injection pressure within a single injection cycle is 131, 105, and 94 MPa (0.169 mm/0.203 mm/0.218 mm), respectively.

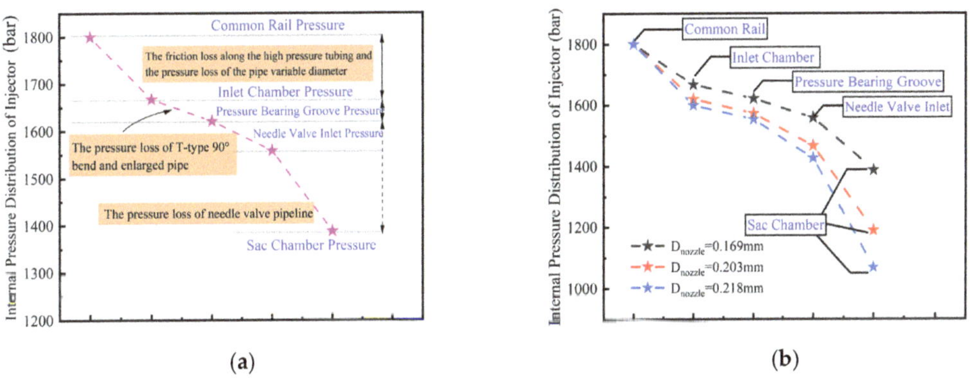

Figure 15. Pressure loss characteristics inside the injector: (**a**) pressure loss forms at various positions inside the injector; (**b**) pressure distribution at various positions inside different diameter injectors.

The change in the nozzle diameter from 0.169 mm to 0.218 mm will result in alterations in both the internal pressure loss of the injector and the cross-sectional area at the nozzle outlet, inevitably affecting the actual fuel injection rate. According to the Bernoulli equation, the fuel flow rate through the nozzle can be expressed as follows:

$$\dot{m}_f = \int Au\cdot\rho dA = A\sqrt{2\Delta p(t)\rho} = \frac{\pi}{4}d^2\sqrt{2\Delta p(t)\rho}, \quad (4)$$

where $\Delta p(t)$ is the pressure difference between the inlet and outlet of the nozzle; A is the cross-sectional area at the nozzle outlet; ρ is for the density of the diesel used in the experiment; and $\dot{m}_f$ is the mass flow rate at the nozzle outlet. Figure 16 illustrates the fuel injection rate corresponding to different nozzle diameters under a fixed injection pulse width. The fuel injection rate curves exhibit a trapezoidal pattern, which is quite similar to the lift curve of the needle valve. This resemblance is primarily due to the influence of the constrained area formed between the needle valve and its seat on the fuel injection rate. When the needle valve reaches its maximum lift, the fuel injection rate peaks. After 0.003 s from the start of injection, the fuel injection rate stabilizes relatively. At this point, the fuel injection rates corresponding to the three nozzle diameters, from smallest to largest, are 5.28 L/min, 6.70 L/min, and 7.23 L/min, respectively. Under the condition of 180 MPa common rail pressure and a cycle fuel mass of 180 mg, the durations of fuel injection for the respective nozzle diameters are 2950 ms, 2370 ms, and 2180 ms.

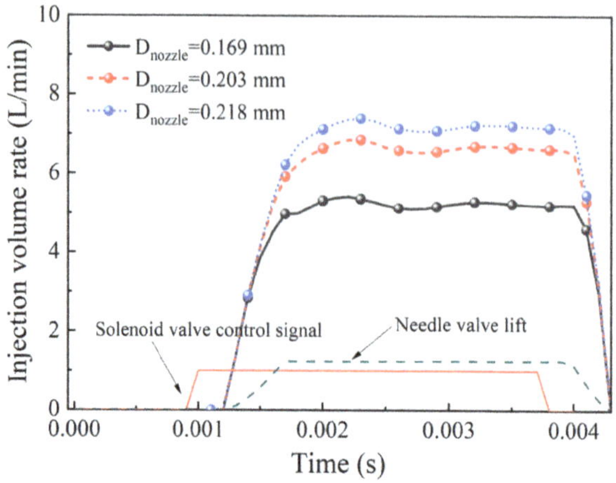

Figure 16. Influence of injector nozzle diameter on actual fuel injection rate.

The momentum of spray ejected will affect the subsequent evaporation and mixing quality of the spray in-cylinder, thereby influencing the final combustion process and emissions. The momentum of the spray at the nozzle outlet can be defined as follows:

$$\dot{M}_f = \int Au^2 \cdot \rho dt = 2A \cdot \int \Delta p(t)dt = \frac{\pi}{2}d^2 \int \Delta p(t)dt, \tag{5}$$

where $\dot{M}_f$ is the momentum of spray at the nozzle outlet within a fixed time step. To calculate the momentum of spray ejection, it is necessary to determine the statistical time step. In this study, a unit time step of 1×10^{-5} s was chosen. Increasing the injector nozzle diameter results in a decrease in actual injection pressure and fuel ejection velocity. However, increasing the injector nozzle diameter significantly improves the fuel injection rate. These two factors combined result in an increase in the average momentum of spray ejection per unit time from 3.25 kg·m/s to 4.21 kg·m/s as nozzle diameter increases when the needle valve is fully open, as shown in Figure 17. However, when cumulative momentum of spray ejection is calculated for a fixed total cycle fuel injection mass, it is found that due to differences in the duration of fuel injection, the cumulative momentum of spray ejection for smaller nozzle diameters actually increases.

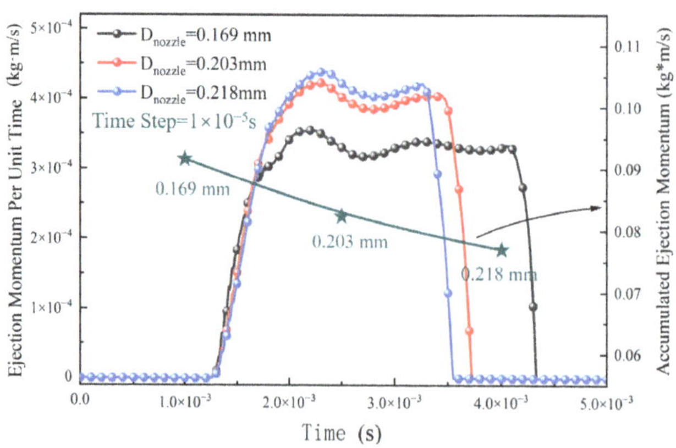

Figure 17. Influence of injector nozzle diameter on fuel ejection momentum.

The phenomenon of air entrainment during the development of fuel spray is a crucial physical process. The high-temperature air entrained into the spray promotes the evaporation of liquid fuel by exchanging energy with the fragmented liquid droplets. Additionally, the motion of air entrainment determines the mixing quality between fuel and fresh air during spray development. Therefore, the quantity and velocity of air entrainment during spray development are important factors affecting the overall combustion performance of diesel engines. The strength of air entrainment during the free development of the spray depends on the rate of momentum exchange between the spray and the ambient gas. Within the range of injector diameters studied, the momentum of spray ejected per unit time increases with the increase in nozzle diameter, and the momentum entering the cylinder space per unit time should also increase synchronously. After high-speed spray enters the cylinder, its momentum density distribution exhibits a spindle-shaped layered structure, with the spray core area near the nozzle outlet having the highest momentum density, as shown in Figure 18. With the increase in nozzle diameter, the axial penetration speed of the spray increases, and the axial range of the high momentum density region expands. While the spray develops axially, the spray edges experience frictional shearing with the air, resulting in intense momentum exchange at the spray edges. To quantify the radial momentum transfer of the spray, the kinetic energy of the core and non-core regions at various axial positions of the spray was statistically analyzed, where the core region is defined as a cylinder with the nozzle axis as the axis and the nozzle diameter as the bottom diameter. As shown in Figure 19, increasing the nozzle diameter significantly increases the total kinetic energy at various axial positions of the spray, enhances the spray penetration rate, intensifies the frictional shearing between the spray and ambient air, improves the momentum exchange capability, and increases the proportion of kinetic energy transferred from the upstream region of the spray to the non-core region from 53% to 59%. As a result, the radial expansion of the spray increases.

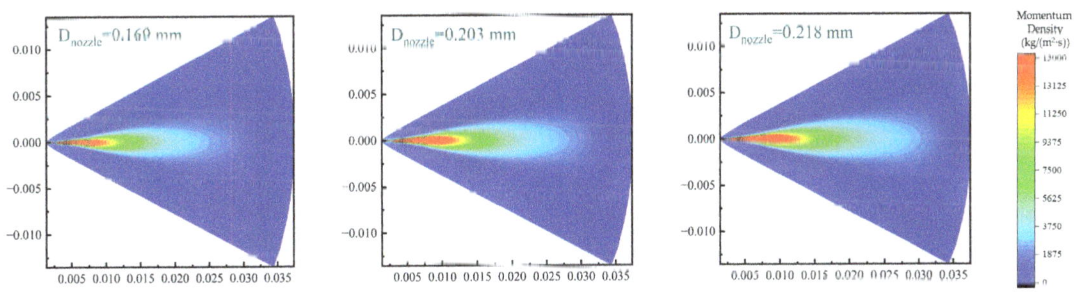

Figure 18. Momentum density distribution contour under three nozzle diameters.

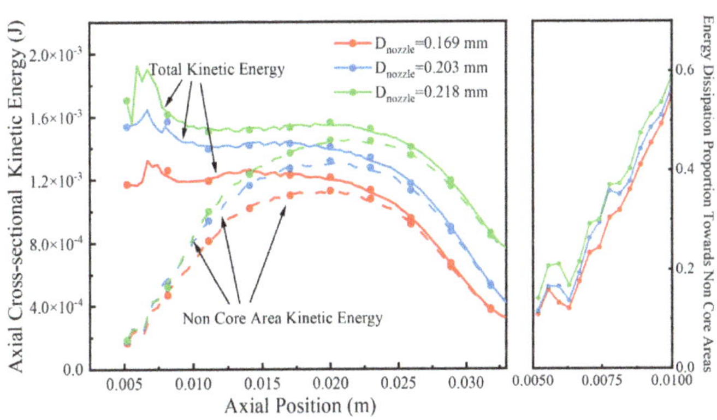

Figure 19. Axial cross-sectional kinetic energy and energy dissipation ratio at various positions under three nozzle diameters.

Figure 20 illustrates the characteristics of flow field distribution in the near and far fields of the spray under different nozzle diameters before spray/wall impingement (5.5 deg. ATDC). From the figure, it can be observed that while diesel spray develops axially, its front end pushes fresh air to both sides of the spray, and under the action of pressure difference, air from the sides enters the interior of the spray. The process of air entrainment by the spray mainly occurs at the middle and upper positions of the spray, and the air entrainment effect at the middle position of the diesel spray is most significant. As the spray develops axially forward, the momentum transferred to the radial air gradually increases, while the corresponding penetrating momentum gradually decreases. Ultimately, under the combined action of upstream spray and ambient gas, the mixture of spray far from the axis stops forward penetration and undergoes lateral movement, and under the action of pressure difference, flows backward upstream, defining this region as the "recirculation zone" where the outflow of fresh charge equals the inflow. Analyzing the velocity field near the edges of the spray under different nozzle diameters, both in the near and far fields, the airflow velocity increases with the increase in nozzle diameter. Statistically, the average air velocities near the spray edges corresponding to nozzle diameters from small to large are 7.67 m/s, 9.69 m/s, and 9.85 m/s in the near field, and 9.75 m/s, 11.71 m/s, and 11.93 m/s in the far field. The airflow velocity of the spray depends on the momentum flux transferred from the spray to the air radially. From the above study, it can be inferred that increasing the injector nozzle diameter increases the momentum ejected per unit time by the spray, increases the total momentum transferred to the surrounding gas, intensifies the mass transfer entrainment process at the spray edge, and increases the lateral airflow velocity near the spray edges.

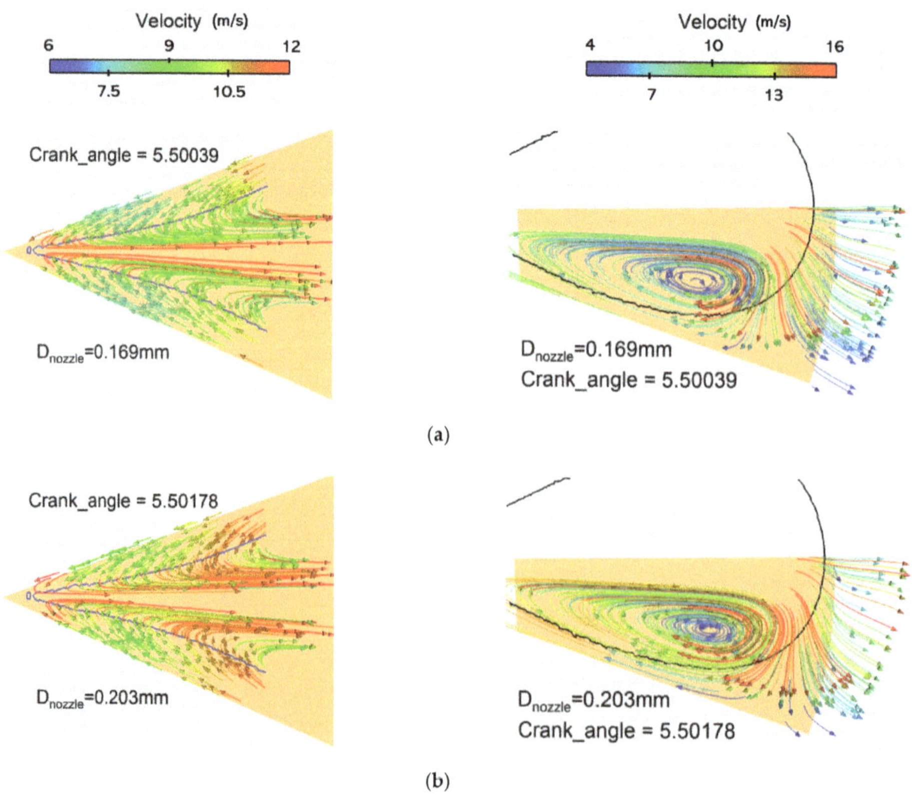

Figure 20. *Cont.*

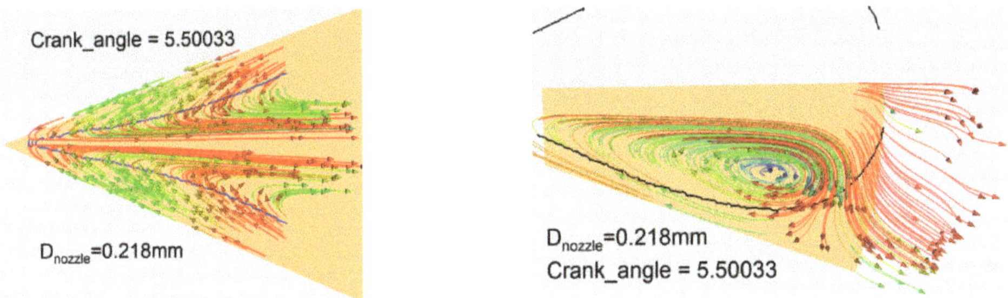

(c)

Figure 20. Distribution flow fields in the near and far fields of the sprays under three injector nozzle diameters (5.5 deg. ATDC): (**a**) 0.169 mm; (**b**) 0.203 mm; (**c**) 0.218 mm.

High airflow velocity does not necessarily imply more efficient oxygen transport. It is also necessary to consider the relationship between flow direction and spray, as well as the concentration gradient of input substances. Therefore, Dr. Zhang from Tianjin University proposed the oxygen transport rate α to measure the strength of oxygen transport during the free development process of the spray [23]. This parameter comprehensively considers the direction of airflow and the concentration gradient of input substances. It is defined as follows:

$$\alpha = \vec{V} \cdot \nabla O_2 = \left|\vec{V}\right| |\nabla O_2| cos\theta, \tag{6}$$

That is a detailed explanation of the oxygen transport rate α, defined as the dot product between the flow direction and the gradient of oxygen mass fraction. θ is the angle between the velocity vector and the gradient vector of oxygen mass fraction. When the angle θ is less than 90° (approximately indicating the same direction), $\alpha > 0$; when the angle is greater than 90°, $\alpha < 0$.

Figure 21 presents the distribution of oxygen transport rate in the cylinder under three different nozzle diameters at 5.5 deg. ATDC. It can be observed from the figure that during the free development process of the spray, oxygen is primarily transported along the concentration gradient into the spray, with a large amount of oxygen entering the interior of the spray. The oxygen transport rate is high in the region near the nozzle outlet, and as the spray moves away from the nozzle outlet, the concentration gradient between the spray edge and the surrounding environment as well as within the spray decreases significantly, leading to a decrease in transport rate. Meanwhile, the distribution of high transport rate regions also changes. Unlike the upstream where the entire spray has globally high transport rates, the oxygen transport in the middle region of the spray mainly occurs between the spray edge and the spray core region due to the large oxygen concentration gradient in that area. The spray core region closer to the axis does not have effective oxygen transport due to low oxygen concentration gradients and high axial penetration speeds. Figure 22 shows the distribution of transport rates at various positions on two radial sections, X = 0.1 m and X = 0.2 m, representing the upstream and midstream of the spray, respectively. The oxygen transport in the upstream of the spray is undergoing intense motion with a bimodal distribution, while the oxygen transport distribution in the midstream differs significantly from the upstream. Although it also shows a bimodal distribution, there is no effective oxygen transport near the nozzle axis. With increasing nozzle diameter, the oxygen transport rate in the upstream and midstream of the spray significantly increases, intensifying the entrainment motion of ambient air into the spray and promoting the mixing and combustion processes.

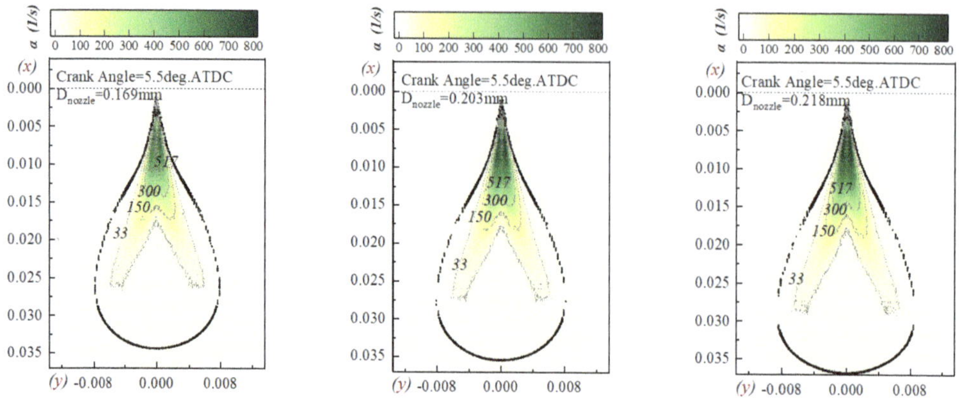

Figure 21. Distribution of oxygen transport rate under three injector nozzle diameters (5.5 deg. ATDC).

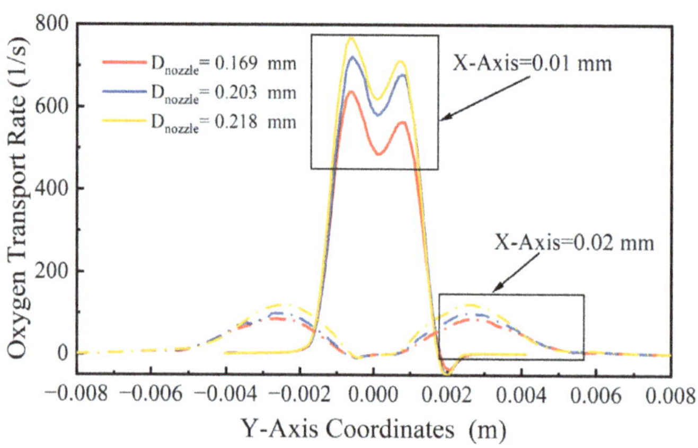

Figure 22. Distribution of oxygen transport rates at various positions on two radial sections of spray under three injector nozzle diameters (5.5 deg. ATDC), X = 0.1 m and X = 0.2 m.

Coupled time series statistically analyzed the cumulative air entrainment mass by the spray at different times, as shown in Figure 23. The air entrainment mass by the spray monotonically increases with the injection time, and the rate of increase continues to rise until the spray/wall impingement. When the spray develops to the piston wall, the effective entrainment area of the spray during the free development stage will no longer increase. Additionally, the fuel impinging on the wall will be guided by the wall to move in the opposite direction from both sides of the spray, which will have a negative effect on the air entrainment during the free development stage. Therefore, the impingement of the spray on the wall is the turning point of the rate of change of air entrainment mass, and its peak appears near the moment of spray/wall impingement. As the nozzle diameter increases from 0.169 mm to 0.218 mm, the air entrainment mass by the spray significantly increases in the same spray development time. Although increasing the nozzle diameter may lead to an earlier spray/wall impingement time, from the perspective of air entrainment mass at the impingement time, the larger nozzle diameter still has advantages. The air entrainment mass at the impingement time corresponding to a nozzle diameter of 0.218 mm is 1.42 times that of 0.169 mm.

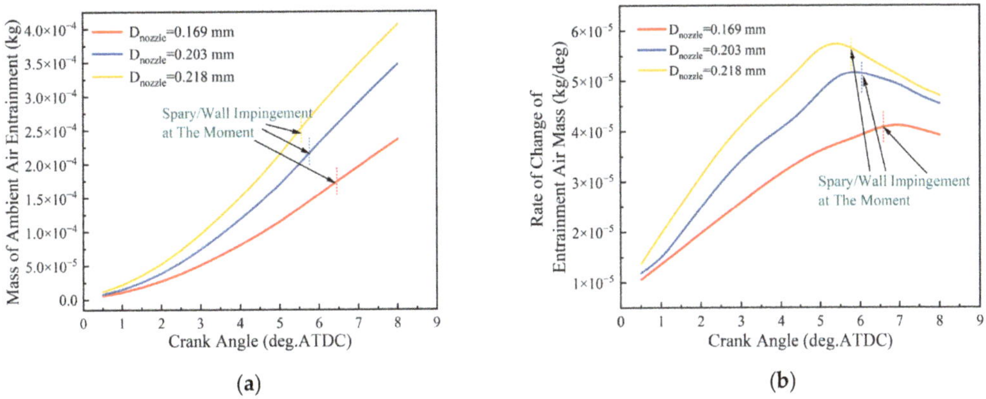

Figure 23. The variation of entrained air mass and its rate of change with injection time under different nozzle diameters: (**a**) air entrainment mass; (**b**) rate of change of air entrainment mass.

Earlier analysis of the combustion path of diesel engines revealed that the proportion of lean mixtures ($0 < \Phi < 1$) plays a crucial role in the overall diesel combustion process, with an ultra-high lean mixture ratio being ideal for efficient and clean combustion. However, the concentration field distribution inside the spray is non-uniform, and different regions within the spray exhibit different mixing characteristics. Taking a nozzle diameter of 0.169 mm as an example (Figure 24), in the region from the nozzle outlet to the ignition boundary (0–0.005 m), the high-temperature air entrained into the spray exchanges energy with the fragmented liquid droplets, thereby promoting droplet evaporation. In this region, the evaporation rate of liquid-phase fuel is significantly higher than the air entrainment rate. Consequently, the phenomenon of $\Phi > 1$ mixtures reaching peak values in ascending order of equivalence ratio intervals is observed. Wherein, the mass fraction of mixtures with $\Phi > 4$ continuously increases and eventually reaches 50%. In the region from the ignition boundary to the maximum liquid length (0.005–0.015 m), air entrainment gradually reaches a balance with fuel evaporation and even dominates. The mass fraction of mixtures with $\Phi > 4$ first stabilizes and then rapidly decreases. Notably, the decrease in mixtures with $\Phi > 4$ does not lead to a significant increase in mixtures with $1 < \Phi < 3$; instead, there is a substantial increase in the proportion of lean mixtures with $\Phi < 1$. In the region from the maximum liquid length to the spray head (>0.015 mm), the liquid-phase fuel has completely evaporated, while air entrainment continues. The equivalence ratio in the core region of the spray gradually decreases. Coupled time series comparison of the mass fractions of lean mixtures with $\Phi < 1$ and rich mixtures with $\Phi > 1$ under different injector nozzle diameters (Figure 25) reveals that in the early stage of injection, the mass fraction of rich mixtures corresponding to larger nozzle diameters is higher, indicating more liquid-phase fuel has evaporated, which is beneficial for subsequent fuel–air mixing. As the injection progresses, the proportion of liquid fuel that has evaporated inside the spray continuously increases, leading to an increase in the mass fraction of lean mixtures. At the moment of ignition, the mass fractions of lean mixtures corresponding to the three nozzle diameters are comparable, with approximately 52%. Larger nozzle diameters change the fuel injection rate, leading to an increase in the total amount of fuel entering the cylinder at the moment of ignition, thereby synchronously increasing the mass fraction of lean mixtures with $\Phi < 1$. The mass fraction of lean mixtures inside the spray reaches its peak value at the moment of impingement. The mass fractions of lean mixtures corresponding to the three nozzle diameters, from smallest to largest, are 60.2%, 62.7%, and 64.0%, respectively. Hence, increasing the nozzle diameter can increase the proportion of lean mixtures inside the spray during the free development stage, thereby enhancing the rate of heat release during the free development stage.

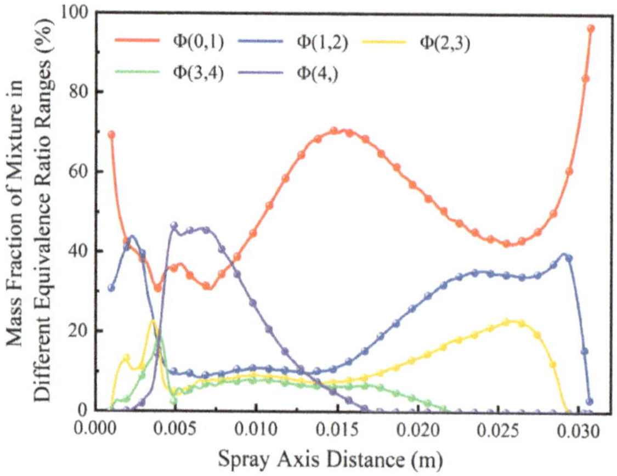

Figure 24. Mass fraction of mixture in different equivalence ratio range (0.169 mm 5.5 deg. ATDC).

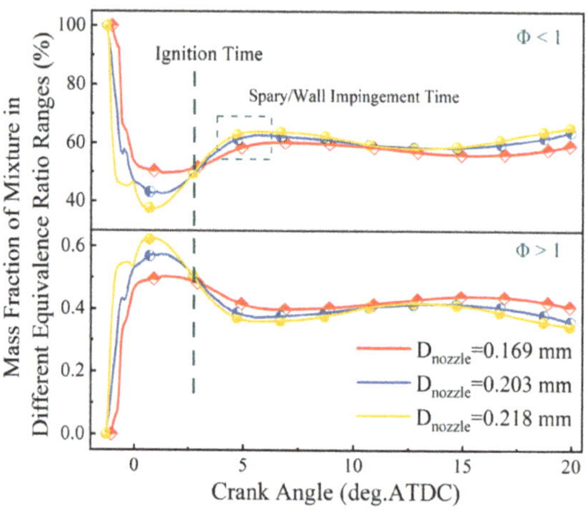

Figure 25. Mass fraction of mixture in different equivalence ratio ranges under three injector nozzle diameters.

Owing to the increase in spray momentum at the nozzle outlet, the friction and shear effects between the spray and fresh air become more pronounced during the axial penetration development process. As the injector nozzle diameter increases, the proportion of momentum transferred radially in the core region of the spray upstream becomes higher. The increase in the total momentum coupled with the increase in the dissipation ratio, results in an increase in the momentum transferred to the fresh air by the spray radially. Consequently, the overall oxygen transport process within the spray becomes more intense, the air entrainment rate significantly increases, and the accumulated air entrainment mass during the free development stage increases. This leads to an improvement in the mixture quality inside the spray, an increase in the proportion of lean mixtures, and an acceleration of the heat release during the free development stage of the spray combustion. However, it is noteworthy that, under the operating conditions considered in this study, over 70% of the combustion process is accompanied by spray/wall impingement. After spray/wall impingement, the mixing rate between the spray and air is significantly reduced due to momentum loss, and the limited fuel–air contact area slows down the fuel heat release rate, affecting the overall combustion quality. Therefore, it is necessary to discuss the influence of the wall-attached fuel mixing rate. In this study, a stepped-lip combustion chamber was used, which replaces the protruding lip of the traditional reentrant ω-shaped combustion chamber with a recessed annular lip. As shown in Figure 26, under conventional injection strategies, spray/wall impingement occurs near the lip, and under the guidance of the combustion chamber wall, the spray is divided into two parts: the majority enters the bottom of the piston and forms a clockwise vortex core, while a minority develops upwards, crosses the step, enters the squish region, and forms a counterclockwise vortex core. Both vortex cores continue to grow over time and form large-scale vortex structures with strong air entrainment capabilities at the leading edge of the wall-attached fuel, especially structure A at the bottom of the piston. With increasing nozzle diameter, the scale and intensity of the vortices at the bottom of the piston increase, while the intensity of the vortices in the squish region decreases. This phenomenon is particularly evident 15 degrees after spray/wall impingement, as the larger nozzle diameter causes the spray/wall impingement point at the end of injection to be closer to the bottom of the piston, resulting in more fuel entering the piston and less fuel entering the squish region. Furthermore, the residual momentum of wall-attached fuel is relatively high, leading to larger scale and intensity of vortices at the bottom of the piston. Although increasing the nozzle diameter weakens the vortices in the squish region, comparing the oxygen concentration in the squish region under different nozzle diameters reveals that even with strong vortex structures in the squish region, it is difficult to entrain air between the piston and cylinder liner gaps. However, the strengthened vortices at the bottom of the piston can entrain a large amount of surrounding air into the vortex structure, facilitating the mixing of air and unburnt fuel within the vortex structure and improving the air utilization rate in the piston center region. Benefiting from the higher mixture quality inside the spray during the free development stage and the larger and stronger vortices induced after spray/wall impingement, the mixing of wall-attached fuel is enhanced, leading to an increase in the mixing quality after the spray/wall impingement stage.

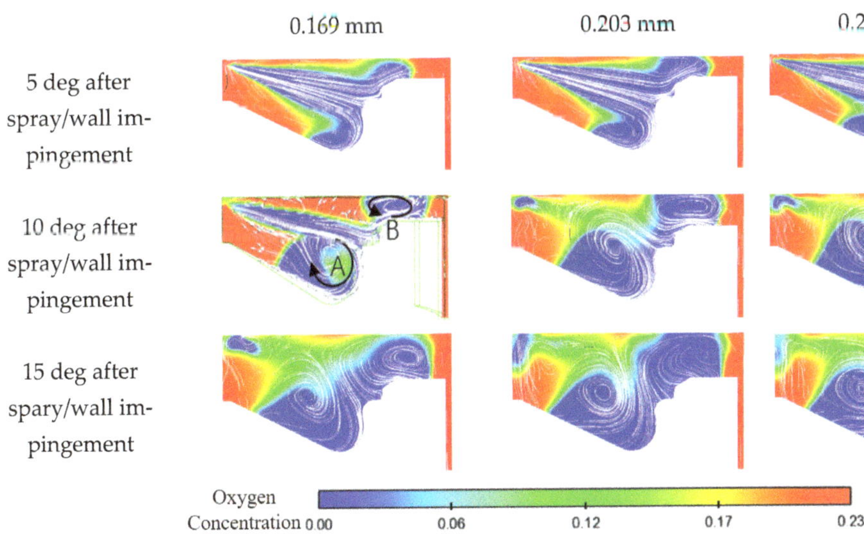

Figure 26. Distribution of in-cylinder flow field and oxygen concentration after spray/wall impingement for three nozzle diameters.

There are significant differences in the reaction pathways corresponding to different equivalence ratio ranges of the mixture, which inevitably lead to significant differences in the contribution to combustion heat release. Instantaneous heat release rates of mixture in different equivalence ratio ranges were statistically analyzed, as shown in Figure 27. When the nozzle diameter increases from 0.169 mm to 0.218 mm, due to the increase in the mass of fuel entering the cylinder per unit time, the heat release rates of mixture in different equivalence ratio ranges during the injector duration all show varying degrees of improvement. The heat release contribution of lean mixtures with Φ < 1 is the highest, and the corresponding increase in heat release rate is most significant. This is also the key reason why increasing the nozzle diameter leads to the most significant increase in heat release rate.

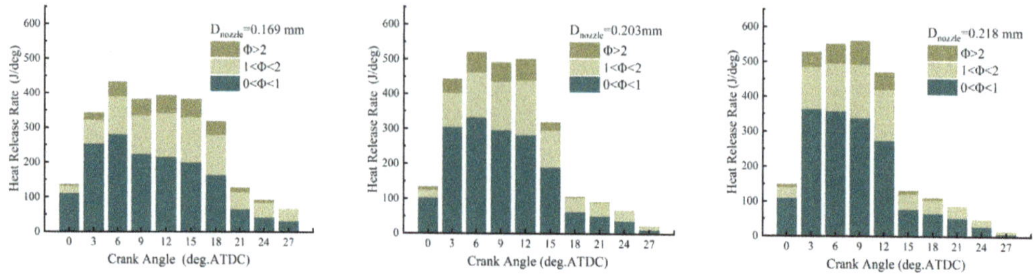

(a) (b) (c)

Figure 27. The instantaneous heat release rates of mixture in different equivalence ratio ranges for three injector nozzle diameters: (**a**) 0.169 mm; (**b**) 0.203 mm; (**c**) 0.218 mm.

In summary, the research results show that in a high charge density environment, increasing the nozzle diameter from 0.169 mm to 0.218 mm not only increases the mass of fuel entering the cylinder per unit time but also improves the overall mixture quality during the entire spray development process. The proportion of lean mixtures inside the spray increases, resulting in more combustible mixture being produced per unit time. As a result, the combustion heat release rate significantly increases, the combustion duration is notably shortened, and the proportion of constant volume combustion increases, allowing for more complete expansion of the working fluid, leading to an indicated thermal efficiency increase to 51.5%.

The above research was conducted within the range of experimental fuel injectors (0.169 mm to 0.218 mm). Given the limitations of the experimental fuel injector nozzle diameter range, numerical simulation methods were employed to expand the research scope of the fuel injector nozzle diameter in order to verify the applicability of the conclusions obtained within a wider range of nozzle diameters. Figure 28 shows the simulation results of the hydraulic system model of the fuel injector. With the increase in nozzle diameter, the actual fuel injection pressure shows a linear decreasing trend, the fuel injection rate continuously increases and the duration of fuel injection continuously shorten. However, when the nozzle diameter exceeds 0.2 mm, the rate of decrease in the duration of fuel injection significantly slows down. This means that the benefits of increasing the fuel injection rate decrease as the nozzle diameter increases, mainly because increasing the nozzle diameter will increases the pressure loss inside the fuel injector, resulting in an excessive decrease in the actual fuel injection pressure. In addition, expanding the research scope of the fuel injector nozzle diameter reveals that the spray momentum at the nozzle outlet per unit time shows a trend of initially increasing and then decreasing with increasing nozzle diameter. This is the result of the interplay between the nozzle flow area and the actual fuel injection pressure. Excessive enlargement of the nozzle diameter will lead to a significant decrease in spray momentum, thereby affecting subsequent droplet atomization, fuel–air mixture, and combustion processes. In this study, the nozzle diameter corresponding to the peak spray momentum at the nozzle outlet was within the range of 0.218 mm to 0.25 mm.

Figure 29 shows the numerical simulation results of the in-cylinder combustion process. When the nozzle diameter exceeds 0.25 mm, the decrease in spray momentum entering the cylinder per unit time leads to a reduction in the air entrainment rate, and the duration of free spray development is significantly shortened, resulting in deterioration of the mixture quality inside the spray at the spray/wall impingement moment. From the perspective of heat release rate, within the range of nozzle diameters from 0.13 mm to 0.25 mm, the combustion duration gradually shortens and the thermal efficiency significantly improves as the nozzle diameter increases. Within the range of nozzle diameters from 0.25 mm to 0.37 mm, the combustion duration remains around 23 degrees and the thermal efficiency shows a slow decreasing trend as the nozzle diameter increases. Therefore, under the engine operating parameters and thermodynamic boundary conditions considered in this study, the optimal nozzle diameter should be within the range of 0.218 mm to 0.25 mm.

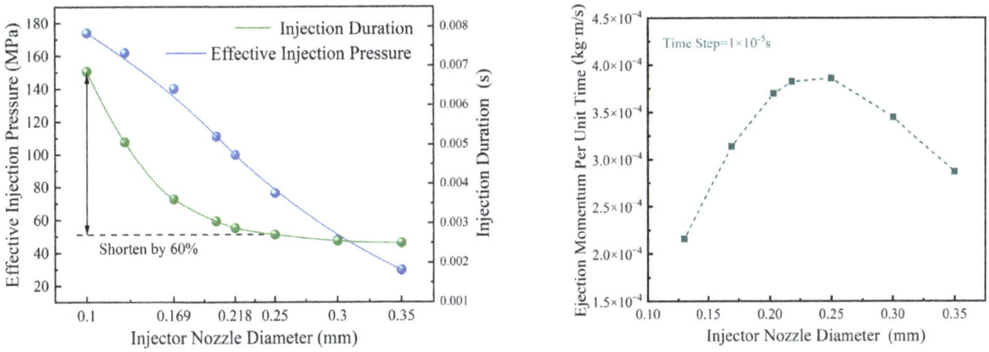

(a) (b)

Figure 28. Simulation results obtained from expanding the research scope of nozzle diameters: (**a**) effective injection pressure and fuel injection duration; (**b**) spray momentum at nozzle outlet.

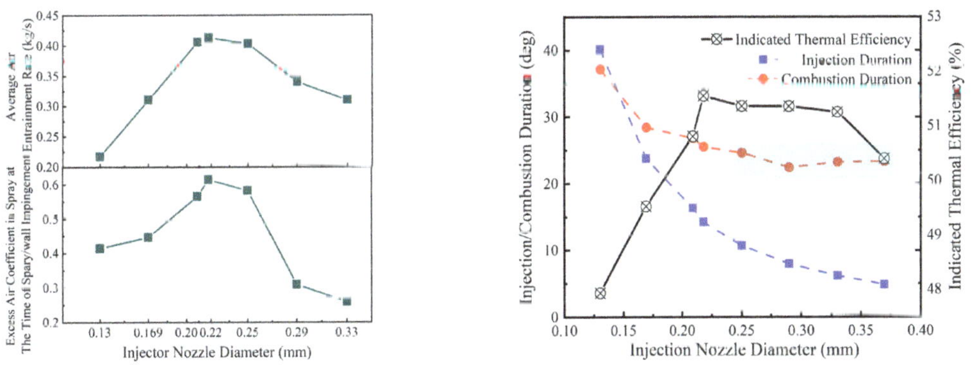

(a) (b)

Figure 29. Numerical simulation results obtained from expanding the fuel injector nozzle diameter range: (**a**) spray mixture quality; (**b**) combustion duration and indicated thermal efficiency.

5. Conclusions

In order to improve the fuel economy of heavy-duty diesel engines under high-load conditions, based on the combustion pathway model, it is proposed that the proportion of lean mixture with $0 < \Phi < 1$ is the most important spray characteristic affecting the overall diesel combustion process. Answering the question of how to increase the proportion of lean mixture inside the spray is the key to achieving efficient and clean combustion of diesel engines. Through experiment and numerical simulation methods, the following conclusions are drawn regarding the impact mechanism of fuel injector nozzle diameter on spray lean mixture characteristics, fuel–air mixture, and combustion process under high charge density conditions, as well as the potential to improve thermal efficiency:

In a high charge density environment, increasing the nozzle diameter from 0.169 mm to 0.218 mm can improve the overall mixture quality throughout the spray development process. The experimental results show that with the increase in nozzle diameter, the peak pressure and instantaneous heat release rate significantly increase, the combustion duration is shortened by about 20%, the heat release becomes more concentrated, and the proportion of constant volume combustion increase. At 1200 rpm and IMEPg~2.3 MPa conditions, the indicated thermal efficiency increases by 1.3%, reaching a maximum of 51.5%.

With the increase in nozzle diameter, the spray ejection momentum per unit time also increases significantly. The increase in spray penetration velocity strengthens the friction and shear effects between the spray and fresh air, resulting in faster energy dissipation rates and a higher proportion of energy dissipation in the radial direction of the spray core, especially in the upstream region where the penetration velocity is highest. The higher spray momentum combined with the increased dissipation proportion means that larger nozzle diameter fuel injectors transfer more momentum to the fresh air radially, resulting in a wider radial distribution range of high-momentum regions in the middle region of the spray, intensifying the mass transfer and entrainment process at the spray edge.

Introducing the oxygen mass transport rate to evaluate the strength of air entrainment motion at various positions of the spray: during the free development process of the spray, oxygen forward transport is dominant; i.e., the oxygen transport rate parameter is greater than 0. During this process, a large amount of oxygen enters the interior of the spray, and the most intense oxygen transport occurs at the inner boundary of the gas phase spray with obvious oxygen concentration stratification. With the increase in nozzle diameter, the higher spray momentum during the free development stage leads to more intense oxygen transport motion, and the radial range of effective oxygen transport in the middle region of the spray increases significantly.

In the region near the spray outlet where liquid-phase fuel evaporation is dominant, high air entrainment leads to a significant increase in the amount of gas-phase fuel generated, which is beneficial for subsequent fuel–air mixing. The proportion of rich mixture increases rapidly, the mass fraction of lean mixture inside the spray reaches its peak until the spray/wall impingement moment. Increasing the nozzle diameter increases the air entrainment by 42% during the free development stage of the spray, and the proportion of lean mixture inside the spray increases from 60.2% to 64.0%. Benefit from the higher mixture quality during the free development stage of the spray and the larger and stronger vortex generated at the front of the spray after spray/wall impingement, increasing the nozzle diameter intensifies the mixing process between the wall-attached fuel and air, leading to an improvement in the mixture quality after the spray/wall impingement stage. Overall, increasing the nozzle diameter from 0.169 mm to 0.218 mm increases the mass of fuel entering the cylinder per unit time, while also increasing the proportion of lean mixture throughout the entire spray development process, resulting in an increase in the heat release rate of the lean mixture, making the overall combustion more intense and concentrated.

By using numerical simulation methods to expand the research scope of fuel injector nozzle diameters, the applicability of the conclusions obtained within a wider range of nozzle diameters is verified. It is found that a nozzle diameter that is too small sacrifices fuel injection duration, leading to a large amount of fuel combustion during the piston descent process, while a nozzle diameter that is too large leads to an excessive reduction in effective injection pressure, affecting spray momentum and subsequent fuel–air mixture processes, with no further reduction in combustion duration but a slow decrease in thermal efficiency. Under the engine operating parameters and thermodynamic boundary conditions considered in this study, the optimal nozzle diameter should be in the range of 0.218 mm to 0.25 mm.

Author Contributions: Conceptualization, W.S. and Y.L.; methodology, Y.L.; software, Y.L.; validation, Y.L.; formal analysis, Y.L.; data curation, Y.L.; writing—original draft preparation, Y.L.; writing—review and editing, W.S.; funding acquisition, W.S. All authors have read and agreed to the published version of the manuscript.

Funding: This research was funded by the National Key Research and Development Program of China (No. 2022YFE0100100).

Data Availability Statement: The data presented in this study are available on request from the corresponding author. The data are not publicly available due to privacy.

Conflicts of Interest: The authors declare no conflicts of interest.

Nomenclature

Item	Definition
NOx	Nitrogen Oxide
UHC	Unburnt Hydrocarbon
ATDC	After Top Dead Center
CA	Crank Angle
ECU	Electronic Control Unit
EGR	Exhaust Gas Recirculation
Φ	Fuel–oxygen Equivalence Ratio
CA10	10% Heat Release Point
Ca50	50% Heat Release Point
Ca90	90% Heat Release Point
HRR	Heat Release Rate
IMEP	Indicated Mean Effective Pressure
ITEg	Gross Indicated Thermal Efficiency

References

1. Huang, M.; Zhai, P. Achieving Paris Agreement temperature goals requires carbon neutrality by middle century with far-reaching transitions in the whole society. *Adv. Clim. Chang. Res.* **2021**, *12*, 281–286. [CrossRef]
2. Gunfaus, M.T.; Waisman, H. Assessing the adequacy of the global response to the Paris Agreement: Toward a full appraisal of climate ambition and action. *Earth Syst. Gov.* **2021**, *8*, 100102. [CrossRef]
3. Senecal, P.; Leach, F. Diversity in transportation: Why a mix of propulsion technologies is the way forward for the future fleet. *Results Eng.* **2019**, *4*, 100060. [CrossRef]
4. Kumano, K.; Iida, N. Analysis of the effect of charge inhomogeneity on HCCI combustion by chemiluminescence measurement. *J. Fuels Lubr.* **2004**, *113*, 974–986.
5. Verma, S.K.; Gaur, S.; Akram, T.; Gautam, S.; Kumar, A. Emissions from homogeneous charge compression ignition (HCCI) engine using different fuels: A review. *Green Energy Environ. Sustain.* **2022**, *29*, 50960–50969. [CrossRef] [PubMed]
6. Liu, Y.; Su, W.; Wu, B.; Wang, J. The Research and Development of a Jet Disturbance Combustion System for Heavy-Duty Diesel Engines. *Energies* **2024**, *17*, 1065. [CrossRef]
7. Zhang, Z.; Liu, Y.; Wu, B.; Nie, J.; Su, W. Effect of Nozzle Diameter on Combustion and Emissions of a Heavy Duty Diesel Engine. *Trans. CSICE* **2022**, *40*, 97–105.
8. Zhai, C.; Jin, Y.; Nishida, K.; Ogata, Y. Diesel spray and combustion of multi-hole injectors with micro-hole under ultra-high injection pressure–non-evaporating spray characteristics. *Fuel* **2021**, *283*, 119322. [CrossRef]
9. Zhao, J.; Grekhov, L.; Yue, P. Limit of fuel injection rate in the common rail system under ultra-high pressures. *Int. J. Automot. Technol.* **2020**, *21*, 649–656. [CrossRef]
10. Wang, L.; Lowrie, J.; Ngaile, G.; Fang, T. High injection pressure diesel sprays from a piezoelectric fuel injector. *Appl. Therm. Eng.* **2019**, *152*, 807–824. [CrossRef]
11. Shi, Z.; Wu, H.; Li, H.; Zhang, L.; Lee, C. Effect of injection pressure and fuel mass on wall-impinging ignition and combustion characteristics of heavy-duty diesel engine at low temperatures. *Fuel* **2021**, *299*, 120904. [CrossRef]
12. Xia, J.; Zhang, Q.; Huang, Z.; Ju, D.; Lu, X. Experimental study of injection characteristics under diesel's sub/trans/supercritical conditions with various nozzle diameters and injection pressures. *Energy Convers. Manag.* **2020**, *215*, 112949. [CrossRef]
13. Dec, J. *A Conceptual Model of DI Diesel Combustion Based on Laser-Sheet Imaging*; SAE Technical Paper 970873; JSTOR: New York, NY, USA, 1997.
14. Pickett, L.M.; Siebers, D.L. Non-sooting, low flame temperature mixing-controlled DI diesel combustion. *SAE Trans.* **2004**, *113*, 614–630.
15. Kook, S.; Bae, C.; Miles, P.C.; Choi, D.; Pickett, L.M. The influence of charge dilution and injection timing on low-temperature diesel combustion and emissions. *SAE Trans.* **2005**, *114*, 1575–1595.
16. Polonowski, C.J.; Mueller, C.J.; Gehrke, C.R.; Bazyn, T.; Martin, G.C.; Lillo, P.M. An experimental investigation of low-soot and soot-free combustion strategies in a heavy-duty, single-cylinder, direct-injection, optical diesel engine. *SAE Int. J. Fuels Lubr.* **2012**, *5*, 51–77. [CrossRef]

17. Manin, J.; Skeen, S.; Pickett, L.M.; Kurtz, E.; Anderson, J.E. Effects of oxygenated fuels on combustion and soot formation/oxidation processes. *SAE Int. J. Fuels Lubr.* **2014**, *7*, 704–717. [CrossRef]
18. Dumitrescu, C.E.; Mueller, C.J.; Kurtz, E. Investigation of a tripropylene-glycol monomethyl ether and diesel blend for soot-free combustion in an optical direct-injection diesel engine. *Appl. Therm. Eng.* **2016**, *101*, 639–646. [CrossRef]
19. Huang, H. Numerical and Experimental Study on the Combustion Process of Diesel HCCI Engines. Ph.D. Dissertation, Tianjin University, Tianjin, China, 2007.
20. Wang, H.; Zheng, D.; Tian, Y. High pressure common rail injection system modeling and control. *ISA Trans.* **2016**, *63*, 265–273. [CrossRef] [PubMed]
21. Kim, J.; Lee, J.; Kim, K. Numerical study on the effects of fuel viscosity and density on the injection rate performance of a solenoid diesel injector based on AMESim. *Fuel* **2019**, *256*, 115912. [CrossRef]
22. Blessing, M.; König, G.; Krüger, C.; Michels, U.; Schwarz, V. Analysis of flow and cavitation phenomena in diesel injection nozzles and its effects on spray and mixture formation. *SAE Trans.* **2003**, *112*, 1694–1706.
23. Zhang, X. Research on Interaction of Physical and Chemical Factors in MULIN-BUMP Compound Combustion. Ph.D. Dissertation, Tianjin University, Tianjin, China, 2007.

MDPI AG
Grosspeteranlage 5
4052 Basel
Switzerland
Tel.: +41 61 683 77 34

MDPI Books Editorial Office
E-mail: books@mdpi.com
www.mdpi.com/books

www.ingramcontent.com/pod-product-compliance
Lightning Source LLC
Chambersburg PA
CBHW061551220926
51850CB00005B/308
* 9 7 8 3 7 2 5 8 1 9 4 3 0 *